# HEALTHCARE SENSOR NETWORKS

## CHALLENGES TOWARD PRACTICAL IMPLEMENTATION

# HEALTHCARE SENSOR NETWORKS

## CHALLENGES TOWARD PRACTICAL IMPLEMENTATION

**Daniel Tze Huei Lai**
**Rezaul Begg**
**Marimuthu Palaniswami**

CRC Press is an imprint of the
Taylor & Francis Group, an **informa** business

CRC Press
Taylor & Francis Group
6000 Broken Sound Parkway NW, Suite 300
Boca Raton, FL 33487-2742

First issued in paperback 2019



ISBN-13: 978-1-4398-2181-7 (hbk)
ISBN-13: 978-0-367-38233-9 (pbk)



**Library of Congress Cataloging-in-Publication Data**

Healthcare sensor networks : challenges toward practical implementation / edited by Daniel Tze Huei Lai, Rezaul Begg, Marimuthu Palaniswami.
p. ; cm.
Includes bibliographical references and index.
ISBN 978-1-4398-2181-7 (hardcover : alk. paper)
1. Biosensors. 2. Wireless sensor networks. 3. Biotelemetry. I. Lai, Daniel T. H. II. Begg, Rezaul. III. Palaniswami, Marimuthu.
[DNLM: 1. Biosensing Techniques--instrumentation. 2. Equipment Design. 3. Monitoring, Ambulatory. 4. Signal Processing, Computer-Assisted. 5. Telemetry--methods. QT 36]

R857.B54H43 2012
610.285--dc23 2011023882

**Visit the Taylor & Francis Web site at**
**http://www.taylorandfrancis.com**

**and the CRC Press Web site at**
**http://www.crcpress.com**

# Contents

Preface ........ vii
Editors ........ ix
Contributors ........ xi

1. **Sensor Networks in Healthcare: A New Paradigm for Improving Future Global Health** ........ 1
   *Daniel T.H. Lai, Braveena Santhiranayagam, Rezaul K. Begg, and Marimuthu Palaniswami*

2. **Healthcare and Accelerometry: Applications for Activity Monitoring, Recognition, and Functional Assessment** ........ 21
   *Andrea Mannini and Angelo Maria Sabatini*

3. **Intrabody Communication Using Contact Electrodes in Low-Frequency Bands** ........ 51
   *Ken Sasaki, Fukuro Koshiji, and Shudo Takenaka*

4. **The Prospect of Energy-Harvesting Technologies for Healthcare Wireless Sensor Networks** ........ 75
   *Yen Kheng Tan*

5. **Addressing Security, Privacy and Efficiency Issues in Healthcare Systems** ........ 111
   *Kalvinder Singh and Vallipuram Muthukkumarasamy*

6. **Flexible and Wearable Chemical Sensors for Noninvasive Biomonitoring** ........ 139
   *Hiroyuki Kudo and Kohji Mitsubayashi*

7. **Monitoring Walking in Health and Disease** ........ 159
   *Richard Baker*

8. **Motion Sensors in Osteoarthritis: Prospects and Issues** ........ 183
   *Tim V. Wrigley*

9. **The Challenges of Monitoring Physical Activity in Children with Wearable Sensor Technologies** ........ 221
   *Gita Pendhakar, Daniel T.H. Lai, Alistair Shilton, and Remco Polman*

**10. Ambulatory and Remote Monitoring of Parkinson's Disease Motor Symptoms** ........ 247
*Joseph P. Giuffrida and Edward J. Rapp*

**11. Nocturnal Sensing and Intervention for Assisted Living of People with Dementia** ........ 283
*Paul J. McCullagh, William M.A. Carswell, Maurice D. Mulvenna, Juan C. Augusto, Huiru Zheng, and W. Paul Jeffers*

**12. Experiences in Developing a Wearable Gait Assistant for Parkinson's Disease Patients** ........ 303
*Marc Bächlin, Daniel Roggen, Meir Plotnik, Jeffrey M. Hausdorff, and Gerhard Tröster*

**13. Designing a Low-Cost ECG Sensor and Monitor: Practical Considerations and Measures** ........ 339
*Ahsan H. Khandoker and Brian A. Walker*

**14. Sensors, Monitoring and Model-Based Data Analysis in Sports, Exercise and Rehabilitation** ........ 375
*Jurgen Perl, Daniel Memmert, Arnold Baca, Stefan Endler, Andreas Grunz, Mirjam Rebel, and Andrea Schmidt*

**15. Robust Monitoring of Sport and Exercise** ........ 407
*Andrew J. Wixted*

**Index** ........ 437

# *Preface*

The booming interest in the medical devices industry despite recent economic fluctuations (2008–2010) can be attributed to the increasing awareness that health is essential for enjoying a good quality of life. The resulting demand for quality healthcare has strained present global healthcare systems, which currently face large workforce shortages and inadequate medical facilities. It is paramount, therefore, to develop widespread and affordable consumer-based devices to augment the current healthcare demands. One emerging technology is healthcare sensor networks (HSNs), a human-oriented network of tiny medical monitoring devices deployed on the body and in the surrounding infrastructure. Body area networks (BANs) are sensor networks worn by the user, while infrastructure area networks (IANs) consist of monitoring networks embedded in the surrounding environment. HSNs are envisioned to present autonomous, consumer-friendly monitoring solutions for disorder prevention, diagnosis and postsurgery monitoring, thereby addressing some of the stringent expectations placed on today's healthcare systems.

This book brings together a collection of chapters highlighting the latest developments in HSNs with an emphasis on the applications and the respective challenges faced in the design phases. Unlike traditional wireless sensor network applications, HSN design focuses on human-network interaction, which poses a new set of technical and human engineering challenges. These engineering challenges include, for example, a higher data bandwidth demand, longer network operational time, robust network design in dynamic environments, lower cost versus higher computational demand and human safety requirements. Some of these problems exist in parallel with other sensor network applications, e.g., military and environmental monitoring, where intense research is still ongoing. Other human-centric challenges, as is seen later, remain unique to the specific healthcare application requirements.

In contrast to textbooks on biomedical engineering or sensor networks, this edited collection provides insights into the practical design and deployment challenges of healthcare sensor networks. Sufficient theory is provided in each chapter to acquaint the reader with the problem complexity before more attention is paid towards developing practical solutions. This allows the reader to appreciate the challenges in this field and cultivate further innovations as newer healthcare applications emerge. The 15 chapters presented in this compilation highlight the key areas requiring further research for HSNs to become a technological and commercially viable reality. The first six chapters focus on particular core issues, while the remaining chapters focus on application and systems design. We hope that this edited volume will be beneficial to researchers, postgraduate students and engineers looking for new, innovative solutions in the fields of healthcare and sports.

# Editors

**Dr. Daniel T.H. Lai** received his bachelor of electrical and computer systems engineering (Hons) and PhD in electrical and computer systems from Monash University, Melbourne, Australia. He was a postdoctoral research fellow at the University of Melbourne and Victoria University (2007–2010). He is currently with the Faculty of Health, Engineering and Science at Victoria University (2011). His research interests include computational intelligence and sensor network technology for healthcare and sports applications. This involves the design of new noninvasive and proactive sensing technologies capable of detecting, diagnosing and predicting health risks. He has more than 50 peer-reviewed publications and is a current reviewer for several international journals, e.g., *IEEE Transactions of Information Technology and Biomedicine, Journal of Biomechanics* and *Sensors and Actuators.* He is also actively involved in the organization of several workshops and international conferences.

**Dr. Rezaul K. Begg** received his BSc and MSc Eng degrees in electrical and electronic engineering from Bangladesh University of Engineering and Technology (BUET), Dhaka, Bangladesh, and his PhD in biomedical engineering from the University of Aberdeen, UK. Currently he is a professor within the Biomechanics Unit at Victoria University, Melbourne, Australia. Previously, he worked with the University of Aberdeen, Deakin University and BUET. His research interests are in biomedical engineering, gait biomechanics and machine learning. He has published more than 160 research papers in these areas. He is on the editorial board for several international journals and regularly reviews submissions for more than 15 international journals. He has also been actively involved in organizing a number of major international conferences. He has received several awards for academic excellence, including the ICISIP 2005 Best Paper Award, the Vice Chancellor's citation award for excellence in research at Victoria University, the BUET Gold Medal, and the Chancellor Prize.

**Prof. Marimuthu Palaniswami** received his bachelor of electrical engineering (Hons) from the University of Madras, ME, from the Indian Institute of Science, India; his MSc in engineering from the University of Melbourne; and his PhD from the University of Newcastle, Australia, before rejoining the University of Melbourne. He has been serving the University of Melbourne for over 16 years. He has published more than 180 refereed papers. His research interests include support vector machines (SVMs), sensors and sensor networks, machine learning, neural networks, pattern recognition, signal processing and control. He was given a Foreign Specialist Award by the Ministry of Education, Japan, in recognition of his contributions to the field of machine learning. He served as associate editor for journals/transactions including *IEEE Transactions on Neural Networks* and *Applied Computational Intelligence for Finance and Economics*. He is also the subject editor for the *International Journal on Distributed Sensor Networks*.

# Contributors

**Juan C. Augusto**
School of Computing and Mathematics
University of Ulster
Newtownabbey, United Kingdom

**Arnold Baca**
Institute of Biomechanics, Kinesiology and Applied Computer Science
University of Vienna
Vienna, Austria

**Marc Bächlin**
Wearable Computing Laboratory
Swiss Federal Institute of Technology Zurich
Zürich, Switzerland

**Richard Baker**
University of Salford
Greater Manchester, United Kingdom

and

Murdoch Childrens Research Institute
Royal Children's Hospital
Parkville, Victoria, Australia

**Rezaul K. Begg**
School of Sport and Exercise Science
Victoria University
Melbourne, Victoria, Australia

**William M.A. Carswell**
School of Computing and Mathematics
University of Ulster
Newtownabbey, United Kingdom

**Stefan Endler**
Institute of Computer Science
University of Mainz
Mainz, Germany

**Joseph P. Giuffrida**
Great Lakes Neuro Technologies, Inc.
Cleveland, Ohio

**Andreas Grunz**
Institute of Cognitive and Team/Racket Sport Research
German Sport University Cologne
Cologne, Germany

**Jeffrey M. Hausdorff**
Laboratory for Gait and Neurodynamics
Tel Aviv Sourasky Medical Center
Tel Aviv, Israel

**W. Paul Jeffers**
Fold Housing Association
Hollywood, United Kingdom

**Ahsan H. Khandoker**
Department of Electrical and Electronic Engineering
The University of Melbourne
Parkville, Victoria, Australia

**Fukuro Koshiji**
Department of Science and Engineering
Kokushikan University
Setagaya-ku, Tokyo, Japan

**Hiroyuki Kudo**
Department of Biomedical Devices and Instrumentation
Tokyo Medical and Dental University
Tokyo, Japan

**Daniel T.H. Lai**
School of Engineering and Science
Victoria University
Melbourne, Victoria, Australia

**Andrea Mannini**
The BioRobotics Institute
Scuola Superiore Sant'Anna
Pisa, Italy

**Paul J. McCullagh**
School of Computing and Mathematics
University of Ulster
Newtownabbey, United Kingdom

**Daniel Memmert**
Institute of Cognitive and Team/Racket Sport Research
German Sport University Cologne
Cologne, Germany

**Kohji Mitsubayashi**
Department of Biomedical Devices and Instrumentation
Tokyo Medical and Dental University
Tokyo, Japan

**Maurice D. Mulvenna**
School of Computing and Mathematics
University of Ulster
Newtownabbey, United Kingdom

**Vallipuram Muthukkumarasamy**
School of Information and Communication Technology
Griffith University
Southport, Queensland, Australia

**Marimuthu Palaniswami**
Department of Electrical & Electronic Engineering
University of Melbourne Parkville
Parkville, Victoria, Australia

**Gita Pendhakar**
School of Engineering (TAFE)
RMIT University
Melbourne, Victoria, Australia

**Jurgen Perl**
Institute of Computer Science
University of Mainz
Mainz, Germany

**Meir Plotnik**
Laboratory for Gait and Neurodynamics
Tel Aviv Sourasky Medical Center
Tel Aviv, Israel

**Remco Polman**
Institute of Sport, Exercise and Active Living
Victoria Univeristy
Melbourne, Victoria, Australia

**Edward J. Rapp**
Cleveland Medical Devices, Inc.
Cleveland, Ohio

**Mirjam Rebel**
Department II, Scientific Consulting
Federal Institute of Sport Science
Bonn, Germany

**Daniel Roggen**
Wearable Computing Laboratory
Swiss Federal Institute of Technology Zurich
Zürich, Switzerland

**Braveena Santhiranayagam**
School of Sport and Exercise Science
Victoria University
Melbourne, Victoria, Australia

**Angelo Maria Sabatini**
The BioRobotics Institute
Scuola Superiore Sant'Anna
Pisa, Italy

**Ken Sasaki**
Department of Human and Engineered Environmental Studies
University of Tokyo
Kashiwa-shi, Chiba-ken, Japan

**Andrea Schmidt**
Institute of Sport Science
University of Bremen
Bremen, Germany

**Alistair Shilton**
Department of Electrical and Electronic Engineering
The University of Melbourne
Melbourne, Victoria, Australia

**Kalvinder Singh**
Australia Development Lab IBM
Southport, Queensland, Australia

**Shudo Takenaka**
Department of Human and Engineered Environmental Studies
University of Tokyo
Kashiwa-shi, Chiba-ken, Japan

**Yen Kheng Tan**
Department of Electrical and Computer Engineering
National University of Singapore
Singapore

**Gerhard Tröster**
Wearable Computing Laboratory
Swiss Federal Institute of Technology Zurich
Zürich, Switzerland

**Brian A. Walker**
School of Electrical and Information Engineering
The University of Sydney
Darlington Campus, New South Wales, Australia

**Andrew J. Wixted**
Centre for Wireless Monitoring and Applications
Griffith University
Nathan, Queensland, Australia

**Tim V. Wrigley**
Faculty of Medicine, Dentistry and Health Sciences
The University of Melbourne
Melbourne, Victoria, Australia

**Huiru Zheng**
School of Computing and Mathematics
University of Ulster
Newtownabbey, United Kingdom

# 1

# *Sensor Networks in Healthcare: A New Paradigm for Improving Future Global Health*

**Daniel T.H. Lai, Braveena Santhiranayagam, Rezaul K. Begg, and Marimuthu Palaniswami**

**CONTENTS**

Introduction .......... 1
Applications of Healthcare Sensor Networks .......... 4
  The Young .......... 4
  Adults .......... 5
  The Elderly .......... 5
Engineering and Technical Challenges .......... 6
  Sensor Hardware .......... 7
  Sensor Fusion Algorithms and Models .......... 9
  Network Architectures and Telecommunications .......... 9
  Power Management .......... 11
  Security .......... 12
  Actuators .......... 12
Human-Centric Challenges .......... 13
Commercialization Challenges: Barriers to Successful Implementation .......... 15
Future Work and Future Directions .......... 17
References .......... 19

## Introduction

The growing demand for a better quality of life is poised to affect every aspect of the economy, especially the healthcare sector, as health is becoming an essential expectation of today's society. Current worldwide shortages in medical personnel have strained present healthcare systems immensely in the midst of a growing global population. A recent WHO report (2006) depicts critical shortages in healthcare providers, with 57 countries possessing less than 80% coverage of healthcare providers for their citizens. It is estimated that training the healthcare workforce to cope with these shortages by 2015 would incur annual costs of US$1.6–2.0 billion per country depending on its size. The cost of healthcare (measured as a percentage of the gross domestic product [GDP]), however, has also risen in tandem with rising costs of living. For example, in Australia the cost of healthcare was AUD87 billion (9% of GDP) in 2005–2006, while a decade ago the cost was 7.5% of GDP (Tegart 2010). The global need for healthcare is rising, as advances in health technology are poised to increase longevity. The escalation of demographic factors and rising costs require new healthcare solutions which are cost-effective and address the various demographic healthcare needs. Immediate plans for new technologies and paradigms must be actively pursued to provide efficient and effective avenues in response to the anticipated increase in healthcare demand.

Recent advances in sensor network technology have given rise to the notion of healthcare sensor networks (HSNs). HSNs are networks of devices designed with the primary function of monitoring human health. Unlike traditional wireless sensor networks (WSNs) which have been deployed for environmental and military applications, HSNs are required to monitor physical and physiological signals in a continually mobile environment, i.e., human activity. These HSNs may be directly deployed on the individual's body via body area networks (BANs; Yang 2006), in which case they move wherever the wearer goes (Figure 1.1), or they can be deployed in infrastructure via infrastructure area networks (IANs), e.g., in hospitals, aged-care facilities and homes, in which case they monitor several people continuously. It is envisioned that HSNs would reduce the strain on the present healthcare workforce by providing new autonomous monitoring services ranging from simple user-reminder systems to more advanced monitoring agents for preventive, diagnostic and rehabilitative purposes. For example, HSNs deployed for high-risk individuals would provide early warning services for the onset of cardiac ischaemia (heart attacks), epileptic seizures, diabetic attacks and falls in the elderly. The networks can be designed to automatically contact emergency services if the individual requires medical attention. HSNs would enable clinicians to evaluate the effectiveness of surgical interventions (rehabilitation), for example, the effectiveness of a new pacemaker. More ambitious applications focus on preventive efforts; HSNs, for example, could be deployed to monitor lower-limb locomotion, which would assist physiotherapy programmes to prevent the onset/progression of osteoarthritis. In research, wireless HSNs introduce the freedom of portability, which promises more accurate studies conducted in natural environments as opposed to the confinements of movement laboratories.

Despite sharing many general aspects of traditional sensor networks, HSNs face new research challenges originating from the focus on human and network interaction. This new focus requires that HSNs continuously sample human data at higher rates, process the data quickly and transmit it through a network that could be in continuous motion. Human data could consist of movement variables—e.g., walking velocity and limb positions—and physiological data—e.g., heart rate (electrocardiogram [ECG]), muscle

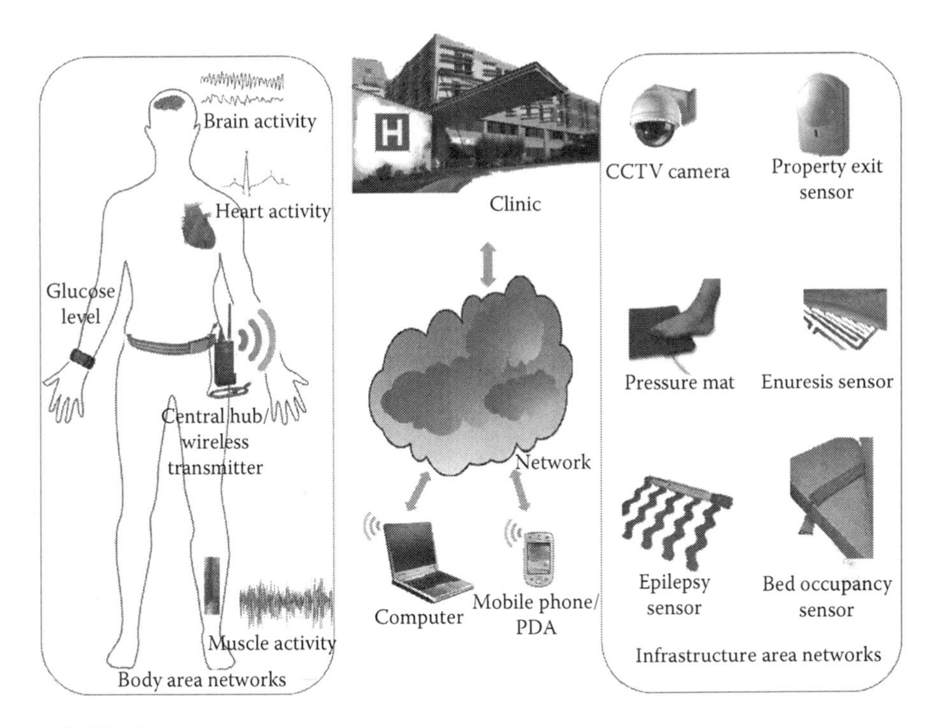


**FIGURE 1.1**
Schematic of a healthcare sensor network (HSN) consisting of body area networks (BANs) and infrastructure area networks (IANs) that send data through a network e.g., Internet to a monitoring station.

performance (electromyogram [EMG]) and cerebral functionality (electroencephalogram [EEG]). In contrast to temperature measurements taken at 10 min intervals, these health variables need at least 20 Hz or 1000 Hz sampling rates in the case of physiological signals. Instead of relaying ambient environmental readings, the network would have to process these data using lightweight signal-processing algorithms and either store or transmit the data or initiate an attached actuator, e.g., functional electrical stimulation units. The requirement that HSNs reside and operate in dynamic environments leads to power management issues for processing and transmission, and the fact that personal health data are collected and processed leads to security and privacy issues. This new set of requirements necessitates research in existing fields that explores new directions such as high-bandwidth low-power transmission protocols and efficient on-chip processing algorithms, as well as in exciting new fields such as biosensors, energy-harvesting schemes, nanoscale networks and lightweight security techniques.

The realization and implementation of HSNs will benefit the broader society, with users ranging from children with congenital disorders to the elderly population at high risk of chronic diseases and disabilities. The technology is set to reduce the strain on the present healthcare workforce by replacing manual monitoring, i.e., carers and nurses, with autonomous monitoring. HSNs will potentially provide continuous online information about the subject's health, together with detection, prediction, biofeedback and actuation capabilities. These will ultimately contribute to more effective medical practices, more efficient emergency response mechanisms and improved rehabilitation programmes. Further advances will ultimately see a reduction in system/deployment costs, thereby making the technology more accessible to society and more commonplace.

However, in order for HSN technology to realize its true potential, a significant amount of research and development is still required to address engineering, human-centric and commercialization challenges. In this opening chapter, we take a peek at the priority areas in healthcare which could see rapid deployment of HSN technology. We then briefly highlight the key engineering issues which remain research foci and the additional human-centric challenges that must be surmounted for successful uptake of the technology. Finally, we touch on several commercialization challenges that remain market barriers for HSNs. References to the 14 other chapters contained in this book are made at appropriate sections to direct the audience to further relevant details.

## Applications of Healthcare Sensor Networks

HSN technology is still an active area of research; however, numerous real-world applications have already been envisioned, providing a rich base for prototyping and testing. Portable BANs are designed to monitor physiological and physical processes in the human body, while infrastructure sensors can monitor human–environment interactions. BAN applications include personalized medical monitoring, fitness and exercise monitoring, and gaming and interactive entertainment. Sensors in infrastructure would be useful for monitoring falls, human physical activity and human–environment interactions. In the domain of health, these applications could be divided according to user groups, in this case, children (0–17 years of age), adults (18–64 years) and the elderly (>65 years).

### The Young

HSNs have been applied to the younger population in a number of ways. As children rely on others for assistance, HSNs can play an important role in providing the necessary safety and support. For infants, and children prone to falls, HSNs have been used as fall-detection devices as well as for monitoring their comfort levels (Alemdar and Ersoy 2010). This has been achieved through the use of wearable body sensor networks, which have been applied as an informative tool to obtain data relating to "children's corporal and intellectual information" (Rodrigues, Pereira, and Neves 2011). An example relating to children's comfort could be the creation of smart homes where the environmental temperature could be controlled using sensor technologies to maintain optimal room conditions. BANs have also been applied to prevent and manage asthma in young people, particularly through the collection and evaluation of localized environmental data (Rodrigues et al. 2011). In addition, monitoring physical activity in children would permit longitudinal studies on physical and mental development, as described further in Chapter 9.

### Adults

The anticipated effect of the use of HSNs on adults is that they would provide greater access to healthcare and make significant progress towards realizing the goal of pervasive healthcare (Varshney 2007). However, for adults, HSNs have so far been predominantly applied to individuals suffering from a particular illness or disability, as this area is in its infancy and currently undergoing lots of major developments. HSNs have been utilized to monitor medication consumption (Alemdar and Ersoy 2010), and this may be particularly useful for adults suffering from a cognitive disability, i.e., post-stroke, as mistakes in handling medication can be identified and prevented. Further, navigation systems have also been developed that use such sensors to aid individuals with cognitive disabilities to make decisions regarding moving to a destination or location of choice (Alemdar and Ersoy 2010).

### The Elderly

HSNs play an important role in assisting the elderly. With respect to day-to-day home activities, HSNs may be used to identify and isolate abnormal behaviours of the elderly to ultimately alert external healthcare supervisors of any risks posed. Body-worn sensors may also be applied to monitor the physiology and movement characteristics of elderly persons to facilitate the detection of falls and fall risks. Through their utilization as a method for locating individuals as well as monitoring medication consumption, HSNs have been applied to supervise the usage of medication by the elderly, specifically, to prevent mistakes in medicine selection. Additionally, HSNs continue to be used to monitor the health and physiological characteristics of elderly individuals (including tracking vital signs) and relaying this data to caregivers (Alemdar and Ersoy 2010; Nyan, Tay, and Murugasu 2008). However, in spite of these benefits, it has been suggested that elderly participants' perceptions regarding independence, privacy and cost could affect the applicability of these sensors (Steele, Lo, Secombe, and Wong 2009; see section on human-centric challenges in this chapter).

The global population is projected to increase, and advances in health have led to increased longevity. This phenomenon has caused an inadvertent increase in the number of older individuals within the community (generally persons over the age of 65) and has consequently led to an increased ageing population. Appropriate technologies are required to address the needs of an ageing society. Developing these technologies is likely to make a substantial contribution to solving the social and economic challenges posed by the change in global demographics.

Already the elderly population in Australia is projected to reach 7.7 million by 2050. In the past, the costs of ageing have not had a major impact on healthcare costs. However, the rising proportion of elderly individuals within the community is set to change this. It has been estimated that in Australia expenditures for aged care will increase from 4% of GDP (2009–2010) to 7.5% (2049–2050). Residential aged-care costs are also set to rise by 1%, and pensions are expected to increase from 1.2 to 3.9% (Tegart 2010).

## Engineering and Technical Challenges

There are several engineering challenges that need to be addressed in the design and implementation of sensor networks for healthcare. Most recent scientific reports have been application-specific, as is also seen in the other chapters of this book; however, the technical challenges can be segmented into six key areas as depicted in Figure 1.2.

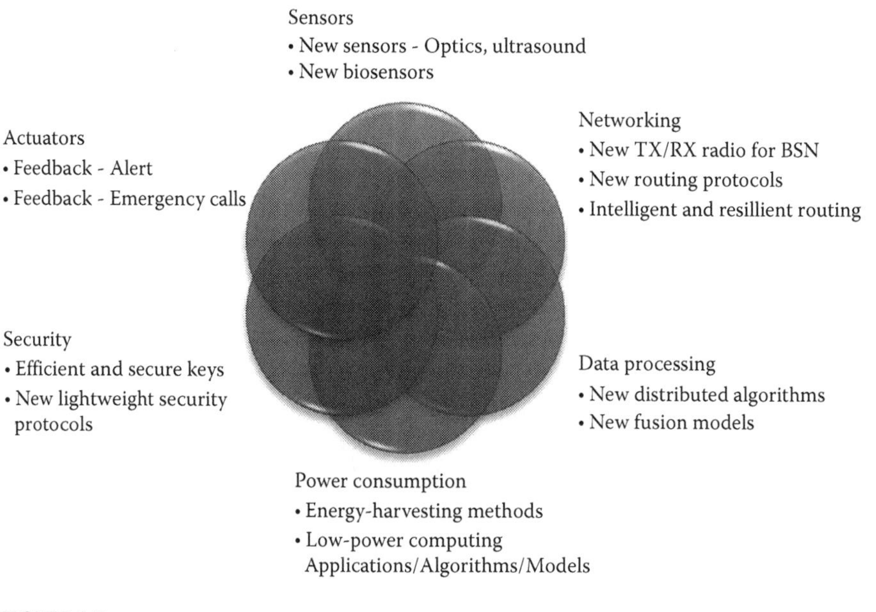


**FIGURE 1.2**
Engineering challenges that need to be addressed in the development of healthcare sensor networks.

## Sensor Hardware

Sensors transduce physical measurements such as length, density and pressure into an equivalent signal, usually an electrical signal which can be easily recorded and processed (Zhou and Hu 2008). During a biological event such as a heartbeat or muscle contraction, the electrical, chemical and mechanical activities that occur often produce signals that can be measured to understand the underlying physiological mechanisms of the biological system. Classification of biomedical sensors is done using fundamental measuring techniques as depicted in Table 1.1.

**TABLE 1.1**

Classifications of Biomedical Sensors

| Type | Description | Examples |
|---|---|---|
| Bioelectric sensors | When a nerve or muscle cell is stimulated beyond a threshold level, the cell generates action potential, which is measured by intracellular or extracellular electrodes | Electrocardiogram (ECG; heart activity), electrogastrogram (EGG; gastrointestinal tract), electroencephalogram (EEG; brain activity), electromyogram (EMG; muscle activity), electrooculography (EOG; retina activity) |
| Biomagnetic sensors | Electric field created by action potential accompanies a very weak magnetic field of a specific tissue or organ, which is measured by precise magnetic sensors/magnetometers | Magnetoencephalography (MEG; brain activity), magnetoneurography (MNG; peripheral nerves activity), magnetogastography (MGG; gastrointestinal tract), magnetocardiography (MCG; heart activity) |
| Biochemical sensors | Change in intracellular and extracellular ion (e.g., calcium and potassium) concentration, measured using appropriate chemicals | Glucose-level sensors, lactate-level sensors, blood oxygen concentration |
| Biomechanical sensors | Mechanical functions of biological systems, such as motion, displacement, tension, force, pressure and flow, are measured using mechanical sensors | Inertial sensors, blood pressure sensors, goniometers |
| Bioacoustic sensors | Acoustic noise caused by the vibrations of biological events is measured using acoustic transducers at the skin surface | Heartbeat monitor, blood flow monitor, ultrasound scanning |
| Biooptical sensors | Natural light produced by a biological event or introduced light properties are sensed by optical sensors | Fetus monitoring using fluorescence characteristics of the amniotic fluid (natural), red and infrared lights to measure blood oxygen level (introduced) |

**TABLE 1.2**

Requirements of a Biomedical Sensor

| Performance Requirements | User Requirements |
|---|---|
| Accuracy | User friendliness, easy calibration, simple interface, suitability |
| Precision | Long lifetime |
| Sensitivity | Cost-effective |
| Selectivity | Portability/wearability |
| Stability | Real-time response |
| High dynamic range | Ruggedness |

Following signal acquisition, biological signals are amplified, filtered, digitized, processed and stored/transmitted using electronic devices for further analysis of the biological event. Even though biomedical sensors serve as the interface between a biological and an electronic system, they should function in such a way as to not adversely affect either system. Table 1.2 tabulates some of the key requirements of a good sensor for biomedical and health sensing.

With the advent of microelectromechanical systems (MEMS) and nanotechnology, the sensors mentioned previously have started to be miniaturized, developed and made more commercially available for a variety of applications. MEMS are made up of microcomponents between 1 and 100 μm in size, and MEMS devices generally range in size from 20 m to 1 mm. MEMS consist of a central unit that processes data, the microprocessor, and several components that interact with the physical world, such as microsensors. *Nanobio* sensors operate on the basis of a structure–function relationship. Molecular recognition and the binding mechanism are the key tasks in detection of analytes such as glucose, cholesterol, amino acids, drugs and various ions ($Li^+$, $Na^+$, $K^+$, $Rb^+$, $Mg^{2+}$, $Ca^{2+}$ etc.; Schultz 1995). Various fabric nanosensors and implantable nanosensors are widely attracting many healthcare applications such as glucose-level monitoring and heart rate monitoring during exercise.

The other major group of sensors is the environmentally embedded sensors used in IANs. As is seen in Chapter 11, these sensors complement the biomedical sensors to enhance monitoring platforms for monitoring active living, rehabilitation and risk prevention. IAN sensors are mounted or installed in the natural environment to monitor the surroundings and the interaction of users within it; for example, bed occupancy sensors detect whether the bed is occupied and activate an alarm call if the user does not return to the bed within a certain time frame. Pressure mats are used to monitor walking activity in the deployed area; optical sensors are used to detect obstacles in the walkway; while infrared (IR) motion sensors are deployed to sense motion and automatically switch on the lights or alarms.

## Sensor Fusion Algorithms and Models

As with traditional wireless networks, one is faced with the challenge of interpreting the data collected or fusing the data. In the case of HSNs, data fusion is required to determine the state of health of the user. At present, research is still focused on building the networks rather than interpreting the data gained from each individual sensor in the network. As is seen in Chapters 7, 8 and 9, detecting movement using BAN technology is still an active field, even though research has been ongoing for almost a decade. Besides movement, the sensing of physiological states such as muscle and brain activity is also being explored, for example, in Parkinson's disease and stroke patients. Chapters 10 and 12 discuss the details of fusion algorithms for combining data from inertial and EMG sensors for detection of freezing gait and tremors in Parkinson's disease.

Most fusion algorithms are currently implemented on computers which have substantially more computing power than previously (10 years ago). It would seem probable that successful HSN implementations would see most BANs transmit raw data to a base-station facility where data processing and fusion can take place. However, newer paradigms of HSNs envision the capability to provide feedback to the user in real time. These proactive technologies would require that a significant amount of data processing be accomplished in the BANs themselves, and possibly in the future the BAN would be required to aggregate and process data from IANs as well. As such, fusion algorithms should be lightweight and require low computational effort. A potential candidate framework would be machine learning algorithms based on computational intelligence paradigms, e.g., artificial neural networks. Chapter 2 explores the use of these algorithms in some detail for movement monitoring.

Some researchers have devised application-specific methods to optimize functioning of HSN systems. Signal-processing models have been used to develop energy-efficient data transmission through burst data transmissions. The transmissions from the nodes were based on an integer optimization programme and a greedy algorithm which transmitted when a certain amount of body movement was detected (Ghasemzadeh, Loseu, Ostadabbas, and Jafari 2010).

## Network Architectures and Telecommunications

Current wireless technologies are based on radio frequency (RF) propagation and include well-known protocols such as IEEE 802.11 a/g/n (Wi-Fi), IEEE 802.15.1 (Bluetooth) and the recent ZigBee based on the IEEE 802.15.4 wireless personal area network (WPAN) standard. The challenge in using these protocols is that they require adaptation for medical applications since they were designed mainly for multimedia data applications. Both Wi-Fi and classical Bluetooth were designed for high data rates, and therefore the protocols were not optimized for efficient power usage.

The first major challenge in defining new wireless RF transmission protocols for health applications is the use of the RF spectrum so that designed HSN solutions can be used globally and legally. The industrial, scientific and medical (ISM), wireless medical telemetry service (WMTS), ultra wideband (UWB) and MedRadio (401–402 MHz and 405–406 MHz) bands are current prospective spectrum bands for implementation of HSN communications. Unfortunately, the ISM band is heavily used by existing 2.4 GHz technologies, while WMTS is restricted to facilities in the United States. The use of the IEEE 802.15.3a UWB puts it in contention with high-data-rate multimedia applications using wireless protocols, though UWB is mainly slated for short-range communications. The increasing potential of and need for networked medical devices communicating wirelessly, e.g., HSNs, has prompted the Federal Communications Commission (FCC) to consider using the 2.36–2.40 GHz band for BANs. An additional 24 MHz is being planned for the 413–457 MHz range, with this spectrum intended for use by medical microstimulator devices which can provide biofeedback capabilities (Patel and Jianfeng 2010).

In addition, new challenges in channel modelling, antenna design and protocol design for physical and Medium Access Control (MAC) layers must be faced. The inherent deployment of some HSN solutions as wearable networks which are frequently in motion gives rise to serious multipath effects which are frequently responsible for communication loss. There are several commercially available and ratified wireless protocols that can be used for the design of HSNs.

Classical Bluetooth (http://www.bluetooth.com) was designed as a short-range personal area network protocol (effective range 10–30 m) for networking up to eight devices in a piconet network. The protocol uses frequency-hopping spread spectrum (FHSS) over 79 channels and requires a master–slave pairing scheme which is slow (~3 s to pair) and not energy-efficient. The protocol supports up to 3 Mb/s in enhanced transmission over a range of 10–100 m. The original design used FHSS so that Bluetooth could coexist with Wi-Fi networks; however, this has also reduced interference with other protocols in the ISM band. Encryption consists of a 64- or 128-bit algorithm.

The new ZigBee protocol defines the network and security layers on top of the IEEE 802.15.4 WPAN standard. The ZigBee alliance recently ratified the ZigBee Health Care standard for low-power medical devices used for patient monitoring, disease management, aged care and asset tracking. This protocol is optimized for low duty-cycle applications where the radio transmitter spends the majority of its time in sleep mode. The ANT protocol is another low-power and low-data-rate communications protocol which uses Nordic's proprietary "burst rate" transmission. It also functions in the 2.4 GHz radio spectrum and avoids interference using a time-division multiple-access scheme and low-level security features. Other wireless protocol candidates include Zarlink (250 μW), 6LowWPAN (IPv6 over PAN) and the soon-to-be-released Peanut (Qualcomm), aimed at delivering a short range, low power and high bandwidth for BAN deployment.

Currently, a new wireless protocol standard encompassing the physical and MAC layers for BANs is being developed by the IEEE 802.15.6 Task Group. The development aims to design a protocol that requires ~10 mW and is capable of 1 Mb/s data rates. Currently, the layers of the protocol stack are being standardized to enable interoperability of plug-and-play devices, i.e., ISO/IEEE 11073 Personal Health Data working group, Continua Health Alliance.

A more novel wireless communications method is explored in further detail in Chapter 3, which is to transmit data using the human body as a communications channel. Intrabody communications is a relatively new technology which was proposed in 1996; data transmissions is sent through the human body (Zimmerman 1996). The advantages of this method are lower power consumption, minimal interference with RF protocols and better security (skin contact with the user is required for data transfer). The limitations, however, are a relatively lower bandwidth (10 Mb/s) and the unknown long-term health effects of continuously injecting electrical signals into the body. Work is still required to define channel models for the human body (i.e., multisegment and posture-related models, empirical studies to characterize the method) and to develop BANs which can reliably use this form of communication.

HSN network architectures currently follow the present wireless RF protocols and consist of star topologies where sensing nodes send information to a hub (aggregator node, base station) which collates the information and sends it to a database connected across other networks, e.g., the Internet. Figure 1.1 depicts a general HSN architecture consisting of several BANs and infrastructure sensors monitoring persons in the home environment.

## Power Management

As mentioned earlier, HSNs consists of BANs which users wear and IANs which are situated in the environment. IANs draw power from the main electrical lines and would be more than capable of continuous monitoring as long as there is electricity. BANs, however, require portable power, i.e., batteries, which have limited energy. This means that the user has to change or charge batteries, giving rise to new challenges with maintaining continuous monitoring, real-time data processing and data reliability. In addition, one has to deal with user irritation (e.g., elderly users would be most encumbered by the need to change batteries).

Research addressing power requirements can be categorized into two major areas: (i) the search for alternative portable energy and (ii) work on low-power electronics. Novel alternatives currently being pursued are addressed in detail in Chapter 4, which looks at harvesting energy from the environment to power BAN electronics. The current industry, however, is looking at the development of low-power electronics aimed at prolonging battery life. Zarlink Semiconductor is developing an ultra low-power (ULP) radio that uses 1.2 V (average power 250 μW), which is a magnitude of power lower than the 802.15.4 protocol (Das 2009). The company has also manufactured an ECG monitor requiring only 600 μW with a peak current of 7 mA. Current coin cell batteries operate at 3 V, which requires a DC regulator for delivering a 1.2 V supply. Besides the additional electronics, the regulator also dissipates 60% of the energy as heat, which presents a further engineering enigma, i.e., how to compare the efficiency of low-power portable devices when the batteries have higher voltage ratings.

## Security

Deployment of HSN technology has given rise to new security issues in addition to the traditional WSN issues of low power, computation and communications availability in the network. Network security implies an additional overhead in the way data packets are processed and transmitted, e.g., encryption and Cyclic Redundancy Check (CRC). This burdens individual network nodes with additional computations which use more power and potentially reduce network throughput. Most WSN security schemes try to find a compromise between good security and the physical node limitations. HSNs, however, add a further dimension of complexity to this problem.

HSNs generate and transmit diverse medical data depending on the deployment scenario such as in hospital, in ambulance and in-home monitoring. Whenever data transmission is conducted over physically insecure mediums, e.g., RF propagation, there is an inherent risk of the data being maliciously corrupted. Corrupted data or, worse yet, data that have been tampered with could be fatal, especially if medical decisions such as medicine dosages are decided based on the compromised medical data. In addition, insecure medical data also compromise the privacy of the patient, e.g., medical data records, and generate the potential for legal issues.

As is seen in Chapter 5, security in a sensor network should encompass aspects such as key establishment and setup, integrity and authentication, secure routing, resilience to network compromise and secure data aggregation. Active or passive attacks could occur on the network, and research into practical security protocols for HSNs is still in its infancy (Ng, Sim, and Tan 2006). The major challenges faced in this aspect include network resource constraints; i.e., BAN nodes may not currently have enough computational power to deal with commercial RF security implementations, dynamic networks which pose security risks, and the potential of data tampering by the users or people in contact with them.

## Actuators

An actuator is a mechanical or electrical device for moving or controlling a mechanism, thus enabling a system to perform a physical function. It is operated by an energy source, usually in the form of an electric current or hydraulic pressure, and converts it into motion. Generally, sensor outputs are used as system inputs to trigger the actuator based on an event; for example, fall-detection sensors, i.e., inertial sensors, detect the position of the centre of mass (COM), and if a fall is detected, an alarm (actuator) is triggered (Lindemann, Hock, Stuber, Keck, and Becker 2005). Another form of actuation is an autonomous connection to an emergency contact point to alert clinicians to the user's condition. These are fundamental types of actuation. Vibrotactile devices (consisting of miniature motors) are another example of actuators, which are used in fall prevention due to balance loss (Wall III, Wrisley, and Statler 2009). Leg actuators based on Functional Electrical Stimulation (FES) are examples of more advanced actuation schemes, while devices such as Medtronic neurostimulator devices assist in pain management. FES actuators are helpful to those who have movement difficulties, as the actuator attempts to artificially trigger muscle movement through electrical stimulation. These are more complicated assistive devices and usually expensive. Research in coupling actuators to HSNs, in particular, sensor networks on the body, is still new. Most of the actuators have been proposed for assisting in cases of movement disorders, e.g., fall risk or Parkinson's disease.

## Human-Centric Challenges

In addition to the previously mentioned technical challenges, HSN development involves the human user more than any of the previous sensor network applications (Hanson et al. 2009). Unfortunately, there has been a lack of study into the needs of users for HSN technology, which has resulted in a technology-push situation rather than a demand-pull centred around consumer requirements (Tegart 2010). At this stage, the likely cause could be the intensive research focus on getting basic WSN technology to work in the healthcare space, which in itself is still a major challenge. The immediate effect is that currently developed HSN solutions do not fulfil consumer needs, resulting in a slow uptake of the technology or complete abhorrence of it (in the case of the elderly population). To complicate matters further, the human-centric factors governing acceptance of HSN technology are user-dependent. For example, young children will likely change their behaviour if they are aware of being monitored (Chapter 9), while the elderly will likely use the technology if it does not restrict their independence in daily activities (Chapter 11).

The fundamental human challenges revolve around cost, operation constraints, system complexity and constant monitoring. We term these key factors the 4Cs, or four challenges/considerations (Table 1.3) that should always be considered in the design of all HSN technologies. It is a well-known fact that consumers want the most for their money, and hence commercial HSN developers are forced to find a compromise between, on the one hand, the

**TABLE 1.3**

The Four Key Human-Centric Challenges

| 4Cs | User Expectations |
|---|---|
| Cost | System is low cost or contains multiple functions commensurate with the price. |
| Constraint | System is portable/wearable, does not restrict daily activities, and can be used anywhere and everywhere. |
| Complexity | System is simple to operate and troubleshoot, does not have too many buttons, and produces outputs that are not confusing. |
| Constant | System operates transparently and continuously, and it is constantly present (ubiquitous). |

research and development costs and, on the other, the amount of functionalities supported by the HSN. HSN technology, particularly the BAN side, should be made as user-friendly as possible and not constrain user movement. This necessitates that the system be wearable, lightweight and easy to deploy yet robust and virtually unbreakable in daily use. This tall order is indeed a call for further advances in materials science and ergonomics design. System complexity, on the other hand, is an antagonistic requirement to the uptake of the technology; i.e., if the system is too complex for the elderly or young children to use, then it would be shunned by a majority of consumers. However, if the system is too simplistic, it may not contain sufficient functionality to complete its application task; e.g., simplistic sensor models may deliver inaccurate data. Worse still, users may find it unattractive for the price to be paid. Finally, as with all other technologies, HSN technologies are expected to monitor constantly and operate transparently to the user. Like the Internet, they are expected to work anywhere, everywhere (ubiquitous) and all the time. User transparency is a useful feature because it guarantees minimal tampering with internal HSN mechanisms and minimizes operation failures. However, this is a challenge, as different user groups pose different problems and requirements. Chapter 9, for example, demonstrates how the monitoring technology can be hidden from children by concealment in the heel of the shoe. Chapters 10, 12 and 13 describe the need for systems which can be used by people with movement, cognitive and cardiac disorders. Chapters 14 and 15 describe the need for the technology to be robust yet transparent to athletes for sports activities or exercise programmes.

The major anticipated users of HSN technology, however, would be the elderly population. A recent survey (Steele et al. 2009) on elderly people listed six major themes of concern for elderly users who were asked if they would be comfortable using HSN technology. These themes included continued independence, impact on their quality of life, robust HSN operation, personal preferences, specialized design requirements and the need for support. The primary concern was that HSN technology would change their ability to live freely (*constraint*) in their own homes, in the community (i.e., with family or friends) or in residential aged-care facilities. It was perceived that the technology should help them stay out of aged-care facilities as long as possible, indicating a declining interest in retiring to these institutions. Acceptance of HSN technology was also dependent on whether it would change their quality of life tremendously. Automated emergency calls were a welcomed function, while alerting family members to their social problems was an undesired function. In addition, the elderly preferred to have social contact with nurses and carers rather than remote monitoring solutions. This strongly suggests that HSNs would find most use as an assistive monitoring technology for the elderly instead of a full monitoring replacement solution for carers. Elderly users would require low-cost technologies and are likely to spend only on healthcare devices that are absolutely necessary. Unfortunately, at this point, most healthcare is treatment-based; that is, action is taken only after a serious ailment has occurred, e.g., a first heart attack. It is apparent that tremendous changes in user mindsets are required before HSNs become widely accepted. The elderly also prefer devices with high functionality, with reports suggesting that high functionality at low cost would be one of the main enticing factors for HSN acceptance.

Whether or not the 4Cs can be fulfilled in future HSN deployments remains solely an issue for debate. The chapters in this book have been structured to address aspects of the 4Cs and demonstrate how engineering compromises must be made in HSN solutions based on the user specifics of the application.

## Commercialization Challenges: Barriers to Successful Implementation

The increasing demand for adequate healthcare along with growing awareness of the power of advanced technology has created a market for providing new efficient healthcare services to the public. HSNs are most likely to find their market niche as monitoring and assistive devices. The global market for assistive devices is valued at US$66.1 billion (Datamonitor 2009a), or 29.8% of the global healthcare equipment industry, estimated to be worth US$148.4 billion in 2008 and growing annually by 4.5% (Datamonitor 2009b). This industry sector is forecast to have a compound annual growth rate of 6.2% during the five years 2008–2013, with an estimated value of US$200.8 billion by 2013 (Datamonitor 2009b). The secondary markets with a potential stake in new healthcare technology include healthcare distributors (market value of US$928.2 billion forecast for 2009; Datamonitor 2005), and industries such as footwear, health and life insurance (US$2696 billion in premiums; Datamonitor 2008), and pharmaceutical (US$750 billion in 2009, estimated by IMS Health (NYSE: RX)). The medical device companies currently majoring in the research, development and manufacture of small portable healthcare equipment are General Electric (GE) Health (11.7% of market share); (Datamonitor 2009b), Medtronic (9.10%) and Boston Scientific (5.4%). These companies are future potential developers of HSN technology, through licensing or via other IP agreements with different research centers, e.g., universities and government research centers.

The potential for local and specialized development of HSN products is also encouraging. For example, Australia currently imports 90% of its demand for medical equipment, with the United States having the largest market share (US$1.02 billion, 37.89%), followed by Germany (US$241.1 million, 8.51%) and Ireland (US$200.5 million, 7.08%) ("Aged Care Market in Australia," 2009). Many suppliers in the Australian industry are subsidiaries of overseas corporations such as Medtronic, Boston Scientific, Bard, Baxter Healthcare, Johnson & Johnson, St. Jude Medical and Stryker. These multinationals, however, face competition from Australian companies, and the A$4 billion Australian market for assistive aids and equipment is, therefore, price-sensitive and competitive. Already, specialized monitoring software and networked alarm systems are manufactured and sold by companies such as Smartlink International, Cural, Ultracare, FlashID International and Care Alert Pty Ltd. Distributors such as Biovation, Melbourne, sell such products through major retail chains such as K-Mart, Big W and Harvey Norman and via the Internet. Sales are also promoted by means of local representatives, by subsidiaries of the overseas companies such as Medtronic Australasia Pty Ltd and also through the websites of the manufacturing companies. Most of the HSN solutions are expected to be costly initially and to be sold through specific distributors. However, this is set to change as the technology becomes more commonplace. It is clear that the industry for these kinds of solutions is currently present, and pathways to market are already available.

The success of an HSN solution in the market depends very much on the reliable availability of data from different sources, particularly in this case data that can be obtained from modern sensor network platforms. The constraint, however, is that the critically needed data may be stored in different repositories and locations. Efficient and secure access to these resources can limit on the commercial exploitation capabilities by a variety of organizations. The exact business model of how this is managed needs further consideration. For example, HSN solutions for fall prevention can be wirelessly connected to existing monitoring systems in private homes and residential aged-care facilities. These monitoring systems, such as Australia's Medi-Link dialer from Smartlink International, GE's Quiet Care system (US) and Intel Health Guide (US), are part of the rapidly expanding home-based health monitoring market, estimated to grow from US$3 billion (2009) to 7.7 billion (2012) ("GE and Intel to Form Healthcare Alliance," 2009). One current successful business model in Australia is Harvey Norman, which markets the Medi-Link dialer for US$375 but offers renting and monitoring service options for A$1.08 per day.

Though HSN technology offers vast opportunities for self-managed care through the integration of appropriate ambulatory monitoring sensor devices and Web-enabled software platforms, uptake has been slow. It appears that this is the result of the stringent requirements for necessary medical compliance, patient confidentiality and security. There is also a huge gap between the expectations of the users and what is provided in terms of user-friendly devices and robust operation of these devices. The lack of clear standards in the field also makes it difficult to provide cost-effective and efficient maintenance. Quite often, maintenance can be so expensive that some of these systems hover around being not competitive commercially.

There is going to be a growing demand for health sensor-based services which use mobile communications technologies and are wirelessly networked. A mobile network operator can offer commercial services activating certain services in the ambience of the patient's home and body, including filling the critically needed gap in emergency services. There is still a bottleneck in appropriately linking advanced infrastructure such as 3G mobile networks or open-source gateways with HSNs, and this is limiting the vast potential for exploiting available communication infrastructures.

In addition, HSN technology is still being developed without proper user consultation, and this is flooding the market with unattractive solutions. Most companies developing the technology are still waiting for viable business models and pathways to commercialization. It is becoming increasingly evident that the original market segmentation and target market may have been misguided. HSN technology might not find a consumer base consisting of patients or people in need of medical monitoring. Rather, it appears to be currently positioned as specialist medical technology and thus would more likely require a clinician's recommendation and be deployed in the hospital environment. Market penetration in this market segment, i.e., selling HSN technology to medical practitioners and clinicians, is a whole new challenge!

## Future Work and Future Directions

At this stage, i.e., in 2010, HSN technology potentially offers a multifunction solution which could be cost-effective in terms of alleviating the pressure on the global healthcare system and reducing medical costs. Though work is still in its infancy and commercial uptake is still a major barrier, HSN technology promises to be a central research topic in the coming years. Research is still ongoing to solve the engineering and human-centric issues mentioned previously; however, this section lists some exciting directions which we envision HSN technology treading in the coming years.

Current HSN implementations are designed for monitoring the health of a single user and relaying it back to a central monitoring station, e.g., a hospital. Networks in the current context consist of several devices relaying information on a single user. However, this one-way interaction could be expanded to monitoring human–human interactions, human–environment interactions and environment–environment interactions resulting from human involvement. For example, monitoring of physiological and movement information could be used for crowd behavioural studies in busy areas, e.g., shopping centers, train stations and office complexes. Data can be used to infer user emotions, movements and health states. Here, HSN information would potentially be combined with information from building monitoring networks to design better building layouts or improve building emergency and evacuation procedures. The study of additional interactions would also aid in the design of the appropriate sensor fusion models and increase the usefulness of HSN-collected data.

The growing advances in artificial intelligence are also set to bestow HSNs with intelligent sensing and feedback capabilities. This field has already started to manifest itself in the wireless communications aspect under the theme of *cognitive radios*. Essentially, the problem is that the available wireless RF spectrum is becoming increasingly scarce as more and more applications compete for it. The observation, however, is that the spectrum is not fully utilized by all these applications, giving rise to the opportunity to share the spectrum among applications, as opposed to oversubscribing the spectrum to many users for a single application. Cognitive radios are RF transmitters equipped with intelligence to sense the RF spectrum for gaps in order to transmit or listen for data. This in turn has been recently proposed for WSNs under the broad theme of cognitive WSNs (Zahmati, Hussain, Fernando, and Grami 2009), where nodes transmit data asynchronously and intelligently in a network to minimize data bandwidth usage. In healthcare, the problem of spectrum scarcity is set to escalate in view of the continuous data monitoring requirements, the increasing demand for low-cost HSN monitoring services and an increasing user base in the future.

The cognitive theme can be extended to cognitive HSNs, which cover not only intelligent data transmissions but also intelligent data processing and intelligent feedback actuation capabilities. For example, a cognitive HSN would be capable of autonomously monitoring a user's heart function and be intelligent enough to detect high-risk events, e.g., heart failure, and contact emergency assistance. A more ambitious scheme would be to have the HSN administer proper dosages of medicine in response to the user's cardiac needs. In order to achieve this, work on lightweight intelligent sensing algorithms must first be accomplished. The algorithms must be robust to a broad range of user scenarios and work seamlessly in an embedded environment. Portable computing power is expected to increase in accordance with Moore's law (e.g., Apple's current iPhone 4 already has a 1 GHz CPU with 512 MB RAM), and hence computing capacity is now seen to be less of a serious issue in the future. The major research challenge, however, lies in the design of these intelligent sensing models, which must increasingly utilize user information to enable cognitive HSNs to effectively assess, detect and predict health risks.

Another idea requiring in-depth research is intelligent actuation, a theme which is only just emerging in HSNs. Intelligent actuation raises the prospect of proactive HSN technology, in line with the popular *prevention is better than cure* methodology. In the future, HSNs would, for example, be able to monitor a user's gait patterns, predict high-risk gait and autonomously actuate a person's limb to correct the gait. The NESS L300 foot drop stimulator (http://www.bioness.com) is an example of a first-generation device which tries to accomplish this. The device monitors lower-limb patterns and applies functional electrical stimulation to the peroneal nerve to actuate the foot dorsiflexor muscles. The device is not currently networked, does not collect any other user health information and does not apply intelligent stimulation (i.e., the foot dorsiflexion is not precisely controlled).

In order to solve the human-centric problems of user acceptance and transparent monitoring (the *constraint* and *constant* issues of the 4Cs), it is likely that we would need to consider nanostructures and biosensors (Chapter 6), which promise more compact and natural avenues for human health monitoring. These directions are indeed exciting and remain fertile research ground for future developments of sensor networks for healthcare. Will HSN technology pass the acid test with real human users? Only time will tell.

## References

Alemdar, H., & Ersoy, C. (2010). Wireless sensor networks for healthcare: A survey. *Computer Networks, 54*, 2688–2710.

Das, R. (2009). Wireless body area networks. Retrieved October 16, 2010, from http://www.energyharvestingjournal.com/articles/wireless-body-area-networks-00001430.asp?sessionid = 1

Datamonitor. (2005). Global health care distributors industry profile. (0199-2293) (http://www.datamonitor.com/).

Datamonitor. (2008). Global insurance: Industry profile. (0199-2087) (http://www.datamonitor.com/).

Datamonitor. (2009a). Global health care equipment & supplies: Industry profile (0199-2067) (http://www.datamonitor.com/).

Datamonitor. (2009b). Global Health Care Equipment: Industry Profile. (0199-2065) (http://www.datamonitor.com/).

GE and Intel to form healthcare alliance. (2009). Retrieved June 26, 2009, from http://www.intel.com/pressroom/kits/healthcare%5Cge_alliance%5Cindex.htm

Ghasemzadeh, H., Loseu, V., Ostadabbas, S., & Jafari, R. (2010). Burst communication by means of buffer allocation in body sensor networks: Exploiting signal processing to reduce the number of transmissions. *IEEE Journal on Selected Areas in Communications, 28*(7), 1073–1082.

Hanson, M. A., Powell, H. C., Barth, A. T., Ringgenberg, K., Calhoun, B. H., Aylor, J. H., et al. (2009). Body area sensor networks: Challenges and opportunities. *Computer, 42*(1), 58–65.

International Business Strategies. (2009). Aged care market in Australia. International Market Research Reports. Retrieved July 30, 2010, from www.internationalbusinessstrategies.com.

Lindemann, U., Hock, A., Stuber, M., Keck, W., & Becker, C. (2005). Evaluation of a fall detector based on accelerometers: A pilot study. *Medical & Biological Engineering & Computing, 43*(5), 548–551.

Ng, H., Sim, M., & Tan, C. (2006). Security issues of wireless sensor networks in healthcare applications. *BT Technology Journal, 24*(2), 138–144.

Nyan, M., Tay, F., & Murugasu, E. (2008). A wearable system for pre-impact fall detection. *Journal of Biomechanics, 41*, 3475–3481.

Patel, M., & Jianfeng, W. (2010). Applications, challenges, and prospective in emerging body area networking technologies. *IEEE Wireless Communications, 17*(1), 80–88.

Rodrigues, J., Pereira, O., & Neves, P. (2011). Biofeedback data visualization for body sensor networks. *Journal of Network and Computer Applications, 34*, 151–158.

Schultz, J. S. (1995). Biotechnology-based recognition elements. *IEEE Engineering in Medicine and Biology Magazine, 14*(2), 210–216.

Steele, R., Lo, A., Secombe, C., & Wong, Y. K. (2009). Elderly persons' perception and acceptance of using wireless sensor networks to assist healthcare. *International Journal of Medical Informatics, 78*(12), 788–801.

Tegart, W. J. M. (2010). Smart technology for health longevity: Report of a study by the Australian Academy of Technological Sciences and Engineering. *Australian Academy of Technological Sciences and Engineering (ATSE).*

Varshney, U. (2007). Pervasive healthcare and wireless health monitoring. *Mobile Networks and Applications, 12*, 113–127.

Wall III, C., Wrisley, D. M., & Statler, K. D. (2009). Vibrotactile tilt feedback improves dynamic gait index: A fall risk indicator in older adults. *Gait & Posture, 30*(1), 16–21.

Yang, G.-Z. (2006). *Body Sensor Networks*. London: Springer Verlag.

Zahmati, A. S., Hussain, S., Fernando, X., & Grami, A. (2009). Cognitive wireless sensor networks: Emerging topics and recent challenges. Paper presented at the Science and Technology for Humanity (TIC-STH), 2009 IEEE Toronto International Conference, September 26–27.

Zhou, H., & Hu, H. (2008). Human motion tracking for rehabilitation—A survey. *Biomedical Signal Processing and Control, 3*(1), 1–18.

Zimmerman, T. G. (1996). Personal area networks: Near-field intrabody communication. *IBM Systems Journal, 35*(3–4), 609–617.

# 2

# *Healthcare and Accelerometry: Applications for Activity Monitoring, Recognition, and Functional Assessment*

**Andrea Mannini and Angelo Maria Sabatini**

**CONTENTS**

Nomenclature ... 21
Introduction ... 22
MEMS Accelerometers ... 23
Estimation of Body Inclination, Balance Control, and Postural Transitions ... 25
The Sit-to-Stand Postural Transition ... 28
Temporal and Spatial Parameters of Gait ... 29
Walking Speed and Incline Estimation ... 29
PDA Assessment and EE Estimation ... 30
Human Activity Classification ... 33
Features for Movement Classification ... 34
Classification Methodologies ... 36
Clinical Applications of Accelerometers ... 37
Monitoring of Motor Fluctuations in Parkinson's Disease ... 38
Objective Skill Evaluation for Rehabilitation ... 38
Fall Detection ... 39
Conclusions and Future Trends ... 40
References ... 41

## Nomenclature

| | |
|---|---|
| **AC** | High-frequency component of the specific force reading |
| **ADL** | Activities of daily living |
| **AM** | Activity monitor |
| **ANN** | Artificial neural network |
| **COM** | Centre of mass |
| **CRF** | Conditional random fields |
| **CV** | Coefficient of variation |

| | |
|---|---|
| **DC** | Zero-frequency component of the specific force reading |
| **ECG** | Electrocardiography |
| **EE** | Energy expenditure |
| **EEG** | Electroencephalography |
| **FES** | Functional electrical stimulation |
| **FVC** | Fast video camera |
| **HMM** | Hidden Markov models |
| **ICA** | Independent components analysis |
| ***k*-NN** | *k*-nearest neighbor algorithm |
| **MEMM** | Maximum entropy Markov model |
| **MEMS** | Microelectromechanical system |
| **PCA** | Principal component analysis |
| **PDA** | Physical daily activity |
| **STS** | Sit-to-stand postural transition |
| **SVM** | Support vector machines classifier |

## Introduction

Currently, acceleration sensors (accelerometers) are being considered in several studies for integration in wearable sensor systems that are conceived for a wide range of applications. The intense research activity that is being undertaken and the strong commercial interests at stake have fostered the development of wearable sensor systems in biomechanics and biomedicine; the potential of wearable sensor technologies for advancing the practice of physical medicine and rehabilitation is thoroughly discussed in Bonato (2003). A truly remarkable point is the possibility of using inertial sensing technology in wearable sensor systems, which offer the opportunity for short- or even long-term patient monitoring in free-living conditions. Microelectromechanical systems (MEMS) *inertial sensors* can be conveniently integrated in wireless data acquisition units and can be easily worn by a human being, by virtue of their small size and weight. Moreover, their low power consumption promotes the development of battery-equipped devices with reasonable autonomy. Inertial sensors are self-contained motion sensors, and these features, along with the features already mentioned, explain their popularity in human body motion capture (Sabatini 2006a; Welch and Foxlin 2002; Yazdi, Ayazi, and Najafi 1998). These facts lead quite naturally to the envisioning of *ubiquitous health* scenarios, in which wearable sensor systems monitor patients at home, in a totally unobtrusive way, while sending information, e.g., through the Internet, to remote clinicians. In this context the scientific community is particularly interested in implementing algorithms and methodologies that are capable of eliciting clinically relevant information from such wearable sensor systems. In this chapter we limit ourselves to reviewing applications of accelerometry alone. Relevant applications that are considered in the following sections include:

- Analysis of human locomotion, posture, and balance
- Assessment of physical daily activities (PDAs) and estimation of energy expenditure (EE), e.g., obesity treatment, assessment of elderly quality of life and fall risk
- Functional assessment in the case of patients affected by different pathologies and treated using specific rehabilitation procedures (e.g., Parkinson's disease, post-stroke rehabilitation)

## MEMS Accelerometers

Over the last decade the interest in MEMS accelerometers has grown dramatically. The earliest MEMS accelerometer was announced at the end of the 1970s, although the first MEMS accelerometer to deeply penetrate the market was the Analog Devices ADXL50, a tiny device that has been used as an airbag sensitive element in automotive applications since 1993 (Bhushan 2010).

A MEMS accelerometer consists of a proof mass suspended by compliant beams and anchored to a fixed frame (Yazdi, Ayazi, and Najafi 1998), as depicted in Figure 2.1. This device measures the projection along its sensitive axis of the specific force $f$ to which it is submitted. The specific force additively

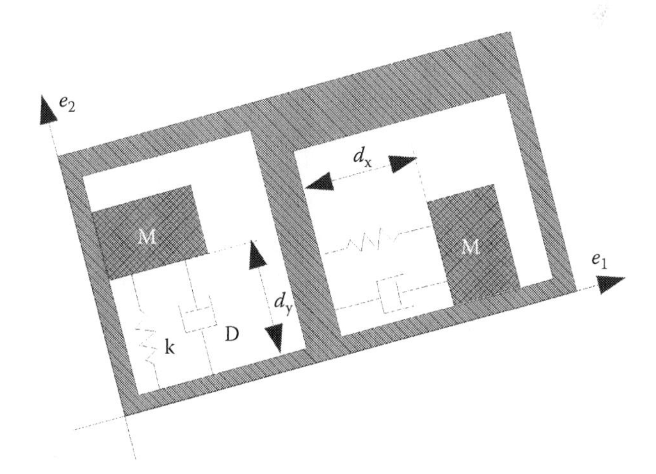

**FIGURE 2.1**
Schematic representation of a biaxial acceleration sensor based on a mass-spring-damper model.

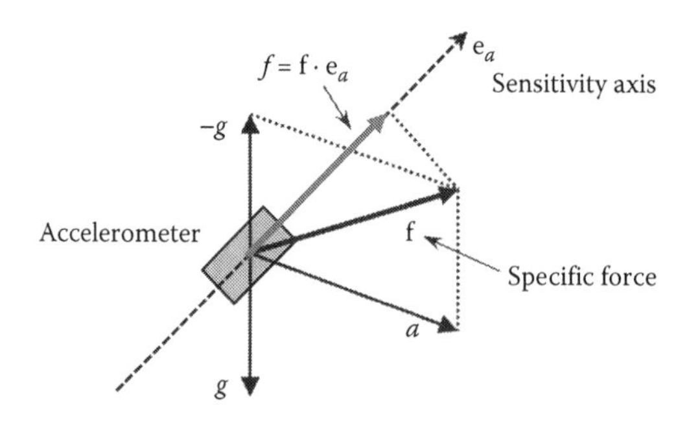


**FIGURE 2.2**
MEMS accelerometers measure the specific force projection $f$ along the sensitive axis.

combines the linear acceleration component, $a$, due to body motion, and the gravitational acceleration component, $-g$, both of which are projected along the sensitive axis of the accelerometer (Figure 2.2). In common parlance, the high-frequency component, also known as the *AC component*, is related to the dynamic motion the subject is performing, e.g., walking, hand weaving, head shaking, while the low-frequency component of the acceleration signal, also known as the *zero-frequency* (DC) component, is related to the influence of gravity and can be exploited to identify static postures.

The presence of these two components in acceleration signals creates some difficulties in the use of these sensors. Usually, we are interested in only one of these components; for instance, the gravitational component is the only one needed when the aim of the analysis is to estimate the inclination of a body part relative to the vertical (inclinometry). Otherwise, in applications where we need to estimate the three-dimensional (3D) position of monitored anatomical points relative to an inertial frame of reference, the component of interest is the inertial acceleration (Bouten, Koekkoek et al. 1997; Hansson et al. 2001).

MEMS accelerometers are immune to occlusions and interferences, which are known to plague standard motion-analysis technology; however, they are heavily influenced by noise and bias drifts, which may lead to gross errors if the position has to be estimated by double-time integrating noisy gravity-compensated acceleration signals: unless sophisticated and complex filtering is applied to raw data, the estimate becomes inaccurate after few seconds (Welch and Foxlin 2002).

The idiosyncrasies of unconstrained sensing by MEMS accelerometers provide an opportunity for assessing human movements without dedicated facilities being required. It is known that testing subjects in laboratory controlled conditions may lead to behavioural artefacts that are typical when standard motion-analysis technology is used (Bao and Intille 2004; Foerster, Smeja, and Fahrenberg 1999). Human accelerations during gait are measured using different measurement sites (head, low back, and tibia; Bouten, Koekkoek et al. 1997). In this latter work it is observed that, generally, the

**TABLE 2.1**

Amplitude and Frequency Range for Different Human Physical Tasks and Body Sites

| Activity | Acceleration Range ($g$) | Frequency Range (Hz) | References |
|---|---|---|---|
| Locomotion | Vertical:<br>–0.3–0.8 (upper body)<br>–1.7–3.3 (tibia)<br>Horizontal:<br>–0.2–0.2 (head)<br>–0.3–0.4 (low back)<br>–2.1–2.3 (tibia) | 0.8–5 | (Bouten, Koekkoek et al. 1997) |
| Running | Vertical:<br>0.8–3.9 (head)<br>0.9–5.0 (low back)<br>3.0–12.0 (ankle) | 1–18 (vertical, ankle) | (Bhattacharya et al. 1980) |
| Trampoline jumping | Vertical:<br>3.0–5.6 (head)<br>3.9–6.0 (low back)<br>3.0–7.0 (ankle) | 0.7–4 (ankle, low back, head) | (Bhattacharya et al. 1980) |
| Daily activities | | 0.3–3.5 | (Sun and Hill 1993) |

acceleration strength is higher in vertical than in the anteroposterior and mediolateral directions, and it shifts towards higher values from the cranial to the more caudal parts of the body. Similar considerations are valid for the bandwidth as well (Bhattacharya et al. 1980; Bouten, Koekkoek et al. 1997). Reference data for the amplitude and frequency ranges of acceleration signals are reported in Table 2.1.

Although high-frequency peak values are observed at the foot, especially during heel strikes, most signal energy is within the bandwidth 0.3–3.5 Hz in the case of activities of daily living (ADL; Sun and Hill 1993). Bouten, Koekkoek et al. (1997) suggest that accelerometers with a range of ±12 $g$ (±6 $g$ if placed at waist level) and bandwidths up to 20 Hz are perfectly suitable for PDA assessment ($g = 9.81$ m/s$^2$).

## Estimation of Body Inclination, Balance Control, and Postural Transitions

An interesting application of accelerometry concerns the determination of the inclination of different body parts with respect to the vertical. This application relies on the ability of accelerometers to measure the gravity component when they are static or slowly moving. An example of a simple

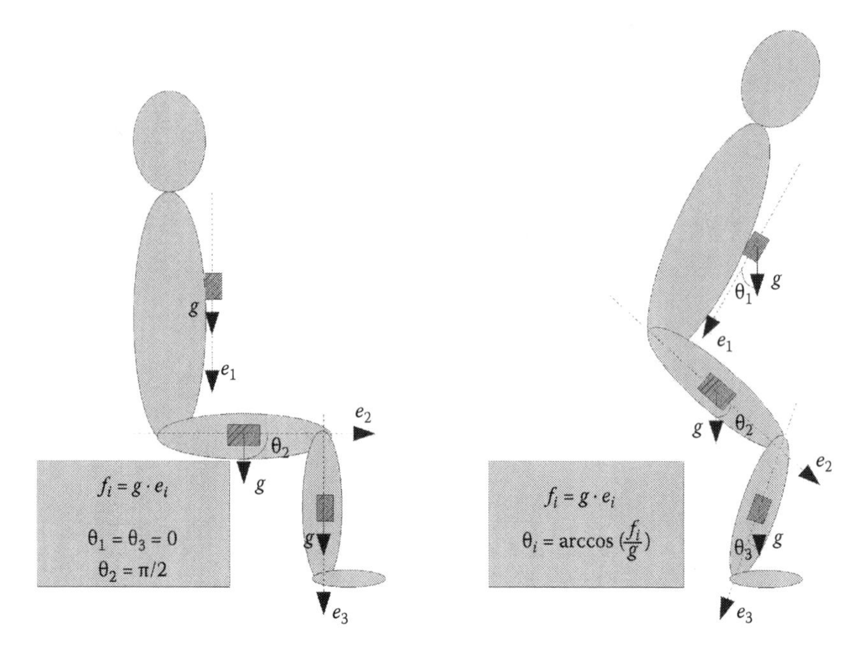


**FIGURE 2.3**
Data from three uniaxial accelerometers feed simple trigonometric functions for estimating the inclination of chest, thigh and shank relative to the vertical in the sit-to-stand postural transition.

biomechanical model for this application is depicted in Figure 2.3: Using uniaxial accelerometers with their sensitivity axes embedded in the sagittal plane, relevant joint angles can be measured. A similar approach is pursued in Lyons et al. (2005), where a method is described for recognizing static postures using two uniaxial accelerometers placed at the sternum and thigh.

A more complete postural characterization can be achieved by using multiaxial accelerometers. Mayagoitia et al. (2002) developed a system based on a triaxial accelerometer fastened to the back, in a position close to the whole-body centre of mass (COM), so as to perform an objective assessment of standing balance: discrimination between different conditions (open/closed eyes, feet together/comfortable, and so forth) is possible with accelerometers, as shown by comparative analyses with results obtained using standard force platforms. Hansson et al. (2001) evaluate the validity and reliability of inclination assessment through a head-mounted triaxial unit. The results confirm the validity of the approach for static and quasi-static conditions.

It should be noted that, since the line of gravity is used as a reference, no information on the rotation angle about the vertical axis (heading) is available, and hence it is not possible to estimate orientations in the three-dimensional (3D) space using accelerometers alone. In the attempt to determine heading, in addition to inclination, additional sensing approaches must be considered. In particular, gyroscopes (*gyros* for short) and magnetic sensors are often integrated in wearable sensor systems for biomechanical applications (Dejnabadi, Jolles, and Aminian 2005; Gallagher, Matsuoka, and Ang 2004; Giansanti, Maccioni, and Macellari 2005; Kemp, Janssen, and van Der Kamp 1998; J. Lee and Park 2009; Luinge, Veltink, and Baten 1999; Sabatini 2006b; Young, Ling, and Arvind 2007; Zhu and Zhou 2004). Gyros measure the angular velocities of rotations around their sensitive axes, while magnetic sensors can sense the Earth's magnetic field, the horizontal component of which, in principle, allows 3D orientation assessment (Kemp, Janssen, and van Der Kamp 1998). Although, conceptually, magnetic sensors seem well suited to solve the stated problem, their use is problematic due to the occurrence of magnetic field variations. These disturbances occur especially within man-made indoor environments, in the presence of ferromagnetic elements or electromagnetic interference near the sensor. The combination of gyros, accelerometers, and magnetic sensors to infer position and orientation leads to the *strap-down integration* approach. The device that integrates a triaxial accelerometer, a triaxial gyro, and a triaxial magnetic sensor together is called an *inertial measurement unit,* referred to as an IMU in the literature. Conceptually, the strap-down approach consists of the following steps:

- Gyro signals are integrated from specified initial conditions, and the rotation matrix (from the body frame to the inertial frame) is estimated; initial conditions are provided by combining gravity and magnetic field data.
- Acceleration signals are gravity-compensated; namely, the inertial component is separated from the gravity component by projecting the measured acceleration signals back to the reference frame.
- Inertial acceleration signals are double-integrated in time for position estimation (if required by the application).

One severe difficulty is the high bias drift that is experienced by inertial sensors, in particular gyros. In an attempt to circumvent the problem of time-integrating noisy signals, Dejnabadi, Jolles, and Aminian (2005) estimate knee flexion using two biaxial accelerometers and two uniaxial gyros fastened to the thigh and shank, without time integration of data. However, the classical approach behind strap-down integration revolves around data fusion and complementary filtering. The idea is to merge the information obtained from accelerometers, gyros, and magnetic sensors, in order to compensate for the limitations of single sensing elements (Table 2.2).

In this regard, the application of Kalman filtering is widespread: the contributions of different sensors can be optimally weighted; in addition, the state vector to be estimated by the filter can also include sensor biases (self-calibration; J. Lee and Park 2009; Luinge, Veltink, and Baten 1999; Sabatini 2006b; Young 2009; Zhu and Zhou 2004). The effects of disturbances (inertial acceleration for accelerometers, magnetic variations for magnetometers)

**TABLE 2.2**

Advantages and Disadvantages of Using Accelerometers, Gyros, and Magnetic Sensors for Position/Orientation Estimation

| | | |
|---|---|---|
| **Accelerometers** | Pros | Useful to build the vertical reference for estimating inclination |
| | Cons | The inertial component must be negligible<br>Moderate to negligible bias drift |
| **Gyros** | Pros | Angular speed is independent of the sensor placement (rigid body) |
| | Cons | High to moderate bias drift |
| **Magnetic sensors** | Pros | Useful to build the horizontal reference for estimating heading<br>Negligible bias drift |
| | Cons | Magnetic variations |

on the orientation estimates can be somewhat mitigated, while the filter is also theoretically capable of deciding the best use of the sensor information at hand.

A different data fusion approach that is applicable in this context is the *complementary filtering* approach. Young (2009) and Gallagher, Matsuoka, and Ang (2004) apply this strategy, which consists in combining different measurements of a signal with different noise properties to produce a single output. The assumption is that noise in the orientation estimation made by accelerometers and magnetometers is principally related to dynamic components (high frequencies), while low-frequency bias drift is the dominant source of error in gyros.

### The Sit-to-Stand Postural Transition

In older people, loss of muscle strength is a strong predictor of falls (Zijlstra et al. 2010), and the sit-to-stand (STS) postural transition is commonly considered as one of the most strength-demanding PDAs (Kralj, Jaeger, and Munih 1990). This postural transition is therefore widely studied, with the aim of quantifying the ability of patients to perform it. Kralj, Jaeger, and Munih (1990) introduced a standardization of the phases occurring in an STS that is commonly adopted in the literature; Janssen, Bussmann, and Stam (2002) discuss the main determinants that have an influence on STS. Janssen et al. (2005) estimate the temporal parameters that characterize an STS using two biaxial accelerometers placed on the trunk and thigh; they also investigate patterns of acceleration data for different STS speeds. The use of accelerometers placed in different body sites for estimating STS time parameters is successfully compared to different strategies such as force platforms (Zijlstra et al. 2010), motion-analysis stereophotogrammetric systems (Boonstra et al. 2006; Janssen et al. 2005), and fast video cameras (FVC; Giansanti et al. 2007). Other studies concentrate on STS detection through wavelet signal decomposition and machine learning techniques (Bidargaddi et al. 2007; Najafi et al. 2003).

## Temporal and Spatial Parameters of Gait

Gait analysis plays a fundamental role for functional assessment of PDAs. The aim of such an analysis is to estimate parameters that are capable of providing an objective description of this functional task. Gait is described by both temporal and spatial parameters. *Temporal parameters* refer to the time localization of specific gait events (gait phases), while *spatial parameters* relate to the description of how the lower limbs are positioned in space (Rueterbories et al. 2010). Typically, the first and second time derivatives of vertical or anteroposterior acceleration waveforms, and/or their time correlations, are computed (Hanlon and Anderson 2009; Moe-Nilssen and Helbostad 2004; Willemsen, Bloemhof, and Boom 1990). The main limitation of these approaches is related to how to prevent sensor jolting during foot impacts with the ground, which may cause the insurgence of noxious artefacts with resulting difficulties in signal interpretation. Nonetheless, several works show the reliability of acceleration recordings and related signal features when machine learning algorithms are created and tested in systems intended for automatic detection of gait phases (Shimada et al. 2005; Williamson and Andrews 2000). Interesting results in terms of detection specificity/sensitivity are reported even for pathological gait, including post-stroke (Shimada et al. 2005; Willemsen, Bloemhof, and Boom 1990) and post-arthroplasty (Aminian et al. 1999). In addition to gait-phase detection, accelerometers can provide some functional indicators that are useful for pathological gait assessment such as, gait symmetry and regularity (Aminian et al. 1999; Moe-Nilssen and Helbostad 2004). Step length can be estimated either through biomechanical models (i.e., inverted pendulum; Brandes et al. 2006; Zijlstra 2004) or by combining estimates of temporal parameters and walking speed (Moe-Nilssen and Helbostad 2004).

### Walking Speed and Incline Estimation

Walking speed and incline are parameters related to locomotion that are interesting to researchers in many areas, including biomechanicists, nutritionists, and sports therapists. Indeed, these parameters influence the estimation of additional biomechanically interesting quantities, such as physical activity and EE. Aminian et al. (1995) propose an approach based on artificial neural networks (ANNs) to estimate each of these quantities during treadmill walking using a uniaxial accelerometer placed at the heel and a triad of uniaxial accelerometers placed at the waist. Twenty features are extracted from the acceleration data. A similar approach is pursued by Y. Song et al. (2007); they propose a system for distance estimation during treadmill walking and running using a single triaxial unit placed at the sternum. Walking and running speed are then calculated by segmenting strides and dividing estimated distances by elapsed times.

Barnett and Cerin (2006) use a commercial device for physical activity assessment, namely the MIT/CSA Actigraph. They find high correlations between its output and the walking speed of the user. Yeoh et al. (2008) correlate walking speed with accelerations averaged over time windows lasting 2 s, using three biaxial units placed at the waist and thighs. In our laboratory, the application of machine learning methods in solving this problem has been tested. Selected features extracted from a triaxial accelerometer at the thigh are used with success in a portable system for measuring walking/running speed (on a treadmill; Mannini and Sabatini 2011).

## PDA Assessment and EE Estimation

Accelerometers are capable of measuring the intensity and frequency of human motion. This capability allows the quantification of PDAs and EE. For proper management of pathologies such as nutrition disorders and cardiovascular diseases, quantitative data about PDAs and EE are fundamental (Plasqui and Westerterp 2007). Rehabilitation, sleep disorders, and attention and personality disorders, notably in children, can also be investigated through accelerometry-based physical activity monitors (AMs; Tryon and Williams 1996; K. Zhang et al. 2003). In occupational and sports medicine, EE estimates are a relevant source of information in evaluating mechanical work efficiency, endurance, respiratory intake, and the thermoregulatory acceptability of working conditions (Capodaglio, Imbriani, and Capodaglio 1993). Estimating metabolic EE during ADLs is quite complex. The gold standard in EE estimation is the doubly labelled water technique, which requires nuclear medicine facilities as well as long time monitoring and yields results with low time resolution. Other approaches are room calorimeters (highly expensive and dedicated chambers) and indirect calorimetry based on pulmonary gas exchange, heart rate monitoring, pedometers, and surveys or questionnaires (LaPorte, Montoye, and Caspersen 1985). With all these techniques a reliable estimation of EE in daily living conditions is impractical (Bouten, Koekkoek et al. 1997). Heart rate monitors and pedometers are known to deliver unreliable estimates (Eston, Rowlands, and Ingledew 1998; Wong et al. 1981), and hence body-fixed motion sensors appear as the best alternative. Early studies conducted in the 1980s by Wong et al. (1981) and Montoye et al. (1983) confirmed that accelerometers outperform pedometers in PDA-related assessments. However, it is known that using motion sensors for estimating EE may lead to AMs that overestimate activity if the force/displacement is low, whereas they underestimate it if this ratio turns out to be high (Montoye et al. 1983). Several systems are now commercially available, as shown in Table 2.3.

Commercial monitors usually assess physical activity levels by *activity counts*, which are generated by applying suitable thresholds to the magnitude

**TABLE 2.3**

Commercial Accelerometers for Physical Activity Assessment and Monitoring

| Commercial Name | Sensors | Placement | Price (US$) | Mass (g) | Website |
|---|---|---|---|---|---|
| ActiGraph (aka MTI/CSA) | 1 × 1D | Midline hip | Out of production | 85 | http://www.theactigraph.com/ |
| ActiGraph GT3X | 1 × 3D | Midline hip | 300 | 27 | http://www.theactigraph.com/ |
| Caltrac | 1D | | Out of production | 78 | NA |
| Tracmor | 1 × 3D | | NA | NA | NA |
| Actiwatch Actical | 1 × 3D | Wrist, hip, leg | 1300 | NA | http://actiwatch.respironics.com/ |
| Philips DirectLife | 1 × 3D | Pocket | 100 | | http://www.directlife.philips.com |
| XSens MVN Biomech | 17 × 3D IMUs | Whole body | NA | NA | http://www.xsens.com/en/mvn-biomech |
| TriTrac RT3 | 1 × 3D | Hip | 500 | 72 | http://www.stayhealthy.com |
| Sensewear armband | 1 × 2D | Posterior midhumeral or lateral triceps | 1200 | 82 | http://www.sensewear.com |
| IDEEA | 5 × 2D | Shanks, thighs, sternum | 5000 | 59 | http://www.minisun.com |
| ActiBelt | 1 × 3D | Hip | NA | NA | http://www.actibelt.com/ |
| Biotrainer, Biotrainer II, Biotrainer Pro | 1 × 1D | Hip | 300 | | http://www.imsystems.net |
| Actitrac | 1 × 2D | Hip | NA | 34 | http://www.imsystems.net |
| Lifecorder | 1 × 1D | Hip | 160 | NA | http://www.thepedometercompany.com |
| StepWatch | 1 × 2D | Ankle | NA | 38 | http://orthocareinnovations.com |
| ActivPAL | 1 × 1D | Chest, thigh, foot | NA | 15 | http://www.paltech.plus.com |
| Apple iPhone | 1 × 3D | Pocket | 500 | 137 | http://www.apple.com |
| Nokia sports phone | 1 × 3D | Arm strap | NA | 103 | http://www.nokia.com |
| Nike + iPod sport kit | 1 × 3D | Shoe | 30 | 7 | http://www.nike.com |
| Nintendo Wii-mote | 1 × 3D | Handheld | 40 | 180 | http://www.nintendo.com |

of acceleration signals from uni-, bi-, and triaxial accelerometers. These quantities showed strong correlations with EE (Eston, Rowlands, and Ingledew 1998; Meijer et al. 1989; Plasqui and Westerterp 2007; Tweedy and Trost 2005; Welk et al. 2000). Commercial devices have been also tested on pathological subjects (Jacobi et al. 2007; Tweedy and Trost 2005). The evidence is that the EE estimation is generally reasonably accurate, although anomalously large errors are sometimes encountered in assessing individual pathological subjects. The early generation of commercial monitors was equipped with uniaxial sensor units. The use of triaxial sensors is more appropriate since the dominant directions of acceleration profiles may change depending on the activity performed (Bouten, Westerterp, and Verduin 1994). The same research group evaluates the effects of the placement and orientation of body-fixed accelerometers in the assessment of EE during walking (Bouten, Sauren et al. 1997). They conclude that, although acceleration profiles with markedly different magnitudes are expected when the sensor placement site changes even a little, their correlation with EE is generally high during walking. Commercial AMs are usually placed at the waist, hip, or back, in a position close to the whole-body COM. Although this aspect does not limit system performance in assessing locomotion tasks, severe underestimation is possible in the presence of significant upper-limb movements (Welk et al. 2000). These movements cannot be sensed unless a specific sensor unit is placed, e.g., at the wrist.

A different approach to improving the accuracy of EE estimation consists in adding information instead of adding signal sources. This would be of great interest in order to avoid the "Christmas tree effect" and its adverse effects on patient compliance. Adding information means devising sophisticated data processing methods in order to enhance the EE estimation process. It is customary to correlate activity counts and EE values, with highly satisfactory results in the case of walking; however, the explained variance in the EE-activity count regression can be significantly lower, using the same activity monitor, when activities other than walking are tested (Swartz et al. 2000). In attempting to improve the estimation accuracy, it would be fundamental to formulate specific activity-dependent regression models, as proposed, e.g., in Dongwoo and Kim (2007) and Hyun Choi et al. (2005). This consideration opens the way to the application of machine learning methods aimed at identifying the activity before applying the specific regression model (see the following section for further details of this approach). At the first level, two regression models can be adopted, one model that is specific to locomotion tasks (walking and running) and a second that is specific to other activities. Crouter, Clowers, and Bassett (2006) propose a simple solution in this direction based on a threshold classifier applied to the coefficient of variation (CV) of activity counts totalled in time windows lasting 10 s. Van Hees, van Lummel, and Westerterp (2009) implement a classifier that is capable of recognizing three static postures and walking, again with a threshold classifier applied to the CVs of windowed activity counts. In the IDEEA device (K. Zhang et al. 2003; K. Zhang, Pi-Sunyer, and Boozer 2004), 32 types of physical activities are classified, while EEs incurred during sedentary postures are estimated using biometric and personal information. Machine learning is applied to directly estimate EE from accelerometer data using ANNs applied to selected signal features (Rothney et al. 2007; Staudenmayer et al. 2009).

Finally, it is worth noting that, at the time being, we are witnessing to a proliferation of high-tech utilities (such as smart phones, game consoles, and other multimedia devices) that incorporate motion sensors, such as accelerometers (see Table 2.3) and gyros. Interesting applications for activity recognition and physical activity assessment are being investigated in these systems, which can exploit their high computational capabilities (C. Chen, Anton, and Helal 2008; Ketabdar and Lyra 2010; Khan et al. 2010; Kwapisz, Weiss, and Moore 2010; Longstaff, Reddy, and Estrin 2010; Yang 2009).

## Human Activity Classification

As already mentioned, signal features extracted from acceleration data can be used to classify physical activities (either static postures or dynamic motions). Apart from the development of smart AMs, automatic classification of physical activities can be highly valuable in pedestrian navigation systems, ubiquitous intelligence systems, and advanced human–machine interfaces. According to its definition by Dey and Abowd (2000) page 1, the *context* is "an implicit situational information that human uses to increase the conversational bandwidth and that cannot be easily transferred to the interaction of human with computers." In this sense, the study of the activity or posture being performed by a human being can be considered as the study of a kind of context.

Human activity can be recognized through ultrasonic or electromagnetic localization systems (Su et al. 2000) or optoelectronic marker-based or markerless motion analysis (Begg and Kamruzzaman 2005; Poppe 2007). The advantage of using motion sensors such as accelerometers for accomplishing this task is that they are sourceless, in contrast with the other methods mentioned, which require external sources operating in relatively structured environments.

In general terms, the architecture of a pattern recognizer revolves around several blocks (Jain and Duin 2000; Mannini and Sabatini 2010), as represented in Figure 2.4. When developing a machine for automatic classification of physical activities, the first critical decision concerns how many accelerometers have to be integrated in the wearable sensor system and at which anatomical points they would be optimally placed. The positioning of accelerometers for activity recognition is still an open research theme. Several

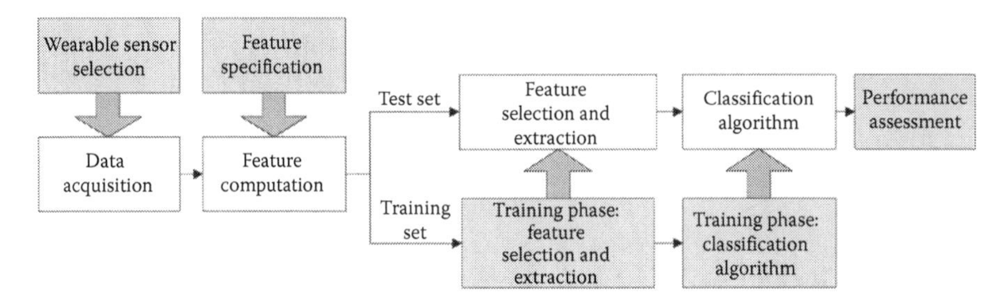


**FIGURE 2.4**
Conceptual scheme of a generic classification system with a supervised learning approach for human activity recognition purposes. (From Mannini, A. and Sabatini, A. M., *Sensors*, 10(2), 1154–1175, 2010.)

studies report the classification accuracies achieved using single-sensor or multisite approaches for a specific set of physical activities (Altun, Barshan, and Tunçel 2010; Mannini and Sabatini 2010; Preece et al. 2009). Intuitively, more complex sensor configurations will have to be devised if the tested activities grow in number and complexity, especially when all parts of the human body are involved in their execution (Bao and Intille 2004).

## Features for Movement Classification

In human activity classification based on accelerometry, many different time and frequency domain features can be proposed and successfully tested (see Table 2.4). They are evaluated using data from partially overlapping sliding windows (Bao and Intille 2004).

To avoid being plagued by the *curse of dimensionality* (Jain and Duin 2000), the dimensionality of the feature space would be reduced before applying the classification algorithm. This is achieved using a *feature selection* or *extraction* algorithm. Feature selection algorithms attempt to select the variables that are endowed with the highest discriminative capabilities. Typical algorithms are based on the following approaches (Jain and Duin 2000):

- Individual search, in which each feature is evaluated separately and those whose discriminative abilities are better are retained.
- Sequential search, in which a fixed or variable number of features are added and/or removed at each step of an iterative procedure. The effects of these modifications are assessed in terms of the resulting discriminative power.

Feature extraction algorithms are aimed at extracting the most relevant information from the data, without necessarily preserving the physical information conveyed by any single feature. *Principal component analysis*

**TABLE 2.4**
Signal Features for Activity-Classification Tasks

| Features | Description | References |
|---|---|---|
| Raw data or low-/high-frequency components | Acceleration data can be used as features or processed in order to extract low-frequency components (gravitational, related to posture) and high-frequency components (related to movement) | (Allen et al. 2006; Bao and Intille 2004; S. W. Lee and Mase 2002; Ravi et al. 2005; Veltink et al. 1996) |
| Signal magnitude and signal magnitude area | Suitable for distinguishing between static and dynamic activities | (Allen et al. 2006; Khan et al. 2010) |
| Signal morphology | Acceleration profiles for different movements can be correlated to template realization of them (template matching) | (Veltink et al. 1996) |
| Jerk and snap | Acceleration first- and second-order time derivatives | (S. W. Lee and Mase 2002) |
| Mean, variance, skewness, and kurtosis | Statistical analysis is often evaluated from sliding windows | (Aminian et al. 1999; Bao and Intille 2004; S. W. Lee and Mase 2002; Ravi et al. 2005; Van Laerhoven and Cakmakci 2000; Veltink et al. 1996) |
| Pearson's correlation coefficients | Cross-correlation coefficients extracted from different acquired signal windows (usually related to pairs of body-fixed sensors to assess motion relationships) | (Bao and Intille 2004; Ravi et al. 2005) |
| Spectrum of signals and spectral energy | Frequency domain information | (Bao and Intille 2004; Bussmann et al. 2001; Foerster, Smeja, and Fahrenberg 1999; Karantonis et al. 2006; S. H. Lee et al. 2003; Ravi et al. 2005; Van Laerhoven and Cakmakci 2000) |
| Frequency domain entropy | Frequency domain complexity feature | (Bao and Intille 2004) |
| Fractal dimensions | Useful feature to infer about movement complexity | (Sekine et al. 2002) |
| Wavelet coefficients | Coefficient in a time frequency representation | (Mantyjarvi, Himberg, and Seppanen 2001; Najafi et al. 2003; Sekine et al. 2000, 2002; K. Song and Wang 2005) |

(PCA) is the most common approach; PCA attempts to rotate the reference frame of the feature space along those directions where the data exhibit the highest variability. Dimensionality reduction is obtained by retaining these directions in the succeeding steps of the analysis. An alternative to PCA is *independent component analysis* (ICA), whose goal is to separate the additive independent components that are present in the data.

## Classification Methodologies

Machine learning algorithms for automatic classification are distinguished according to the nature, supervised or unsupervised, of the training process. Supervised training requires annotating the acquired dataset. Generally, a problem of automatic classification of human physical activities can be tackled using a supervised approach (Mannini and Sabatini 2010). A further distinction can be made based on how the validation process is implemented:

- *Individual validation.* Each subject is considered in isolation from the others; one classifier is then trained and tested for each subject.
- *Cross P-fold validation.* Upon data randomization, the dataset is split into $P$ subsets, each of which contains data from different subjects. Each subset is considered as the validation set, while the remaining $P - 1$ subsets are used for training. The procedure is repeated $P$ times.
- *Leave-1-out.* Data relative to one subject are considered as the validation set of the algorithm, which is trained over the data from the other subjects. The procedure is then repeated for each subject.

Typically, as a consequence of the high inter-individual variability expected in human movements, the last approach is less efficient, especially when the number of tested subjects is limited.

Classification algorithms for human activity can be divided into two main approaches: single-frame and sequential. A *single-frame classifier* works by assigning a label to each data frame it receives as its input, without any regard to previous assignments. Conversely, a *sequential classifier* takes the *past* classifications into account in order to orient the decision on the current feature vector. Single-frame classifiers can also be subdivided into three main classification approaches: probabilistic, geometrical, and template matching (Mannini and Sabatini 2010). All these different approaches to the classification have been pursued for human activity analysis. In the probabilistic approach a feature vector $\mathbf{x}$ is classified as a member of the class $C_i$ that maximizes the conditional probability $p(\mathbf{x}|C_i)$ with $i = 1,\ldots,C$, where $C$ is the total number of classes. Some examples of this approach are the naïve Bayesian classifier (Bao and Intille 2004; Ravi et al. 2005), Gaussian mixture models (Allen et al. 2006; Ketabdar and Lyra 2010; Mannini and Sabatini 2010), logistic regression (Kwapisz, Weiss, and Moore 2010; Mannini and

Sabatini 2010), and the Parzen classifier (Mannini and Sabatini 2010). The geometrical approach focuses on the definition of decision boundaries on the feature space. Examples of this kind of machine learning algorithms that have been specifically applied to human activity recognition are the support vector machines (SVM; Lustrek and Kaluza 2009; Ravi et al. 2005), *k*-nearest neighbor classifier (*k*-NN; Ravi et al. 2005; K. Song and Wang 2005), ANNs (Mantyjarvi, Himberg, and Seppanen 2001; Randell and Muller 2000; Van Laerhoven and Cakmakci 2000), and threshold-based algorithms (Bussmann et al. 2001; Karantonis et al. 2006; S. H. Lee et al. 2003; Najafi et al. 2003; Sekine et al. 2000). Finally, the template-matching algorithm works by comparing acquired data windows of features with predefined activity templates (Foerster, Smeja, and Fahrenberg 1999; Veltink et al. 1996). Single-frame classifiers can also be merged in *co-learning* approaches (Longstaff, Reddy, and Estrin 2010; Lustrek and Kaluza 2009; Ravi et al. 2005). In such strategies different classifiers are trained, and their outcomes are finally combined.

In the sequential approach to classification we orient ourselves to make inferences on, say, *sentences* of activities, namely, a chain of single activities, just as words are chained in sentences in similar problems arising in speech recognition: the sequential approach turns out to be a typical problem of *sequential supervised learning* (Dietterich 2009). Hidden Markov models (HMMs) have already found applications in automatic classification of human physical activities (Mannini and Sabatini 2010; Yang 2009). However, HMMs and related algorithms rely on a set of strong statistical assumptions (discriminative modelling), which motivated the search for alternative algorithmic solutions (conditional modelling): among them, *maximum entropy Markov models* (MEMM) or *conditional random fields* (CRF) are very promising (Chieu, Lee, and Kaelbling 2006; Sminchisescu, Kanaujia, and Metaxas 2006; Vail, Lafferty, and Veloso 2007; Vail, Veloso, and Lafferty 2007).

## Clinical Applications of Accelerometers

As explained in Bonato (2005), wearable sensor systems based on accelerometers have evolved over recent years from *simple* assessment of mobility and level of functional independence to the capability of capturing the kinematics of human motion and even muscle activity patterns (in conjunction with other sensing means, e.g., true inertial sensing and electromyography) that are involved in the execution of given functional motor tasks.

Further investigations have focused on the use of wearable sensor systems to promote some significant advancement in the practice of physical medicine and rehabilitation. An objective and reliable functional assessment tool can in fact greatly aid the clinician in diagnosis and patient management for many pathological conditions (Bussmann et al. 2004).

## Monitoring of Motor Fluctuations in Parkinson's Disease

The severity of motor disorders in Parkinson's disease (e.g., tremor, dyskinesia, bradykinesia) can be assessed through accelerometers. Ambulatory monitoring of tremor over long time observation periods can be achieved using a wrist-worn Actigraph (Van Someren et al. 2006). This system is capable of discriminating tremor from other movements and of estimating the period lengths and the intensity during a monitoring period of up to one week. This information is useful to assess a treatment effect, and the acceptability of such a device can be considered high, because of the accelerometer's low weight and small size. Similar results are reported in Salarian et al. (2007), where attempts are made to quantify tremor and bradykinesia by a triaxial wrist-mounted gyro.

Dyskinesia assessment and monitoring are performed in a similar way. Manson et al. (2000) developed a triaxial accelerometer worn on the shoulder to monitor subject behaviour during dyskinesia-provocation tasks, obtaining high correlations between clinical scores and mean accelerations. Keijsers, Horstink, and Gielen (2003) propose an ANN-based automatic procedure to assess the severity of levodopa-induced dyskinesia using six triaxial accelerometers; reasonable agreements with the clinical scores are also reported. A machine learning approach is applied in Patel et al. (2009); they implement a SVM classifier to estimate the severity of tremor, bradykinesia and dyskinesia using eight uniaxial accelerometers.

Wearable sensors based on accelerometers (Sekine et al. 2004) or gyros (Salarian et al. 2004) are exploited for gait assessment in subjects with Parkinson's disease. In Sekine et al. (2004), an approach based on recognizing the fractal nature of acceleration waveforms is successfully proposed: they track changes of these fractal properties in pathological subjects compared with healthy subjects.

## Objective Skill Evaluation for Rehabilitation

In general, obtaining an objective way to quantify human skills is not a trivial issue. Accelerometry and motion sensing could be useful to this aim. For instance, they can be aimed at assessing the effect of rehabilitation treatment on the functional activity of patients, also in their daily environment (Uswatte et al. 2000). The recovery of upper-limb function in post-stroke patients has been evaluated through accelerometers; clinical scores were predicted by simple data thresholding (Uswatte et al. 2000), by using multiple regressions (Hester et al. 2006), or through feature extraction and machine learning algorithms (Patel et al. 2010). Gait data collected from a trunk-fixed triaxial accelerometer are considered for post-stroke patients: since significant differences are observed in acceleration signal features between healthy and post-stroke subjects, indexes for objectively measuring treatment outcomes during rehabilitation can, in principle, be devised (Mizuike, Ohgi, and Morita 2009).

Activity-type classification and long-term monitoring capabilities are instrumental in the assessment of walking by patients with prosthetic during extended time periods. Bussmann et al. (2004) propose a system to distinguish walking from rest, so as to estimate how long the patient uses her prosthesis during the day. In those periods of time when the user is detected to be walking, simple spatiotemporal parameters of gait are extracted from prosthesis-worn accelerometers. A similar approach is followed for the estimation of the durations and frequencies of daily activities in total hip patients for preclinical testing of prosthetics (Morlock et al. 2001) or in the definition of functional scores for patients with knee problems performing several daily activity tasks (Van Den Dikkenberg et al. 2002).

Functional assessment of knee arthroplasty assessment is validated in Senden et al. (in press); they achieve good results using a triaxial accelerometer to distinguish patients from healthy subjects while analysing the sensitivity to gait changes during rehabilitation. These are assessed using gait parameters such as walking speed, step frequency, step length; further studies are also performed to correlate gait parameters and clinical scores.

From a clinical viewpoint, balance assessment based on inertial sensing (Mayagoitia et al. 2002) can be exploited in evaluating motor control of balance in patients with vestibular disorders or unilateral cerebellar-pontine-angle tumours (Allum et al. 2001).

## Fall Detection

Epidemiological studies estimate that falls have an incidence of approximately 30% in a year in community-dwelling people more than 65 years old (Rubenstein 2006; Tinetti, Speechley, and Ginter 1988). Fall-related injuries have profound physical and psychological consequences. In particular, a phenomenon called *long-lie* is estimated to be associated with a 50% rate of postinjury deaths within 6 months, even if no direct injury from the fall is documented. Long-lie consists of involuntarily remaining on the ground for 1 h or more after a fall and occurs in 20% of falling hospitalized elderly people (Wild, Nayak, and Isaacs 1981). The need for small and lightweight wearable devices that would prevent long-lie by sending an alarm is a pressing problem, in particular because of the rapid ageing of industrialized countries' populations. Detecting falls is not trivial. In fact, the definition itself of human falling is still the object of determination studies (Zecevic et al. 2006). Furthermore, human movement is altered during ageing with respect to young behaviour, and some daily activities can be misleading for an automatic fall-detection system (Bourke, O'Brien, and Lyons 2007). A simple example of how these alterations can introduce detection errors is the way in which a senior man, in attempting to sit down, lets himself fall onto a chair as a result of his reduced strength.

Commonly used systems to detect falls are vibration recognition (detection of floor vibration), image recognition (through markerless posture recognition) and wearable devices (usually accelerometers or gyros; Sposaro and Tyson 2009). The latter solution is more practical; it also requires simpler instrumentation. For both accelerometers and gyros as fall-detection devices, the more common strategy is the simple use of thresholds applied to acceleration or angular velocity magnitudes. Acceleration thresholds to distinguish between falls and activities of daily living were evaluated for different body sites (Kangas et al. 2007), showing better results for the waist (J. Chen et al. 2005; Diaz et al. 2004; Karantonis et al. 2006; Nyan, Tay, and Murugasu 2008), head (Lindemann et al. 2005), and sternum (Bourke, O'Brien, and Lyons 2007; Li et al. 2009) than for the wrist (Degen et al. 2003). To limit false positives in detection a rough posture assessment through the estimation of trunk or lower-limb joint angles was usually implemented (Bourke, O'Donovan, and Olaighin 2008; J. Chen et al. 2005). More complex approaches involve machine learning algorithms for posture and activity recognition such as binary decision trees (Mathie et al. 2004), SVM classifiers, kernel Fisher discriminants, and *k*-NN classifiers (Lustrek and Kaluza 2009; T. Zhang et al. 2006). Recently, a smart-phone application has also been presented that is capable of detecting falls using its embedded triaxial accelerometer (Sposaro and Tyson 2009).

The long-term goal of research in this area is pre-impact fall detection. A reliable system for detecting fall events before they actually occur (or while they are still occurring) could aid the development of protection systems capable of preventing injuries. Wu (2000) measured through motion analysis that the increase in horizontal and downward velocity occurs 300–400 ms before the fall impact. Based on this assumption, Bourke, O'Donovan, and Olaighin (2008) developed a sternum-fixed device (with a triaxial accelerometer and a triaxial gyro) capable of measuring vertical velocity; falls are detected by thresholding with 100% accuracy an average of 323 ms prior to the impact of the trunk. In Shi et al. (2009) a simple waist-worn airbag is presented. The device is activated by an embedded SVM classifier, which processes data from a triaxial gyro and a triaxial accelerometer. The fall detector is capable of sending a signal input to the mechanical actuator that releases the compressed air-inflating bags in 466 ms.

## Conclusions and Future Trends

Accelerometers in biomechanics and biomedical research are widely accepted as a respectable instrument for many applications. At the time being, the most important items on the research agenda concern, on the one hand, the production of better sensors and, on the other hand, the development of computational methods applicable to sensor data.

Accelerometers are about to become pervasive devices, in a sense. Market availability and low cost are certainly motivating factors; this can also explain their growing presence in portable devices and entertainment platforms. In this context, the widespread acceptance of accelerometers is paralleled by the development of ad hoc software packages for healthcare management and monitoring and by the development of specific smartphone applications.

In our opinion, the machine learning approach can be a common answer for most of the issues discussed. Indeed, the technology is mature for embedding machine learning processing in wearable (or portable) devices, and many advances have been achieved from the computational viewpoint. Moreover, machine learning is a powerful tool for data fusion purposes. This capability allows us to merge information from the whole set of different available data sources in a wearable sensor net. The exponential trend in the number of publications on this topic put our considerations into a larger perspective: In the last five years the number of ISI (Institute for Scientific Information) indexed publications on machine learning for biomedical data processing purposes exceeds 3400 papers.

In conclusion, the efforts of the scientific community in this direction could lead, in the short term, to many interesting developments in wearable inertial sensing for health assessment and monitoring. The vision of a smart environment that is capable of monitoring people's health status and activity and of sharing the collected information with a remote clinic centre (or with a robotic homecare system) is about to become reality. There is wide consensus that the technology of MEMS accelerometers and the continuing progress in information and communication technologies will combine to produce innovative practices of healthcare management.

## References

Allen, F. R., E. Ambikairajah, N. H. Lovell, and B. G. Celler. 2006. Classification of a known sequence of motions and postures from accelerometry data using adapted Gaussian mixture models. *Physiol. Meas.* 27, no. 10: 935–951.

Allum, J. H., A. L. Adkin, M. G. Carpenter, M. Held-Ziolkowska, F. Honegger, and K. Pierchala. 2001. Trunk sway measures of postural stability during clinical balance tests: Effects of a unilateral vestibular deficit. *Gait Posture* 14, no. 3: 227–237.

Altun, K., B. Barshan, and O. Tunçel. 2010. Comparative study on classifying human activities with miniature inertial and magnetic sensors. *Pattern Recognit.* 43, no. 10: 3605–3620.

Aminian, K., P. Robert, E. E. Buchser, B. Rutschmann, D. Hayoz, and M. Depairon. 1999. Physical activity monitoring based on accelerometry: Validation and comparison with video observation. *Med. Biol. Eng. Comput.* 37, no. 3: 304–308.

Aminian, K., P. Robert, E. Jequier, and Y. Schutz. 1995. Estimation of speed and incline of walking using neural network. *IEEE Trans. Instrum. Meas.* 44, no. 3: 743–746.

Bao, L., and S. S. Intille. 2004. Activity recognition from user-annotated acceleration data. In *Pervasive Computing, Second Int. Conf. Pervasive 2004 Proc.*, ed. A. Ferscha and F. Mattern 1–17.

Barnett, A., and E. Cerin. 2006. Individual calibration for estimating free-living walking speed using the MTI monitor. *Med. Sci. Sports Exercise* 38, no. 4: 761–767.

Begg, R., and J. Kamruzzaman. 2005. A machine learning approach for automated recognition of movement patterns using basic, kinetic and kinematic gait data. *J. Biomech.* 38, no. 3: 401–408.

Bhattacharya, A., E. P. McCutcheon, E. Shvartz, and J. E. Greenleaf. 1980. Body acceleration distribution and O2 uptake in humans during running and jumping. *J. Appl. Physiol.* 49, no. 5: 881–887.

Bhushan, B. 2010. *Springer Handbook of Nanotechnology*. Springer.

Bidargaddi, N., L. Klingbeil, A. Sarela, J. Boyle, V. Cheung, C. Yelland, M. Karunanithi, and L. Gray. 2007. Wavelet based approach for posture transition estimation using a waist worn accelerometer. *Conf. Proc. IEEE Eng. Med. Biol. Soc.*, 2007: 1884–1887.

Bonato, P. 2003. Wearable sensors/systems and their impact on biomedical engineering. *IEEE Eng. Med. Biol.* 22, no. 3: 18–20.

Bonato, P. 2005. Advances in wearable technology and applications in physical medicine and rehabilitation. *J. Neuroeng. Rehabil.* 2, no. 1: 2.

Boonstra, M. C., R. M. van der Slikke, N. L. Keijsers, R. C. van Lummel, M. C. de Waal Malefijt, and N. Verdonschot. 2006. The accuracy of measuring the kinematics of rising from a chair with accelerometers and gyroscopes. *J. Biomech.* 39, no. 2: 354–358.

Bourke, A. K., J. V. O'Brien, and G. M. Lyons. 2007. Evaluation of a threshold-based tri-axial accelerometer fall detection algorithm. *Gait Posture* 26, no. 2: 194–199.

Bourke, A. K., K. J. O'Donovan, and G. Olaighin. 2008. The identification of vertical velocity profiles using an inertial sensor to investigate pre-impact detection of falls. *Med. Eng. Phys.* 30, no. 7: 937–946.

Bouten, C. V., K. T. Koekkoek, M. Verduin, R. Kodde, and J. D. Janssen. 1997. A triaxial accelerometer and portable data processing unit for the assessment of daily physical activity. *IEEE Trans. Biomed. Eng.* 44, no. 3: 136–147.

Bouten, C. V., A. A. Sauren, M. Verduin, and J. D. Janssen. 1997. Effects of placement and orientation of body-fixed accelerometers on the assessment of energy expenditure during walking. *Med. Biol. Eng. Comput.* 35, no. 1: 50–56.

Bouten, C. V., K. R. Westerterp, and M. Verduin. 1994. Assessment of energy expenditure for physical activity using a triaxial accelerometer. *Med. Sci. Sports Exercise* 26, no. 12: 1516–1523.

Brandes, M., W. Zijlstra, S. Heikens, R. van Lummel, and D. Rosenbaum. 2006. Accelerometry based assessment of gait parameters in children. *Gait Posture* 24, no. 4: 482–486.

Bussmann, H. B. J., K. M. Culhane, H. L. D. Horemans, G. M. Lyons, and H. J. Stam. 2004. Validity of the prosthetic activity monitor to assess the duration and spatio-temporal characteristics of prosthetic walking. *IEEE Trans. Neural Syst. Rehabil. Eng.* 12, no. 4: 379–386.

Bussmann, J. B. J., W. L. J. Martens, J. H. M. Tulen, F. C. Schasfoort, H. J. G. Van Den Berg Emons, and H. J. Stam. 2001. Measuring daily behavior using ambulatory accelerometry: the Activity Monitor. *Behav. Res. Meth. Instrum.* 33, no. 3: 349–356.

Capodaglio, E., M. Imbriani, and P. Capodaglio. 1993. Misure e stime del costo energetico di Attività Motorie in Medicina Riabilitativa e in Medicina Occupazionale. In *Nuovi Approcci alla Riabilitazione Industriale,* ed. G. Bazzini, 137–150. Pavia: Pime Editrice.

Chen, C., S. Anton, and A. Helal. 2008. A brief survey of physical activity monitoring devices. *University of Florida Tech Report MPCL-08-09.*

Chen, J., K. Kwong, D. Chang, J. Luk, and R. Bajcsy. 2005. Wearable sensors for reliable fall detection. *Conf. Proc. IEEE Eng. Med. Biol. Soc.,* 4: 3551–3554.

Chieu, H. L., W. S. Lee, and L. P. Kaelbling. 2006. Activity recognition from physiological data using conditional random fields. *Singapore-MIT Alliance Comp. Sci. Tech. Report.* Boston: MIT-Libraries.

Crouter, S. E., K. G. Clowers, and D. R. Bassett. 2006. A novel method for using accelerometer data to predict energy expenditure. *J. Appl. Physiol. (Bethesda, Md.: 1985)* 100, no. 4: 1324–1331.

Degen, T., H. Jaeckel, M. Rufer, and S. Wyss. 2003. SPEEDY: A fall detector in a wrist watch. *Wearable Comp. Proc. 2003,* 184–189.

Dejnabadi, H., B. M. Jolles, and K. Aminian. 2005. A new approach to accurate measurement of uniaxial joint angles based on a combination of accelerometers and gyroscopes. *IEEE Trans. Biomed. Eng.* 52, no. 8: 1478–1484.

Dey, A. K., and G. D. Abowd. 2000. Towards a better understanding of context and context-awareness. *Workshop What Who Where When How Cont. Aw.,* 4: 1–6.

Diaz, A., M. Prado, L. M. Roa, J. Reina-Tosina, and G. Sanchez. 2004. Preliminary evaluation of a full-time falling monitor for the elderly. *Conf. Proc. IEEE Eng. Med. Biol. Soc.,* 3: 2180–2183.

Dietterich, T. G. 2009. Machine learning for sequential data: A review. *Lect. Notes Comput. Sci.* 2396: 1–15.

Dongwoo, K., and H. Chan Kim. 2007. Activity energy expenditure assessment system based on activity classification using multi-site triaxial accelerometers. *Conf. Proc. IEEE Eng. Med. Biol. Soc.,* 2007: 2285–2287.

Eston, R. G., A. V. Rowlands, and D. K. Ingledew. 1998. Validity of heart rate, pedometry, and accelerometry for predicting the energy cost of children's activities. *J. Appl. Physiol.* 84, no. 1: 362.

Foerster, F., M. Smeja, and J. Fahrenberg. 1999. Detection of posture and motion by accelerometry: A validation study in ambulatory monitoring. *Comput. Hum. Behav.* 15: 571–583.

Gallagher, A., Y. Matsuoka, and W. T. Ang. 2004. An efficient real-time human posture tracking algorithm using low-cost inertial and magnetic sensors. *IEEE Intell. Robot. Sys. Proc.,* 3: 2967–2972.

Giansanti, D., G. Maccioni, F. Benvenuti, and V. Macellari. 2007. Inertial measurement units furnish accurate trunk trajectory reconstruction of the sit-to-stand manoeuvre in healthy subjects. *Med. Biol. Eng. Comput.* 45, no. 10: 969–976.

Giansanti, D., G. Maccioni, and V. Macellari. 2005. The development and test of a device for the reconstruction of 3-D position and orientation by means of a kinematic sensor assembly with rate gyroscopes and accelerometers. *IEEE Trans. Biomed. Eng.* 52, no. 7: 1271–1277.

Hanlon, M., and R. Anderson. 2009. Real-time gait event detection using wearable sensors. *Gait Posture* 30, no. 4: 523–527.

Hansson, G. A., P. Asterland, N. G. Holmer, and S. Skerfving. 2001. Validity and reliability of triaxial accelerometers for inclinometry in posture analysis. *Med. Biol. Eng. Comput.* 39, no. 4: 405–413.

Hester, T., R. Hughes, D. M. Sherrill, B. Knorr, M. Akay, J. Stein, and P. Bonato. 2006. Using wearable sensors to measure motor abilities following stroke. *Proc. Int. Workshop Wearable Impl. Body Sensor Net.*, 5–8.

Hyun Choi, J., J. Lee, H. Tai Hwang, J. Pal Kim, J. Chan Park, and K. Shin. 2005. Estimation of activity energy expenditure: Accelerometer approach. *Conf. Proc. IEEE Eng. Med. Biol. Soc.*, 4: 3830–3833.

Jacobi, D., A. E. Perrin, N. Grosman, M. F. Doré, S. Normand, J. M. Oppert, and C. Simon. 2007. Physical activity-related energy expenditure with the RT3 and TriTrac accelerometers in overweight adults. *Obesity* 4: 950–956.

Jain, A. K., and P. W. Duin. 2000. Statistical pattern recognition: A review. *IEEE Trans. Pattern Anal.* 22, no. 1: 4–37.

Janssen, W. G. M., J. B. J. Bussmann, H. L. D. Horemans, and H. J. Stam. 2005. Analysis and decomposition of accelerometric signals of trunk and thigh obtained during the sit-to-stand movement. *Med. Biol. Eng. Comput.* 43, no. 2: 265–272.

Janssen, W. G. M., H. B. J. Bussmann, and H. J. Stam. 2002. Determinants of the sit-to-stand movement: A review. *Phys. Ther.* 82, no. 9: 866–879.

Kangas, M., A. Konttila, I. Winblad, and T. Jämsä. 2007. Determination of simple thresholds for accelerometry-based parameters for fall detection. *Conf. Proc. IEEE Eng. Med. Biol. Soc.*, 2007: 1367–1370.

Karantonis, D. M., M. R. Narayanan, M. Mathie, N. H. Lovell, and B. G. Celler. 2006. Implementation of a real-time human movement classifier using a triaxial accelerometer for ambulatory monitoring. *IEEE Trans. Inf. Technol. B.* 10, no. 1: 156–167.

Keijsers, N. L. W., M. W. Horstink, and S. C. Gielen. 2003. Automatic assessment of levodopa-induced dyskinesias in daily life by neural networks. *Movement Disord.* 18, no. 1: 70–80.

Kemp, B., A. J. Janssen, and B. van der Kamp. 1998. Body position can be monitored in 3D using miniature accelerometers and earth-magnetic field sensors. *Electroen. Clin. Neurophysiol.* 109, no. 6: 484–488.

Ketabdar, H., and M. Lyra. 2010. System and methodology for using mobile phones in live remote monitoring of physical activities. *IEEE Int. Symp. Tech. Soc.*, 1: 350–356.

Khan, A. M., Y. K. Lee, S. Y. Lee, and T. S. Kim. 2010. Human activity recognition via an accelerometer-enabled-smartphone using kernel discriminant analysis. *Int. Conf. on Future Inf. Tech.*

Kralj, A., R. J. Jaeger, and M. Munih. 1990. Analysis of standing up and sitting down in humans: Definitions and normative data presentation. *J. Biomech.* 23, no. 11: 1123–1138.

Kwapisz, J. R., G. M. Weiss, and S. A. Moore. 2010. Activity recognition using cell phone accelerometers. *ACM SIGKDD Explor. Newslet.*, 12, no. 2: 74–82.

LaPorte, R. E., H. J. Montoye, and C. J. Caspersen. 1985. Assessment of physical activity in epidemiologic research: Problems and prospects. *Public Health Rep. (Washington, D.C.: 1974)* 100, no. 2: 131–146.

Lee, J. K., and E. J. Park. 2009. A minimum-order Kalman filter for ambulatory real-time human body orientation tracking. *IEEE Int. Conf. Robot. Autom.*, 2009: 3565–3570.

Lee, S. H., H. D. Park, S. Y. Hong, K. J. Lee, and Y. H. Kim. 2003. A study on the activity classification using a triaxial accelerometer. *Conf. Proc. IEEE Eng. Med. Biol. Soc.*, 3: 2941–2943.

Lee, S. W., and K. Mase. 2002. Activity and location recognition using wearable sensors. *IEEE Pervasive Comput.*, no. 3: 24–32.

Li, Q., J. A. Stankovic, M. A. Hanson, A. T. Barth, J. Lach, and G. Zhou. 2009. Accurate, fast fall detection using gyroscopes and accelerometer-derived posture information. *Proc. 6th IEEE Int. Workshop Wearable Impl. Body Sensor Net.*,138–143.

Lindemann, U., A. Hock, M. Stuber, W. Keck, and C. Becker. 2005. Evaluation of a fall detector based on accelerometers: A pilot study. *Med. Biol. Eng. Comput.* 43, no. 5: 548–551.

Longstaff, B., S. Reddy, and D. Estrin. 2010. Improving activity classification for health applications on mobile devices using active and semi-supervised learning. *Perv. Comput. Tech. Healthcare*, 2010: 1–7.

Luinge, H. J., P. H. Veltink, and C. T. M. Baten. 1999. Estimating orientation with gyroscopes and accelerometers. *Technol. Health Care* 7, no. 6: 455–459.

Lustrek, M., and B. Kaluza. 2009. Fall detection and activity recognition with machine learning. *Informatica* 33: 205–212.

Lyons, G. M., K. M. Culhane, D. Hilton, P. A. Grace, and D. Lyons. 2005. A description of an accelerometer-based mobility monitoring technique. *Med. Eng. Phys.* 27, no. 6: 497–504.

Mannini, A., and A. M. Sabatini. 2010. Machine learning methods for classifying human physical activity from on-body accelerometers. *Sensors* 10, no. 2: 1154–1175.

Mannini, A., and A. M. Sabatini. 2011. Automatic machine learning methods for analysis of signals from accelerometers: classification of human activity and walking–running speed estimation. *Gait Posture*, 33 no. Suppl: 24.

Manson, A. J., P. Brown, J. D. O'Sullivan, P. Asselman, D. Buckwell, and A. J. Lees. 2000. An ambulatory dyskinesia monitor. *J. Neurol. Neurosurg. Psychiatry* 68, no. 2: 196–201.

Mantyjarvi, J., J. Himberg, and T. Seppanen. 2001. Recognizing human motion with multiple acceleration sensors. In *IEEE Syst., Man, Cybern.* 2: 747–752.

Mathie, M. J., B. G. Celler, N. H. Lovell, and A. C. F. Coster. 2004. Classification of basic daily movements using a triaxial accelerometer. *Med. Biol. Eng. Comput.* 42: 679–687.

Mayagoitia, R. E., J. C. Lötters, P. H. Veltink, and H. Hermens. 2002. Standing balance evaluation using a triaxial accelerometer. *Gait Posture* 16, no. 1: 55–59.

Meijer G. A., K. R. Westerkerp, H. Koper, and T. H. Foppe. 1989. Assessment of energy expenditure by recording heart rate and body acceleration. *Med. Sci. Sports Exercise* 21: 343–347.

Mizuike, C., S. Ohgi, and S. Morita. 2009. Analysis of stroke patient walking dynamics using a tri-axial accelerometer. *Gait Posture* 30, no. 1: 60–64.

Moe-Nilssen, R., and J. L. Helbostad. 2004. Estimation of gait cycle characteristics by trunk accelerometry. *J. Biomech.* 37, no. 1: 121–126.

Montoye, H., R. Washburn, S. Servais, A. Ertl, J. G. Webster, and F. J. Nagle. 1983. Estimation of energy expenditure by a portable accelerometer. *Med. Sci. Sports Exercise* 15: 403–407.

Morlock, M., E. Schneider, A. Bluhm, M. Vollmer, G. Bergmann, V. Müller, and M. Honl. 2001. Duration and frequency of everyday activities in total hip patients. *J. Biomech.* 34, no. 7: 873–881.

Najafi, B., K. Aminian, A. Paraschiv-Ionescu, F. Loew, C. Bula, and P. A. Robert. 2003. Ambulatory system for human motion analysis using a kinematic sensor: Monitoring of daily physical activity in the elderly. *IEEE Trans. Biomed. Eng.* 50, no. 6: 711–723.

Nyan, M. N., F. E. H. Tay, and E. Murugasu. 2008. A wearable system for pre-impact fall detection. *J. Biomech.* 41, no. 16: 3475–3481.

Patel, S., R. Hughes, T. Hester, J. Stein, M. Akay, J. G. Dy, and P. Bonato. 2010. A novel approach to monitor rehabilitation outcomes in stroke survivors using wearable technology. *Proc. IEEE* 98, no. 3: 450–461.

Patel, S., K. Lorincz, R. Hughes, N. Huggins, J. Growdon, D. Standaert, M. Akay, J. Dy, M. Welsh, and P. Bonato. 2009. Monitoring motor fluctuations in patients with Parkinson's disease using wearable sensors. *IEEE Trans. Inf. Technol. Biomed.* 13, no. 6: 864–873.

Plasqui, G., and K. R. Westerterp. 2007. Physical activity assessment with accelerometers: An evaluation against doubly labeled water. *Obesity* 15, no. 10: 2371–2379.

Poppe, R. 2007. Vision-based human motion analysis: An overview. *Comput. Vision Image Understanding* 108, no. 1: 4–18.

Preece, S. J., J. Y. Goulermas, L. P. J. Kenney, and D. Howard. 2009. A comparison of feature extraction methods for the classification of dynamic activities from accelerometer data. *IEEE Trans. Biomed. Eng.* 56, no. 3: 871–879.

Randell, C., and H. Muller. 2000. Context awareness by analysing accelerometer data. *IEEE 4th Int. Symp. Wearable Computers*, 175–176.

Ravi, N., N. Dandekar, P. Mysore, and M. L. Littman. 2005. Activity recognition from accelerometer data. *Am. Assoc. Artif. Intell.* 5: 1541–1546.

Rothney, M. P., M. Neumann, A. Béziat, and K. Y. Chen. 2007. An artificial neural network model of energy expenditure using nonintegrated acceleration signals. *J. Appl. Physiol.* 103, no. 4: 1419–1427.

Rubenstein, L. Z. 2006. Falls in older people: Epidemiology, risk factors and strategies for prevention. *Age Ageing* 35, Suppl. 2: ii37–ii41.

Rueterbories, J., E. G. Spaich, B. Larsen, and O. K. Andersen. 2010. Methods for gait event detection and analysis in ambulatory systems. *Med. Eng. Phys.* 32, no. 6: 545–552.

Sabatini, A. M. 2006a. Inertial sensing in biomechanics: A survey of computational techniques bridging motion analysis and personal navigation. In *Computational Intelligence for Movement Sciences, Neural Networks and Other Emerging Techniques*, ed. R. Begg and M. Palaniswami, 70–100, Hershey: Idea Group Publishing.

Sabatini, A. M. 2006b. Quaternion-based extended Kalman filter for determining orientation by inertial and magnetic sensing. *IEEE Trans. Biomed. Eng.* 53, no. 7: 1346–1356.

Salarian, A., H. Russmann, F. J. G. Vingerhoets, C. Dehollain, Y. Blanc, P. R. Burkhard, and K. Aminian. 2004. Gait assessment in Parkinson's disease: Toward an ambulatory system for long-term monitoring. *IEEE Trans. Biomed. Eng.* 51, no. 8: 1434–1443.

Salarian, A., H. Russmann, C. Wider, P. R. Burkhard, F. J. G. Vingerhoets, and K. Aminian. 2007. Quantification of tremor and bradykinesia in Parkinson's disease using a novel ambulatory monitoring system. *IEEE Trans. Biomed. Eng.* 54, no. 2: 313–322.

Sekine, M., M. Akay, T. Tamura, Y. Higashi, and T. Fujimoto. 2004. Fractal dynamics of body motion in patients with Parkinson's disease. *J. Neural Eng.* 1, no. 1: 8–15.

Sekine, M., T. Tamura, M. Akay, T. Fujimoto, T. Togawa, and Y. Fukui. 2002. Discrimination of walking patterns using wavelet-based fractal analysis. *IEEE Trans. Neural Syst. Rehabil. Eng.* 10, no. 3: 188–196.

Sekine, M., T. Tamura, T. Togawa, and Y. Fukui. 2000. Classification of waist-acceleration signals in a continuous walking record. *Med. Eng. Phys.* 22, no. 4: 285–291.

Senden, R., B. Grimm, K. Meijer, H. Savelberg, and I. C. Heyligers. 2011. The importance to including objective functional outcomes in the clinical follow up of total knee arthroplasty patients. *Knee*, in press.

Shi, G., C. S. Chan, W. J. Li, K. Leung, Y. Zou, and Y. Jin. 2009. Mobile human airbag system for fall protection using MEMS sensors and embedded SVM classifier. *IEEE Sens. J.* 9, no. 5: 495–503.

Shimada, Y., S. Ando, T. Matsunaga, A. Misawa, T. Aizawa, T. Shirahata, and E. Itoi. 2005. Clinical application of acceleration sensor to detect the swing phase of stroke gait in functional electrical stimulation. *Tohoku J. Experiment. Med.* 207, no. 3: 197–202.

Sminchisescu, C., A. Kanaujia, and D. Metaxas. 2006. Conditional models for contextual human motion recognition. *Comput. Vision Image Understanding* 104, no. 2–3: 210–220.

Song, K. T., and Y. Q. Wang. 2005. Remote activity monitoring of the elderly using a two-axis accelerometer. In *2005 CACS Autom. Control Conf. Proc.*

Song, Y., S. Shin, S. Kim, D. Lee, and K. H. Lee. 2007. Speed estimation from a tri-axial accelerometer using neural networks. *Conf. Proc. IEEE Eng. Med. Biol. Soc.*, 2007: 3224–3227.

Sposaro, F., and G. Tyson. 2009. iFall: An Android application for fall monitoring and response. *Conf. Proc. IEEE Eng. Med. Biol. Soc.*, 2009: 6119–6122.

Staudenmayer, J., D. Pober, S. Crouter, D. Bassett, and P. Freedson. 2009. An artificial neural network to estimate physical activity energy expenditure and identify physical activity type from an accelerometer. *J. Appl. Physiol.* 107, no. 4: 1300–1307.

Su, M. C., Y. Y. Chen, K. H. Wang, C. Y. Tew, and H. Huang. 2000. 3D arm movement recognition using syntactic pattern recognition. *Artif. Intell. Eng.*14, no. 2: 113–118.

Sun, M., and J. O. Hill. 1993. A method for measuring mechanical work and work efficiency during human activities. *J. Biomech.* 26, no. 3: 229–241.

Swartz, A. M., S. J. Strath, D. R. Bassett, W. L. O'Brien, G. A. King, and B. E. Ainsworth. 2000. Estimation of energy expenditure using CSA accelerometers at hip and wrist sites. *Med. Sci. Sports Exercise* 32, Supp: S450–S456.

Tinetti, M. E., M. Speechley, and S. F. Ginter. 1988. Risk factors for falls among elderly persons living in the community. *New Engl. J. Med.* 319, no. 26: 1701–1707.

Tryon, W. W., and R. Williams. 1996. Fully proportional actigraphy: A new instrument. *Behav. Res. Meth. Instrum.* 28, no. 3: 392–403.

Tweedy, S. M., and S. G. Trost. 2005. Validity of accelerometry for measurement of activity in people with brain injury. *Med. Sci. Sports Exercise* 37, no. 9: 1474–1480.

Uswatte, G., W. H. Miltner, B. Foo, M. Varma, S. Moran, and E. Taub. 2000. Objective measurement of functional upper-extremity movement using accelerometer recordings transformed with a threshold filter. *Stroke* 31, no. 3: 662–667.

Vail, D. L., J. D. Lafferty, and M. M. Veloso. 2007. Feature selection in conditional random fields for activity recognition. *IEEE Intell. Robot. Sys. Proc.,* 2007: 3379–3384.

Vail, D. L., M. M. Veloso, and J. D. Lafferty. 2007. Conditional random fields for activity recognition. In *Proc. 6th Int. Joint Conf. Auton. Agents Multiagent Systems,* 2351–2358, New York: ACM.

Van Den Dikkenberg, N., O. G. Meijer, R. M. A. van Der Slikke, R. C. van Lummel, J. H. van Dieën, B. Pijls, R. J. Benink, and P. I. J. Wuisman. 2002. Measuring functional abilities of patients with knee problems: Rationale and construction of the DynaPort knee test. *Knee Surg. Sport Tr. A.* 10: 204–212.

Van Hees, V. T., R. C. van Lummel, and K. R. Westerterp. 2009. Estimating activity-related energy expenditure under sedentary conditions using a tri-axial seismic accelerometer. *Obesity* 17, no. 6: 1287–92.

Van Laerhoven, K., and O. Cakmakci. 2000. What shall we teach our pants? *IEEE 4th Int. Symp. Wearable Computers Proc.,* 77–83.

Van Someren, E. J. W., M. D. Pticek, J. D. Speelman, P. R. Schuurman, R. Esselink, and D. F. Swaab. 2006. New actigraph for long-term tremor recording. *Movement Disord.* 21, no. 8: 1136–1143.

Veltink, P. H., H. B. J. Bussmann, W. de Vries, W. L. J. Martens, and R. C. Van Lummel. 1996. Detection of static and dynamic activities using uniaxial accelerometers. *IEEE Trans. Rehabil. Eng.* 4, no. 4: 375–385.

Welch, G., and E. Foxlin. 2002. Motion tracking: No silver bullet, but a respectable arsenal. *IEEE Comput. Graph.* 22, no. 6: 24–38.

Welk, G. J., S. N. Blair, K. Wood, S. Jones, and R. W. Thompson. 2000. A comparative evaluation of three accelerometry-based physical activity monitors. *Med. Sci. Sports Exercise* 32, no. 9 Suppl: S489–S497.

Wild, D., U. S. Nayak, and B. Isaacs. 1981. How dangerous are falls in old people at home? *Brit. Med. J.* 282, no. 6260: 266–268.

Willemsen, T., F. Bloemhof, and H. B. Boom. 1990. Automatic stance-swing phase detection from accelerometer data for peroneal nerve stimulation. *IEEE Trans. Biomed. Eng.* 37, no. 12: 1201–1208.

Williamson, R., and B. J. Andrews. 2000. Gait event detection for FES using accelerometers and supervised machine learning. *IEEE Trans. Rehabil. Eng.* 8, no. 3: 312–319.

Wong, T. C., J. G. Webster, H. J. Montoye, and R. Washburn. 1981. Portable accelerometer device for measuring human energy expenditure. *IEEE Trans. Biomed. Eng.* 28, no. 6: 467–471.

Wu, G. 2000. Distinguishing fall activities from normal activities by velocity characteristics. *J. Biomech.* 33, no. 11: 1497–1500.

Yang, J. 2009. Toward physical activity diary: Motion recognition using simple acceleration features with mobile phones. In *Proc. 1st Int. Workshop Interact. Multimed. Cons. Electronics,* 1–10. New York: ACM.

Yazdi, N., F. Ayazi, and K. Najafi. 1998. Micromachined inertial sensors. *Proc. IEEE* 86, no. 8: 1640–1659.

Yeoh, W. S., I. Pek, Y. H. Yong, X. Chen, and A. B. Waluyo. 2008. Ambulatory monitoring of human posture and walking speed using wearable accelerometer sensors. *Conf. Proc. IEEE Eng. Med. Biol. Soc.*, 2008: 5184–5187.

Young, A. D. 2009. Comparison of orientation filter algorithms for realtime wireless inertial posture tracking. *Proc. 6th Int. IEEE Workshop Wearable Impl. Body Sensor Net.*, 59–64.

Young, A. D., M. J. Ling, and D. K. Arvind. 2007. Orient-2: A realtime wireless posture tracking system using local orientation estimation. *In Proc. 4th Workshop Emb. Net. Sens.*, 53–57. New York: ACM.

Zecevic, A. A., A. W. Salmoni, M. Speechley, and A. A. Vandervoort. 2006. Defining a fall and reasons for falling: Comparisons among the views of seniors, health care providers, and the research literature. *Gerontologist* 46, no. 3: 367–376.

Zhang, K., F. X. Pi-Sunyer, and C. N. Boozer. 2004. Improving energy expenditure estimation for physical activity. *Med. Sci. Sports Exercise* 36, no. 5: 883–889.

Zhang, K., P. Werner, M. Sun, F. X. Pi-Sunyer, and C. N. Boozer. 2003. Measurement of human daily physical activity. *Obes. Res.* 11, no. 1: 33–40.

Zhang, T., J. Wang, P. Liu, and J. Hou. 2006. Fall detection by embedding an accelerometer in cellphone and using KFD algorithm. *IJCSNS* 6, no. 10: 277.

Zhu, R., and Z. Zhou. 2004. A real-time articulated human motion tracking using tri-axis inertial/magnetic sensors package. *IEEE Trans. Neural Syst. Rehabil. Eng.* 12, no. 2: 295–302.

Zijlstra, W. 2004. Assessment of spatio-temporal parameters during unconstrained walking. *Eur. J. Appl. Physiol.* 92, no. 1–2: 39–44.

Zijlstra, W., R. W. Bisseling, S. Schlumbohm, and H. Baldus. 2010. A body-fixed-sensor-based analysis of power during sit-to-stand movements. *Gait Posture* 31, no. 2: 272–278.

# 3

# *Intrabody Communication Using Contact Electrodes in Low-Frequency Bands*

**Ken Sasaki, Fukuro Koshiji, and Shudo Takenaka**

**CONTENTS**

Introduction .......... 51
Transmission Model .......... 52
  Intrabody Communication Using Contact Electrodes .......... 53
  Numerical Simulation .......... 55
  Electric Field Distribution Including an Off-Body Device .......... 57
  Intrabody Communication Using Capacitive Coupling .......... 59
  Comments on Carrier Frequency .......... 59
Configuration and Size of Electrodes .......... 60
  Configuration of Electrodes .......... 60
  Spacing between Two Contact Electrodes .......... 66
  Distance between the Human Body and a Circuit Board .......... 67
  Transmission Characteristics of On-Body Devices .......... 67
  Impedance Matching of Electrodes .......... 68
Summary .......... 72
References .......... 72

## Introduction

Intrabody communication is a wireless communication method that utilizes a part of the human body as a transmission medium. It is considered one of the options for data communications between wearable devices and peripheral devices. Networks containing these devices are called *body area networks* (BANs) or *body-centric networks*. One of the primary application fields for BANs is healthcare monitoring systems, because multiple wearable sensors have to be connected wirelessly to collect the patient's vital data.

Research and development activities on the communication methods for a BAN can generally be divided into two groups in terms of carrier frequency. The lower band is approximately 0.1 MHz to 100 MHz, and the higher band is approximately 300 MHz to several gigahertz. In the higher-frequency band, the characteristics of the communication channel involving the human body are being extensively investigated, because many popular wireless communication standards are in this frequency band, such as Bluetooth and ZigBee (Ruiz et al. 2006; Reusens et al. 2009). The evaluation and selection of these standards for practical applications are also widely being discussed (Monton et al. 2008; Pantelopoulos and Bourbakis 2010; Patel and Wang 2010; Yuce 2010). Although it is a practical choice to use popular communication standards, the presence of a human body usually leads to loss or uncertainty of the transmission in the frequency band of 300 MHz and higher, because the wavelength becomes comparable to or shorter than the size of a human body.

However, a human body can be utilized as a part of the transmission medium in the lower-frequency band. This communication method, called *intrabody communication* or *body-coupled communication* (BCC), is regarded as a new wireless communication method for BANs (Zimmerman 1996; Shinagawa et al. 2004; Doi and Nishimura 2005; Schenk et al. 2008). One of the differences to airborne wireless communication is that intrabody communication uses electrodes instead of antennas for signal transmission. Since part of the signal flows through the human body, the power consumption can be lower than that of ordinary airborne wireless transmission. Enhanced security is also expected because the electromagnetic wave propagation into the air is weaker. This security is beneficial in monitoring systems for healthcare.

Although intrabody communication has existed for more than a decade, there are still some issues to be solved. For example, there is no well-established guideline for carrier frequency selection. A designer may have to look through related literature or fabricate prototypes.

This chapter aims to alleviate some of these deficiencies by providing transmission models and experimental data, so that designers who are considering using intrabody communication can gain a better perspective. The following sections deal primarily with intrabody communication using contact electrodes in the frequency band of 0.1 MHz to 100 MHz. The chapter is organized as follows. First, we review some methods of intrabody communications and some transmission characteristics. Second, issues in practical circuit design, such as electrode arrangement and impedance matching, are discussed in detail based on numerical analyses and experimental results. The final section presents a summary.

## Transmission Model

It is useful to have a qualitative understanding of the transmission mechanism of intrabody communication to evaluate whether it is better to use intrabody communication instead of ordinary wireless communication standards for a given application. In intrabody communication, electrodes are used instead of antennas. These electrodes may be in direct contact with the human body or held close to the surface of the body without contact, depending on the configuration of the communication system. The former interface is called *galvanic coupling*, and the latter is called *capacitive coupling* (Schenk et al. 2008).

## Intrabody Communication Using Contact Electrodes

Handa et al. (1997) developed an electrocardiogram (ECG) monitoring system using an intrabody communication system with contact electrodes. They transmitted ECG signals from an ECG detector on the chest to a relay receiver/transmitter on the wrist, which then sent the data to a remote receiver. The transmitter electrodes of the ECG detector were placed 7 cm apart on the chest, and the receiver electrodes were placed 3 cm apart on the wrist. Although the direction of the receiver electrodes was not explicitly mentioned in Handa et al. (1997), the electrodes depicted in the figure were placed across the wrist. The carrier frequency was 70 kHz. In Handa et al. (1997), all of the electrodes were in contact with the human body.

The transmission characteristics of similar electrode arrangements were investigated by Wegmueller et al. (2007) in the frequency range of 10 kHz to 1000 kHz. The reported attenuation was small for transmission on the thorax and for short distances on the limbs. In these systems, the electrodes of the transmitter and the receiver were both placed across the direction of transmission. Here, the direction of transmission corresponds to the longitudinal direction of the human arms, legs and torso. We refer to this electrode arrangement as a *transversal arrangement* (Fujii et al. 2004).

A transversal arrangement assumes that the signal is transmitted by the current flowing through the human body and that the effect of the electric field around the body is small. This transmission model was analysed as a four-terminal network by Hachisuka et al. (2005). However, this assumption may be valid only for frequencies below several megahertz. Fujii et al. (2004) showed that, at a frequency of 10 MHz, the electric field around the body contributed to the signal transmission, and it was better to arrange the electrodes along the direction of transmission. This electrode arrangement is referred to as a *longitudinal arrangement*. Fujii et al.'s numerical analyses and experiments using a human-arm-size phantom showed that the received signal voltage of the longitudinal arrangement was approximately 10 times larger than that of the transversal arrangement.

Another way to verify the effect of the electric field around the body in intrabody transmission is as follows: If the electric field around the body has no effect on the signal transmission, a phase difference of 180° results when the two electrodes of the transmitter are interchanged. Hachisuka et al. (2004) measured the phase difference between the transmitter and the receiver signal at 10 MHz while rotating the direction of the transmitter electrodes. In their experiments, a battery-operated transmitter was used, and a synchronizing signal for the phase measurement was transmitted to a measurement instrument by an optical fibre to reduce the coupling between the cables and the instruments. A similar method was also reported in other systems (Baldus et al. 2009). Contrary to the prediction based on the four-terminal network model, only a small phase difference occurred when the transmitter electrodes were interchanged in the transversal direction. In other words, there was no inversion of polarity. On the other hand, the phase difference was approximately 150° when the transmitter electrodes were interchanged in the longitudinal direction. Furthermore, the received signal was larger when the transmitter electrodes were aligned in the longitudinal direction. These results agree with the analysis and experimental data of Fujii et al. (2004).

The transmission model of intrabody communication using the contact electrodes arranged in the longitudinal direction is shown in Figure 3.1. In this figure, a person is wearing a receiver on his or her wrist and touching a transmitter box placed on the floor. The two electrodes of the receiver are placed along the person's arm. The transmitter box has a signal source inside; one output terminal is connected to the electrode placed on the top of the box, and the other terminal is connected to the box, which is made of metal. A common explanation of the signal transmission in this type of configuration is as follows. When the person touches the top electrode with his or her hand, electric current flows from the signal source to the hand through the top electrode, and then to the upper arm, torso and legs. Part of the current

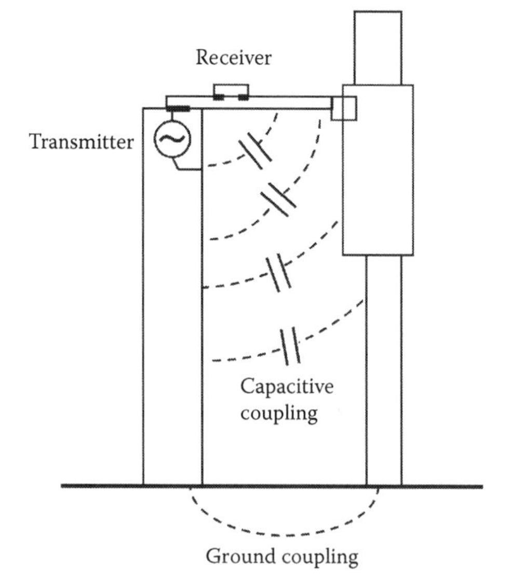


**FIGURE 3.1**
Transmission model based on capacitive coupling between a human body and an off-body device.

flows between the receiver electrodes that are in contact with the arm. This partial current becomes the received signal. The return path is the capacitive coupling between the transmitter box and the human body. In some explanations, the ground loop through the floor is included (Doi and Nishimura 2005). Although this explanation is intuitive, it is somewhat misleading to picture an electric current flowing through a closed circuit that consists of a human body, electronic devices, capacitive coupling between the open ends, and a ground loop. Electromagnetic analyses and experiments have shown that the majority of the current flows through the capacitive coupling in the vicinity of the transmitter and the receiver, and that the capacitive coupling at the farther ends and the ground loop are only a part of the signal transmission path. We explain this in more detail in the following sections.

## Numerical Simulation

Before we show the simulation results, the simulation method and conditions are briefly explained. The human body consists of various body tissues such as skin, muscle, fat, bone, blood, etc., and each tissue has its own unique electrical properties (Gabriel 1996). For numerical electromagnetic analysis of the radio frequency (RF) signals that include a human body, whole-body numerical models with a voxel size of 1 to 2 mm are available commercially or from some public organizations (Nagaoka et al. 2004). Although analyses using these detailed models give better estimations of the transmission characteristics, not all models allow changes in the posture of the human body due to the complexity of the model. However, when the wavelength is longer than the human body, simpler models may be sufficient, depending on the purpose of the simulation (Hall and Hao 2006). We are interested in the transmission characteristics of intrabody communication for frequencies from 0.1 MHz up to approximately 100 MHz, and these frequencies satisfy the wavelength condition.

Figure 3.2a and b show a comparison of numerical simulations using a high-resolution whole-body model (Nagaoka et al. 2004) and a simple whole-body model consisting of several cylinders which have uniform electrical properties. These parameters were adopted from typical electrical properties of muscle (Gabriel 1996). These figures show the electric field distributions within the cross section that contains the human body and the transmitter attached on the wrist. The strength of the electric field is normalized by the maximum value. The conditions for the simulation were set as follows.

The mesh sizes for the high-resolution whole-body model were 1 mm around the electrodes and 2 mm for the rest of the model. The mesh sizes for the simple cylindrical model were 1 mm around the electrodes and 10 mm for the rest of the model. The transmission line matrix (TLM) method was used for the numerical analysis. An absorbing boundary condition was adopted as the boundary condition. A rectangular prism that enveloped

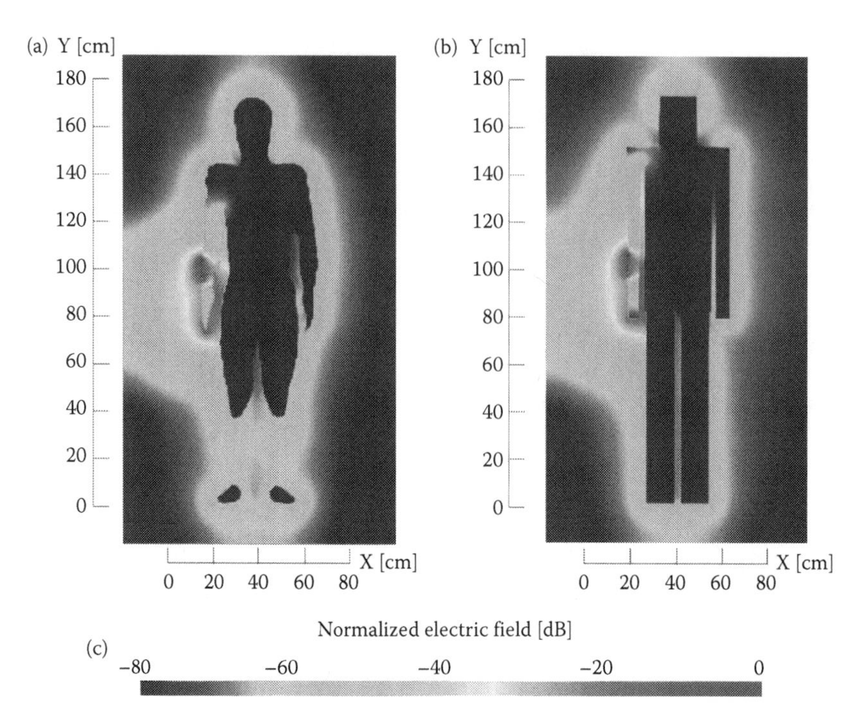


**FIGURE 3.2**
Comparison of the electrical field obtained from the high resolution whole-body model (a), the simplified cylindrical muscle model (b), and scale: normalized electric field [dB] (c).

the human-body model was enlarged by 30% in all three dimensions, and the numerical calculation was carried out over this space. We verified that further enlargement of this space did not change the results. A frequency-dependent Debye dispersive medium was adopted for the dielectric properties of the simple human-body model (Gabriel 1996). Figure 3.3 shows the transmitter model. The overall transmitter size was 30 by 80 mm. This was the same size as the transmitter used in later experiments. The two transmitter electrodes were both 30 by 20 mm in size and were separated by a space of 40 mm. Both electrodes were in contact with the human body at the wrist. The circuit board was modelled as a uniform plane conductor, which was connected to one of the electrodes. A signal source of 10 MHz was inserted between the circuit board and the other electrode. The output impedance was 50 Ω.

We can see that the overall distribution of the electric field of the simple cylindrical model (Figure 3.2b) is similar to that of the high-resolution whole-body model (Figure 3.2a). We also calculated the impedance between the two electrodes, which was 66.2–j8.57 Ω for the simple cylindrical model and 64.2–j7.78 Ω for the high-resolution model. These similarities suggest that the

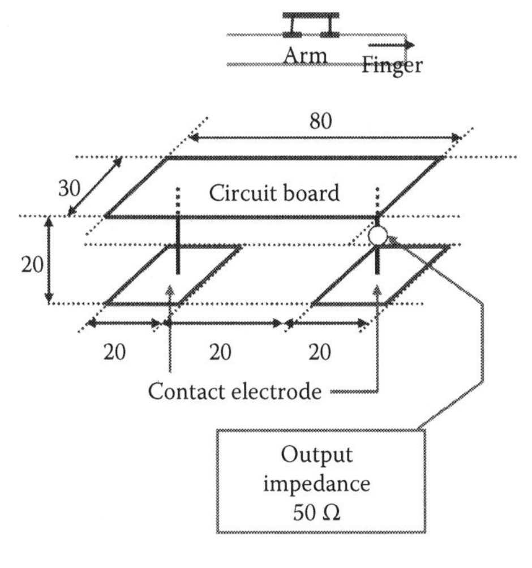


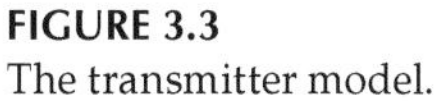

**FIGURE 3.3**
The transmitter model.

simple cylindrical muscle model is sufficient for analyses of the electric field distribution and the impedance at a frequency of 10 MHz. Similar results were obtained for other frequencies in the frequency band of interest to us. The posture of the human body can be easily modified with this simple model. The numerical simulations in the following sections were calculated by using this model.

## Electric Field Distribution Including an Off-Body Device

Figure 3.4a and b show the electric field distributions when a subject is touching transmitter boxes of different heights. The height of the transmitter box is 200 mm for Figure 3.4a and 1200 mm for Figure 3.4b. A receiver is placed on the subject's wrist. The dimensions of the receiver are the same as those for the transmitter model shown in Figure 3.3. The difference is that the signal source and the output impedance of the transmitter were replaced by an input impedance of 2 kΩ. Both electrodes of the receiver are in contact with the arm.

In the case of the short transmitter box, shown in Figure 3.4a, an electric field is formed around the transmitter box and the arm. The electric field between the transmitter box and the subject's body is not discernible in this figure's scale of field intensity. In the case of the tall transmitter box, an electric field is formed around the whole transmitter box as well as around the human body. This shows that there is capacitive coupling between the transmitter box and the human body.

Figure 3.5 shows the received signal voltages for four different transmitter box heights: 200, 400, 800 and 1200 mm. These results were obtained by

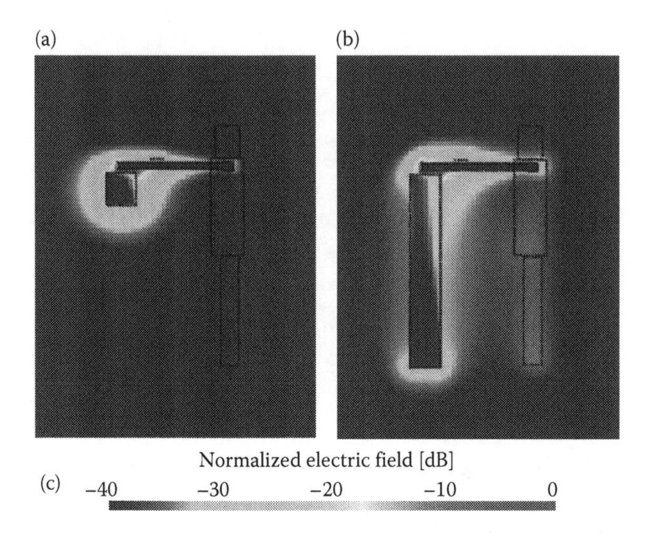


**FIGURE 3.4**
Electric field distributions for transmitter boxes of different heights: (a) box height 200 mm; (b) box height 1200 mm and (c) scale: normalized electric field [dB].

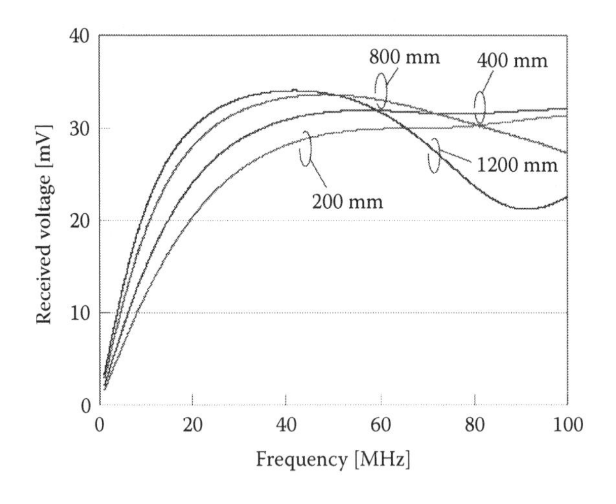


**FIGURE 3.5**
Signal voltages received by a receiver on the wrist when a person is touching transmitter boxes of different heights (numerical analysis).

numerical analysis. Note that the received signal voltage does not vary much for the different heights of the transmitter box. This indicates that the capacitive coupling between the lower part of the transmitter box and the human body is not a major factor in the signal transmission.

For the 1200 mm transmitter box, the received signal voltage decreases as the frequency increases above 50 MHz. This is because the wavelength becomes comparable to the size of the transmitter box and the size of the human body. This suggests that the transmission characteristics depend on the body posture and the size of the device for frequencies above 50 MHz. Similar results were reported by Cho et al. (2007).

### Intrabody Communication Using Capacitive Coupling

In intrabody communication systems using capacitive coupling, all electrodes are detached from the human body (Zimmerman 1996; Shinagawa et al. 2004). One of the electrodes of a transmitter/receiver is placed close to the body, with an insulator between the electrode and the body, and it forms capacitive coupling with the body. The other electrode is placed on the outer side of the transmitter/receiver. The signal transmission model of this system is similar to that of systems using electrodes that have galvanic coupling arranged in the longitudinal direction. The difference is that the electrode closer to the body has capacitive coupling. This is not a fundamental difference because capacitive coupling acts as impedance in RF signals. The transmission mechanism is explained as follows. A small electric current flows through the human body between the two electrodes that are close to the body, and the return path is provided by the capacitive coupling that consists of the outer electrodes, the human body and the environment (Zimmerman 1996). Although there is no quantitative report on the electric current that flows through the body, the receivers used in capacitively coupled systems have higher input impedances (Shinagawa et al. 2004).

Transmission model of Zimmerman (1996) was further investigated by Cho et al. (2007) in the frequency range of 0.1 MHz to 150 MHz. They modelled a human body as a distributed resistor-capacitor (RC) model and compared the results of numerical analyses and experiments. They concluded that the body channel showed high-pass filter characteristics for frequencies below 4 MHz because of the capacitive return path, whereas, for frequencies above 10 MHz, the transmission characteristics varied widely according to the frequency and the distance.

### Comments on Carrier Frequency

The carrier frequency for systems using the transversal electrode arrangement is expected to be lower than several megahertz, because the longitudinal electrode arrangement shows better transmission characteristics at 10 MHz (Fujii et al. 2004). Examples of carrier frequencies for systems using the transversal electrode arrangement are 70 kHz (Handa et al. 1997) and 0.01 MHz to 1 MHz (Wegmueller et al. 2007). However, there is a wider choice of carrier frequencies for systems using the longitudinal electrode arrangement. Examples of carrier frequencies for systems using the longitudinal electrode arrangement and capacitive coupling are 330 kHz (Zimmerman 1996), 455 and 1250 kHz (Doi and Nishimura 2005), 10 MHz (Fujii et al. 2004;

Shinagawa et al. 2004; Hachisuka et al. 2003, 2005; Koshiji and Sasaki 2008; Okamoto et al. 2010), below 60 MHz (Baldus et al. 2009), 15 to 60 MHz (Wang and Nishikawa 2007) and below 100 MHz (Cho et al. 2007). Transmission above 100 MHz becomes more airborne, which means that the contribution of the human body in the transmission decreases.

## Configuration and Size of Electrodes

### Configuration of Electrodes

There are four possible electrode configurations, depending on the combination of the contact and the noncontact electrodes of the transmitter and the receiver. If we denote the number of contact electrodes for the transmitter by the character *T* and those for the receiver by *R*, the four possible configurations are 2T2R, 1T2R, 2T1R and 1T1R. For example, the transversal electrode arrangement described earlier assumes that all four electrodes are in contact with the human body. This configuration is 2T2R. The configuration of the system shown in Figure 3.4 and Figure 3.5 is 1T2R. In this configuration, one electrode on the top of the transmitter box is touching the human body, and the other electrode, which is the transmitter box, is left open. The two receiver electrodes are both in contact with the human body. Fujii et al. (2004) adopted the opposite configuration, which was 2T1R. Both the transmitter electrodes were touching the phantom, while the ground plane of the receiver circuitry was suspended above the surface and was not in contact with the phantom. An intrabody system using capacitive coupling can be viewed as a 1T1R system, in which one electrode of both the transmitter and the receiver is open. Although the electrodes in this system have no physical contact with the body, the electrodes on the body side are electrically connected to the body by capacitive coupling. Therefore, these electrodes can be regarded as contact electrodes in a broader sense. The differences in transmission characteristics with respect to these electrode configurations are described later in this section.

Figure 3.6 shows the received signal voltages of the four possible electrode configurations in the frequency range of 1 to 100 MHz, and Figure 3.7 shows the electric field distributions at the frequency of 10 MHz. Figure 3.8 shows the dimensions and arrangement of the output electrodes on the top of the transmitter box, and Figure 3.9 shows the dimensions and arrangement of the input electrodes of a receiver worn on the wrist.

The most efficient electrode configuration has a single contact electrode for the stationary transmitter box and two contact electrodes for the receiver on the wrist. When two electrodes on the transmitter are in contact with the hand, the signal voltage decreases approximately by half. In contrast,

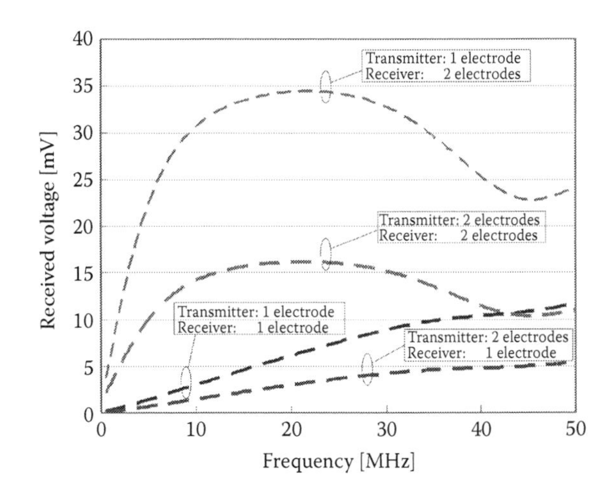


**FIGURE 3.6**
Signal voltages received by a receiver on the wrist for the four configurations of contact electrodes (numerical analysis).

for the receiver, it is better to have both electrodes in contact with the body. For a receiver with a single contact electrode, the received signal voltage increases as the frequency increases. This is because, as the wavelength becomes shorter, the circuit board which is held above the surface of the body becomes a more efficient airborne antenna to capture the electric field formed around the human body.

Figure 3.10 shows the experimentally measured received signals of two configurations, 1T2R and 2T2R. These results correspond to the numerical analyses shown in Figure 3.6. The measured values also show that the received signal of 2T2R is approximately half that of 1T2R. The measurement of the received signal voltage requires special care in the experimental apparatus. The receiver must be a battery-operated stand-alone device without any wires or cables attached to it, because wires connected to the receiver act as antennas. Instead of using displays or wireless transmission, we used a comparator and a small LED in the receiver circuit to visually check whether the received signal was above or below the threshold voltage. The amplitude of the transmitter output was adjusted so that the received signal was equal to the threshold. The received signal voltages shown in Figure 3.10 correspond to the transmitter output of 1 V. In this conversion, we assumed that the transmission characteristic was linear with respect to the amplitude.

The results shown in Figures 3.6 and 3.10 indicate that it is better to touch a single-contact-electrode transmitter than a two-contact-electrode transmitter with the hand when the receiver is worn on the wrist. However, the transmission characteristics change when the contact position of the device with respect to the body changes. Figure 3.11a and b show the electric field distributions of (a) the two-contact-electrode transmitter and (b) the

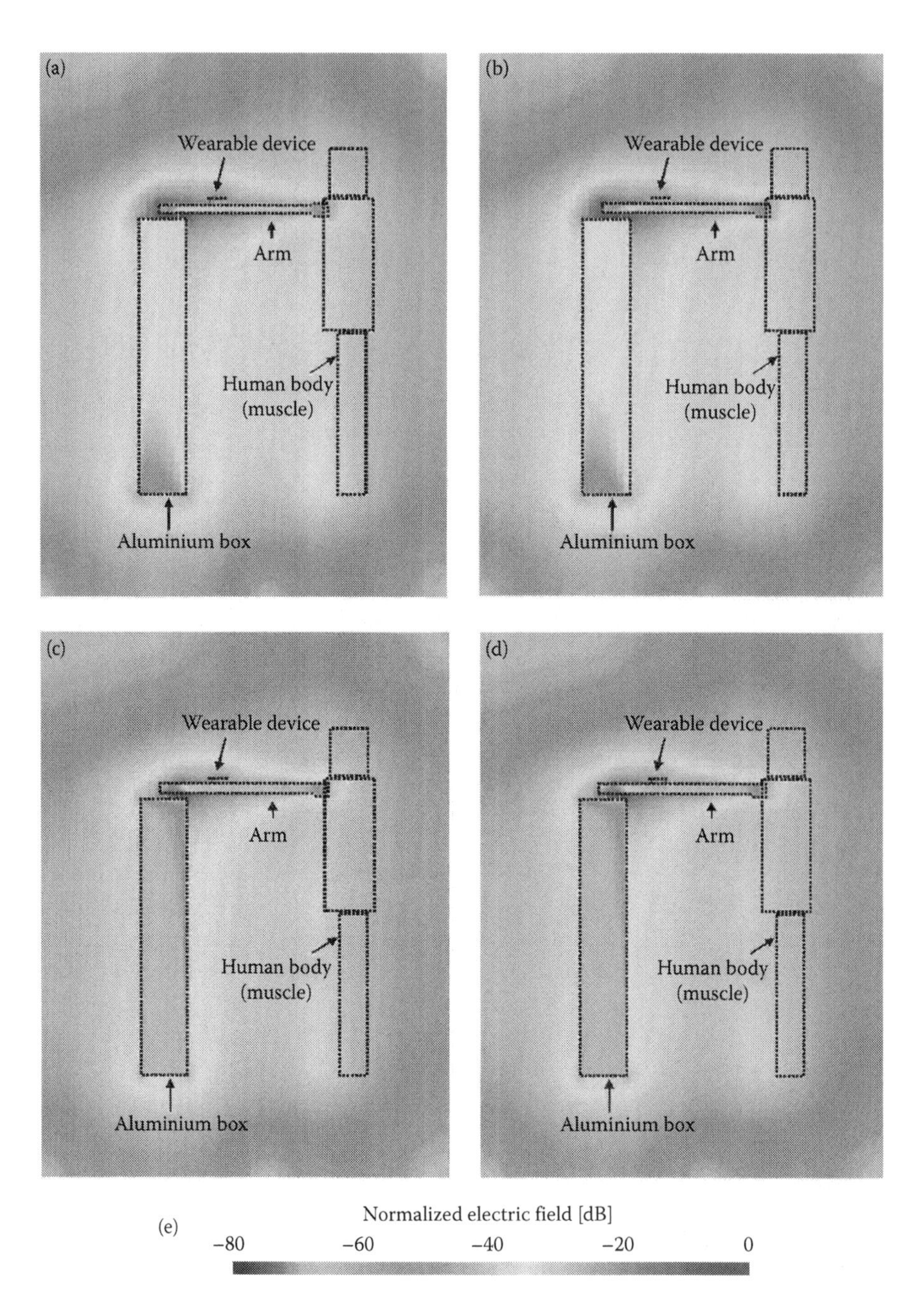


**FIGURE 3.7**
Electric field distributions of the four configurations of contact electrodes. (a) Transmitter (aluminium box): 1 electrode; receiver (wearable device): 1 electrode; (b) Transmitter (aluminium box): 1 electrode; receiver (wearable device): 2 electrodes; (c) Transmitter (aluminium box): 2 electrodes; receiver (wearable device): 1 electrode; (d) Transmitter (aluminium box): 2 electrodes; receiver (wearable device): 2 electrodes; (e) Scale: normalized electric field [dB].

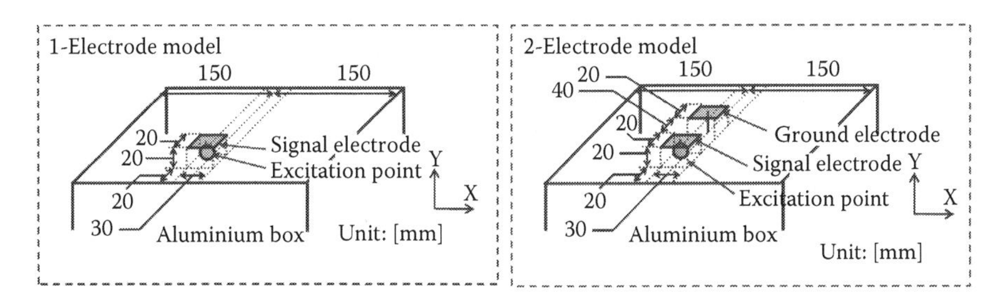


**FIGURE 3.8**
Arrangement of the output electrodes on the top of the transmitter box.

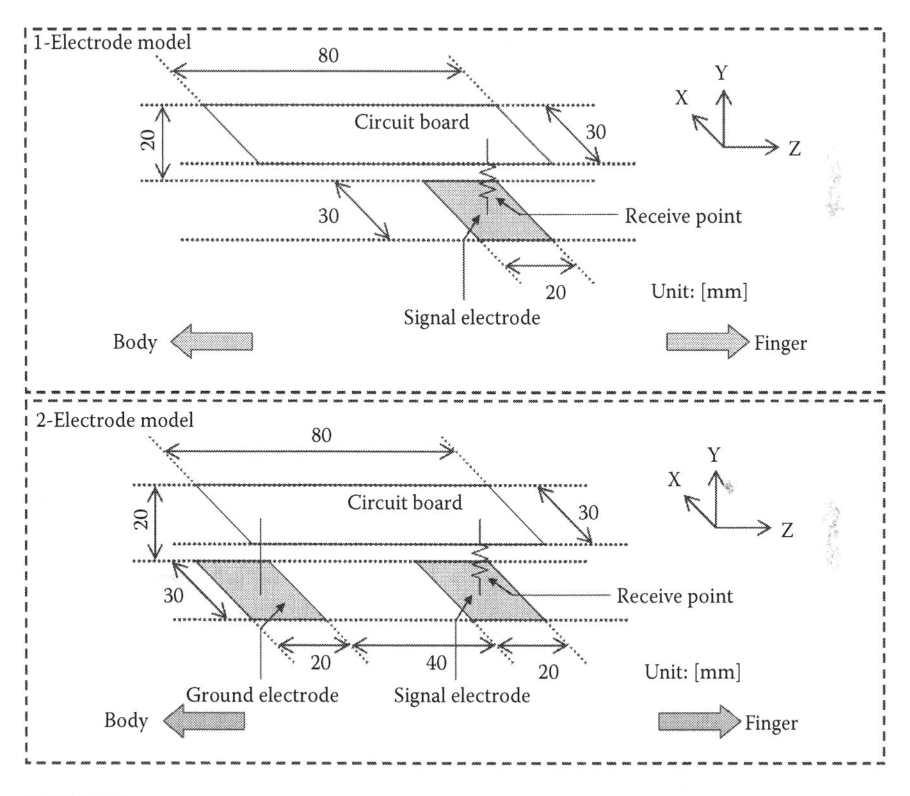


**FIGURE 3.9**
Arrangement of the input electrodes of the receivers.

single-contact-electrode transmitter placed at the centre of a cylindrical phantom. This bar-shaped phantom is a simplified model of a human arm without the torso and the head. Figure 3.12 shows the analytical models used in this simulation. In both cases, the output voltage of the transmitter was the same.

Figure 3.11a resembles the electric field distribution of a dipole antenna. The difference between an intrabody transmitter and a dipole antenna is

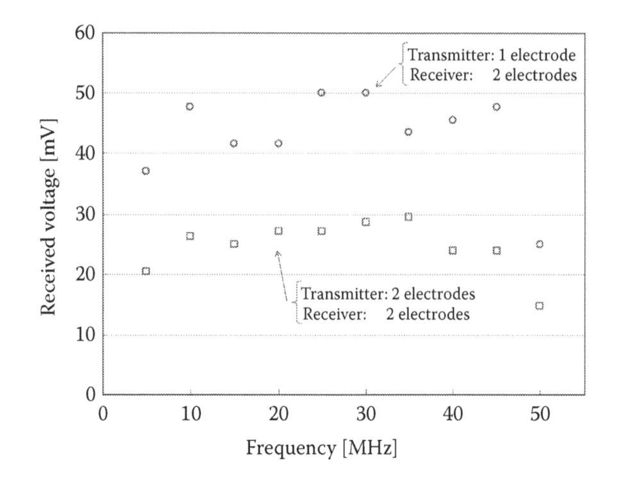


**FIGURE 3.10**
Signal voltages received by a receiver on the wrist (measured).

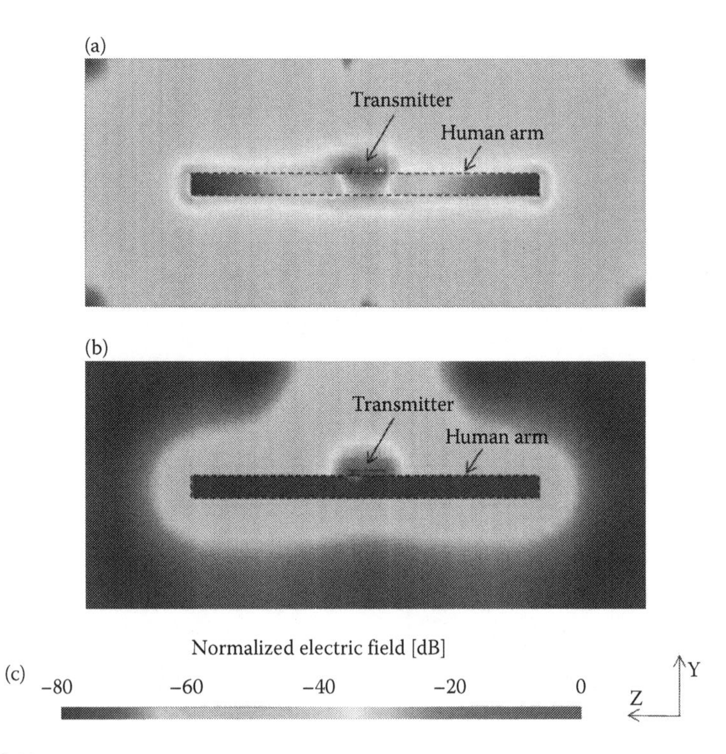


**FIGURE 3.11**
Electric field distributions around a human-arm phantom with a single-contact-electrode transmitter (a), a two-contact-electrode transmitter (b), and scale: normalized electric field [dB] (c).

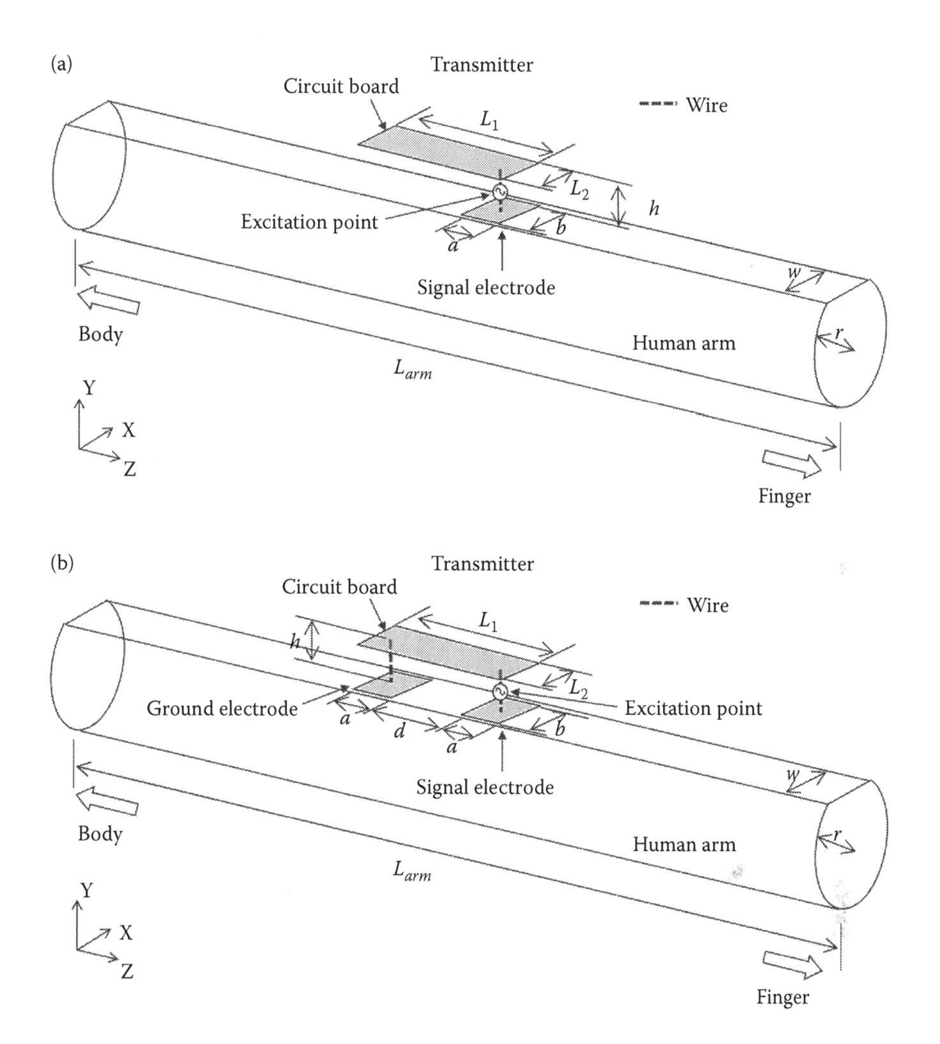


**FIGURE 3.12**
Analytical models of a human-arm phantom with a single-contact-electrode transmitter (a) and a two-contact-electrode transmitter (b).

that the two elements of the dipole antenna are separated, whereas the two electrodes of the intrabody transmitter are electrically connected by the part of the body underneath the electrodes. The current flowing through this part of the body does not contribute to the signal transmission. In contrast, there is no wasted current flowing between the two electrodes of a single-contact-electrode transmitter. The downside is that the electric field at both ends of the phantom is weaker for the single-contact-electrode transmitter than for the two-contact-electrode transmitter, as shown in Figure 3.11b. It is not obvious which configuration requires less transmission power to send the same received power to the receiver. This issue will be investigated in the future.

## Spacing between Two Contact Electrodes

The spacing between electrodes is an important design parameter when the two electrodes of a transmitter and/or a receiver are both contact electrodes. In Figure 3.9, this spacing is 40 mm. Figure 3.13 shows the result of a numerical analysis of the signal voltages received by a receiver on the wrist with different electrode spacing. Figure 3.14 shows the measured values. The analytical model of the receiver was the same as that in Figure 3.9, but the spacing between the electrodes was changed for the analysis and experiment. The whole analytical model was the same as in Figure 3.9: The subject was touching a stationary transmitter box with his or her hand. The results of the numerical analyses and the experiments were similar in terms of the

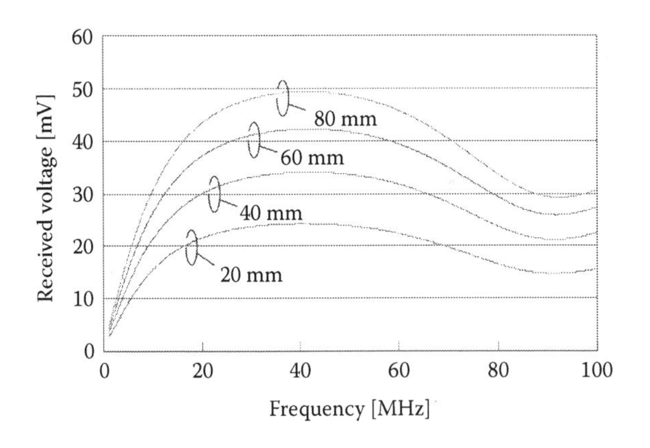


**FIGURE 3.13**
Received signal voltages of a receiver on the wrist with different spacing between the two contact electrodes (numerical analysis).

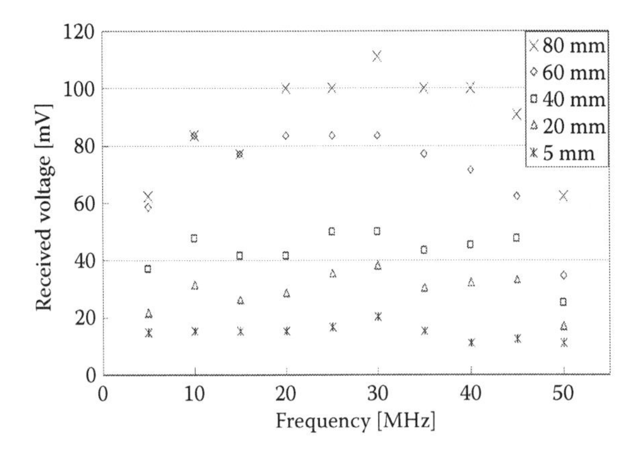


**FIGURE 3.14**
Received signal voltages of a receiver on the wrist with different spacing between the two contact electrodes (measured).

frequency dependence and the relation with the electrode spacing, except for the value of the received signal. This difference is due to the differences in the model parameters. The data show that the received signal increases as the space between the electrodes increases. This suggests that it is better to place the two electrodes as far apart as possible within the limits of the device size.

### Distance between the Human Body and a Circuit Board

Usually, one of the two terminals of a transmitter or a receiver is connected to the ground of the circuit board. Since the ground pattern usually has the largest area on a circuit board, the distance between the body and the circuit board can affect the transmission characteristics. We need to know this influence to determine the position of the circuit board. The received signal voltages for different circuit board heights are shown in Figure 3.15 (numerical analysis) and Figure 3.16 (measured). Both sets of results indicate that the distance between the body and the circuit board has little influence on the received signal voltage.

### Transmission Characteristics of On-Body Devices

In the previous sections, communication between a wristwatch device and an off-body device to be touched with a hand has been described. In this section, the characteristics of transmissions between a wristwatch transmitter and receivers located at several possible locations on the body are shown to gain a more practical perspective on possible applications. The arrangement of devices is shown in Figure 3.17. For the receivers located at positions 5 and 6, we assumed that the receivers were put in clothing pockets and that the electrodes were not in direct contact with the body. In the analytical model,

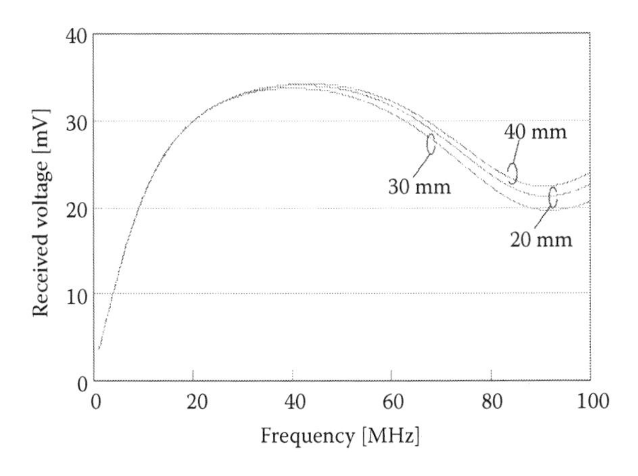


**FIGURE 3.15**
Received signal voltages of a receiver on the wrist with different circuit board heights (numerical analysis).

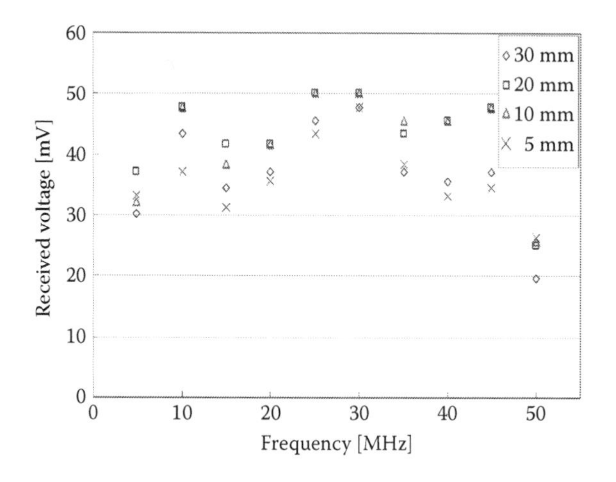


**FIGURE 3.16**
Received signal voltages of a receiver on the wrist with different circuit board heights (measured).

both electrodes were held 4 mm above the surface of the body. The amplitude of the transmission signal was 1 V. Table 3.1 shows the results. For receivers placed at the hand (position 2) and the upper arm (position 3), the received signal level was more than 1 mV. Signals of this strength are very easy to handle. The transmission to the wrist of the other arm (position 4) and to the receivers in pockets (positions 5 and 6) was very weak. Although it is technically possible to receive these weak signals, the merits of intrabody communication may be lost when compared to ordinary airborne transmission.

## Impedance Matching of Electrodes

In RF transmitters and receivers, the output impedance of a driving circuit and the input impedance of a receiver front end are designed to match the impedance of an antenna in order to maximize the transmission efficiency. In this section, we show that impedance matching is also effective in intrabody communication.

In intrabody communication, a transmitter excites the electrodes and transmits an electrical signal through the human body. Therefore, the electrical load of the final driving circuit consists of the electrodes, the human body and the near-field space around the human body. This impedance between the two output electrodes is equivalent to the impedance of the antenna in airborne transmission. In airborne-transmission circuits, common impedances are 50 Ω and 75 Ω, which are derived from the impedances of standard coaxial cables and antennas. Although there is no theoretical reason to adopt these impedances in intrabody transmission systems, these impedances have practical merits: we can use commercially available integrated circuits and RF parts that are specifically designed to match these impedances.

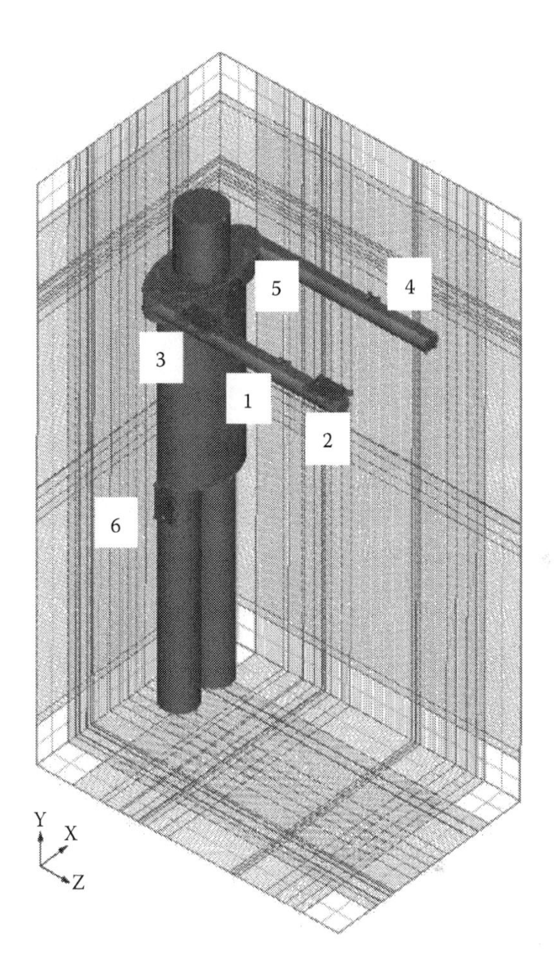


**FIGURE 3.17**
Placement of devices (1, right wrist; 2, right hand; 3, right upper arm; 4, left wrist; 5, left chest pocket; 6, right pants pocket; A transmitter is placed at 1, and receivers are placed at 2–6).

**TABLE 3.1**

Transmission Characteristics between a Wristwatch Transmitter and Receivers on Other Body Parts

| | Transmission Characteristics/Received Power/Received Voltage | | | | | |
|---|---|---|---|---|---|---|
| | Simulated | | | Measured | | |
| Paths | $S_{21}$ (dB) | Power (W) | Voltage (V) | $S_{21}$ (dB) | Power (W) | Voltage (V) |
| 1–2 | –55.8 | 2.63E–09 | 1.62E–03 | –58.1 | 1.55E–09 | 1.24E–0.3 |
| 1–3 | –55.6 | 2.75E–09 | 1.66E–03 | –57.6 | 1.74E–09 | 1.32E–03 |
| 1–4 | –87.3 | 1.86E–12 | 4.32E–05 | –88.2 | 1.51E–12 | 3.89E–05 |
| 1–5 | –102.5 | 5.62E–14 | 7.50E–06 | (–83.2) | (4.79E–12) | (6.92E–05) |
| 1–6 | –98.7 | 1.35E–13 | 1.16E–05 | (–82.1) | (6.17E–12) | (7.85E–05) |

In this section, transmission from a wristwatch transmitter to a handheld personal digital assistant (PDA) or a stationary device that is touched by the user's hand is evaluated as a model case. The impedances between the two transmission electrodes were analysed for a 10 MHz signal by varying the sizes and configurations of the electrodes. Figure 3.18 gives the three electrode designs for comparison. We assumed that the driving circuit had an output impedance of 50 Ω. Figure 3.19 shows the electric field distributions for the three electrode designs. The electrode design that had an

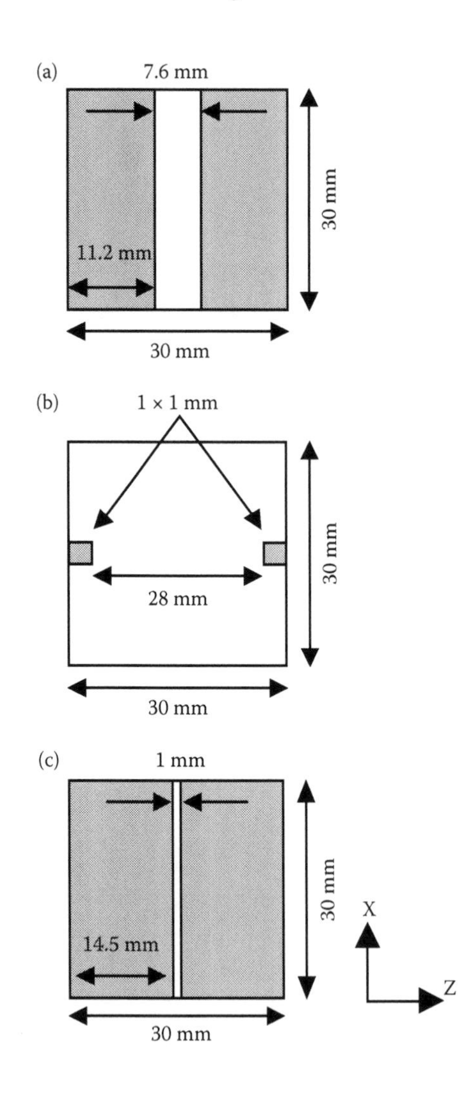


**FIGURE 3.18**
Contact electrode sizes of a wristwatch device for evaluation of impedance matching. (a) Impedance-matched electrode structure; (b) impedance-mismatched electrode structure; (c) impedance-mismatched electrode structure.

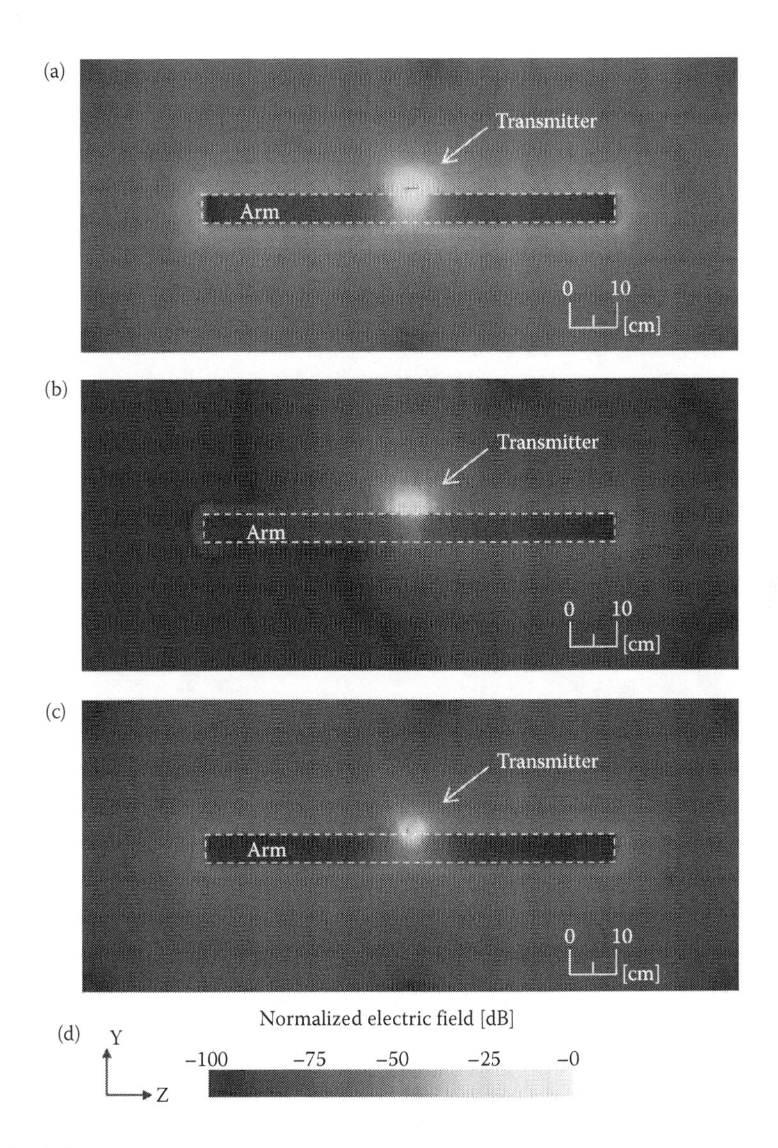


**FIGURE 3.19**
Electric field distributions with electrodes with matched and unmatched impedance: (a) electrode impedance matched with 50 Ω signal source (Zin = 49.6 – j5.9 Ω); (b) electrode impedance not matched with 50 Ω signal source (Zin = 1270 – j267.5 Ω); (c) electrode impedance not matched with 50 Ω signal source (Zin = 30 – j3.8 Ω); (d) scale: normalized electric field [dB].

impedance close to 50 Ω formed the strongest electric field at the end of the phantom, as shown in Figure 3.19a. The differences in the electric field strength from the other two electrode designs were 20 to 30 dB. We need to know how to determine the size and arrangement of electrodes that would satisfy a given specification. However, the relationships between the electrode sizes and the impedance can be obtained only by numerical analysis, so designers have to find solutions by trial and error. To alleviate this process, we calculated the electrode impedances for various sizes and spacings and derived experimental equations that approximate these relationships (Koshiji and Sasaki 2008). With these equations, designers can estimate the electrode impedance and determine the electrode dimensions for a given specification.

## Summary

Intrabody communication using contact electrodes in low-frequency bands was presented with numerical analyses and experimental results. Although the data transmission rates may be lower than the high-frequency communication standards, and the use of contact electrodes may limit some applications, intrabody communication with low-frequency bands has the advantage of simplicity, and many applications can benefit from this communication method.

## References

Baldus H., Corroy S., Fazzi A., Klabunde K., and Schenk T. 2009. Human-centric connectivity enabled by body-coupled communications. *IEEE Communications Magazine,* Vol. 47, No. 6, pp. 172–178.

Cho N., Yoo J., Song S.J., Lee J., Jeon S., and Yoo H.J. 2007. The human body characteristics as a signal transmission medium for intrabody communication. *IEEE Trans. on Microwave Theory and Techniques,* Vol. 55, No. 5, pp. 1080–1086.

Doi K. and Nishimura A. 2005. High-reliability communication technology using human body as transmission medium. *Matsushita Electric Works, Ltd. Technical Report,* Vol. 53, No. 3, pp. 72–76, Panasonic Electric Works.

Fujii K., Ito K., Hachisuka K., Terauchi Y., Sasaki K., and Itao K. 2004. Study on the optimal direction of electrodes of a wearable device using the human body as a transmission channel. *Proc. of the 2004 Int. Symp. on Antennas and Propagation (ISAP'04),* Vol. 2, pp. 1005–1008, Sendai Japan, Institute of Electronics, Information and Communication Engineers (IEICS).

Gabriel C. 1996. Compilation of the dielectric properties of body tissues at RF and microwave frequencies. Brooks Air Force Technical Report AL/OE-TR-1996-0037.

Hachisuka K., Nakata A., Takeda T., Terauchi Y., Shiba K., Sasaki K., Hosaka H., and Itao K. 2003. Development and performance analysis of intra-body communication device. In *Digest of Technical Papers of the 12th Int. Conf. on Solid-State Sensors, Actuators, and Microsystems (Transducers'03),* Vol. 2, pp. 1722–1725, Boston, IEEE.

Hachisuka K., Terauchi Y., Sasaki K., Hosaka H., and Itao K. 2004. Measurement of power transmission and phase characteristics of intra-body communication. In *Proc. of the 18th Japan Institute of Electronics Packaging (JIEP) Annual Meeting*, pp. 195–196 (in Japanese), Tokyo Japan, JIEP.

Hachisuka K., Terauchi Y., Kishi Y., Hirota T., Sasaki K., Hosaka H., and Itao K. 2005. Simplified circuit modeling and fabrication of intrabody communication devices. In *Digest of Technical Papers of the 13th Int. Conf. on Solid-State Sensors, Actuators, and Microsystems (Transducers'05)*, Vol. 1, pp. 461–464, Seoul Korea, IEEE.

Hall P.S. and Hao Y. (Eds.). 2006. *Antennas and Propagation for Body-Centric Wireless Communications*. Boston, Artech House.

Handa T., Shoji S., Ike Shinichi., Takeda S., and Sekiguchi T. 1997. A very low-power consumption wireless ECG monitoring system using body as a signal transmission medium. In *Proc. of Int. Conf. on Solid-State Sensors and Actuators (Transducers'97)*, Vol. 2, pp. 1003–1007, Chicago, IEEE.

Koshiji F. and Sasaki K. 2008. Input impedance characteristics of wearable transmitters for body-centric networks. In *Proc. of the Int. Conf. on Electronic Packaging (ICEP 2008)*, 10B1-1, Tokyo Japan, Japan Institute of Electronic Packaging.

Monton E., Hernandez J.F., Blasco J.M., Herve T., Micallef J., Grech I., Brincat A., and Traver V. 2008. Body area network for wireless patient monitoring. *IET Commun.*, Vol. 2, No. 2, pp. 215–222.

Nagaoka T., Watanabe S., Sakurai K., Kunieda E., Watanabe S., Taki M., Y. Yamanaka. 2004. Development of realistic high-resolution whole-body voxel models of Japanese adult male and female of average height and weight, and application of models to radio-frequency electromagnetic-field dosimetry. *Physics in Medicine and Biology*, Vol. 49, pp. 1–15.

Okamoto E., Sato Y., Seino K., Kiyono T., Kato Y., and Mitamura Y. 2010. Basic study of a transcutaneous information transmission system using intra-body communication. *J. Artif Organs* Vol. 13, pp. 117–120.

Pantelopoulos A. and Bourbakis N.G. 2010. A survey on wearable sensor-based systems for health monitoring and prognosis. *IEEE Trans. on Systems, Man, and Cybernetics-PART C: Applications and Reviews*, Vol. 40, No. 1, pp. 1–12.

Patel M. and Wang J. 2010. Applications, challenges, and prospective in emerging body area networking technologies. *IEEE Wireless Communications*, Vol. 17, No. 1, pp. 80–88.

Reusens E., Joseph W., Latre B., Braem B., Vermeeren G., Tanghe E., Martens L., Moerman I., and Blondia C. 2009. Characterization of on-body communication channel and energy efficient topology design for wireless body area networks. *IEEE Trans. on Information Technology in Biomedicine*, Vol. 13, No. 6, pp. 933–945.

Ruiz J.A., Xu J., and Shimamoto S. 2006. Propagation characteristics of intra-body communications for body area networks. In *3rd IEEE Consumer Communications and Networking Conf.*, CCNC 2006, Las Vegas, pp. 509–513.

Schenk T.C.W., Mazloum N.S., Tan L., and Rutten P. 2008. Experimental characterization of the body-coupled communications channel. In *IEEE Int. Symp. on Wireless Communication Systems (ISWCS '08)*, Reykjavik, Iceland, pp. 234–239.

Shinagawa M., Fukumoto M., Ochiai K., and Kyuragi H., 2004. A near-field-sensing transceiver for intrabody communication based on the electrooptic effect. *IEEE Trans. on Instrumentation and Measurement*, Vol. 53, No. 6, pp. 1533–1538.

Wang J. and Nishikawa Y. 2007. Characterization and performance of high-frequency pulse transmission for human body area communications. *IEICE Trans. Commun.,* Vol. E90-B, No. 6, pp. 1344–1350.
Wegmueller M.S., Kuhn A., Froehlich J., Oberle M., Felber N., Kuster N., and Fichtner W. 2007. An attempt to model the human body as a communication channel. *IEEE Trans. on Biomedical Engineering,* Vol. 54, No. 10, pp. 1851–1857.
Yuce M.R. 2010. Implementation of wireless body area networks for healthcare systems. *Sensors and Actuators A,* Vol. 162, No. 1, pp. 116–129.
Zimmerman T.G. 1996. Personal area networks: Near-field intrabody communication. *IBM Systems Journal,* Vol. 35, No. 3–4, pp. 609–617.

# 4

# *The Prospect of Energy-Harvesting Technologies for Healthcare Wireless Sensor Networks*

**Yen Kheng Tan**

**CONTENTS**

Introduction .......... 76
Motivation for Healthcare WSNs .......... 76
Architecture of WSNs .......... 77
The Protocol Stack of a WSN .......... 78
The Wireless Sensor Nodes of the WSN .......... 79
Problems in Powering Wireless Sensor Nodes .......... 80
High Power Consumption of Sensor Nodes .......... 81
Limits on Energy Sources for Sensor Nodes .......... 82
Energy-Harvesting Solutions for Wireless Sensor Nodes .......... 84
Overview of Energy Harvesting .......... 85
Review of Past Works on EH Systems .......... 87
Solar EH Systems .......... 88
Thermal EH Systems .......... 89
Vibration EH Systems .......... 92
Wind EH Systems .......... 94
Prospect of EH Technologies for Healthcare WSNs .......... 96
Case Study: TEH from Human Warmth for WBANs in a Medical Healthcare System (Hoang, Tan, Chng, and Panda 2009) .......... 97
TEH Structure with TEG .......... 100
Power Management Circuit .......... 102
Fall-Detection Sensor .......... 103
Experimental Test Results .......... 105
Conclusions .......... 106
References .......... 106

## Introduction

The rapid growth in demand for computing everywhere has made the computer a pivotal component of humans' daily lives (Kansal, Hsu, Zahedi, and Srivastava 2006). Whether we use computers to gather information from the Web, to entertain ourselves or to run a business, computers are noticeably becoming more widespread, mobile and smaller in size (Cook and Das 2004). What we often overlook and do not notice is the presence of billions of small pervasive computing devices around us, which provide the intelligence being integrated into the real-world "smart environment" (Cook and Das 2004) to help us to solve some crucial problems in the activities of our daily lives. To achieve this vision of the smart environment with pervasive computing, also known as *ubiquitous computing*, many computational devices are integrated in everyday objects and activities to enable better human–computer interaction. These computational devices, which are generally equipped with sensing, processing and communicating abilities, are known as *wireless sensor nodes*. When these wireless sensor nodes are connected together, they form a network called a *wireless sensor network* (WSN).

### Motivation for Healthcare WSNs

With the recent advances in wireless communication technologies, sensors and actuators, and highly integrated microelectronics technologies, WSNs have gained worldwide attention for their ability to facilitate monitoring and controlling, from remote locations, of physical environments that could be difficult or dangerous to reach. In Massachusetts Institute of Technology's (MIT) *Technology Review* magazine of innovation published in February 2003 (MIT 2003), the editors identified WSNs as the first of the top 10 emerging technologies that would change the world. Mobile, wireless, pervasive computing and communication environments are changing the way medical staffs interact with their patients and the elderly. Through the deployment of self-organized wireless physiological monitoring hardware and software systems, continual patient monitoring in certain types of patient postures becomes convenient to assuring timely intervention by a healthcare practitioner or a physician. For example, cardiac patients wearing electrocardiogram (ECG) sensor systems can be monitored remotely without leaving their residences. Healthcare sensor systems are required to be connected directly or indirectly to the Internet at all times, which allows medical staff to acquire timely data on arrhythmia events and abnormal ECG signals for corrective medical procedures. Moreover, physiological records are collected over a long period of time so that physicians can provide accurate diagnoses and correct treatment.

## Architecture of WSNs

WSNs represent a significant improvement over wired sensor networks with the elimination of the hardwired communication cables and associated installation and maintenance costs. The architecture of a WSN typically consists of multiple pervasive sensor nodes, a sink, public networks, manager nodes and the end users (Akyildiz, Su, Yogesh, and Erdal 2002). Many tiny, smart and inexpensive sensor nodes are scattered in the targeted sensor field to collect data and route the useful information back to the end user. These sensor nodes cooperate with each other via a wireless connection to form a network and collect, disseminate and analyse data coming from the environment. To ensure full connectivity, fault tolerance and long operational life, WSNs are deployed in an ad hoc manner, and the networks use multihop networking protocols to obtain real-world information and perform control ubiquitously (Sohrabi, Gao, Ailawadhi, and Pottie 2000). As illustrated in Figure 4.1, the data collected by node A are routed within the sensor field by other nodes. When the data reach the boundary node E, they are then transferred to the sink. The sink serves as a gateway with higher processing capacity to communicate with the task manager node. The connection between the sink and the task manager node is the public networks in the form of the Internet or satellite. Once the end user receives the data from the task manager node, some processing actions are then performed on the received data.

In Figure 4.1, the sink is essentially a coordinator between the deployed sensor nodes and the end user, and it can be treated like a gateway node. The need for a sink in the WSN architecture is due to the limited power and computing capacity of each of the wireless sensor nodes. The gateway node, typically powered by a readily available power source from the AC main, is equipped with a better processor and sufficient memory space, such that it is able to satisfy the need for extra information processing before data are transferred to the final destination. The gateway node can therefore share the loads placed on the wireless sensor nodes and hence prolong their

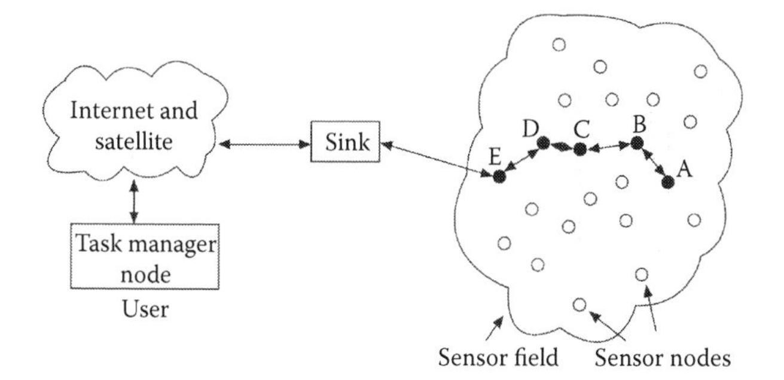


**FIGURE 4.1**
Architecture of wireless sensor networks (WSNs) to facilitate smart environments. (From Akyildiz, I., et al., *IEEE Communications Magazine*, 40(8), 102–114, 2002. With permission.)

working life. To understand how data are communicated within the sensor nodes in a WSN as shown in Figure 4.1, the protocol stack model of the WSN is investigated. With this understanding, the energy-hungry portions of the wireless sensor node can be identified, and then the WSN can be redesigned accordingly for lower power consumption.

## The Protocol Stack of a WSN

Starting from the lowest level, the physical layer is to meet the needs of receiving and transferring data collected from the hardware. It is well known that long-distance wireless communication can be expensive, in terms of both energy and implementation complexity. When designing the physical layer for WSNs, energy minimization is considered significantly more important over and above other factors like propagation and fading effects. Energy-efficient physical layer solutions are currently being pursued by researchers to design tiny, low-power, low-cost transceiver, sensing and processing units (Lin, Kaiser, and Pottie 2004). The next higher layer is the data link layer, which ensures reliable point-to-point and point-to-multipoint connections for the multiplexing of data streams, data frame detection, medium access and error control in the WSN. The data link layer should be power-aware and at the same time minimize the collisions between neighbours' signals because the environment is noisy and sensor nodes themselves are highly mobile. This is also one of the layers in the WSN whereby power-saving modes of operation can be implemented. The most obvious means of power conservation is to turn the transceiver off when it is not required. By using a random wake-up schedule during the connection phase and by turning the radio off during idle time slots, power conservation can be achieved. A dynamic power management scheme for WSNs has been discussed in Sinha and Chandrakasan (2001), where five power-saving modes are proposed, and intermode transition policies are investigated.

The network layer takes care of routing the data supplied by the transport layer. In WSN deployment, the routing protocols in the network layer are important because an efficient routing protocol can help to serve various applications and save energy. By setting appropriate energy and time-delay thresholds for data relay, the protocol can help prolong the lifetime of sensor nodes. Hence the network layer is another layer in the WSN that can reduce power consumption. The transport layer helps to maintain the flow of data if the sensor networks' application requires it.

Depending on the sensing tasks, different types of application software can be built and used on the application layer. In contrast to the traditional networks, which focus mainly on how to achieve high quality-of-service (QoS) provisions, WSN protocols tend to focus primarily on power conservation and power management for sensor nodes (Merrett, Al-Hashimi, White, and Harris 2005; Sinha and Chandrakasan 2001) as well as the design of energy-aware protocols and algorithms for WSNs (Lattanzi, Regini, Acquaviva, and Bogliolo 2007; Sohrabi, Gao, Ailawadhi, and Pottie 2000) in order to reduce the power consumption of the overall WSN. By doing so, the lifetime of the WSN can be extended.

However, there must be some embedded trade-off mechanisms that give the end user the option of prolonging the WSN lifetime but at the cost of lower throughput or longer transmission delay. Conversely, the power consumption of the WSN can be reduced by sacrificing the QoS provisions, i.e., lowering the data throughput or having longer transmission delay. Among the various challenging requirements posed by the design of the underlying algorithms and protocols of the WSNs, it is well known within academia as well as the industry (Chong and Kumar 2003; Kuorilehto, Hannikainen, and Hamalainen 2005; Tubaishat and Madria 2003) that the energy constraint is one of the most significant challenges in the WSN research field (Callaway 2003). The functionalities of the WSN are highly dependent on the amount of energy that is available to be expended by each of the sensor nodes in the network. As such, the energy-constraint challenge of WSNs is substantial enough to be investigated and discussed in this chapter. It is a multi-objective optimization problem concerning various WSN parameters like QoS, transmission delays, lifetime, energy, etc.

## The Wireless Sensor Nodes of the WSN

WSNs, based on the collaborative efforts of a large number of sensor nodes distributed throughout an area of interest, have been proven by many researchers to be good candidates to provide economically viable solutions for a wide range of healthcare applications. These sensor nodes are coordinated based on some network topologies to cooperate with one another within the WSNs to satisfy the applications' requirements. Each sensor node monitors its local environment, locally processing and storing the collected data so that other sensor nodes in the network, via a wireless communication link, can use it. Because of the great potential of WSNs, many groups around the world have invested lots of research efforts and time in the design of sensor nodes for their specific applications. These include Berkeley's Mica motes (Hill and Culler 2002), PicoRadio projects (Rabaey, Ammer, da Silva, Patel, and Roundy 2000) and MIT's μAmps (Massachusetts Institute of Technology, n.d.), as well as many others. In addition, the TinyOS project (TinyOS, n.d.) provides a framework for designing flexible distributed applications for data collection and processing across the sensor network. All of these sensor nodes have similar goals such as small physical size, low power consumption and rich sensing capabilities. A block diagram of a wireless sensor node in a WSN is shown in Figure 4.2. The sensor node typically consists of four subunits, namely, the sensor itself, the data acquisition circuit, a local microcontroller and a radio communication block.

As shown in Figure 4.2, the sensor/transducer converts physical vital parameters of the human body such as the body temperature, heartbeat, movement activities, etc. into an electrical signal. A data acquisition circuit is

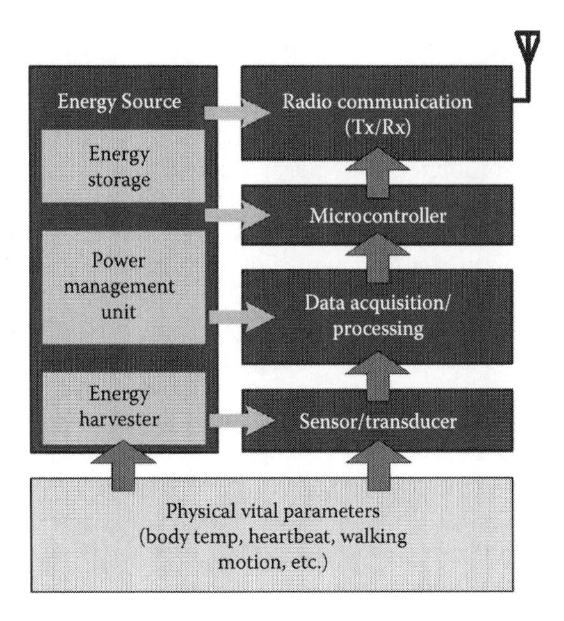


**FIGURE 4.2**
Block diagram of a wireless sensor node.

incorporated in the sensor node to realize amplification and preprocessing of the output signals from sensors, for example, conversion from analogue to digital form and filtering. The conditioned signals are then processed and stored in the embedded microcontroller for other sensor nodes in the network to use them. Other than data processing, the microcontroller also provides some level of intelligence, like time scheduling, to the sensor node. To enable the sensor node to communicate with its neighbour node or the base station in a wireless manner, a radio communication block is included. All four subunits of the sensor node are power sink modules, and they need to consume electrical energy from the power source in order to operate. Because of that, the wireless sensor node would consume all the energy stored in the battery after some time, and the sensor node would then go into an idle state. Once the percentage of nodes that retain some residual energy falls below a specific threshold, which is set according to the type of application (it can be 100% or less; Verdone, Dardari, Mazzini, and Conti 2008), the operational lifetime of the WSN ends.

## Problems in Powering Wireless Sensor Nodes

As the WSN becomes dense with many sensor nodes, the problems in powering the wireless sensor nodes—namely, (1) the high power consumption of the sensor nodes and (2) the limited of energy sources for sensor nodes—become critical, and even worse when one considers the prohibitive cost of providing power to them through wired cables or replacing the batteries. Furthermore, when the sensor nodes must be extremely small, as tiny as several cubic centimeters, so that they can be conveniently placed and used, they are very limited in the amount of energy that the batteries can store. Hence severe limits would be imposed on the nodes' lifetime by the miniaturized battery that is meant to last the entire life of the node.

## High Power Consumption of Sensor Nodes

Based on the breakdown of a wireless sensor node in Figure 4.2, the amount of electrical power a sensor node consumes during operation can be determined. The power consumed by each individual component of a sensor node, i.e., the microcontroller, radio, logger memory and sensor board, is tabulated in Table 4.1. It can been observed from Table 4.1 that all the components in the sensor node consume milliwatts of power during the active mode of operation and then drop to microwatts of power when in sleep or idle mode. If the sensor node is set to operate at full duty cycle, i.e., 100%, the current and therefore the power consumption of the sensor node would be as high as 30 mA.

For most practical healthcare WSN applications, duty cycling of the sensor node's operation is a common method discussed in the literature (Callaway, 2003; Culler, Estrin, and Srivastava 2004; Tubaishat and Madria 2003) to reduce its power consumption and therefore extend the lifetime of the WSN. During one operational cycle, the sensor node remains active for a brief period of time before going into sleep mode. During the sleep period, the current consumption of the sensor node is typically in the microampere range as opposed to the milliampere range during the active period. This results in the sensor node drawing a very small amount of current, in the range of 2–8 μA, for the majority of the time, with short current spikes in the range of 5–17 mA while sensing, processing, receiving and transmitting data, as illustrated in Table 4.1. For the Xbow sensor node in Table 4.1, operating at a duty cycle of 1% as opposed to 100%, the average current consumption of a node with a supply voltage of 3 V is significantly reduced, from 30 mA to around 0.3 mA. Thus, with the help of duty cycling the operation of the sensor node, the power consumed by the wireless sensor node has been decreased tremendously. The power density of its battery is still not high enough to support the high power consumption of the sensor node for a long period of time. Based on an AA alkaline battery of 3000 mA·h, the lifetime of the battery powering the sensor node illustrated in Table 4.1 is computed to be at most 1.1 years. Unless the battery is replaced, the sensor node can be considered as an expired node. The situation is even worse in the case of a coin-type alkaline battery of 250 mA·h, which is smaller in size than the AA battery. The coin battery can sustain the operation of the sensor node for only 1–2 months

**TABLE 4.1**

Battery Life Estimation for an Xbow Sensor Node Operating at 100% and 1% Duty Cycles

| | Current | Duty Cycles | |
|---|---|---|---|
| **System Specifications** | **Value** | **Model 1 (%)** | **Model 2 (%)** |
| *Microprocessor (Atmega 128L)* | | | |
| Full operation | 8 mA | 100 | 1 |
| Sleep | 8 μa | 0 | 99 |
| *Radio* | | | |
| Receiving | 16 mA | 75 | 0.75 |
| Transmitting (3dB) | 17 mA | 25 | 0.25 |
| Sleep | 1 μa | 0 | 99 |
| *Logger* | | | |
| Write | 15 mA | 0 | 0 |
| Read | 4 mA | 0 | 0 |
| Sleep | 2 μa | 100 | 100 |
| *Sensor board* | | | |
| Full operation | 5 mA | 100 | 1 |
| Sleep | 5 μa | 0 | 99 |
| **Computed average current consumed** | | **Model 1 (mA)** | **Model 2 (mA)** |
| μP | | 8.0000 | 0.0879 |
| Radio | | 16.2500 | 0.1635 |
| Flash memory | | 0.0020 | 0.0020 |
| Sensor board | | 5.0000 | 0.0550 |
| **Total current (mA) used** | | **29.2520** | **0.3084** |

*Source:* Adapted from Crossbow Technology, Inc., *MPR-MIB Users Manual*, Crossbow Resources, San Jose, California, Revision A, 2007.

at most. Clearly, an operational lifetime of a year or so or even lower for the wireless sensor node is far from sufficient because the duration of the node's operation needs to last for several years for the WSN to be useful in practical situations. This is a serious limitation to computing paradigms like ubiquitous computing or sensor networks, in which there are dozens or hundreds of small, autonomous sensor node systems with batteries to maintain.

## Limits on Energy Sources for Sensor Nodes

In many invasive and noninvasive healthcare application scenarios, the lifetime of the sensor node typically ranges from 2 to 10 years depending on the requirements of the specific application. The energy supply is one of the major bottlenecks to be addressed. Many different types of energy storage technologies are already available. Alkaline/rechargeable batteries and supercapacitors are the most portable and popular energy supply option for powering the sensor nodes in WSNs. Batteries convert stored chemical energy directly into electrical energy. They are generally classified into two groups: (1) single-use/primary and (2) rechargeable/secondary batteries. The distinction between the two groups is based on the nature of the chemical reactions. Primary batteries are discarded when sufficient electrical energy can no longer be obtained from them. Secondary batteries, on the other hand, convert chemical energy into electrical energy by chemical reactions that are essentially reversible. Thus, by passing the electrical current in the reverse direction to that during discharge, the chemicals are restored to their original state and the batteries are restored to full charge again.

The *supercapacitor*, in short *supercap*, is another electrochemical energy system other than batteries that has been increasing its presence in powering wireless sensor nodes. There are several reasons for this phenomenon. One reason is that the supercap is very scalable, and its performance scales well with its size and weight. Another reason is that the supercap has many desirable characteristics that favour the operations of the sensor nodes such as high power density, rapid charging times, high cycling stability, temperature stability, low equivalent series resistance (ESR) and very low leakage of current (Flipsen 2004). Supercapacitors have much higher peak power density than other energy storage devices like batteries and fuel cells. This means that supercaps can deliver more electrical power than batteries and fuel cells within a short time. The peak power densities of supercaps are well above 1000 W/kg, whereas the power densities of all types of batteries are in the range of 60 to 200 W/kg, and fuel cells are even lower, below 100 W/kg. Hence, for burst power operation, supercaps are a better choice than batteries and fuel cells. Conversely, batteries have much higher energy storage capacities than supercaps. This means that batteries can deliver electrical power for a longer period of time as compared to supercaps. The peak energy densities of all types of batteries are in the range of 20 to 200 Wh/kg, whereas the power density of supercap is below 10 Wh/kg. Hence, for sustaining the extended operational lifetime of wireless sensor nodes, relying solely on supercaps might not be suitable due to their very low energy density as compared to other energy storage devices. Research to increase the energy storage density of both batteries and supercaps has been conducted for many years and continues to receive substantial focus (Blomgren 2002). While these technologies promise to extend the lifetime of wireless sensor nodes, they cannot extend it indefinitely.

Among these nonrenewable energy systems or sources, the rechargeable/alkaline battery is one of the most popular methods for powering the great majority of autonomous sensor nodes in WSNs. The electrical energy necessary for their operation is provided primarily by batteries. Although batteries have been widely used in powering sensor nodes in WSNs presently, the problem is that the energy density of batteries is limited and the batteries may not be able to sustain the operation of the sensor nodes for a long period of time. Referring back to the case scenario of an Xbow sensor node in Table 4.1, with an operating duty cycle of 1%, the average power consumption of the node with a supply voltage of 3 V is around 1 mW. Among current battery technologies, the highest reported energy runs around 3.78 kJ/cm$^3$ (Kansal, Hsu, Zahedi, and Srivastava 2006), so that the ultra-low-power miniaturized wireless sensor node with a volumetric size of around 10 cm$^3$, operating at an average power consumption of 1 mW, would need a battery as large as 100 cm$^3$ to have a 10-year lifespan. The size of the battery is 10 times the sensor node's size. In fact, this calculation is a very optimistic estimate, as the entire capacity of the battery usually cannot be completely used up, depending on the voltage drop. Additionally, it is also worth mentioning that the sensors and electronic circuits of a wireless sensor node could be far smaller than 10 cm$^3$. In this case, the battery makes up a significant fraction of the total size and weight of the overall system and is also the most expensive part of the system. Thus, the energy supply is largely constrained by the size of the battery.

In short, batteries with a finite energy supply must be optimally used for both processing and communication tasks. The communication task tends to dominate over the processing task in terms of energy consumption. Thus, in order to make optimal use of energy, the amount of communication tasks should be minimized as much as possible.

In practical real-life applications, wireless sensor nodes are usually deployed in hostile or unreachable areas, especially for invasive healthcare applications. They cannot be easily retrieved for the purpose of replacing or recharging the batteries, and, therefore, the lifetime of the network is usually limited. There must be some kind of compromise between the communication and processing tasks in order to balance the duration of the WSN lifetime and the energy density of the storage element. In summary, limitations on the device size and energy supply typically mean a restricted amount of resources, i.e., CPU performance, memory, wireless communication bandwidth used for data forwarding and range allowed. The need to develop alternative methods for powering the wireless sensor and actuator nodes is acute. The main research focus of this chapter is to resolve the energy supply problems faced by the wireless sensor nodes in a WSN.

## Energy-Harvesting Solutions for Wireless Sensor Nodes

To overcome the major hindrance to the "deploy-and-forget" nature of WSNs due to the limitation on available energy for the network, constrained by the high power consumption of the sensor nodes and the energy capacity and unpredictable lifetime performance of batteries, energy-harvesting technology has emerged as a promising solution to sustain the operation of WSNs (Niyato, Hossain, Rashid, and Bhargava 2007; Raghunathan, Ganeriwal, and Srivastava 2006).

## Overview of Energy Harvesting

*Energy harvesting* (EH) is a technique that captures, harvests or scavenges a variety of unused ambient energy sources such as solar, thermal, vibration and wind and converts the harvested energy into electrical energy to recharge the batteries. The harvested energy is generally very small (on the order of millijoules) as compared to the large-scale EH using renewable energy sources such as solar farms and wind farms (on the order of several hundred megajoules). Unlike the large-scale power stations which are fixed at a given location, the small-scale energy sources are portable and readily available for usage. Various EH sources that can be converted into electrical energy, excluding the biological type, are shown in Figure 4.3.

Our environment is full of waste and unused ambient energy generated from the energy sources seen in Figure 4.3. These renewable energy sources are ample and readily available in the environment, and so it is not necessary to deliberately expend efforts to create these energy sources like, for example, the burning the nonrenewable fossil fuels to create steam, which in turn would cause the steam turbine to rotate to create electrical energy. Unlike fossil fuels, which are exhaustible, the majority of environmental energy

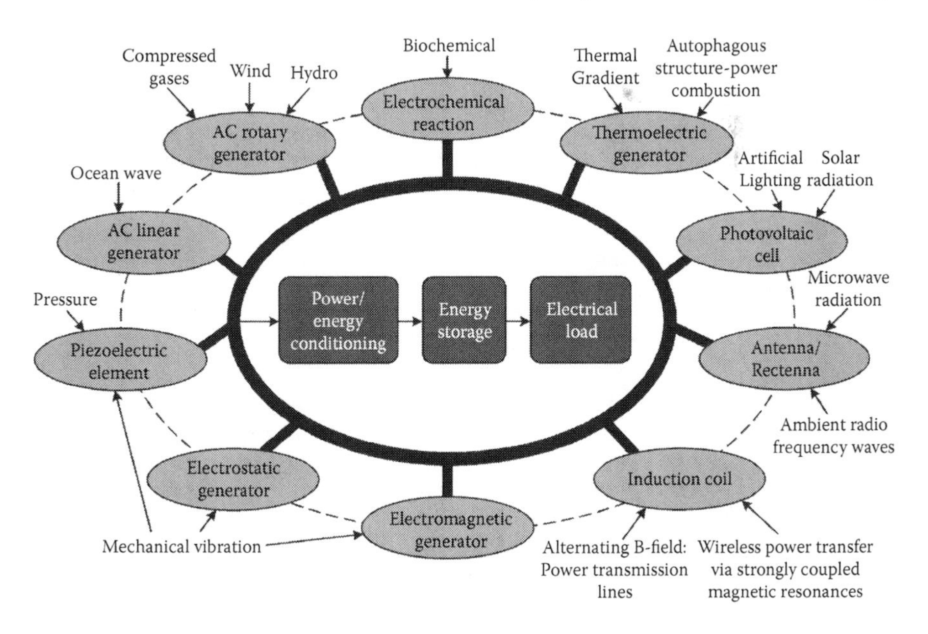


**FIGURE 4.3**
Energy-harvesting sources and their energy harvesters. (Adapted from Thomas, J.P., Qidwai, M.A., and Kellogg, J.C., *Journal of Power Sources*, 159, 1494–1509, 2006.)

**TABLE 4.2**

Energy-Harvesting Opportunities and Demonstrated Capabilities

| Energy Source | Performance | Notes |
|---|---|---|
| Ambient light | 100 mW/cm² (direct sunlight)<br>100 μW/cm² (illuminated office) | Common polycrystalline solar cells are 16–17% efficient, while standard monocrystalline cells approach 20% |
| Thermal | a) 60 μW/cm² at 5 K gradient<br>b) 135 μW/cm² at 10 K gradient | Typical efficiency of thermoelectric generators is ≤ 1% for ΔT < 313 K<br>a) Seiko Thermic wristwatch at 5 K body heat<br>b) Quoted for a ThermoLife® generator at ΔT = 10 K |
| Blood pressure | 0.93 W at 100 mmHg | When coupled with piezoelectric generators, the power that can be generated is on the order of microwatts when loaded continuously and milliwatts when loaded intermittently |
| Vibration | 4 μW/cm³ (human motion-Hz)<br>800 μW/cm³ (machines-kHz) | Predictions for 1 cm³ generators. Highly dependent on excitation (power tends to be proportional to ω, the driving frequency, and $y_o$, the input displacement |
| Hand linear generator | 2 mW/cm³ | Shake-driven flashlight of 3 Hz |
| Push button | 50 μJ/N | Quoted at 3 V DC for the MIT Media Lab Device |
| Heel strike | 118 J/cm³ | Per walking step on piezoelectric insole |
| Ambient wind | 1 mW/cm² | Typical average wind speed of 3 m/s |
| Ambient radio frequency | <1 μW/cm² | Unless near a radio frequency transmitter |
| Wireless energy transfer | 14 mW/cm² | Separation distance of 2 m |

*Source:* Adapted from Paradiso, J. and Starner, T., *IEEE Pervasive Computing*, 4(1), 18–27, 2005.

sources are renewable and sustainable for an almost infinitely long period. Numerous studies and experiments have been conducted to investigate the levels of energy that could be harvested from the ambient environment. Various EH sources and their power/energy densities are listed in Table 4.2.

Table 4.2 shows the performance of each EH source in terms of the power density factor. It can be clearly observed that there is no unique solution suitable for all environments and applications. In Table 4.2, it can be observed that the solar energy source yields the highest power density. However, this may not be always the case. Under illuminated indoor conditions, the ambient light energy harvested by the solar panel drops tremendously. The other

EH sources could provide higher power density, depending on the renewable energy sources available at the specific application areas, such as a bright sunny day outdoors with a rich amount of solar energy, a coastal area with a lot of wind energy, a bridge structure with vehicles travelling that has strong vibrations, etc. In addition, there could also be a possibility of having two or more energy sources available for harvesting at the same time. As such, EH technology can provide numerous benefits to the end user; some of the major benefits of EH that make it suitable for WSNs are stated and elaborated in the following list (Mathna, O'Donnell, Martinez-Catala, Rohan, and Brendan 2008; Thomas, Qidwai, and Kellogg 2006). EH solutions can

1. Reduce the dependency on battery power. With the advancement of microelectronics technology, the power consumption of the sensor nodes is getting lower and lower, and hence harvested ambient/environmental energy may be sufficient to eliminate batteries completely.
2. Reduce installation cost. Self-powered wireless sensor nodes do not require power cables, wiring and conduits, and hence they are very easy to install.
3. Reduce maintenance cost. EH allows for the sensor nodes to function unattended once deployed and eliminates service visits to replace batteries.
4. Provide sensing and actuation capabilities in hard-to-access hazardous environments on a continuous basis.
5. Provide long-term solutions. A reliable self-powered sensor node will remain functional virtually as long as the ambient energy is available. Self-powered sensor nodes are perfectly suited for long-term applications looking at decades of monitoring.
6. Reduce environmental impact. EH can eliminate the need for millions of batteries and the energy costs of battery replacements.

Clearly, it can be deduced from the list of benefits that the EH technology is a viable solution to power WSNs and mobile devices for extended operation with the supplement of the energy storage devices, if not completely eliminating storage devices such as batteries.

## Review of Past Works on EH Systems

There is quite a significant amount of research work being recorded in the literature on harvesting or scavenging small-scale environmental energy for powering wireless sensor nodes. One important point to note is that in order to make the sensor node truly autonomous and self-sustainable in a WSN, the selection of the EH technique is crucial. As such, a review of past work on EH systems is necessary.

### *Solar EH Systems*

The solar energy of an outdoor incident light at midday holds a power density of roughly 100 mW/cm$^2$, which indicates that in a small volume of 1 cm$^2$, 100 mW of electrical power can be harvested from the sun by using a solar panel. Conversely, the lighting power density in indoor environments such as illuminated offices drops tremendously, to almost 100 μW/cm$^2$ (Randall and Jacot 2003). Commercially available off-the-shelf single-crystal solar cells offer efficiencies of about 15%, and efficiencies of up to 20–40% are recorded by Green et al. for state-of-the-art expensive research photovoltaic (PV) cells in a PV progress report (Green, Emery, Hisikawa, and Warta 2007). Thin film polycrystalline and amorphous silicon solar cells are also commercially available and cost less than single-crystal silicon but also have a lower efficiency of only 10–13% (Randall and Jacot 2003).

Recently, a number of solar EH (SEH) prototypes have been presented which perform more and more efficient energy conversion. Two of the first prototypes were Heliomote (Raghunathan, Kansal, Hsu, Friedman, and Srivastava 2005) and Prometheus (Jiang, Polastre, and Culler 2005). In both systems, the solar panels are directly connected with the storage device. The EH system Prometheus is depicted in Figure 4.4a. In this case, the solar panel is directly connected to a supercap. This means that especially for low supercap voltages, the solar panel generates much less power than its maximum power, $P_{MPP}$.

Hence, an efficient SEH system should be able to adapt the electrical operating point of the solar panel to the given light conditions so that $P_{MPP}$ is always maintained. For solar panels of a few square centimetres, particular care has to be taken in order not to waste the few milliwatts generated by the solar panel. Everlast (Simjee and Chou 2008) is an example that uses the fractional short-circuit current technique (see Figure 4.4b) to achieve maximum power point tracking, $P_{MPPT}$. This technique is easy, simple and cheap to implement. The voltage, $V_{MPP}$, is estimated based on the open-circuit voltage, $V_{oc}$, of the solar panel, which is measured periodically by momentarily shutting down the power converter that is connected to the solar panel. However, the drawback with this technique is the transient drop in power during the time when no energy is harvested.

Another SEH system, known as AmbiMax, has been proposed by Park and Chou (2006). The AmbiMax system exploits a small photosensor to detect the ambient light conditions and to force the solar panel to work in its MPP (see Figure 4.4c). In a similar approach to AmbiMax, Dondi et al. presented another circuit that uses a miniaturized photovoltaic module as a pilot panel instead of a photosensor to achieve MPPT for the SEH system (Dondi, Bertacchini, Brunelli, Larcher, and Benini 2008). Indeed, these SEH prototypes have successfully demonstrated that solar energy is a realistic energy source for sensor nodes. However, there is still room for improvement including the system form factor, performance, etc. to suit the power

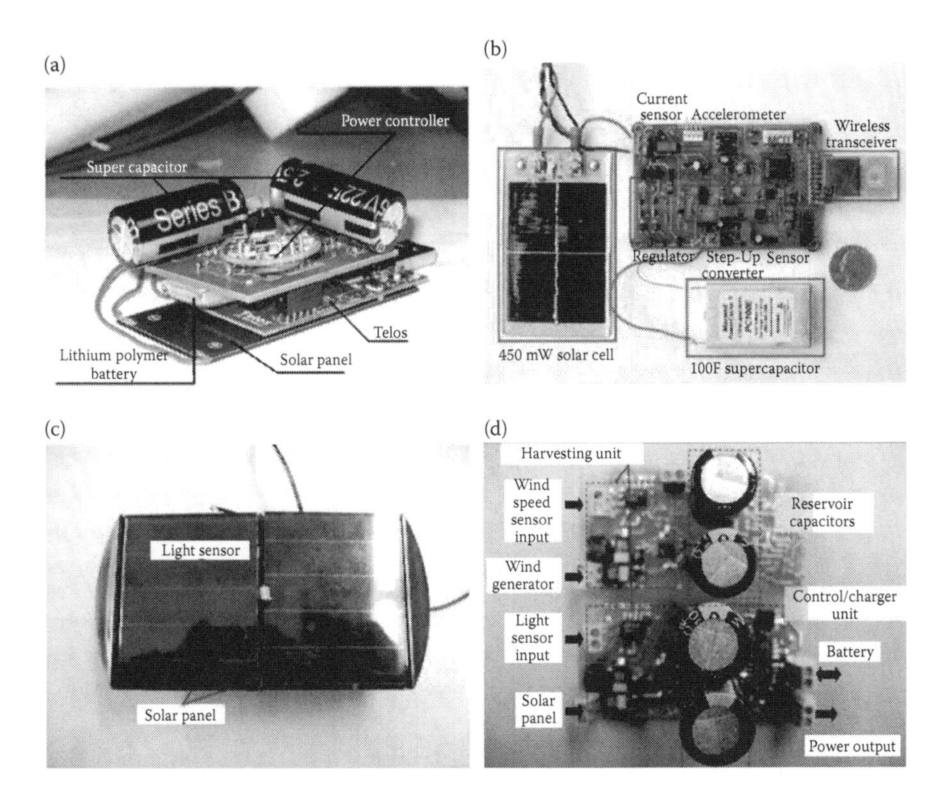


**FIGURE 4.4**
Examples of solar energy–harvesting systems: (a) Prometheus sensor node; (b) Everlast prototype system; (c) AmbiMax solar panel with light sensor; (d) AmbiMax board with supercaps. (Adapted from Jiang, X., Polastre, J., and Culler, D., *Fourth International Symposium on Information Processing in Sensor Networks (IPSN)*, Los Angeles, California, 463–468, 2005; Simjee, F. and Chou, P., *IEEE Transactions on Power Electronics*, 23(3), 1526–1536, 2008; and Park, C. and Chou, P., *3rd Annual IEEE Communications Society on Sensor and Ad Hoc Communications and Networks (SECON)*, Reston, Virginia, 1, 168–177, 2006.)

requirements of embedded wireless sensor nodes deployed in application areas such as indoor and overcast areas where access to direct sunlight is often weak or not available. In some cases where the solar energy source may not be a suitable choice, it is required to search for alternative energy sources either to replace the solar energy source entirely or to supplement the solar energy source when the intensity of the light is low.

### *Thermal EH Systems*

Thermal energy is another example of an alternative energy source. Several approaches to convert thermal energy into electricity are currently under investigation (through the Seebeck effect, thermocouples, or a piezothermal effect; Hudak and Amatucci 2008). The efficiency of these approaches is related to Carnot's law, expressed by the equation $\eta = (T_{max} - T_{min})/T_{max}$. According to Carnot's equation, for a thermal gradient of 5 K with respect to the normal ambient temperature of 300 K, the thermal EH (TEH) efficiency is computed to be around 1.67%. Consider a silicon device with a thermal conductivity of 140 W/mK, as illustrated by Cottone (2008). The heat power that flows through conduction along a 1 cm length for $\Delta T = 5$ K is 7 W/cm$^2$. Hence, the electrical power obtained at Carnot's efficiency is calculated to be 117 mW/cm$^2$. At first sight, this heat power density of 7 W/cm$^2$ seems to exhibit an excellent result, but the TEH devices have efficiencies well below the simple Carnot's rule, so the attainable electrical power density turns out to be a small fraction of that, which is only 117 mW/cm$^2$. Many studies on TEH devices have been discussed in the literature, and the thermoelectric generator (TEG) is one of the popular devices that have been developed to harvest thermal energy based on Seebeck effect. A summary of the implemented TEGs, capable of generating from 1 to 60 μW/cm$^2$ at a 5 K temperature differential, is given in a review paper by Hudak and Amatucci (2008).

A TEH system requires one or more TEGs, heat exchangers on the hot and cold sides of the TEG, a mechanical structure for clamping the heat exchangers to the module and ensuring good thermal contact, thermal insulation to prevent heat losses through the sides and power electronics for impedance load matching (Thomas, Qidwai, and Kellogg 2006). One example of commercial application of a TEG is the Seiko Thermic wristwatch, shown in Figure 4.5a, which is powered by heat generated from the human body.

In the Seiko wristwatch, the TEH system consists of a thermoelectric module, a lithium-ion battery and a simple DC-DC step-up voltage regulator (Hudak and Amatucci 2008). The thermoelectric module of the wristwatch is reported as yielding 60 μW/cm$^2$ at a 5 K temperature gradient with 10 TEGs coupled together in series (Kanesaka 1999). Similarly, Leonov, Torfs, Fiorini and Hoof (2007) have considered TEH through thermoelectric power generation from body heat to power wireless sensor nodes, as shown in Figure 4.5b. The average daytime power generation of about 250 μW corresponds to about 20 μW/cm$^2$ with a temperature difference of 10 K, which is better than solar panels in many indoor situations, especially considering that the TEG power is also available at night. However, these systems do not consider proper matching between the source and the load to ensure maximum power points (MPPs) operation.

In other TEH research, both Stevens (1999) and Lawrence and Snyder (2002) consider the system design aspects for solar TEH via thermoelectric conversion that exploits the natural temperature difference between the ground and air. Later, Sodano, Simmers, Dereux and Inman (2007) presented a solar TEH system placed in a greenhouse with a solar concentrator, as seen in Figure 4.5c. The solar TEH system uses a TEG to recharge a NiMH battery. At an estimated $\Delta T$ of 25 K, the harvested energy was able to recharge an

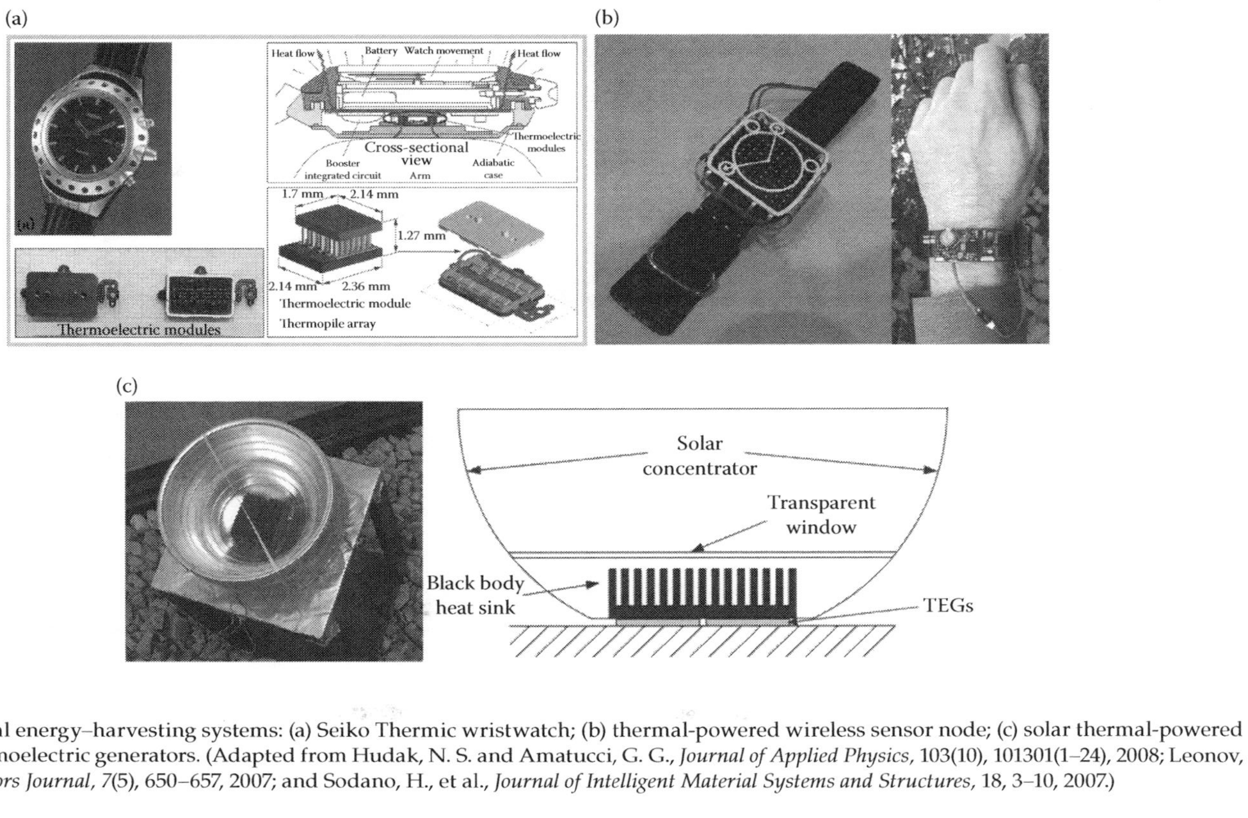


**FIGURE 4.5**
Examples of thermal energy–harvesting systems: (a) Seiko Thermic wristwatch; (b) thermal-powered wireless sensor node; (c) solar thermal-powered system. TEGs, thermoelectric generators. (Adapted from Hudak, N. S. and Amatucci, G. G., *Journal of Applied Physics*, 103(10), 101301(1–24), 2008; Leonov, V., et al., *IEEE Sensors Journal*, 7(5), 650–657, 2007; and Sodano, H., et al., *Journal of Intelligent Material Systems and Structures*, 18, 3–10, 2007.)

80 mAh battery in 3.3 min. The authors have demonstrated that a TEG may be used for solar energy conversion as an alternative to photovoltaic devices. However, as before, there are fewer discussions on the power management aspects of the solar TEH system.

### *Vibration EH Systems*

The first important virtue of random mechanical vibrations as a potential energy source is that they are present almost everywhere. Mechanical vibrations occur in many environments such as buildings, transports, terrains, human activities, industrial environments, military devices, and so on. Their characteristics are various: spectral shape from low to high frequency, amplitude and time duration are manifolds dependently by the surroundings. Theoretical considerations and experiments reported in many research studies show that the power density that can be converted from vibrations is about 300 $\mu W/cm^3$ (Cottone 2008). Devices which convert mechanical motion into electricity can be categorized into electromagnetic, electrostatic and piezoelectric converters (Moser 2009; Roundy, Wright, and Rabaey 2004). In the case of electromagnetic converters, a coil oscillates in a static magnetic field and induces a voltage. In electrostatic converters, an electric charge on variable capacitor plates creates a voltage if the plates are moved. Piezoelectric converters, finally, exploit the ability of some materials like crystals or ceramics to generate an electric potential in response to mechanical stress. A prominent example for the employment of vibrational harvesters is the watch industry, where vibrational energy converters have been used with success to power wristwatches.

Shenck and Paradiso (2001) presented a piezoelectric-powered RFID system for shoes, illustrated in Figure 4.6a, that harvests energy from human walking activity. The developed shoe inserts are capable of generating around 10 mW of power under normal walking conditions. This shows that mechanical vibration from human activity is another promising renewable energy source worth investing effort to investigate. A similar approach has been taken by Roundy (2003), where piezoelectric generators, as seen in Figure 4.6b, have been developed as an attractive method to power wireless transceivers. Other vibration-based EH (VEH) research projects being reported include wearable electronic textiles (Edmison, Jones, Nakad, and Martin 2002) and electromagnetic vibration-based microgenerator devices (see Figure 4.6c) for intelligent sensor systems (Glynne-Jones, Tudor, Beeby, and White 2004). Meninger, Amirtharajan and Chandrakasan (2001) have also demonstrated an electromagnetic vibration-to-electricity converter in their research that can produce 2.5 $\mu W/cm^3$ of electrical power. Similarly, another piece of research discussed by Mitcheson, Green, Yeatman and Holmes (2004) has carried out an analysis that indicated that up to 4 $\mu W/cm^3$ can be achieved from vibrational microgenerators (on the order of 1 $cm^3$ in volume) stimulated by typical human motion (5 mm motion at 1 Hz),

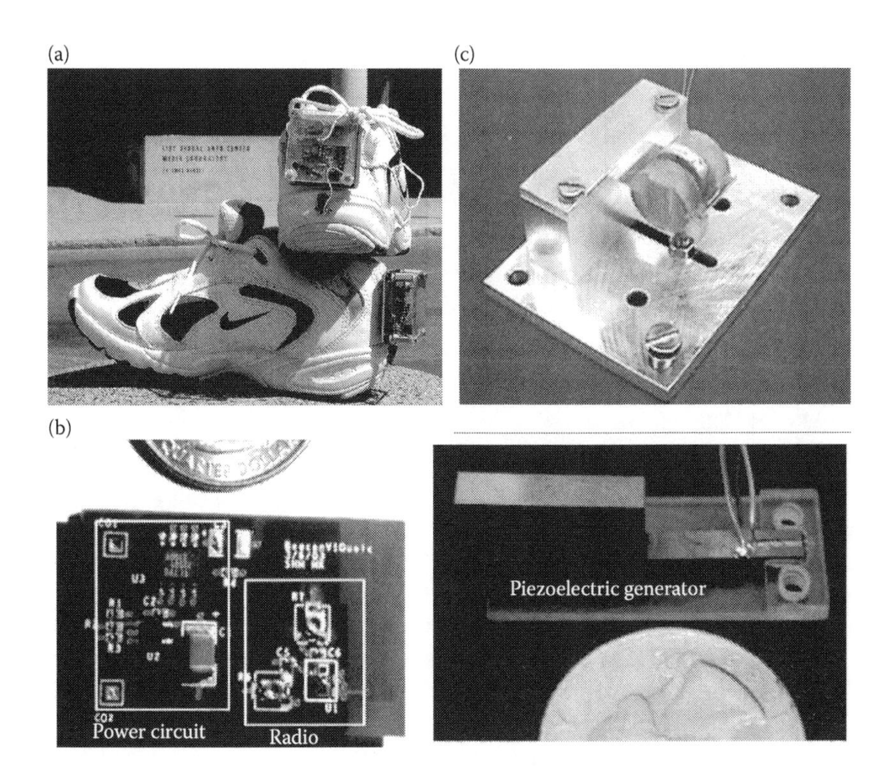


**FIGURE 4.6**
Examples of vibrational energy–harvesting systems: (a) piezoelectric-powered RFID (radiofrequency identification) shoes with mounted electronics; (b) vibration-powered wireless sensor node; (c) electromagnetic vibration-based microgenerator devices. (Adapted from Shenck, N. and Paradiso, J., *IEEE Micro*, 21(3), 30–42, 2001; Roundy, S. J., Energy Scavenging for Wireless Sensor Nodes with a Focus on Vibration to Electricity Conversion, PhD thesis, University of California, Berkeley, 2003; and Glynne-Jones, P., et al., *Sensors and Actuators*, 110(1–3), 344–349, 2004.)

while up to 800 μW/cm$^3$ can be obtained from machine-induced stimuli (2 nm motion at 2.5 kHz). Many meso- and micro-scaled EH generators have been developed in the last 5 years (Cottone 2008); however, there is a lack of adequate ultra-low-power management circuits to condition their micropower generation.

In another VEH research study, Paradiso and Feldmeier (2002) have successfully demonstrated a piezoelectric element with a resonantly matched transformer and conditioning electronics that, when struck by a button, generate 1 mJ at 3 V per 15 N push, enough power to run a digital encoder and a radio that can transmit over 15 m. However, this system requires a large transformer to step up the output voltage generated by the piezoelectric element. The efficiency of the transformer is limited by flux leakage and core saturation when the primary current peaks. Taking an interesting turn, assuming an average blood pressure of 100 mmHg (the normal desired blood pressure is 120/80 above atmospheric pressure), a resting heart rate of 60 beats per minute and a heart stroke volume of 70 mL passing through the aorta per beat (Braunwald 1980), then the power generated is about 0.93 W. Ramsay and Clark (2001) found that when the blood pressure is exposed to a piezoelectric generator, the generator can generate power on the order of microwatts when the load applied changes continuously and milliwatts when the load applied changes intermittently. However, harnessing power from blood pressure would limit the application domains to wearable microsensors.

#### *Wind EH Systems*

Like any of the commonly available renewable energy sources, wind EH (WEH) has been widely researched for high-power applications where large wind turbine generators (WTGs) are used to supply power to remote loads and grid-connected applications (Chen, Guerrero, and Blaabjerg 2009; Siegfried and Rachel 2006). According to a study by the National Renewable Energy Laboratory (NREL) ((RReDC)), wind energy is the fastest-growing electricity-generating technology in the world. In the past 10 years, global installations of wind energy systems have grown at least 10-fold—from a total capacity of 2.8 GW in 1993 to almost 40 GW at the close of 2003 (Chen, Islam, and Priya 2006). In spite of this continuing success of WEH at a large scale, there have been very few attempts to develop small-scale WEH—those that are miniaturized in size and highly portable—to power small, autonomous sensors deployed in remote locations for sensing and/or even to endure long-term exposure to hostile environments such as forest fires. Although very few research works on small-scale WEH are reported in the literature, some efforts to generate power at a very small scale have been made recently. Park and Chou (2006) implemented a MPP tracker for a small windmill. The WEH system, as illustrated in Figure 4.7a, exploits the near-linear relationship between the wind speed and the rotating frequency of the WTG's rotor to force the WTG to work in its MPPs.

Similarly, a small-scale WTG has been presented by Holmes, Hong, Pullen and Buffard (2004), and the power densities harvested from air velocity are quite promising. Later, Weimer, Paing and Zane (2006) presented an anemometer-based solution to perform the WEH and sensing tasks, which are accomplished separately by two different devices. The authors utilize the motion of the anemometer shaft to turn a coupled alternator to generate electrical power for the sensor nodes. Although the sensor nodes incorporating the harvesting solution have an increased operational lifetime, this comes at the price of larger device size, higher overall cost and higher energy-conversion loss.

In another WEH research study, Priya et al. designed a windmill which uses piezoelectric elements to generate electricity from wind energy (see

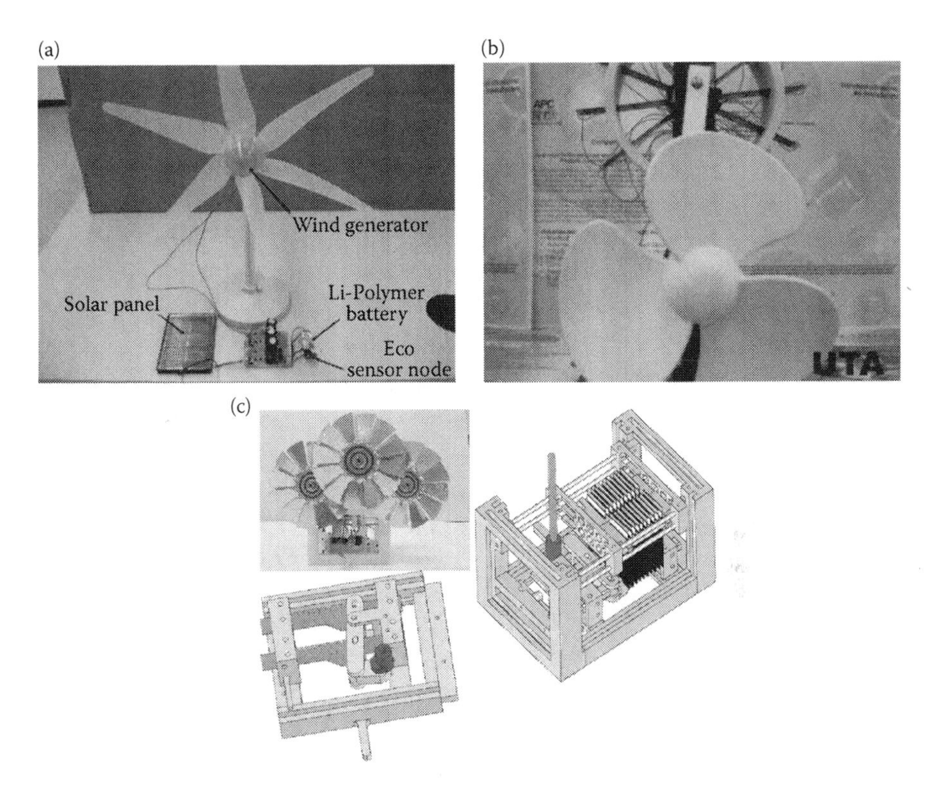


**FIGURE 4.7**
Examples of wind energy–harvesting systems: (a) AmbiMax hardware with a solar panel, wind generator, lithium polymer battery and Eco Node; (b) piezoelectric windmill prototype; (c) optimized design of the small-scale windmill with isometric view and internal crankshaft structure exhibiting translation mechanism. (Adapted from Park, C. and Chou, P., *3rd Annual IEEE Communications Society on Sensor and Ad Hoc Communications and Networks (SECON)*, 1, 168–177, 2006; and Priya, S., et al., *Japanese Journal of Applied Physics*, 44(3), L104–L107, 2005.)

Figure 4.7b). An output power of 10.3 mW was harvested and reported for a wind flow which led to 6 rotations/min. Subsequently, another group of researchers, Myers et al. (2007), developed an optimized small-scale piezoelectric windmill, shown in Figure 4.7c. The whole structure of the windmill is made of plastic, and it utilizes 18 piezoelectric bimorphs to convert wind energy, hence vibration energy, into electrical energy. The windmill was tested to provide 5 mW of continuous power at an average wind speed of 4.5 m/s. Still, the physical size of the WEH systems is too large and bulky as compared to the sensor node, and their harvested power exceeds the sensor node's power requirements. In addition, there is a lack of adequate power management circuits to maximize electrical power transfer from the source to the load.

## Prospect of EH Technologies for Healthcare WSNs

Wireless body area networks (WBANs) can be integrated into various human-related applications like (1) medical healthcare services, (2) assistance to people with disabilities and (3) body interaction and entertainment (Li, Takizawa, and Kohno 2008) to promote healthy lifestyles. There are many benefits with integrated WBANs, which include seamless integration of data into individual personal medical records and research databases as well as discovery of useful medical knowledge through data mining. In a WBAN, there are several sensor nodes, which are comprised of a radio frequency (RF) transceiver, a processing unit and some sensors, deployed on the human being in a star network form as illustrated in Figure 4.2. Various vital signs and healthcare data are collected for medical diagnostic purposes using different types of physiological sensors onboard the sensor nodes like an electrocardiography (ECG) sensor, electromyography (EMG) sensor, electroencephalography (EEG) sensor, blood pressure sensor, tilt sensor, breathing sensor, movement sensor, thermometer, etc.

Unlike a conventional WSN, which has a wide coverage area of tens or hundreds of metres supported by many sensor nodes, the sensor nodes of a WBAN are deployed either outside the human body and form a wearable WBAN or operate within the human body to form an implanted WBAN. The coverage range of the sensor nodes in a WBAN is designed to be within the area of human body of few metres, with a small number of nodes used to monitor the vital signs of the patients. Therefore, WBANs have faced some research challenges (Li, Takizawa, and Kohno 2008) such as power consumption, biocompatibility, reliable data collection, security, context awareness, small size and weight, etc. These research challenges raise the need for developing new wireless communication technologies rather than the available standardized technologies such as the wireless local area network (WLAN), Bluetooth, wireless personal area network (WPAN), etc. Currently, a new standard of wireless communication technology for body area networks, IEEE 802.15.6 (Heinzelman, Chandrakasan, and Balakrishnan 2000), is under construction which focuses on scalable and reliable medium access control, low-power consumption with an EH power source, high quality of service, a medical authorized RF band and safety for human body.

In a typical medical healthcare system with an integrated WBAN, the status of the patient is monitored regularly and reported back to the doctors in a timely manner through the wireless communication media. In order to monitor the patient consistently, it is important to maintain good connectivity between the sensor systems mounted on the human body and the base station connected to the doctors. However, as already mentioned, power consumption is a crucial issue in a WBAN. Depending on the application and the operation of the sensor nodes in the WBAN, the power consumption varies in accordance with the different components of the sensor nodes. Among the various components, the radio transceiver consumes the most energy during its active mode in transmitting or receiving operations. Hence, an efficient energy communication protocol is required to reduce the power consumption of each sensor node and prolong the network lifetime.

In Cheng and Huang (2008), a low-power medium access control (MAC) protocol has been developed to deal with the wasted energy due to inter- and intra-network collisions. The result shows that the carrier sense multiple access/collision avoidance (CSMA/CA) technique achieves lower power consumption than the time-division multiple access (TDMA) technique. When a hybrid mode of the two techniques has been implemented, more energy saving is accomplished, and thus this can be used in WBAN application. Usually, for a WBAN deployed on a human body, sensor nodes are organized in a single-hop star topology where one node becomes a local gateway to collect data from all other nodes and send this information to a base station, which is a hospital's server or a doctor's computer. The local gateway is typically a mobile device like a personal digital assistant (PDA), which is very costly, and it requires a separate communication device to be compatible with other body sensor nodes as well as a special communication service to send data to the system's server. More important, the mobile device consumes a huge amount of energy to operate, and hence the rechargeable battery needs to be recharged regularly in order to sustain its operation. Therefore, instead of using a fixed local gateway, this chapter proposes having all the sensor nodes in the WBAN take turns being the gateway to communicate the collected information to the main WSN system.

## Case Study: TEH from Human Warmth for WBANs in a Medical Healthcare System (Hoang, Tan, Chng, and Panda 2009)

Taking the radio communication model described by Heinzelman, Chandrakasan and Balakrishnan (2000) as a reference, where the radio dissipates $E_{elec} = 50$ nJ/bit to run the transmitter or receiver circuitry and $E_{amp} = 100$ pJ/bit/m$^2$ is the transmit amplifier, the two scenarios of using one fixed sensor node as the gateway and of rotating all sensor nodes as the gateway are simulated, and their results are compared. The energy required to transmit a *k*-bit message for a distance, *d*, is given as (Heinzelman, Chandrakasan, and Balakrishnan 2000)

$$E_{Tx}(k,d) = E_{Tx\text{-}elec}(k) + E_{Tx\text{-}amp}(k,d)$$

$$E_{Tx}(k,d) = E_{elec} * k + E_{amp} * k * d^2,$$

and the energy spent to receive this message is expressed as

$$E_{Rx}(k) = E_{Rx\text{-}elec}(k)$$
$$E_{Rx}(k) = E_{elec} * k.$$

Considering a WBAN with $n$ nodes, one of them is the gateway. The energy consumption of each node to transmit $k$-messages to this gateway is given as

$$E_{node} = E_{Tx}(k,r) = E_{elec} * k + E_{amp} * k * r^2,$$

where $r$ is the distance between the node and the gateway. The energy consumed by the gateway to receive data from all nodes and send them to a base station is

$$\begin{aligned} E_{node} &= E_{Tx}(n * k, d) + E_{Rx}((n-1) * k) \\ &= E_{elec} * n * k + E_{amp} * n * k * d^2 \\ &\quad + E_{elec} * (n-1) * k \\ &= k * [(2n-1)E_{elec} + n * E_{amp}]. \end{aligned}$$

To solve the problem of relying on just one energy-hungry gateway to communicate with the main WSN system which is connected with the hospital's server or the doctor's computer, a selective gateway method has been proposed that selects one of the body sensor nodes to be the local gateway based on the nodes' residual energy. Whenever the residual energy of the gateway falls below a threshold value, $E_{th}$, another of the nodes in the WBAN with a higher energy level than $E_{th}$ is chosen as the current gateway. After one round of selection in which all the nodes' residual energy drops below the threshold, the original threshold value is adjusted lower. It is required to make sure that at least one of the sensor nodes has enough energy supply to become the gateway.

The selection process is repeated again, as already mentioned, until the residual energy stored in all the sensor nodes is used up. The information about each individual node's residual energy is added to the sending data and is compared at the gateway at each round of data gathering. The decision to change the gateway that is made by the current gateway is sent to all the other nodes through the messages acknowledging received data in the next round.

The simulation result, shown in Figure 4.8, illustrates the residual energy of the gateway based on the conventional fixed-gateway case and the last node running out of energy when using proposed selective gateway method with 10 nodes deployed in an area of $0.4 \times 1.7$ m, a 200-bit message (i.e., header, payload, metadata, etc.) is transferred from the sensor node to the gateway every round. Assuming that the distance between the gateway and the base station

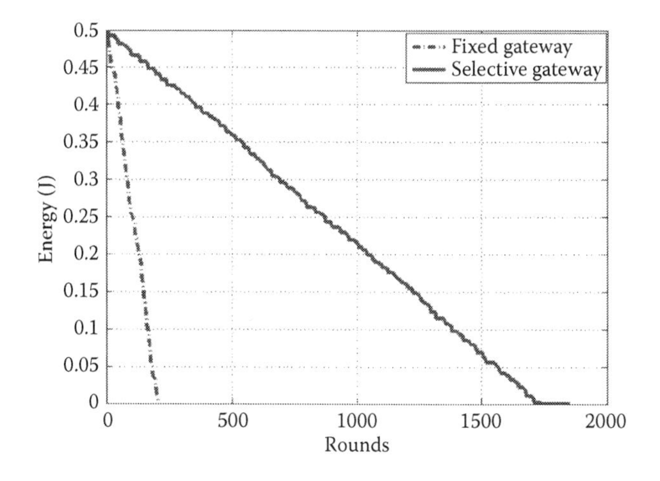


**FIGURE 4.8**
Residual energy levels of the conventional fixed gateway and the selective gateway.

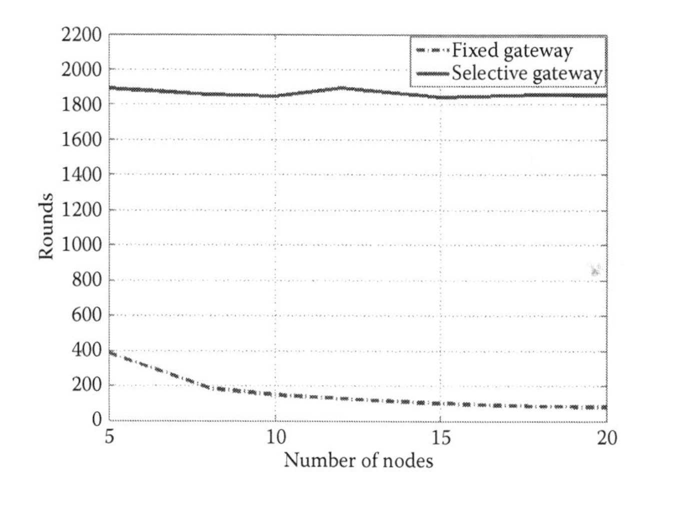


**FIGURE 4.9**
Gateway lifetime with different numbers of nodes in the network.

is less than 200 m and that all nodes have the same initial energy of 0.5 J, the gateway in the first method has spent all of its energy after 200 rounds of receiving and transmitting data. Meanwhile, in the second method, all nodes have run out of energy after an average of 1700 rounds.

The comparison between the lifetime of the single unique gateway and the time until the last node runs out of energy in the selective gateway method with respect to the number of nodes deployed in the WBAN is illustrated in Figure 4.9. When the gateway is fixed, increasing the number of nodes shortens the life of the gateway and thus shortens the network lifetime. In another case, where the gateway is selected based on the residual energy of the sensor nodes, the average lifetime of each node in the network is much longer, and its performance is independent of the number of sensor nodes. Therefore, the network lifetime, which does not depend on a unique local gateway, is much improved.

By changing the gateway based on residual energy, the energy among all nodes is balanced, and the time that the gateway exists in the network is longer; thus, it guarantees the connection with the base station. Furthermore, when an EH source is added, the proposed method provides a benefit of utilizing energy scavenged from all of the nodes as well as letting the gateway have enough time to recharge and restore its energy capacity, therefore prolonging the network lifetime much more. Thus, it is necessary to harvest energy from the ambient environment; in the case of a WBAN, it can be scavenged from the human body.

The EH approach provides the sensor nodes and network with an unlimited power supply. When incorporated with an energy-efficient communication protocol, an EH source helps to enhance the operation of the WBAN, whereas the design of sensor nodes can be optimized in terms of weight, size and power efficiency. Ambient energy sources which can be converted to electrical energy include solar cell, motion and vibration, thermoelectric sources, wind, etc. Among these, thermal energy is one of the potential energy sources: a TEG is used to extract energy from human warmth. The possibility of applying TEH to power a sensor node of a WBAN is investigated.

The application scenario is fall detection for patients in medical healthcare services. Fall detection is very significant to support elderly people's safety in a home-dwelling environment and to guarantee that patients with disabilities in hospitals are assisted by informing doctors and nurses in a timely fashion. The deployment of sensors in WBANs and some methods to identify fall events are described by Kangas, Konttila, Winblad and Jms (2007) and Karantonis, Narayanan, Mathie, Lovell and Celler (2006), where accelerometer sensors mounted on the human body or on clothes are used for fall detection. The sensing signals are transmitted to a base station to process and send to the hospital network and doctors' computers.

## TEH Structure with TEG

According to Stark (2006), the warmth of a human body (and also an animal body) can be used as a steady energy source for powering a sensor node in a WBAN. The amount of energy released by the metabolism (traditionally measured in Met; 1 Met = 58.15 W/m$^2$ of body surface) mainly depends on the amount of muscular activity. A normal adult has an average surface area of 1.7 m$^2$, so that such a person in thermal comfort with an activity level of 1 Met will have a heat loss of about 100 W. The metabolism can range from 0.8 Met (46 W/m$^2$) while sleeping up to about 9.5 Met (550 W/m$^2$) during sports activities, such as running at 15 km/h. A commonly used Met rate is

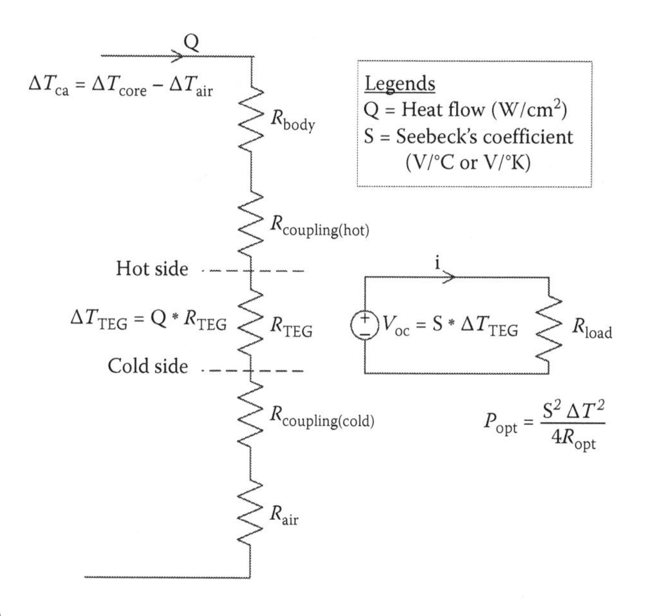


**FIGURE 4.10**
Thermal analysis of the thermoelectric generator (TEG).

1.2 (70 W/m$^2$), corresponding to normal work when sitting in an office, which leads to power dissipation of about 119 W, burning about 10.3 MJ a day. Once the input thermal energy is known, the equivalent electrical circuit of the TEH structure with the TEG, given in Figure 4.10, is analysed.

The left-hand side of Figure 4.10 shows the thermal equivalent circuit representation of a TEG in contact with the human skin. The heat flow, $Q$, takes place between the body, with a core temperature, $T_{core}$, and the ambient air, with a lower temperature, $T_{air}$, through the following thermal resistances: the body, $R_{body}$; the interface between the body and the TEG (hot side), $R_{coupling(hot)}$; the TEG, $R_{TEG}$; the interface between the TEG (cold side), $R_{coupling(cold)}$, and the surrounding air, $R_{air}$. This results in the following equation that describes the relationship among the temperature difference between the body and the air, $\Delta T_{ca}$; the heat flow; and the various thermal resistances of the TEH structure. Knowing this relationship, the important factors that affect the overall system efficiency can be identified for improving its performance

$$Q = \frac{\Delta T_{ca}}{R_{body} + R_{coupling(hot)} + R_{TEG} + R_{coupling(cold)} + R_{air}}.$$

By keeping the ratio

$$\frac{R_{TEG}}{R_{body} + R_{coupling(hot)} + R_{coupling(cold)} + R_{air}},$$

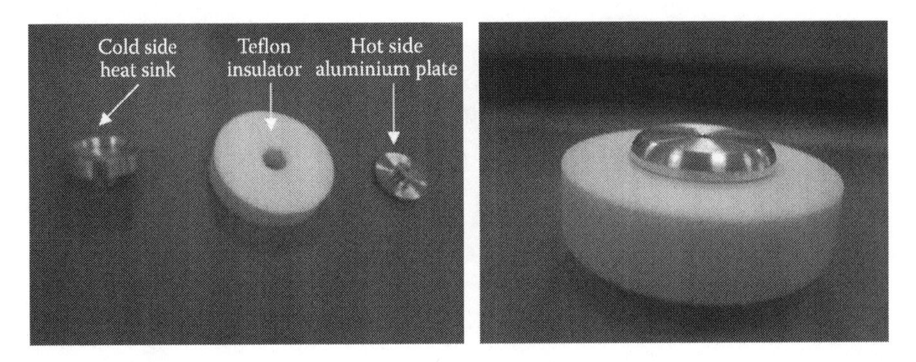


**FIGURE 4.11**
Prototype of the thermoelectric generator (TEG).

$\Delta T_{ca}$ and $Q$ as large as possible, better performance of the overall TEH system is achieved. When there is a temperature difference across the TEG structure, the heat resistances residing in the TEH system generate a certain amount of heat energy loss. These thermal resistances are due to the mechanical structure used to contain the TEG. When heat flows from the body to the TEG and from the TEG to the air, the material used and the design of the mechanical structure greatly affect the performance of the system. These critical factors are taken into consideration during the design and development of the TEH system.

The concept of TEH is not new to most people and is based on one of the thermodynamic concepts, known as *Seebeck's theory*. The theory states that when there is a temperature difference across two dissimilar materials, electric voltage is generated. A TEG, which is based on Seebeck's theory, is used as the energy converter to transform the thermal energy into electrical energy. In this design, the TEG is fabricated using aluminium and Teflon. Aluminium is used to act as the hot plate, designed with a small surface area in order to collect heat fast, and as the cold plate, designed in a shape to act as a good heat diffuser. Teflon is used as the insulator sandwiched between the hot and the cold plate so as to effectively reduce the convection and radiation of heat from the hot plate and the cold plate, preventing the latter from warming up, which would be highly undesirable as it would reduce the thermal gradient between the plates, thus affecting the heat flow and power output. The prototype design of the TEH structure with the TEG is shown in Figure 4.11.

## Power Management Circuit

The power management circuit of the TEH system contains an energy storage and supply circuit, as described by Tan and Panda (2007), and a voltage regulator circuit, illustrated in Figure 4.12. The operation of the power management circuit is as follows: The electrical DC power harvested from the

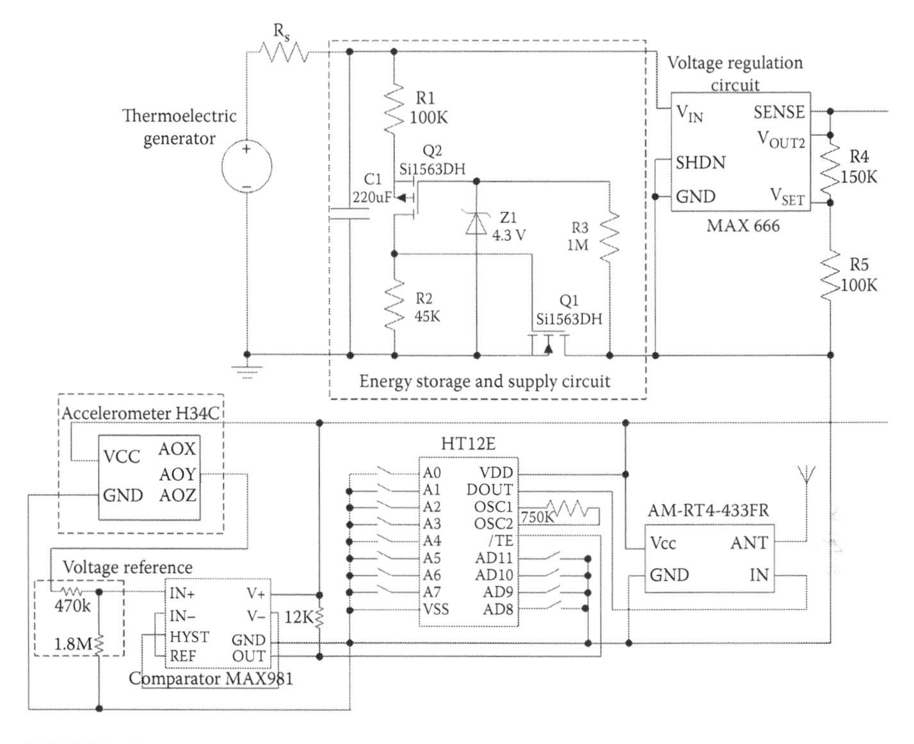


**FIGURE 4.12**
Schematic diagram of thermal energy–harvesting sensor node for fall detection.

TEG is stored within a capacitor to a level sufficient to power the loads. The process of storing and releasing energy is controlled by the supply circuit with 2 MOSFET switches, i.e., Q1 and Q2.

During the time when the capacitor is being charged, Q1 and Q2 are turned off to isolate the TEG source and the RF load. When the built-up voltage across the capacitor reaches the preset voltage of 4.9 V, Q1 is turned on and then in turn activates the control switch Q2. The energy accumulated in the capacitor is then discharged and fed to the voltage regulator. The voltage regulator steps down the input voltage of 4.9 to 3.3 V to supply to the connected load for its sensing and communications operations.

## Fall-Detection Sensor

The TEH system is designed to power a fall-detection system. The body sensor node is designed and implemented to be mounted on a human body to detect any falling event. If a falling event is detected, a signal is sent via the wireless communication system to the base station, where it is postprocessed and forwarded to the doctors or nurses for monitoring of the patient's status or timely responses like activation of an emergency ambulance. In this chapter, the fall-detection event is sensed by using an accelerometer. Based on the application requirements, the accelerometer chosen must be able to sense and differentiate, via its internal nomenclature, between an upright standing posture and a fallen posture and subsequently give different output voltage levels to signify the sensed information accordingly. The accelerometer, H34C, obtained from Hitachi, is small in size, has high sensitivity in three axes (i.e., x-, y- and z-axes), has a very low power consumption of 1 mW with a 3.3 V supply and is capable of sensing both dynamic and static (tilt) acceleration. In this research, the static sensing mode is used to indicate the body posture, standing or fallen position, illustrated in Figure 4.13. The output voltage (Y) is assigned as the indicated signal for detecting a fall. The design principle revolved around the fact that Y is always at its preset voltage level of 1.833 V in a "standing" posture, and always at its $V_{ref}/2$ level in a "fallen" posture.

To make a comparison between a standing and a fallen condition, a low-power comparator is utilized. The output voltage, $V_{out}$, of the comparator is determined by

$$V_{in} < (V^- + 1.182\ V),\ V_{out} = 0\ V$$

$$V_{in} \geq 1.182\ V,\ V_{out} = V^+.$$

Based on the sensing conditions mentioned in the preceding, the signal conditioning circuit is designed to process the accelerometer signal voltage. As already mentioned, radio transmission consumes the greatest amount of energy among the various components of the sensor node, and hence a low-power transmitter–receiver pair, i.e., AM RT4-433 and AM HRR30-433, which consumes around 10 mW at 3.3 V, has been chosen. A matching encoder, HT12E, with very low power consumption is chosen to encode the communication address and patient identification data for transmission. In the scenario when the patient has fallen, the comparator output becomes 0 V, therefore enabling transmission of the communicating address and the patient information via the transmitter to relay the encoded bits to the receiver in a wireless manner. The received signal is then decoded by HT12D to alarm for an emergency call of a patient, recognized by his/her identification number.

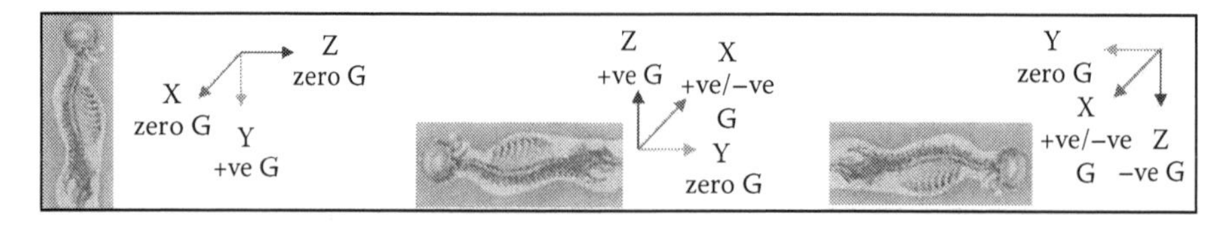


**FIGURE 4.13**
Sensing of body posture (upright and fallen) using H34C.

## Experimental Test Results

Various experiments were conducted to verify the proposed TEH system to power the sensor node in a WBAN. The TEH structure is first characterized to determine the amount of harvested electrical power. When different thermal gradients between 3 and 15°C are applied across the thermal energy harvester, the maximum electrical power generated ranges from 40 to 520 μW, respectively, at the same load resistance of 16 kΩ. The harvested electrical power of 520 μW at 15°C is definitely not sufficient to directly drive the sensor node, which requires around 14 mW power. Hence, an energy storage and supply circuit has been implemented to bridge between the source and the load (see Figure 4.12).

Successful transmissions of the fall-detection signal can be observed in Figure 4.14. The charging time of the storage capacitor is very short, simply intervals of less than 30 s with a thermal gradient of approximately 15°C across the harvester. The actual packets transmitted were five digital packets over approximately 120 ms; the actual energy used equated to 50 ms of active transmission time, using 660 μJ, and 120 ms of operating time for the other loads, using 292 μJ, and, therefore, 952.4 μJ of energy is required to be stored in the capacitor.

In a conventional WBAN, a single local gateway is used to communicate with the base station connected with the doctor's computer or hospital's

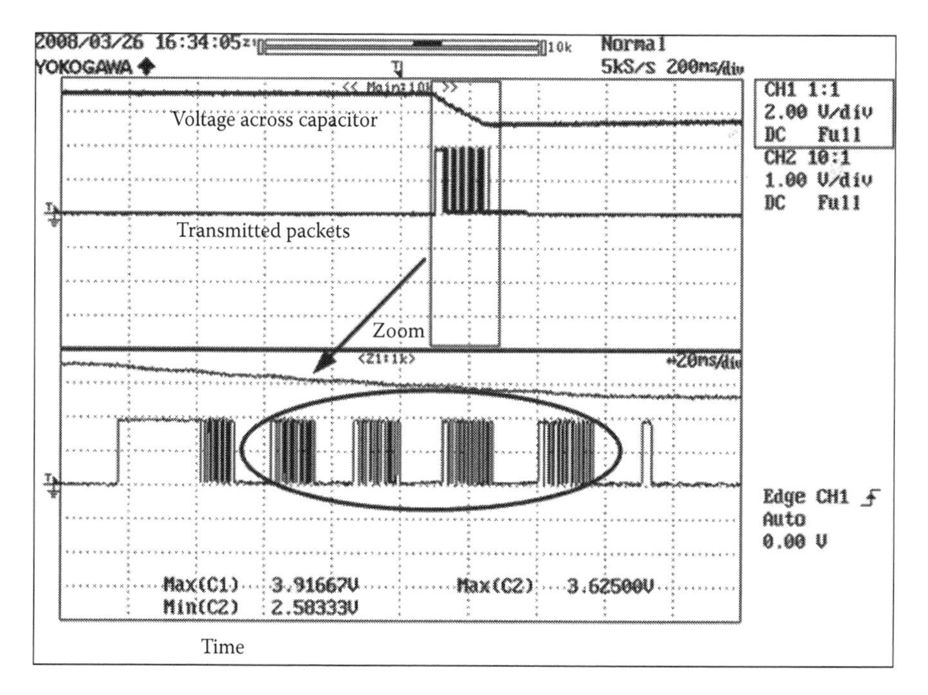


**FIGURE 4.14**
Fall-detection signal received at base station.

servers. Since the local gateway is fixed, the simulation results demonstrate that increasing the number of nodes causes a short gateway life and thus shortens the network lifetime. On the other hand, when the proposed method of selecting the gateway based on the residual energy of the sensor nodes is applied, the average lifetime of each node in the network is much longer, and its performance is independent of the number of sensor nodes. On top of that, the lifetime of the sensor node is further increased by the EH concept. This chapter demonstrates that the sensor node of a WBAN is able to be powered by the thermal energy harvested from human warmth (Hoang, Tan, Chng, and Panda 2009). The sensor node is equipped with fall-detection ability to support elderly people's safety in a home-dwelling environment and to guarantee that patients with disabilities in hospitals are assisted by informing doctors and nurses of falls in a timely fashion. Experimental results show that the body posture (upright or fallen) is sensed and then transmitted wirelessly to a remote patient-monitoring platform that displays the patient identification number and sounds an alarm for the respective personnel.

## Conclusions

In this chapter, the state of the art of EH technologies for WSNs has been reviewed. The chapter has also explored and illustrated the challenges involved in deploying this EH technology in the area of healthcare sensor networks. Take note that not all of the EH techniques are suitable for the healthcare type of WSN. Only some of the EH sources, like indoor solar energy from artificial lighting, thermal energy from human warmth and kinetic energy from movement, are applicable. Proposed solutions and market trends are highlighted in the chapter, followed by a detailed discussion on the prospect of EH technologies that can be practically deployed for healthcare sensor networks. The chapter also records a case study based on a research study published in *IEEE* (Hoang, Tan, Chng, and Panda 2009) to demonstrate how EH technologies can be applied to healthcare WSNs.

## References

(RReDC), R. R. (n.d.). Retrieved November 2010 from http://www.nrel.gov/rredc/.

Akyildiz, I., Su, W., Yogesh, S., & Erdal, C. (2002). A survey on sensor networks. *IEEE Communications Magazine, 40* (8), 102–114.

Blomgren, G. (2002). Perspectives on portable lithium ion batteries liquid and polymer electrolyte types. In *Seventeenth Annual Battery Conference on Applications and Advances* (pp. 141–144).

Braunwald, E. (1980). *Heart Disease: A Textbook of Cardiovascular Medicine*. Philadelphia: W. B. Saunders.

Callaway, E. H. (2003). *Wireless Sensor Networks: Architectures and Protocols*. Boca Raton, FL: Auerbach.

Chen, C., Islam, R., & Priya, S. (2006). Electric energy generator. *IEEE Transactions on Ultrasonics, Ferroelectrics and Frequency Control, 53* (3), 656–661.

Chen, Z., Guerrero, J., & Blaabjerg, F. (2009). A review of the state of the art of power electronics for wind turbines. *IEEE Transactions on Power Electronics, 24* (8), 1859–1875.

Cheng, S., & Huang, C. (2008). Power model for wireless body area network. In *IEEE Biomedical Circuits and Systems Conference* (pp. 1–4).

Chong, C., & Kumar, S. (2003). Sensor networks: Evolution, opportunities, and challenges. *Proceeding of the IEEE, Sensor Networks and Applications, 91* (8), 1247–1256.

Cook, D., & Das, S. (2004). Smart environments: Technologies, protocols and applications. In *Wireless Sensor Networks*. New York: John Wiley.

Cottone, F. (2008). Nonlinear piezoelectric generators for vibration energy harvesting. PhD thesis, University of Perugia, Italy.

Crossbow Technology, Inc. (2007). *MPR-MIB Users Manual*. Crossbow Resources, Revision A, San Jose, California.

Culler, D., Estrin, D., & Srivastava, M. (2004). Overview of sensor networks. *IEEE Computer, 37* (8), 41–49.

Dondi, D., Bertacchini, A., Brunelli, D., Larcher, L., & Benini, L. (2008). Modeling and optimization of a solar energy harvester system for self-powered wireless sensor networks. *IEEE Transactions on Industrial Electronics, 55* (7), 2759–2766.

Edmison, J., Jones, M., Nakad, Z., & Martin, T. (2002). Using piezoelectric materials for wearable electronic textiles. In *Proceedings of Sixth International Symposium on Wearable Computers (ISWC).*

Flipsen, S. (2004). *Alternative Power Sources for Portables and Wearables Part 1 - Power Generation and Part 2 - Energy Storage*. Delft University of Technology.

Glynne-Jones, P., Tudor, M., Beeby, S., & White, N. (2004). An electromagnetic, vibration-powered generator for intelligent sensor systems. *Sensors and Actuators, 110* (1–3), 344–349.

Green, M., Emery, K., Hisikawa, Y., & Warta, W. (2007). Solar cell efficiency tables (version 30). *Progress in Photovoltaics: Research and Applications, 15* (5), 425–430.

Heinzelman, W. R., Chandrakasan, A., & Balakrishnan, H. (2000). Energy-efficient communication protocol for wireless microsensor networks. *Proceedings of the 33rd Hawaii International Conference on System Sciences, 2*, 1–10.

Hill, J. L., & Culler, D. E. (2002). Mica: A wireless platform for deeply embedded networks. *IEEE Micro, 22* (6), 12–24.

Hoang, D., Tan, Y., Chng, H., & Panda, S. (2009). Thermal energy harvesting from human warmth for wireless body area network in medical healthcare system. In *Eighth International Conference on Power Electronics and Drive Systems*. Taipei, Taiwan. (pp. 1277–1282).

Holmes, A., Hong, G., Pullen, K., & Buffard, K. (2004). Axial-flow micro-turbine with electromagnetic generator: Design, CFD simulation, and prototype demonstration. In *17th IEEE International Conference on Micro Electro Mechanical Systems*. Maastricht, The Netherlands. (pp. 568–571).

Hudak, N. S., & Amatucci, G. G. (2008). Small-scale energy harvesting through thermoelectric, vibration, and radio frequency power conversion. *Journal of Applied Physics, 103* (10), 101301(1–24).

Jiang, X., Polastre, J., & Culler, D. (2005). Perpetual environmentally powered sensor networks. In *Fourth International Symposium on Information Processing in Sensor Networks (IPSN)*. Los Angeles, California. (pp. 463–468).

Kanesaka, T. (1999). Development of a thermal energy watch. In *Proceedings of the 64th Conference on Chronometry (Socit Suisse de Chronomtrie)* (pp. 19–22). Switzerland: Le Sentier.

Kangas, M., Konttila, A., Winblad, I., & Jms, T. (2007). Determination of simple thresholds for accelerometry-based parameters for fall detection. In *29th Annual International Conference of the IEEE Engineering in Medicine and Biology Society*, France. (pp. 1367–1370).

Kansal, A., Hsu, J., Zahedi, S., & Srivastava, M. (2006). Power management in energy harvesting sensor networks. *ACM Transactions on Embedded Computing Systems*, 35.

Karantonis, D., Narayanan, M., Mathie, M., Lovell, N., & Celler, B. (2006). Implementation of a real-time human movement classifier using a triaxial accelerometer for ambulatory monitoring. *IEEE Transactions on Information Technology in Biomedicine, 10* (1), 156–167.

Kuorilehto, M., Hannikainen, M., & Hamalainen, T. (2005). A survey of application in wireless sensor networks. *EURASIP Journal on Wireless Communications and Networking, 2005* (5), 774–788.

Lattanzi, E., Regini, E., Acquaviva, A., & Bogliolo, A. (2007). Energetic sustainability of routing algorithms for energy-harvesting wireless sensor networks. *Computer Communications, 30* (14–15), 2976–2986.

Lawrence, E., & Snyder, G. (2002). A study of heat sink performance in air and soil for use in a thermoelectric energy harvesting device. In *Proceedings of 21st International Conference on Thermoelectrics (ICT 02)*. California. (p. 446).

Leonov, V., Torfs, T., Fiorini, P., & Hoof, C. V. (2007). Thermoelectric converters of human warmth for self-powered wireless sensor nodes. *IEEE Sensors Journal, 7* (5), 650–657.

Li, H., Takizawa, K., & Kohno, R. (2008). Trends and standardization of body area network (BAN) for medical healthcare. In *European Conference on Wireless Technology, EuWiT* (pp. 1–4).

Lin, T.-H., Kaiser, W., & Pottie, G. (2004). Integrated low-power communication system design for wireless sensor networks. *IEEE Communications Magazine, 42* (12), 142–150.

Massachusetts Institute of Technology (MIT). (2003, February). 10 emerging technologies that will change the world. *Technology Review.* Retrieved November 2010 from http://www.technologyreview.com/Infotech/13060/?a = f.

Mathna, C., O'Donnell, T., Martinez-Catala, R. V., Rohan, J., & Brendan, O. (2008). Energy scavenging for long-term deployable wireless sensor networks. *Talanta, 75* (3), 613–623.

Meninger, S., Amirtharajan, A., & Chandrakasan, R. (2001). Vibration-to-electric energy conversion. *IEEE Transaction on VLSI System, 9*, 64–71.

Merrett, G., Al-Hashimi, B., White, N., & Harris, N. (2005). Resource aware sensor nodes in wireless sensor networks. *Journal of Physics, 15* (1), 137–142.

Mitcheson, P., Green, T., Yeatman, E., & Holmes, A. (2004). Architectures for vibration-driven micropower generators. *Journal of Microelectromechanical Systems, 13* (3), 429–440.

Moser, C. (2009). Power management in energy harvesting embedded systems. PhD thesis, Swiss Federal Institute of Technology, Zurich, Switzerland.

Niyato, D., Hossain, E., Rashid, M., & Bhargava, V. (2007). Wireless sensor networks with energy harvesting technologies: A game-theoretic approach to optimal energy management. *IEEE Wireless Communications, 14* (4), 90–96.

Paradiso, J., & Feldmeier, M. (2002). *A Compact Wireless Self Powered Pushbutton Controller.* Massachusetts Institute of Technology Media Laboratory.

Paradiso, J., & Starner, T. (2005). Energy scavenging for mobile and wireless electronics. *IEEE Pervasive Computing, 4* (1), 18–27.

Park, C., & Chou, P. (2006). AmbiMax: Autonomous energy harvesting platform for multi-supply wireless sensor nodes. *3rd Annual IEEE Communications Society on Sensor and Ad Hoc Communications and Networks (SECON).* Reston, Virginia. *1,* 168–177.

Myers, R., Vickers, M., Kim, H., & Priya, S. (2007). Small scale windmill. *Applied Physics Letters, 90* (5), 3.

Rabaey, J. M., Ammer, M. J., da Silva, J. L. Jr., Patel, D., & Roundy, S. (2000). Pico-Radio supports ad hoc ultra-low power wireless networking. *IEEE Computer, 33,* 42–48.

Raghunathan, V., Ganeriwal, S., & Srivastava, M. (2006). Emerging techniques for long lived wireless sensor networks. *IEEE Communications Magazine, 44* (4), 108–114.

Raghunathan, V., Kansal, A., Hsu, J., Friedman, J., & Srivastava, M. (2005). Design considerations for solar energy harvesting wireless embedded systems. In *Fourth International Symposium on Information Processing in Sensor Networks (IPSN)* (pp. 457–462).

Ramsay, M., & Clark, W. (2001). Piezoelectric energy harvesting for bio-MEMs applications. *Proceedings of the SPIE—The International Society for Optical Engineering, 4332,* 429–439.

Randall, J. F., & Jacot, J. (2003). Is AM1.5 applicable in practice? Modelling eight photovoltaic materials with respect to light intensity and two spectra. *Renewable Energy, 28* (12), 1851–1864.

Roundy, S. J. (2003). Energy scavenging for wireless sensor nodes with a focus on vibration to electricity conversion. PhD thesis, University of California, Berkeley.

Roundy, S., Wright, P., & Rabaey, J. (2004). *Energy Scavenging for Wireless Sensor Networks with Special Focus on Vibrations.* Boston, MA: Kluwer Academic Press.

Priya, S., Chen, C., Fy, D., & Zahnd, J. (2005). Piezoelectric windmill: A novel solution to remote sensing. *Japanese Journal of Applied Physics, 44* (3), L104–L107.

Shenck, N., & Paradiso, J. (2001). Energy scavenging with shoe-mounted piezoelectrics. *IEEE Micro, 21* (3), 30–42.

Siegfried, H., & Rachel, W. (2006). *Grid Integration of Wind Energy Conversion Systems.* 2nd ed. Chichester, West Sussex, UK: John Wiley & Sons.

Simjee, F., & Chou, P. (2008). Efficient charging of supercapacitors for extended lifetime of wireless sensor nodes. *IEEE Transactions on Power Electronics, 23* (3), 1526–1536.

Sinha, A., & Chandrakasan, A. (2001). Dynamic power management in wireless sensor networks. *IEEE Design & Test of Computers, 18*(2), 62–74.

Sodano, H., Simmers, G., Dereux, R., & Inman, D. (2007). Recharging batteries using energy harvested from thermal gradients. *Journal of Intelligent Material Systems and Structures, 18* (1), 3–10.

Sohrabi, K., Gao, J., Ailawadhi, V., & Pottie, G. (2000). Protocols for self-organization of a wireless sensor network. *IEEE Personal Communications, 7* (5), 16–27.

Stark, I. (2006). Invited talk: Thermal energy harvesting with ThermoLife. In *International Workshop on Wearable and Implantable Body Sensor Networks* (pp. 19–22).

Stevens, J. (1999). Heat transfer and thermoelectric design considerations for a ground-source thermo generator. In *Proceedings of 18th International Conference on Thermoelectrics,* Baltimore. (pp. 68–71).

Tan, Y., & Panda, S. (2007). A novel piezoelectric based wind energy harvester for low-power autonomous wind speed sensor. In *33th Annual IEEE Conference of Industrial Electronics Society* (pp. 2175–2180).

Massachusetts Institute of Technology (n.d.). *μAmps projects.* Retrieved November 2010 from Microsystems Technology Laboratories: http://wwwmtl.mit.edu/researchgroups/icsystems/uamps/.

Thomas, J. P., Qidwai, M. A., & Kellogg, J. C. (2006). Energy scavenging for small-scale unmanned systems. *Journal of Power Sources, 159,* 1494–1509.

TinyOS. (n.d.). *The TinyOS Project.* Retrieved November 2010 from TinyOS Community Forum at http://www.tinyos.net.

Tubaishat, M., & Madria, S. (2003). Sensor networks: An overview. *IEEE Potentials, 22,* 20–23.

Verdone, R., Dardari, D., Mazzini, G., & Conti, A. (2008). Chapter.5: Network lifetime. In *Wireless Sensor and Actuator Networks: Technologies, Analysis and Design* (pp. 115–116). Academic Press.

Weimer, M. A., Paing, T. S., & Zane, R. A. (2006). Remote area wind energy harvesting for low-power autonomous sensors. In *37th IEEE Power Electronics Specialists Conference* (pp. 2911–2915).

# 5

# *Addressing Security, Privacy and Efficiency Issues in Healthcare Systems*

**Kalvinder Singh and Vallipuram Muthukkumarasamy**

**CONTENTS**

Introduction ........ 112
Healthcare Sensor Systems ........ 115
  Compatibility Issues between Different Environments ........ 117
  Limitations with Power and Security ........ 118
Information Assurance, Security and Privacy Threats ........ 120
  Impersonation of the User ........ 120
  Impersonation of the Service ........ 121
  Modification of Software ........ 121
  Modification of Data ........ 121
  Disclosure of Sensitive Data ........ 121
  Denial of Service ........ 122
  Repudiation ........ 122
  Difficulty in Using Complex Technology ........ 122
  Inability to Keep Track of Changing Technology ........ 122
  Lack of Trust in the System ........ 122
  Expectation of Reliability ........ 123
  Expectation of Real-Time Communication ........ 123
Countermeasures to the Threats ........ 123
  Privacy ........ 124
  Mapping the Countermeasures ........ 124
  Key Management ........ 125
  Pairwise Key Establishment ........ 127
  Random Key Establishment ........ 127
  KDC Schemes ........ 128
  Environment Information ........ 129
  Using Physiological Data to Establish Keys ........ 130
Efficiency Issues and Experimental Evaluations ........ 132
Future Directions ........ 135
Acknowledgements ........ 135
References ........ 135

## Introduction

This chapter explains the issues related to the security of a home healthcare system, from assurance and administration to authorization, accountability and availability. We give a brief introduction to each of the security countermeasures. To gain an understanding of the level of difficulty in implementing security mechanisms, we provide a detailed description of how to implement a key management protocol for sensors. The chapter begins with a brief description of the general architecture and service platforms used for healthcare sensor systems.

The ageing population and the increase in chronic diseases have placed a considerable financial burden on healthcare services. Body sensors may play a vital role in reducing healthcare costs significantly. Sensors can be used to remotely monitor elderly patients suffering from chronic diseases and allow them to have relatively independent lives. However, the use of body sensor networks is inherently complex. For instance, an increase in blood pressure during exercise is normal. But blood pressure increases while at rest could mean a serious medical condition. Sensors may measure not only physiological values but also a range of other parameters such as body motions, which can lead to a number of different sensors needing to communicate with each other. As the number of heterogeneous sensors increases, so will the complexity of interactions between the sensors. Different sensors have different associated costs and privacy acceptance levels. For example, a sensor simply detecting light will have different costs to a sensor recording sound. However, less costly and more acceptable sensors can be used to detect a phenomenon before alerting the more costly sensors to start their monitoring.

In some cases large areas can either be covered by relatively few sensors, in other cases large areas can be covered by many sensors. It depends on the sensor application. (Kuorilehto, Hännikäinen, and Hämäläinen 2005). Different environments have a range of varying characteristics. For instance, physical access to a sensor is easier for an adversary for a sensor placed in an open area than for a similar sensor implanted in an individual's body. In this chapter we discuss a number of protocols that are designed for home healthcare systems. However, these protocols can also be applied to other environments that have similar security characteristics. Figure 5.1 gives a diagrammatic representation of a home automation system that can be used to monitor the elderly. The diagram shows the communication of the sensors with the central home health controller (HHC). The home controller is a specialized device that is situated at home and is connected to the Internet. When this system is part of a health system, the HHC is a specialized device that sends the necessary information to the hospital. Depending on the patient's health risks and privacy concerns, all of the information may not be transmitted to a hospital. For instance, home sensors may be activated only if other sensors detect that there may be a medical emergency, such as the

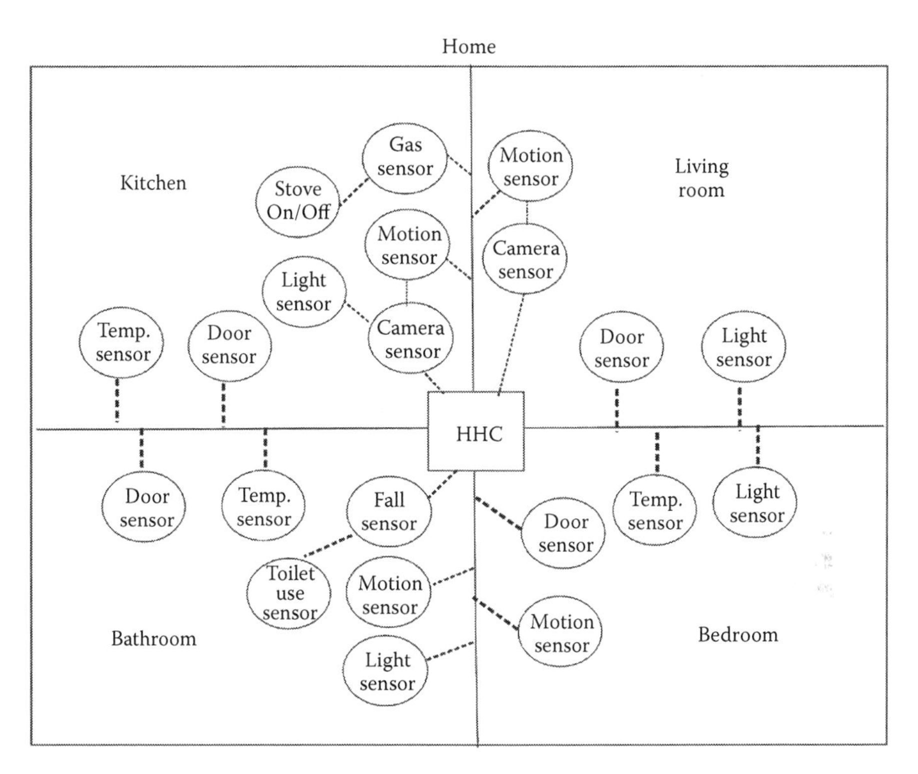


**FIGURE 5.1**
A sensor-enabled smart home. HHC, home health controller.

patient lying horizontal in the kitchen. Surveillance software can be used to detect if the patient is cleaning the kitchen or getting something from the ground, or if there is actually an emergency. If the software does detect an emergency, the hospital staff are notified; they examine the information and decide on the best course of action.

Other sensors that may work in conjunction with home sensors are body sensors. Some of the data recorded by body sensors include the heart rate, blood pressure, temperature and blood oxygen level. The number of messages sent and the size of messages require the data rate to be around 2 bps (Balomenos 2001; Yeatman and Mitcheson 2006). However, other information sent with the message, such as the location of the sender node (8 bits), a message authentication code (we have specified the size to be the same as the size of the physiological data) and a counter to stop replay attacks (32 bits), raises the data rate to around 10 bps. Therefore, due to the number and size of messages, the data rate is measured in bits rather than bytes. In our case the bit rate we require is 10 bps to send secure data. Another type of sensor is a surveillance camera, and the data rate requirement for video streams (Axis 2007) is much greater than that for body sensors. A single camera normally requires a data rate of 1–4 Mbps (with compression).

Providing secure data transfer between sensors is a requirement for our home healthcare system.

An additional advantage of enabling security in the home healthcare system is that secure smart data masking enables data sharing among institutions and reduces costs for duplicate tests. Privacy laws may prevent the complete medical records of an individual from being transferred between institutions. Health information collected from sensors needs to be secured, and in some countries (for example, the US) security is mandated (USA 2003). Securing a home healthcare system becomes difficult, mainly because of the different requirements for various components. For instance, the sensors have considerably more resource constraints than do mobile phones, cameras or desktop computers. With differences in computing power, as well as differences in communication costs, different security protocols may be required throughout the entire system. For instance, an efficient key establishment mechanism specifically for body sensors was created using physiological data (Singh and Muthukkumarasamy 2007). However, the home health system may send physiological data to medical staff or to an analytics engine (Espina, Falck, and Mülhens 2006). The physiological data may also be sent to an actuator to release medicine into the body (Espina et al. 2006).

When the same physiological data are used for more than one purpose (and taking into consideration the complexity of a heterogeneous environment), it becomes important from a security or information assurance point of view to have a formal methodology. A formal methodology is also important to ensure that the information sent to medical staff and actuators to dispense medicine is accurate and that appropriate actions are taken. The formal methodology has a requirement that it can model both the security and the privacy aspects as well as the application correctness. A number of researchers have used environmental data as the only source of secret information to establish keys between body sensors (Bao, Zhang, and Shen 2006; Poon, Zhang, and Bao 2006; Venkatasubramanian and Gupta 2006). The major benefit of using environmental data is that body sensors can use this information to authenticate that the other sensor is also on the same person and not another individual. However, these researchers have cited a number of problems with that approach. The problems include that

- Only cryptographically strong environmental data can be used.
- The environmental values can become compromised, in which case the new session key is also compromised.

These problems limit the use of environmental data for establishing keys to only a few cases. The other difficulty our system encounters is in the key establishment scheme of the PDA with sensors at home and in the body. It is envisaged that the patients will simply be users of the system and will not be able to set up security certificates or keys.

In this chapter, we show protocols to address these problems. We show that password protocols can be used to establish keys between body sensors, if passwords are replaced by physiological data. A new protocol is developed to allow a patient to connect a PDA to the home healthcare system and thus be able to view information about each of the sensors (ranging from cameras to body sensors). The protocol does not require traditional encryption to transport the new session key. We show that the sensor nodes can establish keys even if no previous shared keys exist between them.

## Healthcare Sensor Systems

Healthcare sensor networks may contain many different types of sensors. The sensors in such a system can range from powerful high-definition cameras to low-powered Electrocardiography (ECG) body sensors. Figure 5.2 shows a patient at home with a number of body sensors that can communicate with a camera sensor, the HHC and a PDA. The cameras may start recording only if the body sensors detect that there may be a medical anomaly, such as a dramatic increase in blood pressure. Surveillance systems, such as S3 (Hampapur et al. 2004), can be used to detect if the patient is involved in activity that would increase the blood pressure (such as exercise). If the patient is not involved in any activity that would account for the medical

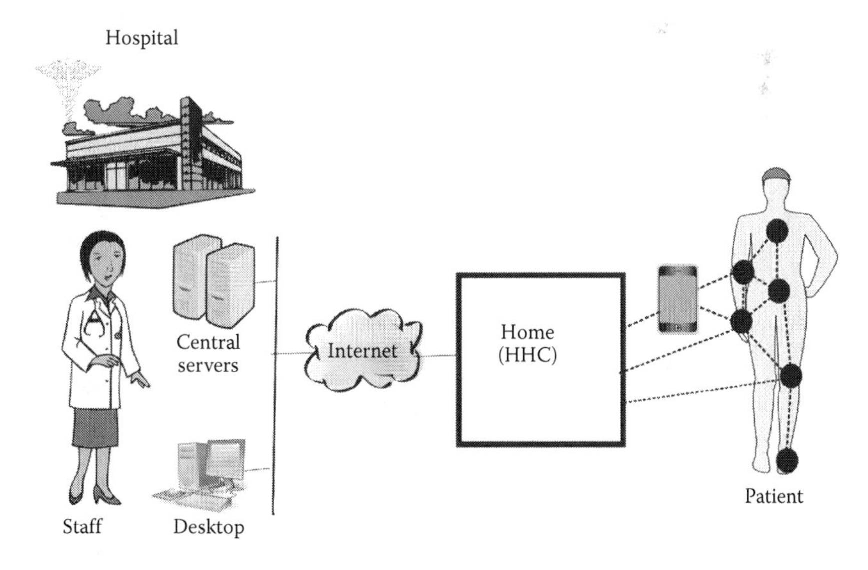


**FIGURE 5.2**
Architecture and service platform of a body sensor network (BSN) for telemedicine and home healthcare.

anomaly, the hospital staff are notified. They examine the information and decide on the best course of action. The PDA is used to give feedback to the patient about the condition of his or her body, as well as the status of the sensors. The PDA can notify the patient of any detected emergency, allowing the patient to report back a false alarm if one has occurred. The PDA can be replaced with a mobile phone or any other handheld communication device.

There are numerous types of communication technologies that are used in the home healthcare system. The communication technology can be

- Bluetooth
- Low-power Bluetooth
- ZigBee
- 802.11 (Wi-Fi)
- 802.3 (Ethernet)

The list shows the need to interoperate between the different technologies, there are many security risks. Each technology has its own security risks, and combining technologies can increase the security risks.

Other software and hardware architectures have been proposed and implemented to evaluate and address the needs of a home healthcare system. Table 5.1 summarizes some of the major research projects into body sensor networks and monitoring (Kuryloski et al. 2009; Malan, Fulfrod-Jones, Welsh, and Moulton 2004; Oliver and Flores-Mangas 2006; Wood et al. 2006).

The wireless protocols used in existing home healthcare systems include a number of security provisions and options. Each protocol has many options and security features. For instance, there are several vulnerabilities

**TABLE 5.1**

Sample Implementations of Home Healthcare Systems

| Platform | Base Device | Wireless Protocol | Function |
|---|---|---|---|
| **HealthGear** | Phone | Bluetooth | Wearable system for connecting sensors and mobile phones |
| **CodeBlue** | PC, PDA | 802.15.4 | Provision of medical monitoring in a hospital environment |
| **ALARM-NET** | PC, PDA, Stargate | Bluetooth 802.11 | Wireless sensor network for assisted-living and residential monitoring |
| **DexterNet** | PC, PDA, phone | 802.15.4 | Body sensor network for indoor and outdoor monitoring |

and pitfalls in 802.15.4 (Sastry and Wagner 2004). They fall into three categories:

- Initialization vector (IV) management
- Key management
- Integrity protection

Bluetooth also has several vulnerabilities (Gehrmann, Persson, and Smeets 2004), which include

- Bluejacking
- Bluecasting
- Bluesnarfing

Several vulnerabilities have also been discovered in the 802.11 protocol over the last decade (Borisov, Goldberg, and Wagner 2001; Housley and Arbaugh 2003). They include

- WEP cipher
- Key management
- 40-bit secret key

The wireless protocols themselves can have security vulnerabilities. Even with secure wireless protocols, the home health systems described in the preceding do not cover attacks such as impersonation of the user, modification of software or other higher-level attacks that are independent of the type of network that is used.

### Compatibility Issues between Different Environments

The possible number of different networks and different types of environments is large. There are various types of sensor environments, ranging from sensors covering large areas (Boulis, Han, and Srivastava 2003; Mainwaring, Culler, Polastre, Szewczyk, and Anderson 2002) to many sensors covering a small area (Heinzelman, Murphy, Carvalho, and Perillo 2004; Marcy et al. 1999; Schwiebert, Gupta, and Weinmann 2001). In a home health system, there are two major types of wireless sensor networks:

- Large-area wireless sensor networks (home sensor networks, HSNs)
- Small-area wireless sensor networks (body sensor networks, BSNs)

Table 5.2 compares HSNs and BSNs. The table is based on a comparison between a BSN and a generic wireless sensor network (WSN; Aziz, Lo,

**TABLE 5.2**

Comparison of Home Sensor Networks (HSNs) and Body Sensor Networks (BSNs)

| Network | HSNs | BSNs |
|---|---|---|
| **Scale** | The size of the home can be over several hundred square metres | As large as the human body |
| **Number of nodes** | Greater number of nodes required for wide area coverage | Fewer, more accurate sensor nodes required |
| **Node function** | Sensors perform a single dedicated task | A single sensor performs multiple tasks |
| **Node accuracy** | Many nodes compensate for accuracy and allow result validation | Each sensor is required to be robust and accurate |
| **Size of node** | Small size preferable but not needed in many cases | Need for miniaturization |
| **Power supply** | Accessible and likely to be changed more easily and frequently | Inaccessible and difficult to replace in implanted sensors |
| **Access** | Sensors can be more easily replaced | Implanted sensors are difficult to replace |
| **Biocompatibility** | Not considered in most applications | Needed for implanted sensors |
| **Data transfer** | Loss of data compensated by number of sensors | Loss of data more significant |

Darzi, and Yang 2006). Some of the more noticeable differences are the larger scale that exists in HSNs, compared to the severe constraints on the BSNs. The different sensor environments have a wide range of different characteristics. For instance, sensors placed in a large open area are not as physically secure as sensors implanted in an individual's body.

This chapter discusses specific protocols for the BSN and HSN environments.

The protocols designed for BSNs can also be applied to other environments that have similar security characteristics to those found when using BSNs. The same is true for the protocols designed for the HSNs: they can be applied to environments that exhibit similar security characteristics.

## Limitations with Power and Security

We refer to a *sensor network* as a heterogeneous system combining small, smart and cheap sensing devices (sensors) with general-purpose computing elements. A sensor network consists of a potentially large number of sensors. Sensor network applications (Bulusu 2005; Kuorilehto et al. 2005; Venkatasubramanian and Gupta 2006) include monitoring health, tracking bushfires, monitoring wildlife, conducting military surveillance and monitoring public exposure to contaminants. A WSN may also contain smaller number of control nodes, which may have more resources. The functions of the control nodes include

- Connecting the sensor network to an external network
- Possibly aggregating results before passing them on
- Controlling the sensor nodes
- Providing services (otherwise not available) to a resource-constrained environment

Examples of control nodes (Singh and Muthukkumarasamy 2008) are the Stargate platform, the GNOME platform, the Medusa MK-2 platform and the MANTIS platform. These platforms may use higher-level operating systems such as the Linux operating system. The platforms themselves may have additional communication mechanisms. For instance, the GNOME platform also has an Ethernet connection. It should be noted that a sensor node may also be a control node. Controller nodes are not limited to custom-made hardware. For instance, the UbiMon project (Lo and Yang 2005) uses a PDA as a controller node, while other options for controller nodes include mobile phones (Espina et al. 2006).

Communication is the most expensive operation in sensor networks, where the received power drops off as the fourth power of distance mainly due to multipath propagation. If the distance between the nodes is 10 m, the energy required to send a single bit set is equivalent to the energy required to perform 5000 operations. Therefore, when the distance between the nodes is 100 m, the energy required to send a single bit is equivalent to the energy required to perform 50 million operations. When information encrypted using symmetric cryptography is transmitted and received, only 2% of the energy consumed is used to encrypt or decrypt the data (Perrig, Szewczyk, Wen, Culler, and Tygar 2001).

Mica sensor nodes run TinyOS (TinyOS 2007) an event-driven operating system specifically designed for wireless sensor environments. The memory footprint for TinyOS is small; a minimum installation (the core components) uses 400 bytes of data and instruction memory. TinyOS is developed in nesC and supports other hardware platforms. The TinyOS network packet wraps the payload during sensor node communication. To reduce the message size, the packet header contains a minimal amount of information. For instance, it does not contain the source address of the sender.

To enable security in a home healthcare system, the sensors require more power. Most of the power consumption is due to the extra communication costs. For instance, without security, a message in the 802.15.4 protocol contains a Cyclic Redundancy Check (CRC) value of 2 bytes. When using an 802.15.4 protocol security suite, the CRC value is replaced by a Message Authentication Code (MAC) value. The smallest MAC size in the 802.15.4 protocol is 4 bytes. Every secured message has an additional overhead. Also, additional security management protocols, such as key establishment protocols, need to be added and run.

## Information Assurance, Security and Privacy Threats

Security and privacy concerns are impediments to health systems. The lack of security and privacy could limit the adoption of a home healthcare system, either because of legal consequences or because of patient misgivings about allowing devices such as cameras into their homes. Even if a home healthcare system was adopted with a limited security framework, the patient would be put at risk. For instance, a device in the home healthcare system that dispenses medicine or measures important vital signs should be secured against attacks. Not only are security and privacy important, but information assurance is also just as important. Information assurance guarantees that the data are correct not only from an adversary but also from malfunctioning devices. Security and privacy threats in the system include

- Impersonation of the user
- Impersonation of the service
- Modification of software
- Modification of data
- Disclosure of sensitive data
- Denial of service
- Repudiation

Threats may stem not only from adversaries but also from the users' expectations or the users themselves:

- Difficulty in using complex technology
- Inability to keep track of changing technology
- Lack of trust in the system
- Expectation of reliability
- Expectation of real-time communication

### Impersonation of the User

An adversary may attack the system by taking on the identity of one of the users in the system. The users may range from patients to medical staff to computer administrators. Each different type of user has specific privileges and permissions, which an adversary may utilize to gain access to the system.

If the adversary assumed the identity of a patient, incorrect information could be passed through to the medical staff. The adversary might also be able to access sensitive information that should be available only to the patient. An adversary who obtained the identity of a medical staff person or computer administrator could be even more dangerous. The adversary might then be able to obtain information about many patients or affect their treatment.

#### Impersonation of the Service

A complex system such as a home healthcare system has many remote services. Services may include analyses of data, authentication server or sending an emergency call. An adversary successfully impersonating a service can have major consequences for the integrity of the system. For example, if an adversary was able to impersonate an authentication server which leased session keys out to devices, then the adversary would then have access to all the session keys within the system, and the devices in the home healthcare system could easily become compromised.

#### Modification of Software

Viruses, Trojan horses, root kits and software update failures are only several methods to modify the software in a home healthcare system. Once the integrity of the system is compromised, any data or information derived from the system are also compromised. As devices become more intelligent and complex, there will be more opportunities for an adversary to create more complex viruses and Trojan horses to manipulate the system.

#### Modification of Data

Another attack is to modify the data. The modification of data can either be at rest (stored in memory) or in motion (transmitted on the network). If the data obtained by the home healthcare system can be modified, then the medical staff and the end user will have misgivings about how accurate the figures are. The data may also be manipulated to produce false emergencies or even contribute to an emergency.

#### Disclosure of Sensitive Data

Privacy is a major concern for patients. In a traditional healthcare system, medical records have sensitive information that patients would not like the general population to know. Home healthcare systems add an extra dimension to the problem, as more sophisticated devices can record a patient's every move. Some devices are not very intrusive, such as temperature sensors, but other devices, such as cameras, are far more intrusive. Patient control over what data are disclosed is an important feature of a home healthcare system.

### Denial of Service

Denial of service can be initiated by an adversary. Denial-of-service activities can range from wormhole attacks to draining the battery of a sensor by initiating requests. If an adversary can easily create a denial-of-service attack and make the system unusable for an extended period of time, the end user could be placed in serious danger.

### Repudiation

Repudiation is the act where a participant in the system has forged that they have performed an action. For instance, a patient may forge that they had taken their medicine, and state that there must be something wrong with the name healthcare system.

The hospital obtaining a guarantee a patient has taken his or her medicine and the patient obtaining a guarantee that a call sent has been received by the hospital are both important infrastructure requirements. In both cases the information assurance of the system increases, as well as its usability.

### Difficulty in Using Complex Technology

A complex system where a user needs to install security certificates is infeasible. Setting passwords on tens to hundreds of devices is technically difficult due to the lack of user interfaces such as keyboards and monitors on many of the sensors. Forcing a patient to remember a new password or PIN will be a deterrent to the take-up of a home healthcare system. It is envisioned that the elderly could benefit from a home healthcare system. However, they would have the greatest difficulty in managing and handling complex technology.

### Inability to Keep Track of Changing Technology

Changing the components in an ever-changing environment and then having the users learn that new technology is infeasible. As technology improves, new devices will be added to an already-complex system. The initial security set-up should be simple so that any new devices can be easily added.

### Lack of Trust in the System

The user needs to trust the system to use it. The user should be in control of the system, as well the data and devices. For instance, devices such as cameras capture sensitive data and therefore should turn on if there is an emergency or when there is a request from the user. The types of control and feedback that are given to the user will influence the adoption of a home healthcare system.

### Expectation of Reliability

The quality of security is a very important criterion of the home healthcare system. Security services should be available to facilitate the smooth operation of the system even if a component malfunctions or is compromised. As the number of components in a home healthcare system increases, so does the likelihood of failure. The system should be designed so that the failure of a component does not jeopardize the security of the system. For instance, if the authentication server is unavailable, then the system should be able to cope, either with a redundant authentication server or with another device that would take on the new load until the authentication server was replaced.

### Expectation of Real-Time Communication

The security mechanisms should enable real-time communication. When an emergency occurs there is an expectation that information will be sent to the relevant authorities as soon as possible. In a medical emergency, time is a critical factor for the well-being of a patient.

## Countermeasures to the Threats

In this section, we describe security mechanisms to mitigate the threats and deterrents to the home healthcare system. When looking at countermeasures, we examined Certified Information Systems Security Professional (CISSP) guidelines. CISSP is an independent information security certification governed by the not-for-profit International Information Systems Security Certification Consortium. According to CISSP there are 10 domains to security (ISC 2007):

- Information security and risk management
- Access control
- Telecommunications and network security
- Cryptography
- Security architecture and design
- Operations security
- Application security
- Business continuity planning and disaster recovery planning
- Legal, regulatory, compliance and investigation issues
- Physical security

The 10 domains are based on three fundamental security characteristics of information: confidentiality, integrity and availability. Cryptographic keys are a cornerstone to confidentiality and integrity. The quality of security is a very important criterion of the home healthcare system.

Ensuring the quality of security services in a home healthcare system is difficult, mainly because of the number of different components, each with different characteristics. The home healthcare system's security services should be able to handle the failure of a single or multiple components. Singh and Muthukkumarasamy (2007) proposed three multiserver protocols that were designed to handle failures in the authentication security service. The proposed protocols show that security in a home health care system is difficult because of the different requirements for various components. Sensors, mobile phones, cameras and desktop computers have dramatically different resource constraints. The differences in computing power and communication costs require different security protocols throughout the entire system.

### Privacy

Privacy is the ability for users to control their personal data. The users should be able to view and control their data. Privacy is enabled by securing data, data transmission and storage. There needs to be due diligence to ensure that no sensitive data are sent to the incorrect groups or individuals unless these have permission to see the information. Any failure will cause a lack of trust by the end users.

Also, it is important that sensitive data (such as from cameras) are recorded only when necessary. The system should allow the end user an opportunity to control the type of data that are sent. Input from the end user is an important consideration when looking at the overall privacy of the system.

### Mapping the Countermeasures

Table 5.3 maps the security and privacy threats with the appropriate countermeasures. An audit is an important countermeasure for most of the threats, and it enables accountability. The countermeasures range from low-level cryptography algorithms to high-level policy management. The difficulty with the countermeasures is that different components have different security requirements. For instance, key management on the HHC is different from the key management of an implanted health sensor found in the body. The HHC may need to manage a wide variety of keys, from small keys found in body sensors to certificates found in some of the cameras. Integrating the different security implementations is a major problem in a home healthcare system.

The security countermeasures can be grouped into the five As: *authorization, accountability, availability, administration* and *assurance*. Figure 5.3 maps the countermeasures associated with the five As. Also, the countermeasures are broken down further. For instance, the audit countermeasure is

**TABLE 5.3**

Threats and Countermeasures in a Home Healthcare System

| Threats | Countermeasures |
|---|---|
| Impersonation of the user | Authentication, audit, user administration, trust policy, key management, intrusion detection |
| Impersonation of the service | Authentication, audit, service ID management, trust policy, key management, intrusion detection, system integrity |
| Modification of code | Code signature, access control, audit, system installation, configuration, remediation, antivirus software, intrusion detection, system integrity |
| Modification of data | Data protection (integrity), access control, audit, data protection policy, intrusion detection |
| Disclosure | Data protection (confidentiality), access control, audit, data protection policy |
| Denial of service | Firewall, service continuity, disaster recovery, routing control, topology control, rate/resource limits, system integrity |
| Repudiation | Audit, non-repudiation, audit policy, non-repudiation policy, system integrity |

an important part of accountability, but two important subsets of audits are entity authentication and policy enforcement. Entity authentication is used to distinguish different services, nodes, components and users. When auditing an action it is important to know who performed that action.

Administration is another vital countermeasure. For instance, the installation and configuration of the system, if done incorrectly, can introduce zero-day security holes. This is an example of requiring a secure mechanism. However, if a secure mechanism is implemented in a user-unfriendly way, where the user needs to set up certificates, keys and passwords on each of the components, then the user will find the system hard to use. With a more complex system the likelihood is higher that the user will produce an error. Without cryptographic keys and key management, most of the countermeasures described in Figure 5.3 will be ineffectual.

## Key Management

Key establishment protocols are used to set up shared secrets between sensor nodes, especially between neighbouring nodes. When using symmetric keys, we can classify the key establishment protocols in WSNs into three main categories: *pairwise schemes*, *random key predistribution schemes* and a *key distribution centre (KDC)*. Recently, a fourth category, in which key establishment protocols use environmental values, has been developed (Poon et al. 2006; Venkatasubramanian and Gupta 2006). This section discusses limitations and problems with existing key establishment protocols when applied to WSNs.

Many of the existing key establishment protocols are designed for open environments where a sensor node can be easily compromised (D. Liu and

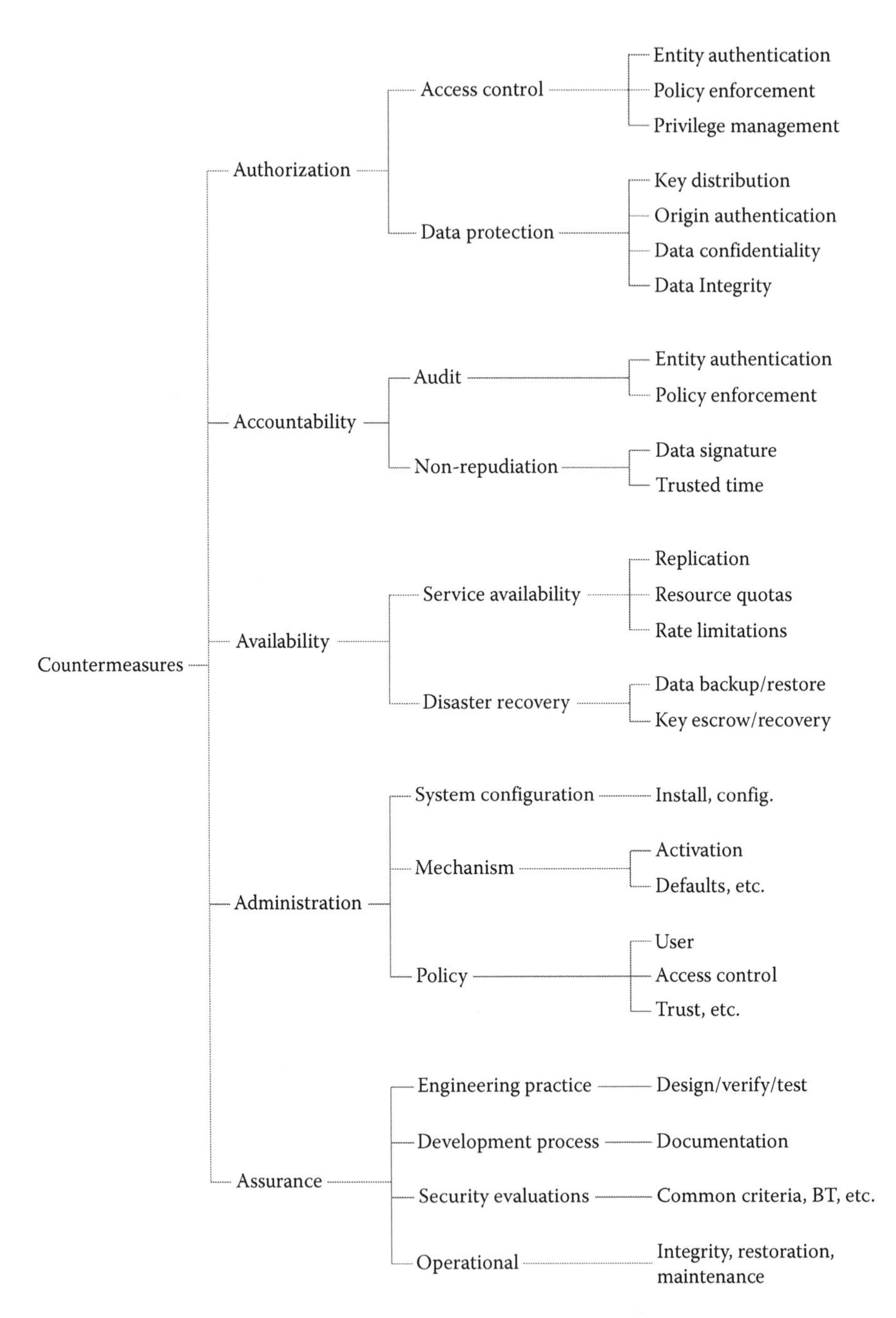


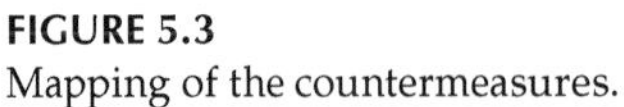

**FIGURE 5.3**
Mapping of the countermeasures.

Ning 2007). Also, sensors may be short-lived, and thus updating keys may not be important. We examine traditional sensor key establishment protocols in the BSN environment as well and show their limitations and problems.

### Pairwise Key Establishment

The first category of key establishment protocol is called the pairwise scheme. The simplest version is the full pairwise scheme (Chan, Perrig, and Song 2003), where each node in a network of total $n$ nodes shares a unique pairwise key with every other node in the network. Each sensor node will have a memory overhead of $(n - 1)$ cryptographic keys. Other pairwise schemes (Blundo et al. 1993) also have Order $n$ which is denoted as $O(n)$ memory cost. In a pairwise scheme, the sensor network will not be compromised even if a fraction of the sensors are compromised.

However, because of the memory constraints found in sensors, predefined keys may not be practical in large sensor networks. Also, pairwise key establishment is difficult to maintain if the sensor network is dynamic in nature, with additions and deletions of sensors. Furthermore, there is no secure mechanism available to update the keys. Keys in sensor networks are usually 64 bits in size, and they may easily become compromised. A server-less or server-based key establishment protocol can be used to update the key (ISO11770-2 1996). If the key between the sensor nodes is compromised before the update protocol is run, then the new key will also be compromised. If the key between the sensor nodes is compromised after the update protocol is run, then the new key will be compromised provided that the messages from the update protocol were saved by the adversary.

Sensors found in BSNs may be operational for years, especially if the sensors are implanted in the body of an individual. Pairwise key establishment is, therefore, not suitable if the keys need to be updated.

### Random Key Establishment

Random key predistribution schemes are the second category of key establishment protocols (Ren, Zeng, and Lou 2006). This is a major class of key establishment protocols for sensor networks. They rely on the fact that a random graph is connected with high probability if the average degree of its nodes is above a threshold. After the connected secure network is formed, the protected links can be further used for agreeing on new keys, called *path-keys*. The random key establishment schemes have a security concern when a certain number of sensor nodes become compromised. If so, the entire network can then be compromised.

Another issue is that random key predistribution schemes require $O(n)$ of predefined data, and it is a major research area to limit the amount of predefined data (Ren et al. 2006). The random key establishment schemes are designed for large sensor networks and assume that sensors can be compromised. These characteristics do not match those of BSNs. The BSN topology may not be suitable for random key establishment schemes. In BSNs, some sensors may be able to communicate with only one other sensor node. Another drawback is that the shared keys cannot be used for entity authentication, since the same key can be shared by more than a single pair of nodes (Hämäläinen, Kuorilehto, Alho, Hännikäinen, and Hämäläinen 2006). Random key establishment protocols also have no mechanism to update the keys.

### KDC Schemes

In a KDC scheme, if two entities, sharing no previous secret, want to communicate securely with each other, they generally do so with the assistance of a third party. In WSNs the two entities are typically resource-constrained sensor nodes, and the third party is a resource-heavy base station (Perrig et al. 2001). The base station provides an authentication service that distributes a secure session key to the sensor nodes. The base station is sometimes referred to as *a trusted third party,* since every client has to trust the base station by sharing a secret with it. The level of security of a typical key distribution protocol depends on the assumption that the authentication server is trustworthy (Boyd and Mathuria 2003).

KDC schemes use the least amount of memory compared with the other two categories, and they have the extra advantage of providing authentication for the sensor nodes (Hämäläinen et al. 2006). Examples of KDCs in WSNs were first proposed in the Security Protocols for Sensor Networks (SPINS) protocol (Perrig et al. 2001). A simple example of a sensor environment may consist of a large number of motion sensors, a camera and a base station. When the motion detectors are triggered, they notify the camera to start recording. If the motion sensors need to communicate with the camera via the base station, this may place undue stress on the network and cause undesirable latency. To enable secure communication, the motion detection sensors will need to establish a shared secret between themselves and the sensor on the camera. Motion detectors may also be added, removed or replaced, thus adding the requirement to have a scheme that allows for a dynamic reconfiguration of the sensor environment.

Another use of a KDC scheme is when a sensor network is created using a random key predistribution scheme. If there is an increase in the area that the sensor network needs to cover, a new set of sensors is distributed. To enhance the security of the expanded network, new values are used in the random key predistribution scheme for the added sensors. The new sensors will have to use a KDC scheme if they want to create a key between themselves and the old sensor nodes. The old sensors may have deleted the random key predistribution scheme from memory, so that they could increase the amount of memory available to the node (and increase security). If new nodes are added, they will not be able to use the old scheme to create a key. Thus, the SPINS protocol may not be suited for every WSN topology. For instance, it does not easily scale to a large WSN, since the nonuniform communication will focus the load onto the KDC. This may cause the battery life of the network to diminish considerably. Updating keys using a KDC mechanism can also cause concerns if the keys between the sensor nodes and the KDC become compromised.

Hybrid schemes have been created to offset this problem by combining different key establishment categories. The Peer Intermediaries Key Establishment (PIKE) scheme (Chan and Perrig 2005) is one such implementation. It combines a pairwise scheme with a KDC scheme, in which one or more sensor nodes acts as a trusted intermediary to facilitate the key establishment. The scheme was developed to limit the amount of memory used by the pairwise and random key predistribution schemes and also to limit the communication load of the KDC scheme.

However, some of the limitations with using a sensor node as the trusted third party are that

- The KDC scheme relies upon other schemes to create the trusted intermediary.
- The key sizes in sensor nodes are not large enough, so over a period of time, the key between the sensor and the trusted intermediary may become compromised. If the KDC protocol messages were captured and saved by an adversary, then the adversary could calculate the new keys created.
- Some sensor networks may not need an encryption algorithm; however, KDC protocols require an encryption algorithm to encrypt the new key.

## Environment Information

*Secure environmental value* (SEV) refers to sensed data that can be obtained by sensors from their environment. These data are usually hard to obtain through other means. Examples of environments where SEVs may be found include

- The human body, where it is difficult to attach a device without the person's knowledge
- A secured location, for instance, a military base or unmanned vehicle, such as Unmanned Aerial Vehicles (UAVs)
- Hard-to-reach places, for instance, a satellite in orbit

The sample environment used in this chapter is the human body, where BSNs have been developed to measure the physiological values found in individuals (Aziz et al. 2006). Recent research on BSNs has shown that environmental information found in the body can be used to secure communication between sensor nodes (Bao and Zhang 2005; Bao, Zhang, and Shen 2005; Bao et al. 2006;

Poon et al. 2006; Venkatasubramanian and Gupta 2006). Health sensors can use the interpulse interval (IPI) (Poon et al. 2006) or heart rate variance (HRV; Bao et al. 2005) as good sources for cryptographically random numbers, and the physiological values can be used as a one-time pad. Protocols that use these physiological values to encrypt a new key between a sensor pair have been developed (Bao and Zhang 2005; Venkatasubramanian and Gupta 2006). For instance, the Venkatasubramanian BSN protocol used a single message to send a new key to the neighbouring sensor node, as shown in Protocol 5.1.

$$A \rightarrow B: N_A, [N_A]_{RANDKEY}, RANDKEY \oplus SEV$$

Protocol 5.1 Venkatasubramanian BSN protocol

The new key, *RANDKEY*, is encrypted with the physiological value *SEV*, which is known only to sensors on a particular person. The sensor node *B* validates that *RANDKEY* is correct by verifying the MAC of $N_A$ where *A* is another sensor node. $N_A$ is a node generated by *A*. Venkatasubramanian and Gupta noted that finding additional cryptographically sound physiological values is still an open research problem. Another problem is that all the protocols developed with physiological values require all the sensor nodes to be able to measure the same phenomenon. Only cryptographically strong physiological values, such as IPI and HRV, can be used. Also, modern wireless technology (ultra wideband [UWB], radar; Staderini 2002) may be used to remotely capture the heart rate and could cause security risks when using only IPI and HRV to secure the communication. Other cryptographically weaker physiological values, such as blood pressure and iron count, are less susceptible to these remote attacks. Key establishment protocols that increase the number of available environmental or physiological values will enhance this research area.

## Using Physiological Data to Establish Keys

Even though PINs and passwords may not be used in body sensors, we show that password protocols can be used. Passwords have low randomness and therefore have similar characteristics to many SEVs. A four-digit PIN contains less than 14 bits of randomness and can be used in a password protocol. A typical password length of eight characters has less than 48 bits of randomness, if we randomly choose upper- and lower-case letters as well as the digits 0 to 9. We investigate the suitability of password protocols for the sensor environment. Password protocols have the special property of allowing secrets with small entropy to be used for key establishment. Password protocols are designed so that both offline and online attacks are not feasible. A feature or by-product of most password protocols is that if the password is compromised, any keys created before the password was compromised are not compromised.

Key sizes in sensor networks are small, normally 64 bits, so that the encryption or integrity tests do not consume a large amount of energy (Karlof,

Sastry, and Wagner 2004). Small key sizes lead to the need to update keys on a regular basis. Researchers have shown that password protocols can be implemented in sensor networks (Singh, Bhatt, and Muthukkumarasamy 2006). The password protocols used either a human-entered password on the sensor or an existing 64-bit key. Instead, we propose that environmental data can be used instead of using small keys or passwords. Also, previous approaches to sensor environments used only elliptic curve cryptography.

However, Rivest, Shamir and Adleman (RSA) password protocols that can be converted to Elliptic Curve Cryptography (ECC) have a large exponent such as 1024 bits. When the protocol is converted to use ECC, then 160-bit arithmetic is required. We instead investigated the Encrypted Key Exchange (EKE) protocol, which is an RSA-based password protocol where the exponent only needs to be 160 bits (Bellovin and Merritt 1992). The EKE protocol was chosen because other variants of password protocols require exponents of 1024 bits. The EKE protocol is diagrammatically shown in Protocol 5.2. A drawback of the EKE protocol is that it cannot use ECC (Boyd and Mathuria 2003).

| *Shared Information* | *Generator* $g$ of $G$ *where* | $p-1=qr$ |
|---|---|---|
| $A$ | $\leftrightarrow$ | $B$ |
| $r_A \in_R z_p$ | | |
| $t_A = g^{r_A}$ | $\rightarrow A,[[t_A]SEV_1]\rightarrow$ | $r_B \in_R z_p$ |
| | | $K_{AB} = t_A^{r_B}$ |
| $K_{AB} = t_B^{r_A}$ | $\leftarrow[[t_B]]_{SEV_2}[[n_B]]_{K_{AB}}\leftarrow$ | $t_B = g^{r_B}$ |
| | $\rightarrow[[n_A,n_B]]_{K_{AB}}\rightarrow$ | Verify $n_B$ |
| Verify $n_B$ | $\leftarrow[[n_A]]_{K_{AB}}\leftarrow$ | |

Protocol 5.2 Diffie–Hellman-based EKE protocol

The EKE protocol contains four messages. Node $A$ sends the first message to node $B$; the message contains the location of $A$ (the location value is in the clear), and the first part of Diffie–Hellman, $t_A$, is encrypted by the weak key $SEV_1$. After the first message is sent, node $B$ will calculate the second part of the Diffie–Hellman scheme and hence be able to calculate the session key $K_{AB}$. Node $B$ then sends the second part of the Diffie–Hellman scheme, encrypted by the weak key $SEV_2$, to node $A$. The nonce $n_B$ is also sent, encrypted by the session key $K_{AB}$. The last two messages authenticate both $A$ and $B$, as well as confirming that they have the session key $K_{AB}$. The encryption of $t_A$, $t_B$, $n_A$ and $n_B$ can be implemented with an exclusive-or function, as originally described by Bellovin and Merritt (1992).

Depending on which environmental value is measured, and how long the protocol will run, different SEVs may be used for the request and response. If the SEV stays constant, then both $SEV_1$ and $SEV_2$ will remain the same throughout the lifetime of the protocol. The EKE protocol is designed for a constant password throughout the running of the protocol, so similar or identical data for both $SEV_1$ and $SEV_2$ will not adversely affect the protocol.

The EKE protocol was originally designed to handle small-entropy secrets, so that offline and online dictionary attacks are infeasible for an adversary. Another useful feature is that even if the secrets $SEV_1$ or $SEV_2$ are compromised or available freely after the running of the key establishment protocol, the session key $K_{AB}$ will remain secure and safe.

Both nonces $n_A$ and $n_B$ are cryptographically strong random numbers, allowing the exclusive-or function to be used for encryption. If any nonce was not cryptographically strong, then either the $[[n_A]]_{K_{AB}}$ or $[[n_B]]_{K_{AB}}$ operation would allow an adversary to significantly reduce the number of valid $K_{AB}$ values. A characteristic of the EKE protocol is that the nonces are never sent out in the clear, since the nonces are used to encrypt the new key $K_{AB}$. The value of $p$ should be chosen wisely (Bellovin and Merritt 1992). It should be as close to $2N - 1$ as possible for the best security.

## Efficiency Issues and Experimental Evaluations

This section describes some of the efficiency requirements faced when implementing security on sensor networks. We implemented and compared different cryptographic primitives that can be used in body sensor security protocols on a Crossbow mica2 MPR2600 mote (Crossbow 2010).

Before comparing the different cryptographic primitives and the benefits that one implementation has over another, we created skeleton code based on TinyOS 2.x (TinyOS 2007). The skeleton code initialized the sensor node, and after the sensor was initialized, we obtained the initial time in milliseconds. We then ran a cryptographic primitive in a loop for 2000 iterations, before obtaining a new time. We subtracted the new time from the initial time to obtain the time (in milliseconds) that it took to run our cryptographic primitive for 2000 attempts. The elapsed time was then sent via the serial connection to a PC running a Linux® distribution where we had a Java® application reading the TinyOS packet from the serial port, and report that data to the user.

The key establishment protocols use exclusive-or (xor) to encrypt the new session key. We compared this method with other methods of encrypting the new session key for BSNs. Singh and Muthukkumarasamy (2007, 2008) have shown how RC5, SKIPJACK, HMAC-MD5, RSA and ECC cryptographic primitives can be used in BSNs; however, their work and comparisons were based on simulations and on TinyOS 1.x. We implemented these cryptographic primitives on real hardware, for TinyOS 2.x. To our knowledge these cryptographic primitives had not (until now) been ported to the latest version of TinyOS. Previously, Singh et al. did not separate the square root function

**TABLE 5.4**

Comparison of Cryptographic Algorithms

| Algorithm | Time | Lines of Code | Size (bytes) |
|---|---|---|---|
| XOR | 2 ms | 80 | 6340 |
| RC5 | 500 ms | 506 | 7168 |
| SKIPJACK | 700 ms | 697 | 8138 |
| HMAC-MD5 | 20 s | 507 | 19,054 |
| RSA | 43 s | 1456 | 7814 |
| SQRT | 80 s | 3366 | 8610 |
| ECC | 78 min | 5038 | 16,328 |

from the elliptic curve cryptography. However, in our comparison we found significant information when separating them.

Table 5.4 shows the time it took to run 2000 iterations of each of the algorithms. We have ordered the algorithms by the time elapsed. The column "Lines of Code" indicates the complexity for the coder to implement the algorithm.

The RC5 application took considerably more effort than the exclusive-or (xor) application. We found an RC5 implementation for TinyOS 1.x in the TinySEC library (Karlof et al. 2004); however, it had not yet been ported to TinyOS 2.x. Most of our effort was spent porting the code to the new platform.

The SKIPJACK application had similar problems to the RC5 application. There was an implementation for TinyOS 1.x in the TinySEC library but not for TinyOS 2.x. Once again, most of our effort was spent porting the code to the platform.

For the HMAC-MD5 application we could not find any previous implementations of it in any version of TinyOS. In this case we obtained code from RFC1321 (Rivest, 1992) and RFC2104 (Krawczyk, Bellare, and Canetti 1997) and ported the code to first the nesC language and then the TinyOS application. This took considerably more effort than either the RC5 or SKIPJACK implementations.

The RSA application also had similar problems to the RC5 and SKIPJACK implementations. We found code in the Deluge System (Dutta, Hui, Chu, and Culler 2006); however, the RSA code was based on TinyOS 1.x. Effort was required to port this code to TinyOS 2.x. We used a 160-bit exponent as required by the EKE protocol.

The SQRT application had the most difficulties since we implemented it from pseudocode rather than porting any code. We used Newton's method (Press, Teukolsky, Vetterling, and Flannery 2007) for finding square roots to implement the SQRT application.

The ECC application also had similar problems to the RSA, RC5 and SKIPJACK implementations. We ported an ECC library (A. Liu, Kampanakis, and Ning 2007) developed for TinyOS 1.x to TinyOS 2.x. The ECC application used 160-bit points, since password protocols that could be converted to use ECC require stronger keys (Singh and Muthukkumarasamy 2007).

The xor application is the quickest by several orders of magnitude compared to the other cryptographic primitives. The size of the application is smaller, and the number of lines of code is lower than for the other applications. The slowest application was the ECC application. This verifies existing research into the differences in speed for password protocols of RSA and ECC implementations in TinyOS simulators (Singh and Muthukkumarasamy 2007). The HMAC-MD5 application is the largest; however, the application was a straight port from the RFCs, where the code was not intended for sensors.

Even though exclusive-or and block cipher symmetric cryptography are suitable in an RSA environment, it is not suitable when converting to elliptic curves (Boyd and Mathuria 2003). The EKE (RSA) protocol is compared with an ECC-based password protocol (Singh et al. 2006). Using the RSA implementation (Dutta et al. 2006) from the Deluge system and porting it to the mica2 mote system, and using only 160-bit exponents, we found that the total number of CPU cycles was 147,879. Password protocols that can use an ECC implementation inherently require 1024-bit exponents when in RSA mode (Boyd and Mathuria 2003). When we measured the number of CPU cycles by using the ECC implementation for sensors (A. Liu et al. 2007), including an implementation of the square root function, we obtained a total of 18,790,689. The key size was 160 bits, which is equivalent to 1024 bits in RSA. When moving to the ECC protocols, more secure keys are required. There is a significant number of extra cycles in an ECC implementation over an RSA implementation.

We also examined the memory of the application, as shown in Table 5.5. The combination of .bss and .data segments uses SRAM, and the combination of .text and .data segments uses ROM. The .text contains the machine instructions for the application. The .bss contains uninitialized global or static variables, and the .data section contains the initialized static variables.

We used the values provided by the TOSSIM simulator (part of the TinyOS installation) to obtain an indication of the power consumption when sending a message. In our calculations we did not take into account any collision-avoidance times. On the mica2 mote, the cost of sending an extra 20 bytes was 28.1 μJ. There is a substantial start-up cost for each message sent, and then there is an added cost for every bit that is sent.

**TABLE 5.5**

Memory Comparison of RSA and ECC

| Memory | RSA | ECC |
|---|---|---|
| **ROM** | 1942 | 9720 |
| **RAM** | 177 | 859 |
| **.data** | 60 | 8 |
| **.bss** | 117 | 851 |
| **.text** | 1882 | 9712 |

## Future Directions

This chapter gave a brief introduction to each of the security countermeasures, with a detailed description of a way to implement a key management protocol, in order to gain an understanding of the difficulties in implementing security on sensors. There has been rapid growth in wireless technology and the creation of new protocols, such as low-power Bluetooth, ANT+ and ZigBee. Each device may use a different protocol to another device. Interoperability between the protocols will create new security and privacy vulnerabilities. With the emphasis put on protocol security, there has been little thought regarding the higher-level security issues.

Considerable work is still needed in each area of security, from assurance and administration to authorization, accountability and availability. The authors believe that the biggest improvements can be made in the assurance and privacy of the information in the entire system. The design, verification and testing of the security of a home healthcare system seem to have been neglected. As more implementations of systems become available, it will become more important to have assurance that the systems are secure.

Key management is another example of an immature area for home healthcare systems. Key management in a home health system has a requirement that elderly patients should be able to easily use the system. Key management should also be able to handle the loss of devices and provide the ability to update small keys or update keys on newly added devices. A suitable key management framework for a home healthcare system needs further investigation.

## Acknowledgements

Linux is a registered trademark of Linus Torvalds in the United States, other countries or both. Oracle and Java are registered trademarks of Oracle and/or its affiliates in the United States, other countries or both. Other company, product or service names may be trademarks or service marks of others.

## References

Axis. (2007, March). Setting up an IP-Surveillance system using Axis cameras and AXIS Camera Station software. http://www.axis.com/

Aziz, O., Lo, B., Darzi, A., & Yang, G.-Z. (2006). Introduction. In G.-Z. Yang (Ed.), *Body Sensor Networks* (pp. 1–40). New York: Springer-Verlag.

Balomenos, T. (2001). *User requirements analysis and specification of health status analysis and hazard avoidance artefacts*. DC FET project ORESTALA, Delieverable D02.

Bao, S.-D., & Zhang, Y.-T. (2005). A new symmetric cryptosystem of body area sensor networks for telemedicine. *Proceedings of the Conference the Japan Society of Medical Electronics & Biological Engineering, 44*, 654.

Bao, S.-D., Zhang, Y.-T., & Shen, L.-F. (2005). Physiological signal based entity authentication for body area sensor networks and mobile healthcare systems. In *27th Annual International Conference of the Engineering in Medicine and Biology Society, 2005* (pp. 2455–2458). Shanghai: IEEE Press.

Bao, S.-D., Zhang, Y.-T., & Shen, L.-F. (2006). A design proposal of security architecture for medical body sensor networks. In *BSN '06: Proceedings of the International Workshop on Wearable and Implantable Body Sensor Networks (BSN'06)* (pp. 84–90). Washington, DC: IEEE Computer Society.

Bellovin, S. M., & Merritt, M. (1992). Encrypted key exchange: Password-based protocols secure against dictionary attacks. In *IEEE Symposium on Research in Security and Privacy* (pp. 72–84). Oakland: IEEE Computer Society Press.

Blundo, C., Santis, A. D., Herzberg, A., Kutten, S., Vaccaro, U., & Yung, M. (1993). Perfectly-secure key distribution for dynamic conferences. *Lecture Notes in Computer Science, 740*, 471–486.

Borisov, N., Goldberg, I., & Wagner, D. (2001). Intercepting mobile communications: The insecurity of 802.11. In *Proceedings of the 7th Annual International Conference on Mobile Computing and Networking (MobiCom '01)* (pp. 180–189). New York: ACM.

Boulis, A., Han, C.-C., & Srivastava, M. B. (2003). Design and implementation of a framework for efficient and programmable sensor networks. In *Proceedings of the 1st International Conference on Mobile Systems, Applications, and Services (MobiSys '03)* (pp. 187–200). New York: ACM.

Boyd, C., & Mathuria, A. (2003). *Protocols for Authentication and Key Establishment*. Berlin and Heidelberg: Springer.

Bulusu, N. (2005). Introduction to wireless sensor networks. In N. Bulusu & S. Jha (Ed.), *Wireless Sensor Networks: A Systems Perspective*. Artech House.

Chan, H., & Perrig, A. (2005). PIKE: Peer intermediaries for key establishment in sensor networks. *INFOCOM* (pp. 524–535). IEEE Computer Society Press.

Chan, H., Perrig, A., & Song, D. (2003). Random key predistribution schemes for sensor networks. In *SP '03: Proceedings of the 2003 IEEE Symposium on Security and Privacy* (p. 197). Washington, DC: IEEE Computer Society.

Crossbow. (2010). Crossbow. http://www.xbow.com

Dutta, P. K., Hui, J. W., Chu, D. C., & Culler, D. E. (2006). Securing the deluge network programming system. In *Fifth International Conference on Information Processing in Sensor Networks (IPSN'06)*.

Espina, J., Falck, T., & Mülhens, O. (2006). Network topologies, communication protocols, and standards. In *Body Sensor Networks* (pp. 145–182). New York: Springer.

Gehrmann, C., Persson, J., & Smeets, B. (2004). *Bluetooth Security*. Boston: Artech House.

Hämäläinen, P., Kuorilehto, M., Alho, T., Hännikäinen, M., & Hämäläinen, T. D. (2006). Security in wireless sensor networks: Considerations and experiments. In S. Vassiliadis, S. Wong & T. D. Hämäläinen (Ed.), *SAMOS* (pp. 167–177). Springer.

Hampapur, A., Brown, L., Connell, J., Haas, N., Lu, M., Merkl, H., et al. (2004). S3-R1: The IBM smart surveillance system-release 1. In *ETP '04: Proceedings of the 2004 ACM SIGMM Workshop on Effective Telepresence* (pp. 59–62). New York: ACM Press.

Heinzelman, W. B., Murphy, A. L., Carvalho, H. S., & Perillo, M. A. (2004). Middleware to support sensor network applications. *IEEE Network, 18*, 6–14.

Housley, R., & Arbaugh, W. (2003). Security problems in 802.11-based networks. *Communications of the ACM, 46*(5), 31–34.

ISC. (2007). CISSP Organization. https://www.isc2.org/

ISO11770-2 (Artist). (1996). *Information Technology—Security Techniques—Key Management—Part 2: Mechanisms Using Symmetric Techniques ISO/IEC 11770-2.* http://www.iso.org/iso/iso_catalogue/catalogue_tc/catalogue_detail.htm?csnumber=25290

Karlof, C., Sastry, N., & Wagner, D. (2004). TinySec: A link layer security architecture for wireless sensor networks. In *SenSys '04: Proceedings of the 2nd International Conference on Embedded Networked Sensor Systems* (pp. 162–175). Baltimore, MD: ACM Press.

Krawczyk, H., Bellare, M., & Canetti, R. (1997, February). HMAC: Keyed-Hashing for Message Authentication. http://www.ietf.org/rfc/rfc2104.txt

Kuorilehto, M., Hännikäinen, M., & Hämäläinen, T. D. (2005). A survey of application distribution in wireless sensor networks. *EURASIP Journal on Wireless Communications and Networking*, 2005(4), 774–788.

Kuryloski, P., Giani, A., Giannantonio, R., Gilani, K., Gravina, R., Seppa, V. P., et al. (2009, 3–5 June). DexterNet: An open platform for heterogeneous body sensor networks and its applications. In *Proceedings of the 2009 Sixth International Workshop on Wearable and Implantable Body Sensor Networks (BSN '09)* (pp.92–97). Washington, DC: IEEE Computer Society.

Liu, A., Kampanakis, P., & Ning, P. (2008). TinyECC: A configurable library for elliptic curve cryptography in wireless sensor networks. In *Proceedings of the 7th International Conference on Information Processing in Sensor Networks (IPSN '08)* (pp. 245–256). Washington, DC: IEEE Computer Society.

Liu, A. and Ning, P. (2007). TinyECC: A configurable library for elliptic curve cryptography in wireless sensor networks. Technical Report TR-2007-36, North Carolina State University, Department of Computer Science.

Lo, B. P. L., & Yang, G.-Z. (2005). Key technical challenges and current implementations of body sensor networks. *International Workshop on Wearable and Implantable Body Sensor Networks* (pp. 1–5). London.

Mainwaring, A., Culler, D., Polastre, J., Szewczyk, R., & Anderson, J. (2002). Wireless sensor networks for habitat monitoring. In *WSNA '02: Proceedings of the 1st ACM International Workshop on Wireless Sensor Networks and Applications* (pp. 88–97). Atlanta, GA: ACM Press.

Malan, D., Fulfrod-Jones, T., Welsh, M., & Moulton, S. (2004, April). CodeBlue: An ad hoc sensor network infrastructure for emergency medical care. Paper presented at the Workshop on Wearable and Implantable Body Sensor Networks.

Marcy, H. O., Agre, J. R., Chien, C., Clare, L. P., Romanov, N., & Twarowski, A. (1999). Wireless sensor networks for area monitoring and integrated vehicle health management applications. In *AIAA Guidance, Navigation, and Control Conference and Exhibit* (Vol. 1) (pp. 1–11). Portland, OR.

Oliver, N., & Flores-Mangas, F. (2006, 3–5 April). HealthGear: A real-time wearable system for monitoring and analyzing physiological signals. Paper presented at the International Workshop on Wearable and Implantable Body Sensor Networks, 2006 (BSN 2006).

Perrig, A., Szewczyk, R., Wen, V., Culler, D., & Tygar, J. D. (2001). SPINS: Security protocols for sensor networks. In *Seventh Annual International Conference on Mobile Computing and Networks (MobiCOM 2001)*. Rome, Italy.

Poon, C. C. Y., Zhang, Y.-T., & Bao, S.-D. (2006). A novel biometrics method to secure wireless body area sensor networks for telemedicine and m-health. *IEEE Communications Magazine, 44*, 73–81.

Press, W. H., Teukolsky, S. A., Vetterling, W. T., & Flannery, B. P. (2007). Root finding and nonlinear sets of equations. In *Numerical Recipes: The Art of Scientific Computing*. (pp. 442–486), Cambridge, UK: Cambridge University Press.

Ren, K., Zeng, K., & Lou, W. (2006). A new approach for random key predistribution in large-scale wireless sensor networks: Research articles. *Wireless Commununications and Mobile Computing, 6*(3), 307–318.

Rivest, R. (1992, April). The MD5 Message-Digest Algorithm. http://tools.ietf.org/html/rfc1321

Sastry, N., & Wagner, D. (2004). Security considerations for IEEE 802.15.4 networks. *Proceedings of the 3rd ACM Workshop on Wireless Security*.

Schwiebert, L., Gupta, S. K. S., & Weinmann, J. (2001). Research challenges in wireless networks of biomedical sensors. In *MobiCom '01: Proceedings of the 7th Annual International Conference on Mobile Computing and Networking* (pp. 151–165). Rome, Italy: ACM Press.

Singh, K., Bhatt, K., & Muthukkumarasamy, V. (2006). Protecting small keys in authentication protocols for wireless sensor networks. In *Proceedings of the Australian Telecommunication Networks and Applications Conference* (pp. 31–35). Australia: The University of Melbourne.

Singh, K., & Muthukkumarasamy, V. (2007). Authenticated key establishment protocols for a home health care system. In *Proceedings of the Third International Conference on Intelligent Sensors, Sensor Networks and Information Processing (ISSNIP)*. (pp. 353–358). IEEE Press.

Singh, K., & Muthukkumarasamy, V. (2008). Performance analysis of proposed key establishment protocols in multi-tiered sensor networks. *Journal of Networks, 3*(6), 13–28.

Staderini, E. M. (2002). UWB radars in medicine. *IEEE Aerospace and Electronic Systems Magazine, 21*, 13–18.

TinyOS. (2007). An operating system for sensor motes. http://www.tinyos.net/

USA. (2003, May). Summary of HIPAA Health Insurance Probability and Accountability Act. US Department of Health and Human Service.

Venkatasubramanian, K. K., & Gupta, S. K. S. (2006). Security for pervasive health monitoring sensor applications. In *ICISIP '06: Proceedings of the 4th International Conference on Intelligent Sensing and Information Processing* (pp. 197–202). Bangalore, India: IEEE Press.

Wood, A., Virone, G., Doan, T., Cao, Q., Selavo, L., Wu, Y., et al. (2006). *ALARM-NET: Wireless Sensor Networks for Assisted-Living and Residential Monitoring*. Department of Computer Science, Virginia: University of Virginia.

Yeatman, E., & Mitcheson, P. (2006). Energy scavenging. In *Body Sensor Networks*. (pp. 183–218). New York: Springer-Verlag.

# 6

# *Flexible and Wearable Chemical Sensors for Noninvasive Biomonitoring*

**Hiroyuki Kudo and Kohji Mitsubayashi**

**CONTENTS**

Introduction .......... 139
Flexible Devices for Healthcare Networks .......... 140
Biomonitoring for Information Systems .......... 140
Flexible Devices for Biomonitoring .......... 140
Flexible Oxygen Sensors for Transcutaneous Oxygen Monitoring .......... 142
Transcutaneous Gas at Body Surfaces .......... 142
Flexible Oxygen Sensors .......... 143
Transcutaneous Oxygen Monitoring with a Flexible Oxygen Sensor .......... 144
Transcutaneous Oxygen Monitoring at the Conjunctiva .......... 145
Flexible Biosensors .......... 148
Continuous Glucose Monitoring .......... 148
Flexible Glucose Sensors .......... 149
Tear Glucose Monitoring at the Eye .......... 152
References .......... 154

## Introduction

The field of advanced bioinstrumentation has in recent years been responsible for the development of sensor technology which is capable of providing highly reliable biological information. These devices have also been improved to maximize comfort and convenience. Current research has been focusing on further improving the flexibility and biocompatibility of biosensor devices for the purpose of ongoing patient monitoring. Such devices would allow for noninvasive bioinstrumentation and consequently reduce their impact on day-to-day activities. In particular, noninvasive chemical sensors for the preceding purposes are strongly requested. To this end, a number of oxygen sensors and enzyme-based biosensors based on functional polymers such as polypropylene and polydimethylsiloxane have been developed. This chapter describes various flexible and wearable chemical sensors and their application in noninvasive biomonitoring.

## Flexible Devices for Healthcare Networks

### Biomonitoring for Information Systems

As evident from the rapid spread of information systems and two-way communication infrastructure that began in the early 1990s, a global information age has come upon us. Advances in information technologies have led to drastic changes in social systems including logistics, emergency services and other essential utilities within the developed world. From 2000 onwards, contactless integrated circuit (IC) smart cards began to gradually replace conventional magnetic cards due to their large data storage capacity and robust security features. These changes afforded greater security to electronic information and have enabled us to carry our own personal data for administrative, medical or commercial network services. More recently, there has been a growing trend towards the utilization of biological information in information systems, as seen in the automobile, entertainment, military and healthcare industries.

Despite technological advances, many developed nations have seen declining birth rates and an overall ageing of the population, which constitute a serious problem in these countries. This social phenomenon has contributed to a shrinking workforce that continues to threaten the sustainability of social welfare systems, particularly healthcare infrastructure. For this reason, advances in medical technologies to improve public health have been strongly sought. Noninvasive biomonitoring is a key technology used in preventive healthcare that has both economic and health benefits.

Modern bioinstrumentation techniques were developed in response to demands for *on-site* monitoring. Fingerprint identification systems have been one of the most successful biometric systems developed to this end. They provide real-time, in vivo, on-site recognition. Moreover, fingerprint identification is a network-friendly system. These features are also necessary for technologies aimed at improving healthcare, including those that monitor chemical information such as fluids and volatile substances from the body.

### Flexible Devices for Biomonitoring

Recently, many kinds of mobile information devices suitable for everyday wear have been developed. These devices are often computerized and called *wearable* devices. Although some of these utilize biological information (e.g., voice or fingerprint recognition), there are still many restrictions in the development of chemical sensors. Many physical sensors make excellent wearable devices because they are microfabricated, reliable and durable. Wearable chemical sensors for biomonitoring similarly need to be safe and comfortable to wear, and they additionally need to be highly accurate. Some chemical sensors, such as Clark's oxygen electrode, also require the presence of other solutions. This reduces the wearability of these devices. Flexibility is one of the more important features required for comfort and safety needs in on-site biomonitoring. Flexible sensors are ones that follow the transformation and expansion of movement of the body without compromising their ability to measure biological information. Additionally, wearers can use these devices with minimum awareness of their presence. These types of sensors are usually fabricated using polymer microfabrication techniques. Another important feature is biocompatibility, which allows the sensors to be attached on any site of the human body. This is expected to build a network system such as that shown in Figure 6.1.

As shown in this figure, sensors and other devices (storage, a controller, etc.) are expected to create a local network in the body. However, it is not reasonable to build a wired network in these sorts of systems. Several types of radio technology for transmitting information such as biological information and control signals can be considered. These include Bluetooth, ultra-wide band (UWB) and those that use the human body as a transmission path. In particular, human body communication technology (baud rate: several tens of Kbps) using an electric field in proximity to the human body is one technique that has been proposed so far.

Various related technologies have also been developed for flexible sensor devices. In order to construct a flexible device, an effective wiring method for a flexible substrate is also necessary. Gray et al. (2004) developed a wavy gold interconnection (line width: 10 μm) formed in a 400 μm thick polydimethylsiloxane (PDMS) membrane. The gold microstructure was formed on a sacrificial photoresist layer by electrodeposition and was sandwiched with PDMS. The resulting material could be stretched up to 50 per cent, as shown in Figure 6.2 (Gray et al. 2004). Khang et al. (2006) also reported on a stretchable wavy transistor and a diode that was transferred to a flexible PDMS

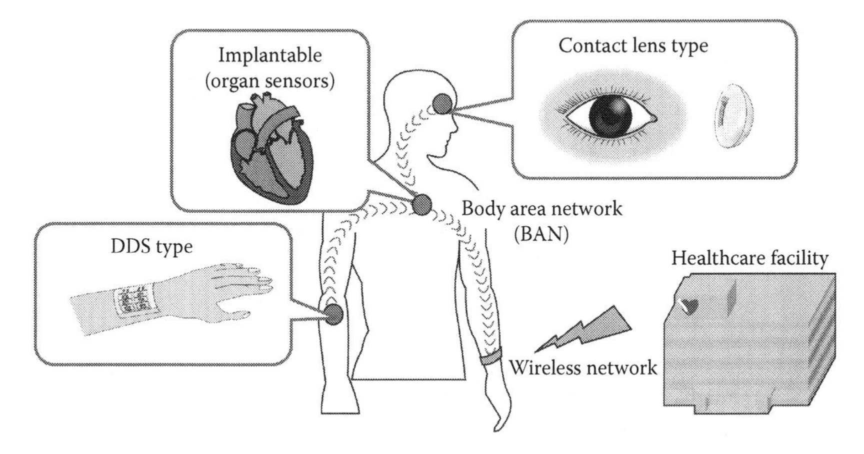


**FIGURE 6.1**
Flexible and wearable sensors in a body area network.

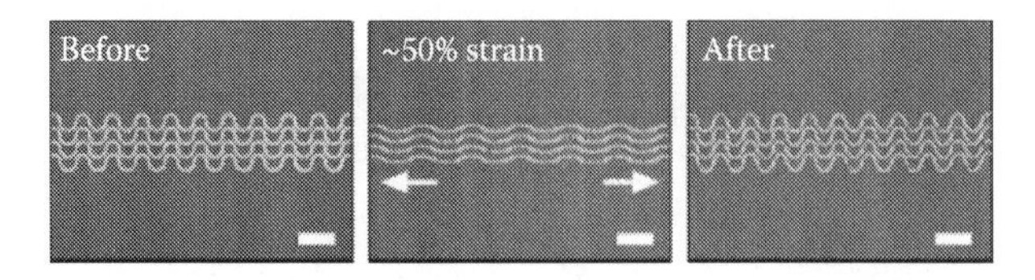


**FIGURE 6.2**
Flexible wiring formed in polydimethylsiloxane (PDMS) membrane. (From Gray, D.S., et al., *Adv. Mater.*, 16, 393–396, 2004.)

membrane. These types of soft devices represent a new concept in biodevices that is different from an extension of solid-state sensors.

## Flexible Oxygen Sensors for Transcutaneous Oxygen Monitoring

### Transcutaneous Gas at Body Surfaces

Humans consume oxygen and release carbon dioxide by the exchange of gases in the lungs. Gas exchange takes place between millions of alveoli in the lungs and the capillaries that envelop them. The partial pressure of oxygen in arterial blood ($PaO_2$) is known to reflect the severity of lung disorders such as pulmonary embolism or atelectasis. It is also used in the diagnosis, treatment and management of respiratory depression. This parameter is usually measured by obtaining blood from an artery. This involves puncturing an artery, often at the wrist, and drawing a small volume of blood with a syringe. Normally, $PaO_2$ is kept within a range of 80 to 100 mmHg, and abnormal $PaO_2$ values are seen in hypoxemia or hyperoxemia. At a $PaO_2$ of less than 60 mmHg, supplemental oxygen should be administered. In premature neonates, it is especially important to monitor arterial oxygen levels continuously to ensure they are maintained at normal levels as these infants have immature cardiopulmonary systems.

The transcutaneous partial pressure of oxygen ($TcPO_2$) can be monitored (Baumberger and Goldfriend 1951; Clark 1956; Evans and Naylor 1967; Huch et al. 1973; Rithalia 1991) using a polarographic technique that is noninvasive and provides a continuous measurement of blood oxygen through the skin. This method of monitoring arterial oxygen pressure is usually used to prevent retinopathy in premature neonates in neonatal intensive care units (Huch et al. 1974; Imura and Baba 1975; Duc 1975). Although it is a relatively painless method, this involves attaching rigid cylindrical cells to the body surface with an adhesive plaster and heating the measurement site to 43.5°C. This induces skin rashes and general discomfort. A less invasive alternative involves utilizing a thin and flexible oxygen sensor rather than a rigid electrode cell. A flexible oxygen sensor that contains electrolyte at the sensing region has been developed using microelectromechanical systems (MEMS) technologies and applied in $TcPO_2$ monitoring.

## Flexible Oxygen Sensors

The wearable oxygen sensor is a filmlike version of Clark's oxygen electrode that is made up of flexible functional polymer membranes such as polypropylene (PP) and fluorinated ethylene-propylene (FEP), on which film electrodes and the electrolyte layer are formed (Mitsubayashi et al. 2003). Figure 6.3a shows the structure of a flexible oxygen sensor used in $TcPO_2$ monitoring. Since the electrolyte layer is sandwiched between gas-permeable (sensitive) and nonpermeable (not sensitive) membranes, oxygen discharged from the body surface can be measured as a reduction current at the working electrode.

A gas-permeable, 25 μm thick FEP membrane was fixed on a dummy substrate, and electrode patterns (working and reference/counter electrodes) were formed by photolithography. Pt (thickness: 200 nm, width: 2 mm) and Ag (thickness: 300 nm, width: 5 mm) were sputtered on the FEP membrane. In order to improve adhesion, Pt was also used as an interlayer between Ag and FEP. They were patterned using a lift-off process, and the Ag layer was chloridized via electrochemical methods (110 mV, 4.5 min) with 0.1 mmol/L HCl solution (Suzuki et al. 1999) to form the Ag/AgCl counter/reference electrodes. The membrane was then cut into 15 × 50 mm strips that each contained a pair of Pt and Ag/AgCl electrodes. A membrane filter (pore size: 12 μm) was soaked with electrolyte solution (0.1 mmol/L KCl) and placed on the electrode. Following this, the KCl membrane was attached by covering it with a nonpermeable thermoplastic membrane (thickness: 50 μm) using a heat-sealing system. As shown in Figure 6.3b, the oxygen sensor was sufficiently flexible to attach to a human body.

The flexible oxygen sensor measures oxygen levels by a two-electrode amperometric method. A constant potential of –600 mV vs. Ag/AgCl was applied to the Pt electrode and tested by measuring dissolved oxygen (DO).

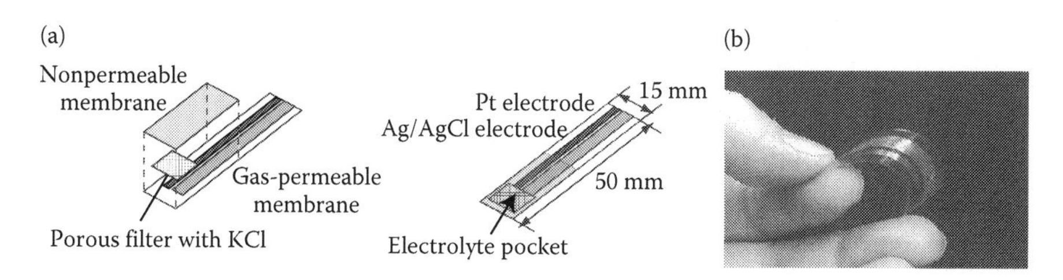


**FIGURE 6.3**
(a) The structure of a flexible oxygen sensor. (Modified from Mitsubayashi, K., et al., *Sens. Actuators B* 95, 373–377, 2003.) and (b) the photograph of the flexible oxygen sensor being bent.

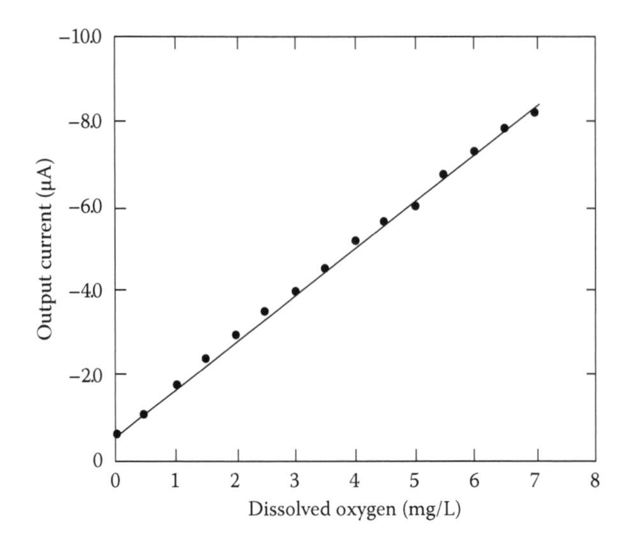


**FIGURE 6.4**
Calibration curve for dissolved oxygen. (From Kudo, H., et al., *Biomed. Microdev.*, 9, 1–6, 2007. With permission.)

The calibration curve of the flexible oxygen sensor is shown in Figure 6.4. In this figure, the sensor output is presented as a displacement of current value. The current output of the flexible sensor device was linearly related to the DO concentration from 0.0 to 7.0 mg/L, with a correlation coefficient of 0.998 deduced by regression analysis, as derived from

$$\text{Sensor output } (\mu\text{A}) = 0.60 + 1.11[\text{DO (mg/L)}].$$

The performance of the thin flexible sensor was similar to that of the DO meters currently on the market.

## Transcutaneous Oxygen Monitoring with a Flexible Oxygen Sensor

Transcutaneous oxygen monitoring with a flexible oxygen sensor device was conducted in healthy male volunteers (Mitsubayashi et al. 2003; Kudo et al. 2007). Subjects were instructed as to how the determination was to be performed and were asked to maintain natural behaviour during the procedure. Informed consent was obtained. Figure 6.5a shows the experimental set-up for $TcPO_2$ monitoring with the flexible oxygen sensor. The flexible oxygen sensor was attached to the skin of the forearm, together with a temperature sensor and a skin warmer (Figure 6.5b). Skin temperature was controlled at 40°C, which is lower than that for general $TcPO_2$ monitoring. The output of the flexible oxygen sensor and the skin temperature were monitored continuously during inhalation of oxygen (21–60–21%). Oxygen (60%) was applied to the subject using a mouthpiece.

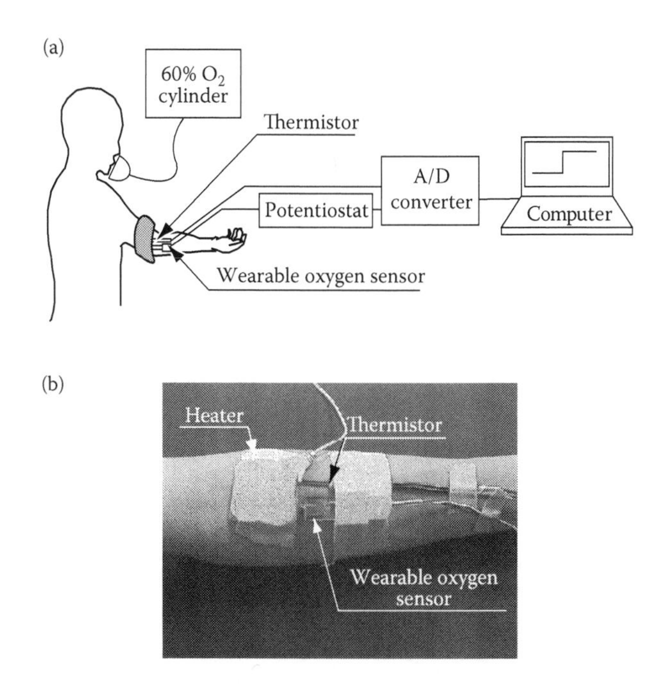


**FIGURE 6.5**
Transcutaneous oxygen monitoring with a flexible oxygen sensor. (a) experimental setup and (b) photograph of the forearm on which the flexible oxygen sensor was attached. (From Kudo, H., et al., *Biomed. Microdev.*, 9, 1–6, 2007. With permission.)

The output of the flexible oxygen sensor and the temperature of the forearm are shown in Figure 6.6. As shown, the output current was sufficiently stable during the test. The output current of the flexible oxygen sensor significantly increased when 60% oxygen was applied and decreased to baseline levels on room air (21%). These changes indicate that $TcPO_2$ increased during inhalation of a high concentration of oxygen. As indicated in the figure, the skin surface temperature was controlled at approximately 40°C. Owing to its flexibility, the application of the sensor did not cause physical discomfort to the subjects such as a skin rash or burn. The physical burden of monitoring $TcPO_2$ is thus reduced when using flexible biometric sensors.

The transmission of arterial blood oxygen depends on the region of body (Imura and Baba 1975; Hagihara 1984). Measurement of arterial oxygenation using a flexible oxygen sensor provides similar results at other body sites. Higher outputs were observed at the wrist compared to the upper arm (and elbow). The forearm was found to be a particularly suitable site for applying the device.

## Transcutaneous Oxygen Monitoring at the Conjunctiva

Sensor flexibility allows biomonitoring at any surface of the body. Although the forearm was well suited to $TcPO_2$ monitoring, the conjunctiva is a known alternate site for monitoring arterial blood oxygen (Kwan and Fatt 1971; Isenberg

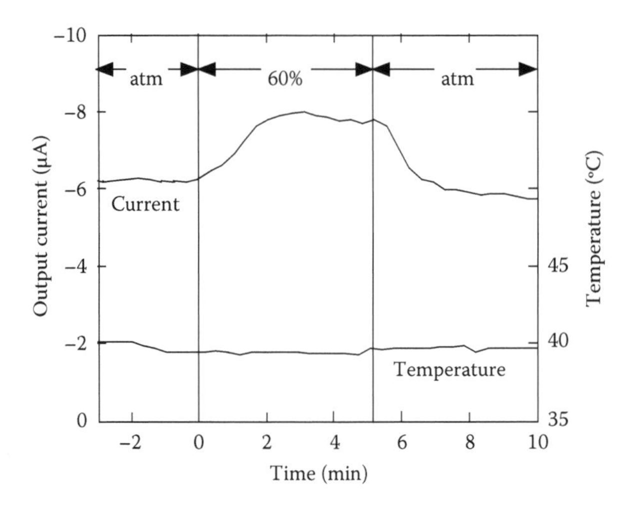


**FIGURE 6.6**
Typical output of the transcutaneous oxygen measurement by the flexible oxygen sensor attached to the forearm of a subject inhaling controlled oxygen (21–60–21%, sequentially). (From Kudo, H., et al., *Biomed. Microdev.*, 9, 1–6, 2007. With permission.)

and Shoemaker 1983; Webb et al. 1985; Brown and Vender 1988; Haljamae et al. 1989; Podolsky et al. 1989; Isenberg et al. 2002). The conjunctiva is a clear mucous membrane that supplies oxygen to the cornea (Weissman et al. 1981; Shoemaker and Lawner 1983). Conjunctival oxygen monitoring using a flexible oxygen sensor (Iguchi et al. 2005) was carried out using a Japanese white rabbit (16 months of age, female). In order to attach the device to the conjunctiva, the width of the flexible oxygen sensor was reduced to 3 mm. The rabbit was placed in a fixing apparatus, and the flexible oxygen sensor was held between the conjunctiva and the eyeball (Figure 6.7). Note that the sensing region was oriented to face the conjunctiva. No sedatives or anaesthetics were used in this experiment. In order to prevent misalignment, the terminal part of the sensor was fixed with a custom holder, which was attached to the fixing apparatus. Due to the high levels of gas transmission that occur at the conjunctiva, oxygen monitoring was conducted without heating. The oxygen supply was increased from 60 to 90% through a special respiratory mask at an interval of 4 min.

The output current significantly increased while high-concentration oxygen (60% and 90%) was applied. Figure 6.8b shows a typical current response for conjunctival oxygen monitoring using the flexible sensor. Similar to $TcPO_2$ monitoring at the forearm, the current was sufficiently stable. Figure 6.8a and c show the sensor's response to standard oxygen before (a) and after (c) its application to the conjunctiva. There was no significant change in characteristics after conjunctival monitoring. The result demonstrates the possibility of $PaO_2$ assessment without skin warming by conjunctival oxygen monitoring. In particular, it would be useful in monitoring adults with a low transmission of $TcPO_2$ because of a thick horny layer.

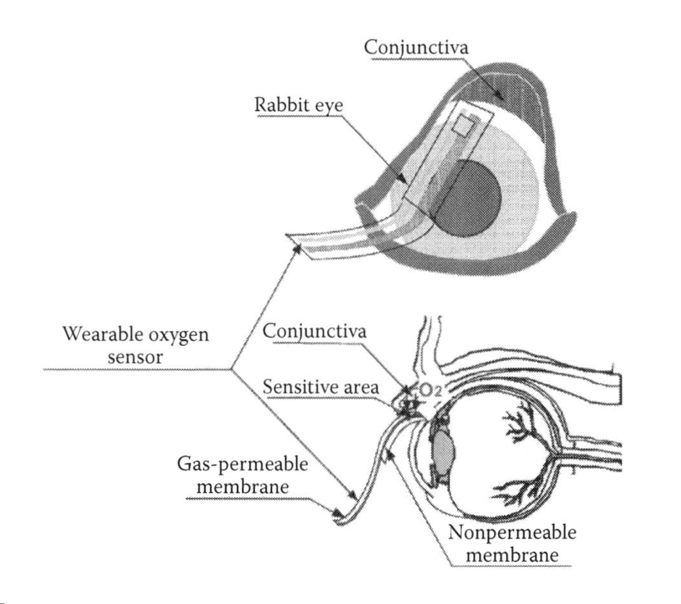


**FIGURE 6.7**
Sensor attachment for rabbit conjunctival oxygen monitoring. (From Iguchi, S., et al., *Sens. Actuators B*, 108, 733–737, 2005. With permission.)

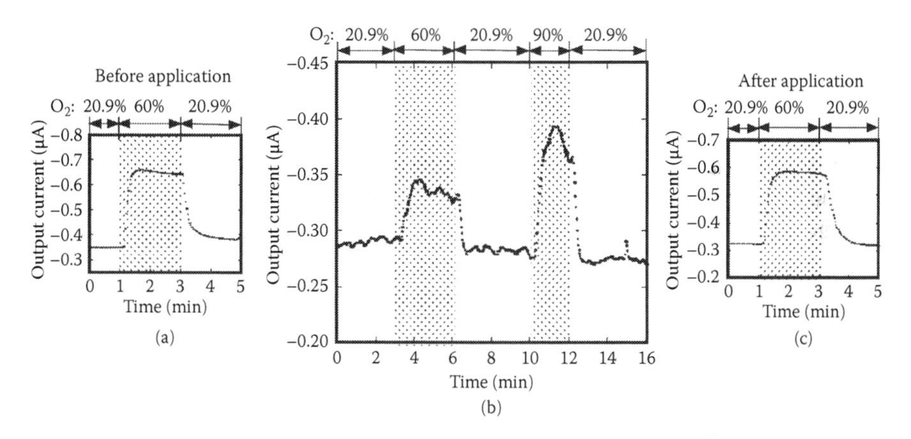


**FIGURE 6.8**
Typical response of transcutaneous oxygen monitoring at the rabbit conjunctiva by the wearable sensor. (a) Sensor characteristics before application to the conjunctiva. (b) Sensor response for rabbit conjunctival transcutaneous oxygen monitoring during inhalation of a high concentration of oxygen. (c) Sensor characteristics after the experiment. (From Iguchi, S., et al., *Sens. Actuators B*, 108, 733–737, 2005. With permission.)

Two types of flexible oxygen sensors were used for $TcPO_2$ monitoring and for conjunctival oxygen monitoring. They were comparable except for differences in the sensor width. Application at both sites demonstrates that the thin and flexible structure of the sensor is compatible with most surfaces of the body.

## Flexible Biosensors

### Continuous Glucose Monitoring

Diabetes mellitus type 2 is well known as a lifestyle-related disease, and it is becoming a serious health problem worldwide. Diabetes occurs when there is a defect in insulin secretion (type 1) or there are both a defects in insulin secretion and insulin resistance (type 2). Regardless of the type, patients with diabetes are suffered from chronic elevated blood-sugar levels (hyperglycemia; Leahy 1996). Type 2 diabetes often results from excess body weight and physical inactivity. Although several symptoms of diabetes have been recognized, such as excessive thirst or urination, hunger, weight loss and acetone-smelling breath, patients often do not recognize these until the disease has reached an advanced stage. Both type 1 and type 2 diabetes can cause serious microvascular complications (diabetic retinopathy, nephropathy and neuropathy; Fuller et al. 1983; Campbell and White 2002). Insulin therapy is a widespread treatment for diabetes and is delivered by injection. Since patients with type 1 diabetes have an absolute insulin deficiency, they depend on injected insulin for survival. Individuals with type 2 diabetes, which is also known as non-insulin-dependent diabetes mellitus (NIDDM), do not require exogenous insulin to survive in most cases. However, the failure of antidiabetic drugs may warrant commencement of insulin therapies to maintain normal glucose levels.

Self-monitoring of blood glucose with portable glucose meters is widely used for maintaining appropriate blood glucose levels (euglycemia). This is usually performed by piercing the pulp of the finger to draw blood, which is applied to a reagent strip for analysis in a portable meter (Pickup et al. 2005). Although it is a relatively painless and reliable measurement method, it is inconvenient and requires patient compliance (Badugu et al. 2003). Since insulin injections must be delicately balanced with meals and exercise, frequent blood glucose monitoring is required in these patients. Patients with NIDDM are also recommended to record their blood glucose levels at least once a day. While the costs of such testing are relatively high, they are reasonable, as the total medical costs of treating diabetes-related complications can be reduced by such strategies.

Many newer types of blood sugar-monitoring techniques have been reported to improve the convenience of monitoring (March et al. 1982; Rabinovitch et al. 1982; D'Costa et al. 1986; Brooks et al. 1988; Clarke et al. 1988; Meadows and Schultz 1988; Schier et al. 1988; Trettnak and Wolfbeis and 1989; Heise et al. 1994; Tolosa et al. 1997, 1999; D'Auria et al. 2000). A continuous glucose monitor (Bailey et al. 2007; Garg et al. 2007) is one of the most practical solutions. Continuous glucose meters measure the glucose level of interstitial fluid, which can be obtained by the minimally invasive placement of tiny sensors under the skin. This principle also makes it possible to monitor blood glucose continuously in other biological fluids such as tears, mucus, sweat and saliva (Man et al. 1979; Daum and Hill 1982; Romano and Rolant 1988; Mitsubayashi et al. 1994). This indicates that glucose monitoring on body surfaces may be possible. In particular, flexible chemical sensors measuring tear fluid at the eye have been considered feasible because of the correlation between glucose levels in tears and blood (Sen and Sarin 1980; Chen et al. 1996; Jin et al. 1997). In order to measure glucose in tear fluid continuously, sensors would have to be in direct contact with the cornea. Flexible devices for biomonitoring at the eye have been developed (Kudo et al. 2006; Mitsubayashi et al. 2006; Kudo et al. 2008). Preclinical experiments in this field have also been reported (Iguchi et al. 2007). These have shown that it may be possible to monitor not only glucose at the eye but also various other biological information at various sites of the human bodies.

## Flexible Glucose Sensors

The most promising method for continuous glucose monitoring has been enzyme-based biosensors. The glucose biosensor uses the enzyme glucose oxidase (GOD), which is responsible for catalyzing the conversion of β-D-glucose and oxygen to D-glucono-1,5-lactone and hydrogen peroxide. Glucose is usually measured by quantifying the production of hydrogen peroxide or the consumption of oxygen by the GOD reaction using electrochemical or spectrophotometric methods. A flexible glucose sensor can therefore be designed by immobilizing GOD at the sensing region of the flexible oxygen sensor. Figure 6.9 illustrates the enzyme-immobilization method used to construct a flexible glucose sensor. GOD was immobilized on the surface of a flexible oxygen sensor that had been treated with an aminopropylsilane monolayer. A mixture of phosphate buffer containing GOD and water-soluble photosensitive resin (AWP: Azide-unit pendant Water-soluble Photopolymer, Toyo Gosei Kogyo Co., Ltd., Japan) was then cured by illuminating it with ultraviolet light.

The flexible biosensor demonstrated a sufficient capability for measurement of glucose in tear fluids. The calibration curve of the flexible biosensor is shown in Figure 6.10. The output current was linearly related to the glucose concentration from 0.025 to 1.475 mmol/L, with a correlation coefficient of 0.998 as deduced by regression analysis. This is represented by

$$\text{Output current } (\mu\text{A}) = -\,0.016 + 0.491\ [\text{glucose (mmol/L)}].$$

As indicated in the figure, the flexible biosensor had an appropriate calibration range including normal glucose levels in the tear fluid (0.14 mmol/L).

In regard to biomonitoring at the eye, the utilization of special materials in flexible biosensors has been reported (Mitubayashi et al. 2006). Since the tear glucose sensor is attached to the eye, it is expected to hinder vision. However, flexible biosensors with an optically transparent working electrode

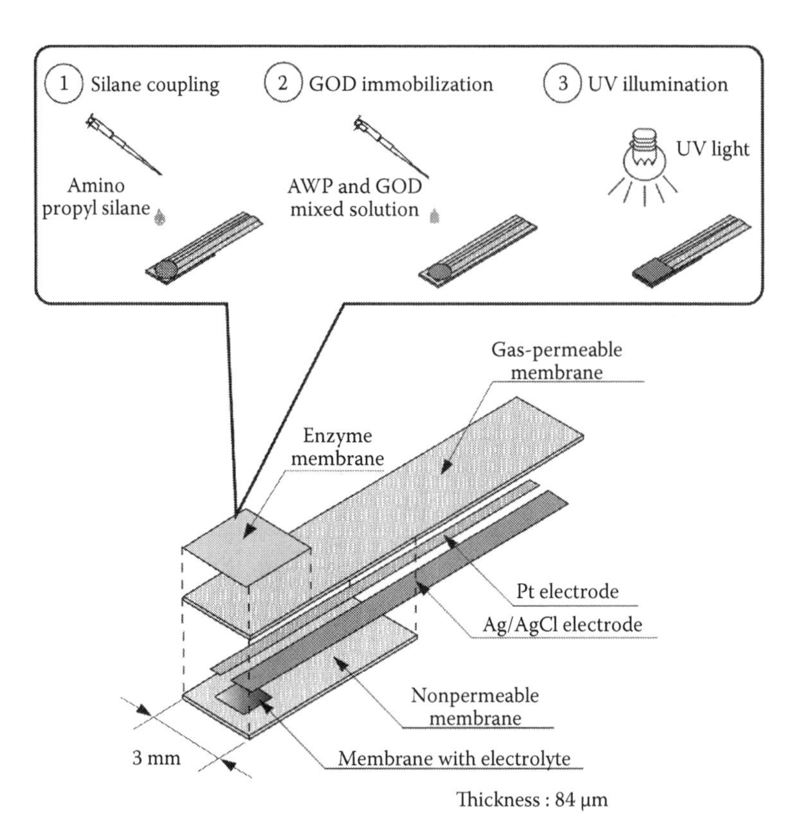


**FIGURE 6.9**
Structure of the flexible glucose sensor. (From Iguchi, S., et al., *Biomed. Microdev.*, 9, 603–609, 2007. With permission.)

have been reported. The transparent biosensor utilizes an indium-tin-oxide (ITO) electrode formed on a polyethylene-terephthalate (PET) film with a thickness of 0.1 mm. GOD was immobilized by a covalent binding method with glutaraldehyde on the oxygen-sensing region (Figure 6.11). The thinner glucose sensor (total thickness: 225 μm) was sufficiently flexible to be applied to the skin surface and was also optically transparent as shown in the photograph. By spectrophotometric analysis, the absorbance was confirmed to be lower than 0.6 Abs at the visible wavelength (400–700 nm).

Polymers that are more biocompatible have been investigated for flexible biosensor applications (Kudo et al. 2006). Figure 6.12 shows a flexible biosensor fabricated using PDMS and poly(2-methacryloyloxyethyl phosphorylcholine (MPC)–co-dodecyl methacrylate (DMA)) (PMD). PDMS (also known as silicone) is a polymeric organosilicon compound that is widely used due to its biocompatible properties and ease of processing (Ostuni et al. 2000; De Silva et al. 2004; Peterson et al. 2005). PDMS is also used for micro total analysis systems (Thiébaud et al. 2002; Arakawa et al. 2007) in bioengineering.

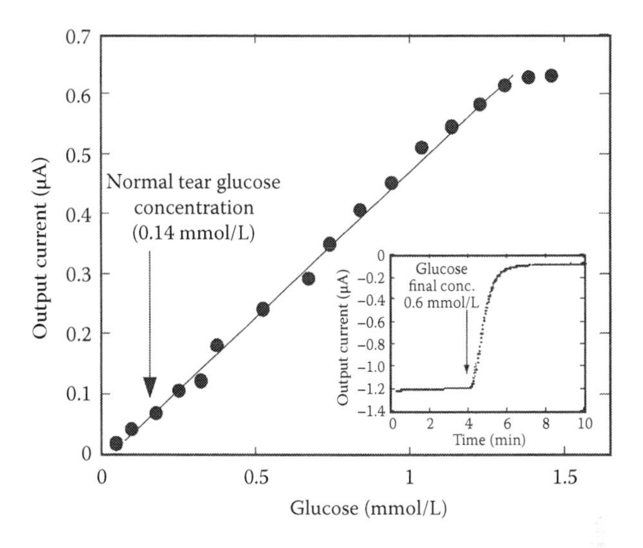


**FIGURE 6.10**
Calibration curve of the flexible glucose sensor (arrowhead, glucose concentration [0.14 mmol/L] in normal tear fluid). (From Iguchi, S., et al., *Biomed. Microdev.*, 9, 603–609, 2007. With permission.)

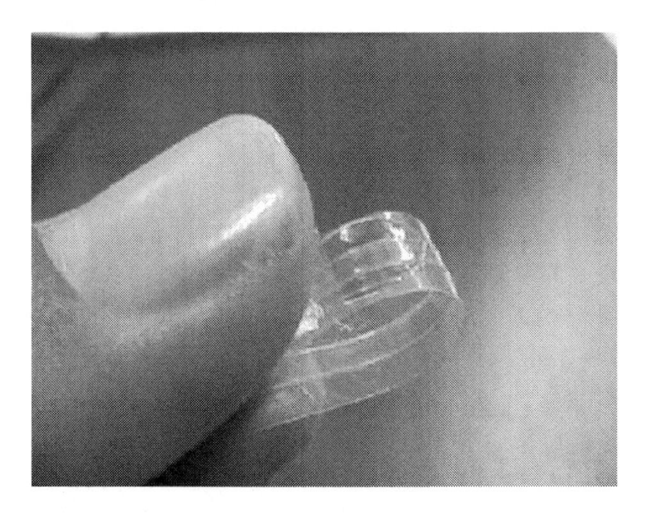

**FIGURE 6.11**
Photograph of the transparent glucose sensor. (From Mitsubayashi, K., et al., *Biosens. Bioelectron.*, 19, 67–71, 2003. With permission.)

In addition, modification of the PDMS surface using MPC polymers has been reported (Iwasaki et al. 2007). Additionally, PMD is a highly biocompatible phospholipid polymer that has a cell-membrane-like chemical configuration (Ishihara et al. 1990, 1998). This sensor measures glucose as determined by the hydrogen peroxide concentration, a by-product of the GOD reaction.

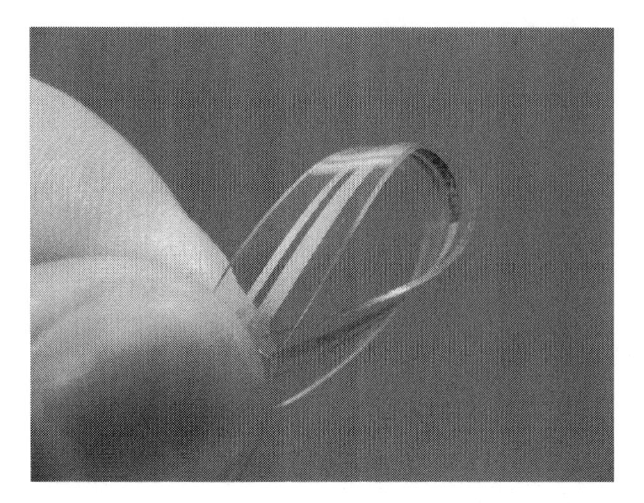

**FIGURE 6.12**
Photograph of the flexible and biocompatible glucose sensor.

In order to form the flexible electrode on PDMS, the ion beam sputtering (IBS) system (EIS-220, Elionix Co., Ltd., Japan) was employed, and Pt deposition was conducted with an acceleration energy of 1900 V. The electrodes showed high adhesion to the PDMS substrate and did not crack or peel off after bending. Enzyme immobilization was carried out by curing the mixture of PMD and GOD. These methods provided a sensor that was not only thinner but also soft and biocompatible, with an arbitrary shape.

## Tear Glucose Monitoring at the Eye

Continuous monitoring of glucose levels in tear fluids with the flexible glucose sensor (Figure 6.9) was conducted in a Japanese white rabbit (18 months of age, female; Iguchi et al. 2007). The rabbit was immobilized in the fixing apparatus. The sensing region of the flexible biosensor was attached to the pupil of the rabbit, and the sensor was fixed using fixing tape. Figure 6.13 illustrates the experimental method used for continuous tear glucose monitoring. A glucose solution was then orally administered to the rabbit with the quantity of glucose being determined as 2 g (1 g of glucose per 1 kg of weight). Blood glucose levels were measured using a commercial glucose meter (MEDISAFE, TERUMO Co., Japan) as a control. The blood samples were drawn from the rabbit's ear.

Figure 6.14 shows the dynamics of glucose levels in the tear fluid and blood. As indicated by the solid line, the glucose level was monitored continuously with the flexible biosensor. The tear glucose level increased after a delay of approximately 10 min compared to the blood glucose level. The physiological lag in the glucose level in the tear fluid was not far from

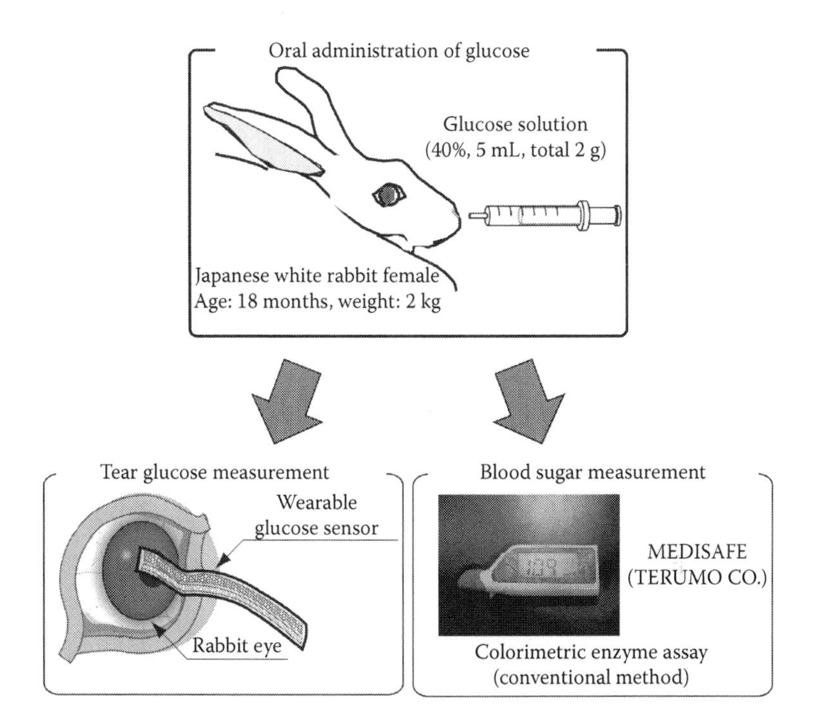


**FIGURE 6.13**
Measurement methods of tear glucose level (left hand) and blood sugar level (right hand) that are varied with oral glucose administration. (From Iguchi, S., et al., *Biomed. Microdev.*, 9, 603–609, 2007. With permission.)

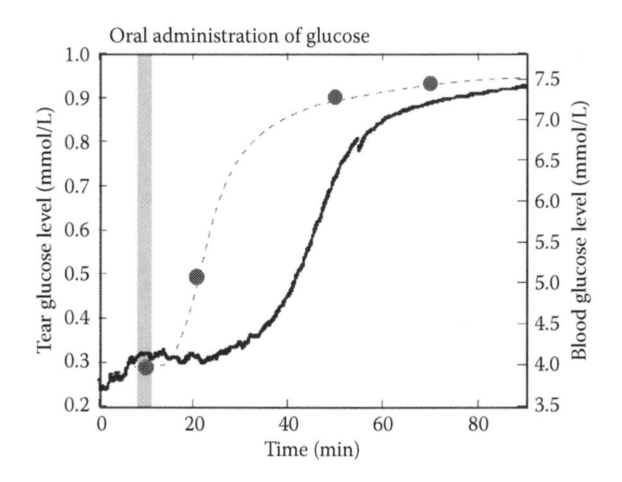


**FIGURE 6.14**
Blood and tear glucose dynamics after oral glucose tolerance test. The solid line indicates the tear glucose level. Solid circles show the blood glucose level as measured by a commercial glucose meter. (From Iguchi, S., et al., *Biomed. Microdev.*, 9, 603–609, 2007. With permission.)

nondiabetic human responses that were reported previously (Lane et al. 2006). Consequently, considerable correlation between blood and tear glucose levels was seen. Further improvements in the flexible sensor (e.g., biocompatibility, assembly with a contact lens) are expected for ophthalmic biomonitoring in the future.

## References

Arakawa, T., Sameshima, T., Sato, Y., Ueno, T., Shirasaki, Y., Funatsu, T., Shoji, S. 2007. Rapid multi-reagents exchange TIRFM microfluidic system for a single biomolecular imaging. *Sens. Actuators B* 128: 218–225.

Badugu, R., Lakowicz, J.R., Geddes, C.D. 2003. A glucose sensing contact lens: A noninvasive technique for continuous physiological glucose monitoring. *J. Fluoresc.* 13: 371–374.

Bailey, T.S., Zisser, H.C., Garg, S.K. 2007. Reduction in hemoglobin A1c with real-time continuous glucose monitoring: Results from a 12-week observational study. *Diabetes Technol. Ther.* 9: 203–210.

Baumberger, J.P., Goldfriend, R.B. 1951. Determination of arterial oxygen tension in man by equilibration through intact skin. *Federation Proc.* 10: 10–21.

Brooks, S.L., Ashby, R.E., Turner, A.P.F., Calder, R.M., Clarke, D.J. 1988. Development of an on-line glucose sensor for fermentation monitoring. *Biosensors* 3: 45–56.

Brown, M., Vender, J.S. 1988. Noninvasive oxygen monitoring. *Crit. Care Clin.* 4: 493–509.

Campbell, R.K., White, J.R., Jr. 2002. Insulin therapy in type 2 diabetes. *J. Am. Pharm. Assoc.* 42: 602–611.

Chen, R., Jin, Z., Colon, L.A. 1996. Analysis of tear fluid by CE/LIF: A noninvasive approach for glucose monitoring. *J. Capillary Electrophor.* 3: 243–248.

Clark, L.C. 1956. Monitor and control of blood and tissue oxygen tension. *Trans. Am. Soc. Art. Int. Organs* 2: 41–52.

Clarke, W., Becker, D.J., Cox, D., Santiago, J.V., White, N.H., Betschart, J., Eckenrode, K., Levandoski, L.A., Prusinki, E.A., Simineiro, L.M., Snyder, A.L., Tideman, A.M., Yaegar, T. 1988. Evaluation of a new system for self blood glucose monitoring. *Diabetes Res. Clin. Pract.* 4: 209–214.

Daum, K.M., Hill, R.M. 1982. Human tear glucose. *Invest. Ophthalmol. Vis. Sci.* 22: 509–515.

D'Auria, S., Dicesare, N., Gryczynski, Z., Gryczynski, I., Rossi, M., Lakowicz, J.R. 2000. A thermophilic apoglucose dehydrogenase as nonconsuming glucose sensor. *Biochem. Biophys. Res. Commun.* 274: 727–731.

D'Costa, E.J., Higgins, I.J., Turner, A.P. 1986. Quinoprotein glucose dehydrogenase and its application in an amperometric glucose sensor. *Biosensors* 2: 71–87.

De Silva, M.N., Desai, R., Odde, D.J. 2004. Micro-patterning of animal cells on PDMS substrates in the presence of serum without use of adhesion inhibitors. *Biomed. Microdev.* 6: 219–222.

Duc, G. 1975. Is transcutaneous PO2 reliable for arterial oxygen monitoring in newborn infants? *Pediatrics* 55: 55–63.

Evans, N.T.S., Naylor, P.F.F. 1967. The oxygen tension gradient across human epidermis. *Resp. Physiol.* 3: 38–42.

Fuller, J.H., Shipley, M.J., Rose, G., Jarrett, R.J., Keen, H. 1983. Mortality from coronary heart disease and stroke in relation to degree of glycaemia: The Whitehall study. *Br. Med. J.* 287: 867–870.

Garg, S.K., Schwartz, S., Edelman, S.V. 2007. Improved glucose excursions using an implantable real-time continuous glucose sensor in adults with type 1 diabetes. *Diabetes Care* 27: 734–738.

Gray, D.S., Tien, J., Chen, C.S. 2004. High-conductivity elastomeric electronics. *Adv. Mater.* 16: 393.

Hagihara, B. 1984. Development and application of blood gas monitoring. *Jpn. J. Clin. Pathol.* 32: 1065–1075.

Haljamae, H., Frid, I., Holm, J., Holm, S. 1989. Continuous conjunctival oxygen tension (PcjO2) monitoring for assessment of cerebral oxygenation and metabolism during carotid artery surgery. *Acta Anaesthesiol. Scand.* 33: 610–616.

Heise, H.M., Marbach, R., Koschinsky, T.H., Gries, F.A. 1994. Noninvasive blood glucose sensors based on near-infrared spectroscopy. *Ann. Occup. Hyg.* 18: 439–447.

Huch, R., Huch, A., Lubbers, D.W. 1973. Transcutaneous measurement of blood $PO_2$ ($tcPO_2$). *J. Perinat. Med.* 1: 183–191.

Huch, R., Lubbers, D.W., Huch, A. 1974. Reliability of transcutaneous monitoring of arterial $PO_2$ in newborn infants. *Arch. Dis. Child.* 49: 213–218.

Iguchi, S., Kudo, H., Saito, T., Ogawa, M., Saito, H., Otsuka, K., Funakubo, A., Mitsubayashi, K. 2007. A flexible and wearable biosensor for tear glucose measurement. *Biomed. Microdev.* 9: 603–609.

Iguchi, S., Mitsubayashi, K., Uehara, T., Ogawa, M. 2005. A wearable oxygen sensor for transcutaneous blood gas monitoring at the conjunctiva. *Sens. Actuators B* 108: 733–737.

Imura, S., Baba, K. 1975. Continuous transcutaneous $PO_2$ measurement in newborn infant. *Resp. Circulation* 23: 53–59.

Isenberg, S.J., Neumann, D., Fink, S., Rich, R. 2002. Continuous oxygen monitoring of the conjunctiva in neonates. *J. Perinat.* 22: 46–49.

Isenberg, S.J., Shoemaker, W.C. 1983. The transconjunctival oxygen monitor. *Am. J. Ophthalmol.* 95: 803–806.

Ishihara, K., Nomura, H., Mihara, T., Kurita, K., Iwasaki, Y., Nakabayashi, N. 1998. Why do phospholipid polymers reduce protein adsorption? *J. Biomed. Mater. Res.* 39: 323–330.

Ishihara, K., Ueda, T., Nakabayashi, N. 1990. Preparation of phospholipid polymers and their properties as polymer hydrogel membranes. *Polym. J.* 22: 355–360.

Iwasaki, Y., Takamiya, M., Iwata, R., Yusa, S., Akiyoshi, K. 2007. Surface modification with well-defined biocompatible triblock copolymers: Improvement of biointerfacial phenomena on a poly(dimethylsiloxane) surface. *Colloids Surf. B* 57: 226–236.

Jin, Z., Chen, R., Colon, L.A. 1997. Determination of glucose in submicroliter samples by CE-LIF using precolumn or on-column enzymic reactions. *Anal. Chem.* 69: 1326–1331.

Khang, D.-Y., Jiang, H., Huang, Y., Rogers, J.A. 2006. A stretchable form of single-crystal silicon for high-performance electronics on rubber substrates. *Science* 311: 208–212.

Kudo, H., Iguchi, S., Yamada, T., Kawase, T., Saito, H., Otsuka, K., Mitsubayashi, K. 2007. A flexible transcutaneous oxygen sensor using polymer membranes. *Biomed. Microdev.* 9: 1–6.

Kudo, H., Sawada, T., Kazawa, E., Yoshida, H., Iwasaki, Y., Mitsubayashi, K. 2006. A flexible and wearable glucose sensor based on functional polymers with Soft-MEMS techniques. *Biosens. Bioelectron.* 22: 558–562.

Kudo, H., Yagi, T., Chu, M.X., Saito, H., Morimoto, N., Iwasaki, Y., Akiyoshi, K., Mitsubayashi, K. 2008. Glucose sensor using phospholipid polymer-based enzyme immobilization method. *Anal. Bioanal. Chem.* 391: 1269–1274.

Kwan, M., Fatt, I. 1971. A noninvasive method of continuous arterial oxygen tension estimation from measured palpebral conjunctival oxygen tension. *Anaesthesiology* 35: 309–314.

Lane, J.D., Krumholz, D.M., Sack, R.A. 2006. Tear glucose dynamics in diabetes mellitus. *Curr. Eye Res.* 31: 895–901.

Leahy, J.L. 1996. Impaired B cell function with chronic hyperglycemia: "Overworked B cell" hypothesis. *Diabetes Rev.* 4: 298–319.

Man, S.F.P., Adams, G.K., Proctor, D.F. 1979. Effects of temperature, relative humidity, and mode of breathing on canine airway secretions. *J. Appl. Physiol. Respir. Environ. Exercise Physiol.* 46: 205–211.

March, W.F., Rabinovitch, B., Adams, R., Wise, J.R., Melton, M. 1982. Ocular glucose sensor. *Trans. Am. Soc. Art. Int. Organs* 28: 232–235.

Meadows, D., Schultz, J.S. 1988. Fiber-optic biosensors based on fluorescence energy transfer. *Talanta* 35: 145–150.

Mitsubayashi, K., Suzuki, M., Tamiya, E., Karube, I. 1994. Analysis of metabolites in sweat as a measure of physical condition. *Anal. Chim. Acta.* 289: 27–34.

Mitsubayashi, K., Wakabayashi, Y., Murotomi, D., Yamada, T., Kawase, T., Iwagaki, S., Karube, I. 2003. Wearable oxygen sensor with a flexible and thinner structure for transcutaneous oxygen monitoring. *Sens. Actuators B* 95: 373–377.

Mitsubayashi, K., Wakabayashi, Y., Tanimoto, S., Murotomi, D., Endo, T. 2003. Optical-transparent and flexible glucose sensor with ITO electrode. *Biosens. Bioelectron.* 19: 67–71.

Ostuni, E., Kane, R., Chen, C.S., Ingber, D.E., Whitesides, G.M. 2000. Patterning mammalian cells using elastomeric membranes. *Langmuir* 16: 7811–7819.

Peterson, S.L., McDonald, A., Gourley, P.L., Sasaki, D.Y. 2005. Poly(dimethylsiloxane) thin films as biocompatible coatings for microfluidic devices: Cell culture and flow studies with glial cells. *J. Biomed. Mater. Res. A* 72: 10–18.

Pickup, J.C., Hussain, F., Evans, N.D., Sachedina, N. 2005. In vivo glucose monitoring: The clinical reality and the promise. *Biosens. Bioelectron.* 20: 1897–1902.

Podolsky, S., Wertheimer, J., Harding, S. 1989. The relationship of conjunctival and arterial blood gas oxygen measurements. *Resuscitation* 18: 31–36.

Rabinovitch, B., March, W.F., Adams, R.L. 1982. Noninvasive glucose monitoring of the aqueous humor of the eye: Part I. Measurement of very small optical rotations. *Diabetes Care* 5: 254–258.

Rithalia, S.V. 1991. Developments in transcutaneous blood gas monitoring: A review. *J. Med. Eng. Technol.* 15: 143–153.

Romano, A., Rolant, F. 1988. A non-invasive method of blood glucose evaluation by tear glucose measurement, for the detection and control of diabetic states. *Metab. Pediatr. Syst. Ophthalmol.* 11: 78–85.

Schier, G.M., Moses, R.G., Gan, E.T., Blair, S.C. 1988. An evaluation and comparison of Reflolux II and Glucometer II, two new portable reflectance meters for capillary blood glucose determination. *Diabetes Res. Clin. Pract.* 4: 177–181.

Sen, D.K., Sarin, G.S. 1980. Tear glucose levels in normal people and in diabetic patients. *Br. J. Ophthalmol.* 64: 693–699.

Shoemaker, W.C., Lawner, P.M. 1983. Method for continuous conjunctival oxygen monitoring during carotid artery surgery. *Crit. Care Med.* 11: 946–947.

Suzuki, H., Watanabe, I., Kikuchi, Y. 1999. Measurement of $PO_2$ using a micromachined oxygen electrode. *Chem. Sens. B* 15: 34–36.

Thiébaud, P., Lauer, L., Knoll, W., Offenhäusser, A. 2002. PDMS device for patterned application of microfluids to neuronal cells arranged by microcontact printing. *Biosens. Bioelectron.* 17: 87–93.

Tolosa, L., Gryczynski, I., Eichorn, L.R., Dattelbaum, J.D., Castellano, F.N., Rao, G., Lakowicz, J.R. 1999. Glucose sensor for low-cost lifetime-based sensing using a genetically engineered protein. *Anal. Biochem.* 267: 114–120.

Tolosa, L., Malak, H., Rao, G., Lakowicz, J.R. 1997. Optical assay for glucose based on the luminescnence decay time of the long wavelength dye Cy5™. *Sens. Actuators B* 45: 93–99.

Trettnak, W., Wolfbeis, O.S. 1989. Fully reversible fiber-optic glucose biosensor based on the intrinsic fluorescence of glucose oxidase. *Anal. Chim. Acta* 221: 195–203.

Webb, A.I., Daniel, R.T., Miller, H.S., Kosch, P.C. 1985. Preliminary studies on the measurement of conjunctival oxygen tension in the foal. *Am. J. Vet. Res.* 46: 2566–2569.

Weissman, B.A., Fatt, I., Rasson, J. 1981. Diffusion of oxygen in human corneas in vivo. *Invest. Ophthalmol. Vis. Sci.* 20: 123–125.

# 7

# Monitoring Walking in Health and Disease

**Richard Baker**

**CONTENTS**

Why Monitor Walking? ........ 159
 Walking for Health ........ 159
 Walking with Disease ........ 161
 Health Conditions Affecting Walking ........ 162
 What Aspects of Health Conditions Might We Be Interested In? ........ 165
What to Monitor? ........ 165
 The International Classification of Function, Disability and Health (ICF) ........ 165
 Gait Analysis and Monitoring Walking ........ 167
  Gait Analysis: Technology ........ 167
  Gait Analysis: The Clinical Paradigm ........ 169
  Gait Analysis and Monitoring Walking ........ 171
How to Monitor Walking? ........ 173
 Uptimers ........ 173
 Pedometers ........ 173
 Accelerometers ........ 174
 Gyroscopes, Magnetometers and Integrated Sensors ........ 175
 Global Positioning System and Other Position Sensors ........ 176
Conclusion: Using Sensors to Monitor Walking ........ 176
References ........ 177

## Why Monitor Walking?

### Walking for Health

Since Hippocrates first stated that "walking is man's best medicine", the health benefits of walking have been appreciated. It was not, however, until the 1990s that official public health recommendations began to promote walking as a means to improve health as distinct from more vigorous exercise to improve fitness (Lee and Buchner 2008; Pate et al. 1995; US Department of Health and Human Services 1996). High-level evidence suggests a strong beneficial effect of moderate levels of physical activity in preventing cardiovascular disease, stroke, type 2 diabetes, osteoporosis and colon cancer in adults (Department of Health 2004). Interestingly, given the publicity relating obesity to lack of exercise, the evidence suggests that controlling weight through exercise is only moderately successful at best (Institute of Medicine 2002).

Current guidance recommends 30 min of moderate-intensity physical activity (brisk walking at 5–6 km/h) each day (Department of Health 2004; Pate et al. 1995) and that this should be in bouts of no shorter than 10 min duration. A different emphasis, which originally evolved with the introduction of widely available pedometers to Japan in the 1960s, has focused on the number of steps taken per day. It is now widely acknowledged that an active lifestyle is characterized by more than 10,000 steps per day for adults (and 12,000 and 15,000 steps per day for schoolgirls and schoolboys, respectively; Tudor-Locke, Hatano, Pangrazi, and Kang 2008). There is growing acceptance that step count alone is not sufficient to improve health and that a substantial proportion of the steps (perhaps as high as 70% in children; Jago et al. 2006) need to be at moderate intensity at least.

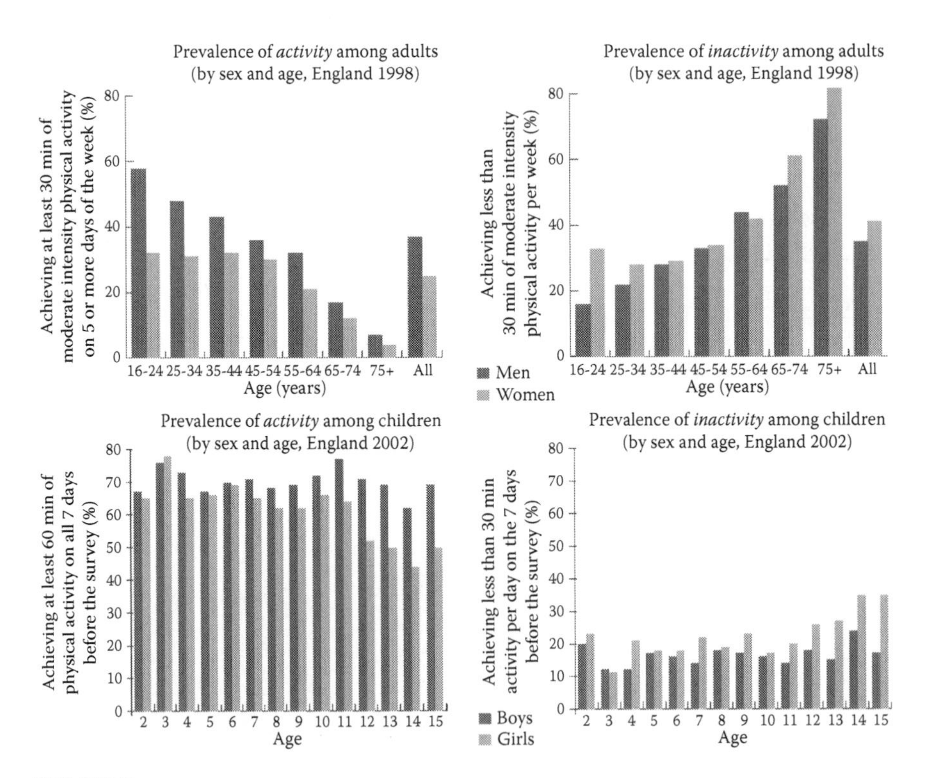


**FIGURE 7.1**
Activity and inactivity across the age range for English adults and children. (From Department of Health, *At Least Five a Week: Evidence on the Impact of Physical Activity and Its Relationship to Health,* Department of Health, London, 2004.)

Just under a third of all adults in the United Kingdom achieve 30 min of moderate-intensity physical exercise at least 5 days a week, and just over a third have a total of less than 30 min per week (Department of Health 2004). There are strong age effects, with the proportion of active people falling and that of inactive people increasing with age. By the age of 65 less than 20% of the population qualifies as active, and over half are inactive. There is also a strong gender effect on activity up to the age of 65 (with males being more active) but not on inactivity. Over two-thirds of British children are active (more than 60 min of physical activity on all 7 days prior to the survey), with just 20% inactive (less than 30 min activity per day over the same period). Substantial portions of both the adult and child population are not involved in sufficient physical activity to promote good health. Clearly, monitoring physical activity, and walking in particular, in the general population may have a role in encouraging healthier lifestyles as reflected by the popularity of the 10,000 steps a day slogan (Tudor-Locke et al. 2008).

## Walking with Disease

Whilst walking is something that many of us take for granted, over 10% of the entire noninstitutionalized US adult population reports having some difficulty with walking just 400 m (Iezzoni, McCarthy, Davis, and Siebens 2001). This equates to a total of 19 million individuals. Another study found that over 18% of individuals between 50 and 65 years of age in the United Kingdom had some difficulty walking a similar distance (Gardener, Huppert, Guralnik, and Melzer 2006). Even in the 18–50 age category, 4.9% of respondents reported some walking difficulties (Iezzoni et al. 2001). Walking difficulties are clearly correlated with age, with rates in the elderly rising to over 50% (Gardener et al. 2006; Iezzoni et al. 2001) as depicted in Figure 7.2. Given

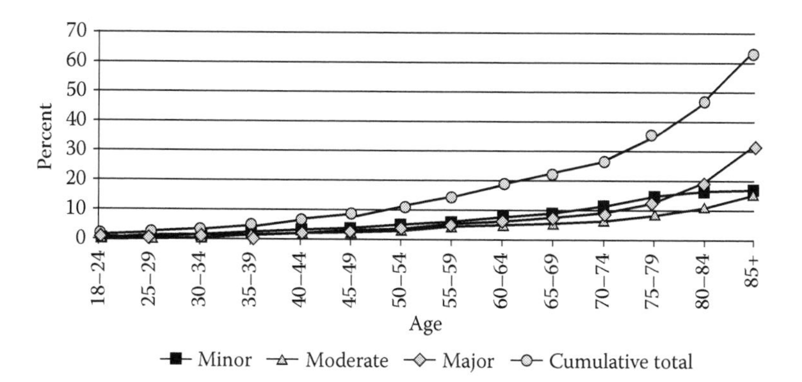


**FIGURE 7.2**
Percentage of population reporting mobility problems by age category determined from reweighted population estimates for noninstitutionalized, civilian US residents. (From Iezzoni, L. I., McCarthy, E. P., Davis, R. B., and Siebens, H., *J Gen Intern Med*, 16, 235–243, 2001. With permission.)

the increasing proportion of elderly people in the populations of most countries, it is clear that the number of people experiencing walking difficulties will increase further in the near future.

Factors which at least double the odds of an individual having difficulty walking are chronic obstructive lung disease, angina, stroke, recently treated cancer, comorbidity and lower-limb and back pain, and these vary little across the age range (Melzer, Gardener, and Guralnik 2005). In the middle-aged population, 38% of mobility disability is related to lower-limb pain and 15% to low back pain. The increasing incidence of mobility problems in the elderly broadly follows the incidence of these underlying conditions. It is interesting that many of the conditions most studied in relation to walking, such as cerebral palsy and Parkinson's disease, are not represented in these statistics and represent only a small proportion of the total population that experiences difficulty with walking, which is dominated by those with pain (mostly arthritis) and "shortness of breath" (Gardener et al. 2006).

The specific socio-economic cost of walking disabilities is not well documented, as most analyses tend to concentrate on specific diseases or the broader spectrum of musculoskeletal disorders (particularly as a consequence of the Bone and Joint Decade; Bone and Joint Decade 2009). Half of all lost work days and hospital bed days in the United States arise from musculoskeletal conditions (American Academy of Orthopaedic Surgeons 2008). In 2004 the cost of healthcare for musculoskeletal conditions in the United States was estimated to be $510 billion (4.6% of GDP) and the indirect costs a further $339 billion. In the same year the annual number of total hip and knee replacement surgeries in the United States passed one million for the first time.

Monitoring walking can give an important indication of health status, disease progression and the effect of interventions on this wide range of conditions which have such a profound effect on people's health and lead to considerable health expenditures. Clearly, there will be an interaction between factors reducing walking ability and the important role that walking, and other physical activities, has in ensuring good general health. This interaction becomes increasingly important in the ageing population, and having good methods of understanding this interaction through measurement studies will be important in addressing the challenge of promoting good health within an ageing population.

## Health Conditions Affecting Walking

It is not possible in a short space to give a full description of the health conditions affecting walking (*Wikipedia* provides excellent introductory material), but a brief overview of the health conditions may be useful. Iezzoni et al. (2001) provide data which can be used to estimate the number of non-institutionalized US adults reporting walking difficulties from a particular cause (see Figure 7.3).

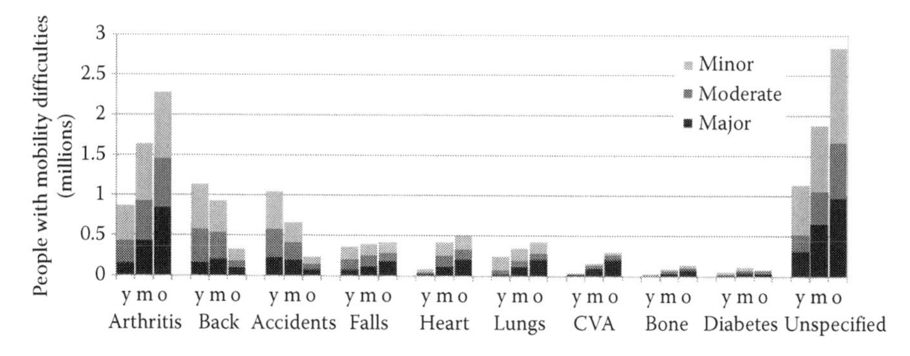


**FIGURE 7.3**
Number of noninstitutionalized American citizens with mobility difficulties by causative condition. CVA, cerebrovascular accident; y, young (18–49 years); m, middle-age (50–69 years); o, old (>70 years). (Data abstracted from Iezzoni, L. I., McCarthy, E. P., Davis, R. B., and Siebens, H., *J Gen Intern Med*, 16, 235–243, 2001.)

*Arthritis* literally means "inflammation of the joints" and is associated with degradation of the cartilage. It is the corresponding pain, particularly during activity, that limits walking. It can occur in any joint but affects walking most directly when it occurs in the hip, knee, ankle or joints of the foot. It can have a variety of causes, but by far the most common is osteoarthritis, which arises from mechanical degradation of the cartilage. Primary osteoarthritis is a consequence of the deterioration in cartilage quality with age, while secondary osteoarthritis can arise as a consequence of previous injuries, obesity or a number of other causes. It is estimated that 60% of the US population will have symptomatic osteoarthritis by the age of 65.

Back pain is also extremely common, with 90% of people reporting pain at some stage in their lives. Despite this, the causes of back pain are poorly understood, with up to 85% of cases not having a clear physiological cause. Disk herniation, osteoarthritis, muscular strain or spasm and nerve compression (including sciatica) may all be causative factors. Walking is limited by the pain, although in many cases keeping mobile are found to be beneficial and is encouraged.

Accidents of one form or another are an important cause of walking difficulties. Although these are a more important cause amongst younger adults, the proportion of different types of accidents leading to walking problems is broadly similar across the age range. About 45% of such injuries arise from motor vehicle accidents, 20% from overexertion and muscle strains, and about 15% from other specific causes such as machinery or firearms (and about 15% from unspecified causes). Accidents can result in a wide range of impairments, but brain and spinal cord injury or damage to the muscles, nerves, bones or blood supply to the lower limbs are the most likely to affect walking.

Falls become increasingly important in limiting walking as people age (Lord, Sherrington, Menz, and Close 2007). Falls may result from poor motor control, weakness and fatigue, which are generally secondary to other conditions. It is generally the fear of falling again that actually limits walking. Cardiovascular problems also become more important with age. Ischaemic heart disease and a variety of lung conditions such as bronchitis, emphysema and asthma all reduce general levels of fitness and hence walking ability. Cerebrovascular accident (CVA; strokes) is caused by localized disruption of the blood supply to the brain, which can lead to persistent brain damage. It is generally loss of motor control of the muscles of the leg that limits walking. There is generally some recovery of function over the days and weeks following an acute attack, but many people do not recover fully, being left with chronic limitations of their walking ability. Cerebral palsy is also caused by localized disruption of the blood supply to the brain, but the term is used only if this occurs while the brain is still developing (Rosenbaum, Paneth, Leviton, Goldstein, and Bax 2007), typically before the age of two. The direct effect is on the brain's ability to coordinate muscle activity, but this can lead to abnormal development of the bones and muscles, resulting in complex neuromusculoskeletal disorders that can profoundly limit walking ability (Bache, Selber, and Graham 2003; Gage, Schwartz, Koop, and Novacheck 2009). Although not specifically included in Figure 7.3, this is the cause of about 1.1% of walking difficulties (Iezzoni et al., 2001).

Diabetes affects nearly 3% of people worldwide (Wild, Roglic, Green, Sicree, and King 2004), but the percentage is much higher in some parts of the developed world. Over 90% of diabetics have type 2 diabetes, or insulin resistance in which the cells fail to use insulin properly. Its occurrence is linked to a range of interrelated lifestyle factors including obesity, smoking, poor diet and lack of exercise, and the incidence increases with age. Type 1 diabetes results from a failure in insulin production and is unrelated to these factors. Chronic diabetes can result in degradation of the blood supply to the extremities (peripheral vascular disease), leading to ulceration of the feet. Severe ulceration can limit walking and may ultimately require amputation.

Parkinson's disease is not included specifically in Figure 7.3 but can have a profound effect on walking. It is a result of degenerative changes in the basal ganglia, the part of the brain responsible for the automatic control of patterned movements such as walking. This can lead to muscle tremor and rigidity and slowing of movements including walking. In extreme cases it can lead to a blocking of muscular activity. Walking is often affected particularly by this.

Less specific factors can also limit walking. These may be consequences of the conditions listed in the preceding paragraphs or may be independent of these. Sensory loss, particularly loss of vision, can be an important factor, as can dizziness or other balance problems. Obesity tends to be more common in people who have difficulty walking, as are depression and anxiety, although it is often not clear whether these are causes or effects. Fear of falling arising from a number of underlying conditions can also limit walking.

### What Aspects of Health Conditions Might We Be Interested In?

Within each of the health conditions mentioned in the preceding paragraphs, there are several aspects that we might be interested in and which might be measured by monitoring walking. The most obvious is that we might wish to describe the relationship between the health condition and walking. In static and relatively homogeneous conditions, this might simply require us to monitor walking within a group with a particular condition. Many of the conditions, however, are progressive or present in a wide variety of ways (and in different levels of severity). In such conditions it can be important to divide the population into subgroups who are similarly affected by the conditions to understand how walking is affected by these different presentations of the disease. A related approach is to look for correlations between walking and specific descriptors of the disease condition.

Monitoring walking can also be used to assess disease progression. Designs can be either cross-sectional, in which different people at different stages of the disease are monitored at a similar time, or longitudinal, in which case the same people are monitored at different times. Monitoring walking can also be used to assess the effects of various clinical interventions (treatments). There are a variety of trial designs for doing this, but the most robust is the randomized clinical trial (RCT). In this approach, baseline measures are taken from all individuals, and then a random process is used to choose a subset of the individuals to have a particular intervention. At a later stage, measurements are taken from all the individuals, and the change in the subset that has had the intervention is compared with the change in the subset that has not had the intervention (placebo group). A variant on this is the comparison of two different treatments in which the subgroups are randomly allocated to receive one treatment or another.

## What to Monitor?

### The International Classification of Function, Disability and Health (ICF)

As modern international medicine has made progress in addressing the issue of preventable life-threatening conditions, the focus of the World Health Organisation (WHO) has shifted somewhat to include non-life-threatening conditions which cause disability and limit function. The *International Classification of Impairments, Disabilities and Handicaps* (*ICIDH*; WHO 1980) was first published in 1980 and has now been superseded by the *International Classification of Functioning, Disability and Health* (*ICF*; WHO 2001), which was endorsed by the World Health Assembly in 2001. Mobility and walking are important features of the classification system, and a consideration of how

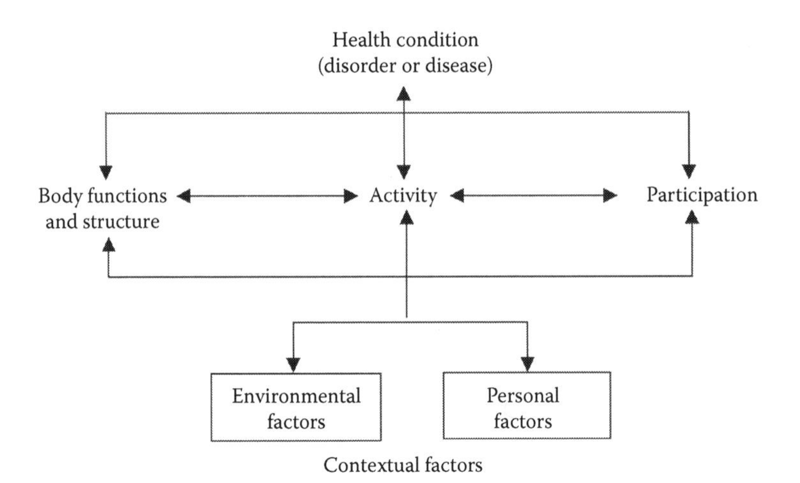


**FIGURE 7.4**
Schematic outline of terms used in the International Classification of Functioning, Disability and Health. (From World Health Organisation, International Classification of Functioning, Disability and Health, World Health Organisation, Geneva, 2001. With permission.)

it operates can be useful in defining exactly what activity monitors have the capacity to measure.

The *ICF* considers that the description of any health condition has two *parts**: the level of *functioning and disability* and *contextual factors*. Both can be further subdivided into *components*. Functioning and disability can be defined in terms of *body structures* (anatomical parts of the body), *body functions* (physiological functions of those body structures), *activities* (tasks that can be executed by an individual) and *participation* (involvement in life situations). Each component can be expressed in both positive and negative terms (e.g., ability or disability). It is tempting to assume some hierarchy that impairments of body functions and structures lead to activity limitations and restrictions on participation, but in many areas the converse can also be true that restricted participation might limit activity and that this can affect body structures and functions. The components are thus considered simply to interact.

Within activity and participation two further concepts of *performance* and *capacity* are introduced. Performance describes what an individual does in his or her current environment or "life situation". Capacity describes that individual's ability to execute that task or action (particularly under standardized or idealized conditions). The level and extent of a person's performance for a given capacity will be determined by contextual factors, which are divided into *personal factors* (which are intrinsic to the individual such as

* Italicized words here have formal definitions within the *ICF* document.

gender, ethnicity, fitness, attitudes, coping styles.) and *environmental factors* (which are extrinsic to the individual and range from factors in the physical environment to legal structures and societal attitudes).

These concepts can be useful in designing measurement systems in general and mobility monitors in particular. The framework allows a specific focus on which elements of the health condition are being investigated. Different devices will be needed to assess body structures, body functions, activities and participation. Similar devices might be used to distinguish performance from capacity, but different testing regimes will be required. Performance needs to be measured within the context of everyday life, whereas capacity needs to be measured under standardized (and possibly idealized) conditions. The process of designing useful devices and experimental regimes needs to be guided by an understanding of the specific health issue being assessed.

A considerable part of the *ICF* is devoted to issues surrounding mobility in general and walking in particular. One chapter out of the eight describing body functions is devoted to "Neuromusculoskeletal and Movement Related Functions" and one out of eight on body structures to "Structures Related to Movement". Activity and participation are grouped, and the fourth of nine chapters describes "Mobility". This includes sections on "walking and moving" and "moving using transportation". Table 7.1 provides the hierarchy of subsections of "Mobility". Using this hierarchy to specify exactly what is to be measured can be very useful in designing measuring devices and testing regimes.

## Gait Analysis and Monitoring Walking

The overriding current paradigm for the study of walking is instrumented gait analysis. In this approach, a subject is required to walk up and down a walkway, while some combination of measurements is recorded, including joint movement (kinematics), the components of the ground reaction, loading of the joints (kinetics) and electrical activity in the muscles (electromyography [EMG]) is recorded. How the use of a new generation of sensors can augment and improve gait analysis measurements requires a consideration of both current technology and the clinical paradigm within which measurements are made and used. Baker (2006) provides a fuller overview of the current state of the art and challenges for the future if required.

### *Gait Analysis: Technology*

Although other systems are available, the most common gait analysis systems use optical systems to track the position of small markers attached to the skin. Some systems use "active" markers emitting light, but most use "passive" systems with a retro-reflective coating to reflect light from light sources built into the camera body around the lens. By using high-intensity

**TABLE 7.1**

Hierarchy of Terms describing Activity and Function within the Chapter "Mobility" in the *International Classification of Functioning.*

- d4 Chapter 4 Mobility
  - d410--d429 Changing and maintaining body position*
  - d430--d449 Carrying, moving and handling objects*
  - d450--d469 Walking and moving
    - d450 Walking
      - d4500 Walking short distances
      - d4501 Walking long distances
      - d4502 Walking on different surfaces
      - d4503 Walking around obstacles
      - d4504/5 Walking, other
    - d455 Moving around
      - d4550 Crawling
      - d4551 Climbing
      - d4552 Running
      - d4553 Jumping
      - D4554 Swimming
      - D4555/6 Moving around, other
    - d460 Moving around in different locations
      - D4600 Moving around within the home
      - D4601 Moving around within buildings other than the home
      - D4602 Moving around outside the home and other buildings
      - D4608/9 Moving around in different locations (other)
    - d465 Moving around using equipment
    - d479 Moving and walking, other
  - d470--d489 Moving around using transportation*

*Source:* From World Health Organisation, *International Classification of Functioning, Disability and Health,* World Health Organisation, Geneva, 2001. With permission.

* Sub-sections have not been expanded for clarity.

short-pulse-width strobe illumination from light-emitting diodes (in the visible or infrared spectrum), good contrast to the background image and minimal blurring from movement of the markers can be achieved. Early systems modified standard video cameras, but more recently cameras designed specifically for this purpose have been developed. Pattern recognition techniques are used to identify the two-dimensional location of the image of the centre of each marker in the image plane of each camera. If the position and orientation of each camera is known, then information from a number of cameras can be combined to determine the position of the markers in three dimensions. Recent reviews of this process should be consulted for further details of this process (Cappozzo, Della Croce, Leardini, and Chiari 2005; Chiari, Della Croce, Leardini, and Cappozzo 2005).

The body is assumed to move as a number of rigid segments linked by joints. If each marker is assumed to be fixed in relation to a particular segment, then a variety of inverse kinematic techniques can be used to determine the orientation of the different joints at any particular instant in time (Kadaba, Ramakrishnan, and Wootten 1990; Lu and O'Connor 1999; Reinbolt et al. 2005). This technique is limited by the assumption that the markers remain fixed in relation to the segment (and variations from this result in "soft-tissue artefact"; Peters, Baker, and Sangeux 2009). A further limitation of most systems has been the difficulty in defining where markers are in relation to bony landmarks defining the rigid segments (Della Croce, Leardini, Chiari, and Cappozzo 2005). This has led to the development of various functional calibration techniques to define the relationship of the rigid segments to the markers (Baker 2007; Ehrig, Taylor, Duda, and Heller 2006, 2007; Lu and O'Connor 1999; Reinbolt et al. 2005; Schwartz and Rozumalski 2005). The validity of these techniques, however, has yet to be established.

The joints of the body can be modelled to have different degrees of freedom, but most current models assume three rotational degrees of freedom at most. These three degrees of freedom are generally described by a set of three joint angles using either a Cardan/Euler decomposition or a joint coordinate system (Baker 2001, 2003; Grood and Suntay 1983; Wu et al. 2002, 2003). The variation of these angles with time is the primary output of any gait analysis system. If the components of the ground reaction can be measured using a force plate, then it is also common to compute the moments that must be acting at each joint using inverse dynamics on the basis of the kinematic data (Kadaba et al. 1989; Ounpuu, Davis, and Deluca 1996). If electromyographic data are collected, then these can be displayed as well.

### *Gait Analysis: The Clinical Paradigm*

Gait analysis is based around the concept that walking is a cyclic activity and that measurements from a small number of cycles are representative of the gait pattern of an individual. The cycle can be divided into two phases. By convention the gait cycle starts at the time when a foot makes contact with the ground. The stance phase then lasts until the same foot is lifted off the ground, at which time the swing phase starts. This lasts while the foot is being advanced, until the next cycle and stance phase starts when the foot again makes contact with the floor. Gait variables are almost always plotted against time expressed as a percentage of the gait cycle. It is most common to have data from several cycles from the left and right sides superimposed on some representation of reference data collected from individuals with no neuromusculoskeletal condition that might affect walking.

Most clinical gait analysis is conducted to aid clinical decision-making. Discrepancies between the data for the patient and the representative data (gait features) are identified and related to underlying impairments at the level of body structures and functions which may be causing them. Once

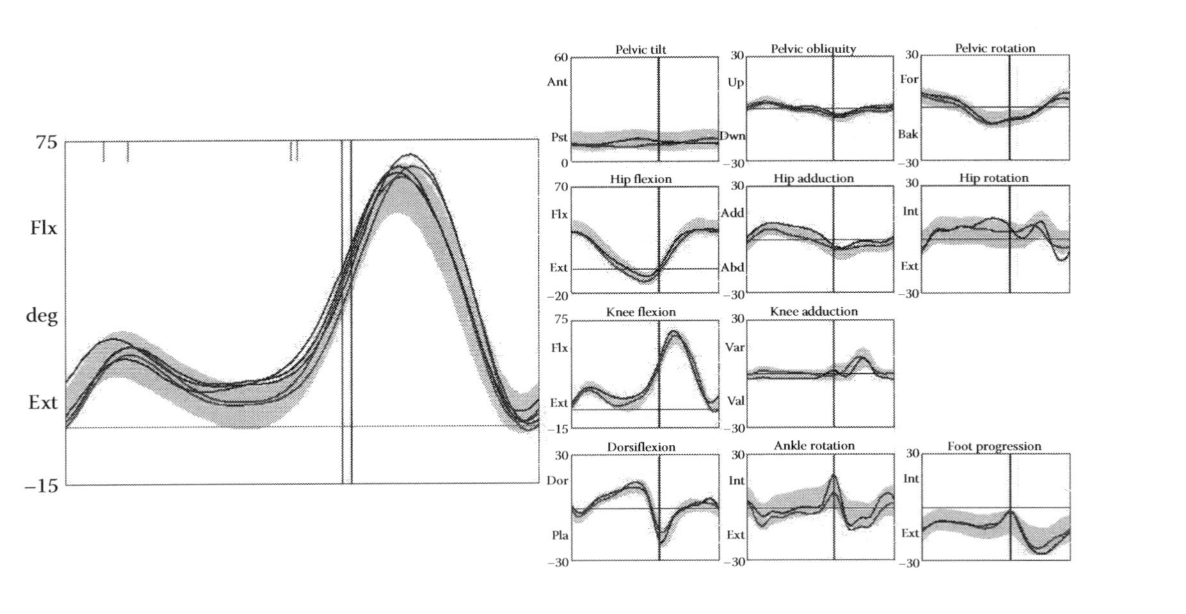


**FIGURE 7.5**
Representation of kinematic data. The left-hand graph shows knee flexion angle plotted across the gait cycle for five strides. The grey area in the background represents the mean ± 1 standard deviation range from reference data for subjects with no neuromusculoskeletal pathology. The array of graphs on the right is typical of gait analysis, representing similar data from a range of different joints (rows) and planes (columns).

these impairments have been identified, then clinicians can decide which of a range of treatment options are best suited to the patient. This process is most commonly applied to the management of children with cerebral palsy (Gage et al. 2009), although use in other conditions is increasing (Morris et al. 2005). Gait analysis data are almost always interpreted in the context of a wider assessment of the patient's neuromusculoskeletal condition including medical and surgical history, a physical examination and relevant medical imaging.

One of the fundamental limitations of gait analysis arises from the variability of gait patterns within the general population. Even people with no neuromusculoskeletal pathology have very different gait patterns (and these have even been suggested as a biometric quality that could be used to identify individuals; Boyd and Little 2005). This is hidden to some extent by plotting just one standard deviation of the reference data—35% of data from people with no pathology will still lie outside these limits. A paradigm which aims to distinguish differences attributable to underlying pathology from the variability amongst people without such pathology is clearly fundamentally limited by the size of that variability.

### Gait Analysis and Monitoring Walking

There is often considerable confusion regarding the relationship between conventional clinical gait analysis and the monitoring of walking. Clinical gait analysis clearly measures walking under highly standardized and idealized conditions and thus measures capacity as defined by the *ICF* (WHO 2001). This might be regarded as a limitation, but it allows the comparison of data from the patient with those from other patients or some reference population in the knowledge that both have been captured under essentially the same conditions. Differences in the gait traces can then be attributed to differences in body structures and functions and can be used as the basis for understanding the disease condition.

Simply using continuous monitoring techniques to provide essentially the same measurements over a longer time period and outside the regulated conditions of the gait laboratory does not necessarily help this process. If differences between the patient data and the reference data are observed, then these might be a consequence of a requirement to avoid tripping over a small obstacle or to manoeuvre around a large one rather than be indicative of some underlying pathology. It can be seen that conducting tests under standardized and observed conditions is an important part of the gait analysis process. Gait analysis, however, can tell very little about activity and participation or how the capacity measured by gait analysis is translated into performance in the real world. It is in these domains that continuous monitoring of walking has most to offer.

Clinical gait analysis generates considerable volumes of data through the large number of variables measured or calculated. This is manageable for a small number of gait cycles and within the well-defined conditions under which the data are captured. When extended to tens of thousands of gait cycles and particularly when data are captured in real-life conditions, such extensive data collection becomes unmanageable. Continuous monitoring of simpler measures representative of overall function and activity such as step count or walking speed can provide extremely useful data under these conditions.

Continuous monitoring is also useful to track the considerable variability that is inherent in functional walking. Variations in walking speed (Orendurff, Bernatz, Schoen, and Klute 2008) and direction (Glaister, Bernatz, Klute, and Orendurff 2007; Huxham, Baker, Morris, and Iansek 2008a, 2008b; Huxham, Gong, Baker, Morris, and Iansek 2006), crossing and manoeuvring about obstacles (Galna, Peters, Murphy, and Morris 2009; Said, Goldie, Patla, and Sparrow 2001; Said, Goldie, Patla, Sparrow, and Martin 1999) and stopping (Jaeger and Vanitchatchavan 1992; Sparrow and Tirosh 2005; Tirosh and Sparrow 2005) and restarting (Halliday, Winter, Frank, Patla, and Prince 1998) are all essential parts of functional mobility which are difficult to assess in the gait laboratory. Orendurff, Schoen, Bernatz, Segal and Klute (2008) have drawn attention to just how much functional walking comprises very short bouts of activity, with 17% of bouts consisting of fewer than 5 steps, 40% of fewer than 13 steps and 75% of fewer than 41 steps (see Figure 7.5).

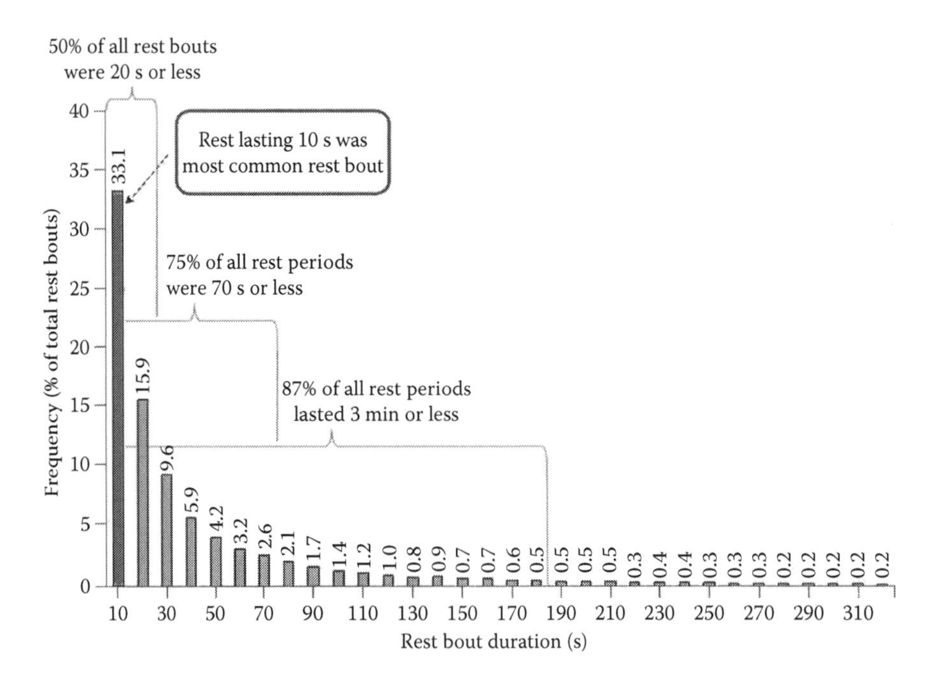


**FIGURE 7.6**
Frequency of bouts as a function of number of steps in each bout. (From Orendurff, M. S., Schoen, J. A., Bernatz, G. C., Segal, A. D., and Klute, G. K., *J Rehabil Res Dev*, 45(7), 1077–1089, 2008. With permission.)

## How to Monitor Walking?

### Uptimers

Uptimers (Diggory, Gorman, Schwartz, and Helme 1994; Follick, Ahern, Laser-Wolston, Adams, and Molloy 1985; Sanders 1980, 1983; White and Strong 1992), strictly speaking, do not monitor walking, but they are the simplest devices giving information on mobility. An uptimer is a small unit that is strapped to the outside of the thigh with a tilt switch that detects whether the thigh is horizontal (which is taken to indicate lying or sitting) or vertical (which is taken to indicate standing or walking). Earlier devices (Diggory et al. 1994; Follick et al. 1985; Sanders 1980, 1983; White and Strong 1992) simply recorded the total amount of uptime over the period of measurement, whereas a more recent device can log the pattern of that activity over periods between several hours and a week (Chastin et al. 2009; Grant, Ryan, Tigbe, and Granat 2006; Ryan, Grant, Tigbe, and Granat 2006).

Early studies used uptimers to investigate chronic pain (Follick et al. 1985; White and Strong 1992). Over the last decade Eldridge and her co-workers have conducted exemplary work documenting normative values for uptime in Australian children (Eldridge, Galea, McCoy, Wolfe, and Graham 2003) and the variability between children (Eldridge, McCoy, Galea, Wolfe, and Graham 2003) and then extended their work to a study of how uptime varied after appendectomy (Eldridge, Kimber, Wolfe, Galea, and Hutson 2003). Pirpiris and Graham extended this work to consider how uptime values for children with cerebral palsy compared with the normative data (Pirpiris and Graham 2004) and also to examine how this changed following complex orthopaedic surgery (Pirpiris 2002). More recent studies have used similar techniques in patients with hip fractures (Bernhardt et al. 2005). Whilst uptime has proved a useful measure in the past, it has generally been superseded by equally reliable but more sophisticated measures of activity.

### Pedometers

Pedometers measure the number of steps taken. Modern pedometers were first introduced in Japan in the 1960s and quickly sparked a movement claiming that 10,000 steps a day were essential for a healthy lifestyle (Tudor-Locke 2003). These devices were purely mechanical, but modern devices are microelectromechanical. They are now widely available at very low prices and are incorporated into many modern personal electronic devices. There are at least three basic types of mechanism: a spring-suspended lever arm with metal-on-metal contact, a magnetic reed proximity switch and an accelerometer.

Crouter, Schneider and colleagues (Crouter, Schneider, and Bassett 2005; Crouter, Schneider, Karabulut, and Bassett 2003; Schneider, Crouter, and

Bassett 2004; Schneider, Crouter, Lukajic, and Bassett 2003) have shown that the accuracy of pedometers can be quite variable but that the better ones can detect more than 99% of steps in straight line walking in adults at speeds above 1.3 m/s (Crouter et al. 2003). Accuracy drops with walking speed, but the better pedometers can still record over 88% of steps at slower speeds (0.9 m/s). Accuracy also drops to similar levels when free walking is monitored over a day (Schneider et al. 2004). In frail people walking slowly, up to 74% of steps can be missed (Cyarto, Myers, and Tudor-Locke 2004), and indeed under such conditions it may be difficult to define exactly what constitutes a step and therefore how many should be measured.

### Accelerometers

Accelerometers can be used simply to measure step count, but much more complex processing of the signal can also be used; this section relates to this more sophisticated use. Most accelerometers comprise a small mass supported on a beam which when deformed produces a change in impedance (resistive or capacitive) or a charge generation (piezoelectric; Zijlstra and Aminian 2007). All accelerometers detect acceleration due to gravity as well as that from movement. This can be useful in that it offers the potential for them to be used as inclinometers (Luinge and Veltink 2004), but it creates a specific problem if the application requires separation of the gravitational effect and the movement effect. The other problem with accelerometers is that the quantities most often required are either speeds or distances. Integration errors arise from DC offsets, signal fluctuations due to factors other than accelerations (e.g., temperature changes) or incorrect removal of gravitational effects. Errors of some magnitude are inevitable and accumulate with time, and they are thus particularly problematic if the accelerometer is operating over long periods of time. Uni- or triaxial goniometers are now readily available. Almost all accelerometer-based activity monitors now use complex signal processing from multiple channels to obtain useful measurements, with various Kalman filters (Kalman 1960) being particularly common.

Two approaches to using single accelerometer units are common. In the first the accelerometer is placed on the pelvis or trunk (Auvinet, Chaleil, and Barrey 1999; Bussmann, Veltink, Koelma, and Stam 1995; Zijlstra and Hof 2003). From measuring the acceleration of the trunk it is possible to distinguish gait cycle events and hence to derive step count and a number of temporal parameters (Rueterbories, Spaich, Larsen, and Andersen 2010). By measuring the vertical component of acceleration and assuming an inverted pendulum model of walking, it is also possible to estimate step length. Such systems typically incorporate some calibration process as a basis for step-length calculation for any individual. Clearly, the combination of cadence and step length allows walking speed and distance walked to be calculated. The other approach is to place the accelerometer on the foot or ankle (e.g., AMP 331, Dynastream Innovations, Alberta, Canada). The signal can be analysed to detect when the foot is moving and when it is stationary (Rueterbories et al. 2010), allowing temporal parameters to be calculated, and the acceleration during swing can be used to estimate step length.

Both designs show high levels of accuracy in measuring walking and running in a straight line or in a regular pattern. Indeed, several devices (particularly ankle- or foot-worn accelerometers) are now commercially available to record running speed and distance run at prices that make them accessible to amateur runners. The more sophisticated of these have Bluetooth connectivity to wristwatches and even the capacity to relay results via the mobile phone network. Unfortunately, such devices cope much less well with the sort of irregular walking activity that is such a feature of everyday life (Kuo, Culhane, Thomason, Tirosh, and Baker 2009). The devices also perform less well in patient populations whose gait patterns may differ from those on which the signal processing is based (Kuo et al. 2009) or whose gait patterns may vary over time.

Several systems are now available using multiple accelerometer units attached to different segments of the body (Zhang, Werner, Sun, Pi-Sunyer, and Boozer 2003). Signal processing allows the relative movement of the body segments to be recorded and can thus be used for continuous monitoring of joint kinematics (Huddleston et al. 2006).

Algorithms have also been developed to recognize different patterns of activity based on these data (Preece, Goulermas, Kenney, and Howard 2009; Zhang et al. 2003). A problem with these devices, however, is that even if very small sensors are used, the wires connecting them can be cumbersome to wear and may be prone to damage if care is not taken while dressing and undressing. Monitoring activity over considerable periods of time may thus be difficult.

## Gyroscopes, Magnetometers and Integrated Sensors

Solid-state gyroscopes are based on the transfer of energy between different modes of a vibrating system in response to Coriolis forces, and they are sensitive to changes in angular velocity. Such devices are insensitive to gravity but are generally much more sensitive to integration error than accelerometers (partly through marked temperature dependence) if used to calculate absolute angles. Magnetometers detect the strength of the earth's magnetic field and hence can give information on the orientation of the sensor. They are, however, susceptible to variations in the magnetic field, particularly around substantial metal objects.

Gyroscopes have been used to give reasonable gait event detection (Rueterbories, Spaich, Larsen, and Andersen 2010). There is, however, little evidence in the scientific literature of gyroscopes or magnetometers being used separately within sensors to monitor segment or joint angles, but a number of integrated sensors which use them to supplement data from accelerometers (often multiple accelerometers) are becoming available

(Roetenberg, Luinge, Baten, and Veltink 2005; Roetenberg, Slycke, and Veltink 2007; Schepers, Roetenberg, and Veltink 2010). Many of these use Kalman filters (Kalman 1960) to combine the information from the different sensors, and others incorporate joint or other physical constraints. Other work has integrated signals from force transducers built into the soles of shoes with accelerometer measurements to estimate the ground reaction and hence the displacement of the whole-body centre of mass (Schepers, van Asseldonk, Buurke, and Veltink 2009). Since the earth's magnetic field is not vertical, it is possible to combine accelerometer and magnetometer outputs to determine the direction (heading) in which the subject is walking. At least one manufacturer has assembled a number of integrated sensors into a suit, allowing complex measurements of intersegmental kinematics for use in the entertainment (computer games) and biomechanics markets.

### Global Positioning System and Other Position Sensors

Another approach to monitoring walking is through the use of position sensors such as the global positioning system (GPS), mobile phone networks or other systems. Maddison and Ni Mhurchu (2009) provide a review of studies of such devices, which are becoming more and more used in sporting applications. Commercially available GPS systems have been found to be capable of measuring location to around 3 m in open space, but tree canopies or urban environments can reduce accuracy to between 10 and 20 m (Rodriguez, Brown, and Troped 2005). The GPS signal is often lost entirely within buildings. Such techniques are thus incapable of measuring most walking which occurs in short bouts inside but may have some role in measuring specific activities outside. Although limited, these devices do have the advantage that, by being based on the location of the individual rather than how the individual moves, they are generally equally reliable across healthy and disabled populations.

## Conclusion: Using Sensors to Monitor Walking

Sensor-based systems for monitoring walking offer potential for use in two areas of the health domain. The first stems from the observation that walking and other moderate types of physical activity have a strong beneficial effect on a number of aspects of human health. Monitoring walking clearly allows for quantification of whether individuals or groups are likely to experience these effects. Various sensors are already available. Pedometers are readily available to count steps and to log the number of steps taken over considerable periods. Accelerometer-based units are becoming more and more widely available at a reasonable cost to record the running speed of professional and amateur athletes, and some are also available which can measure walking speed in the healthy population to good levels of accuracy over walks of medium and long distances (greater than 10 m). Studies show, however, that a substantial part of functional walking occurs in bouts of distances shorter than this, and there are no sensors that will accurately monitor anything other than step count for such walks. Monitoring walking has the potential to give an indication of health status, disease progression or the effect of interventions (an outcome measure).

The other area involves using sensors to monitor walking in people with a variety of health conditions that might limit walking ability. Because such people generally walk slower than their healthy peers and often with a different pattern of movement, the various sensor devices are generally less accurate. Even simple pedometers may fail to record walking in such groups, and accelerometer-based measurements have also been shown to be unreliable. GPS systems are independent of the speed and pattern of walking but, at present, are limited to open outdoor environments.

There is thus considerable potential for further development of sensor-based devices specifically designed to monitor walking in such people. To comprehensively monitor walking, such sensors will need to be capable of detecting short bouts of slow walking achieved with a range of different patterns of walking and to record the speed of walking (and hence distance walked). Devices need to be simple to use, easy to wear and unobtrusive. For monitoring over periods of more than 1 day, sensors need to be simple to put on or take off or be able to be attached to the subject for long periods without restricting the individual's lifestyle. Tamper-proof designs may be essential for applications with children and the confused. Modern sensors are almost always integrated with data loggers, and the combined devices need to have a battery life and data storage capacity to suit the intended purpose. For widespread application, such devices also need to be affordable. Some commercially available step counters satisfy these criteria, but, at present, no devices capable of also measuring speed and distance walked do. Development of such sensor-based devices would be extremely useful.

## References

American Academy of Orthopaedic Surgeons. (2008). *Burden of Musculoskeletal Diseases in the United States: Prevalence, Societal and Economic Cost.* Rosemount, Illinois: American Academy of Orthopaedic Surgeons.

Auvinet, B., Chaleil, D., & Barrey, E. (1999). Accelerometric gait analysis for use in hospital outpatients. *Rev Rhum Engl Ed, 66*(7–9), 389–397.

Bache, C., Selber, P., & Graham, H. (2003). The management of spastic diplegia. *Curr Orthopaed, 17*, 88–104.

Baker, R. (2001). Pelvic angles: A mathematically rigorous definition which is consistent with a conventional clinical understanding of the terms. *Gait Posture, 13*(1), 1–6.

Baker, R. (2003). ISB recommendation on definition of joint coordinate systems for the reporting of human joint motion. Part I: Ankle, hip and spine. *J Biomech, 36*(2), 300–302; author reply 303–304.

Baker, R. (2006). Gait analysis methods in rehabilitation. *J Neuroeng Rehabil, 3*, 4.

Baker, R. (2007). An engineering approach to reducing measurement error in clinical gait analysis. *J Biomech, 40*(S2), s26.

Bernhardt, J., Borschmann, K., Crook, D., Hill, K., McGann, A., & DeGori, M. (2005). Stand up and be counted: Measuring time spent upright after hip fracture and comparison with community dwelling older people. *Physiotherapy, 91*, 215–222.

Bone and Joint Decade. (2009). *Made to Last: Looking Back at a Decade of Action in Musculoskeletal Health.* Bone and Joint Decade.

Boyd, J. E., & Little, J. J. (2005). Biometric gait recognition. In M. Tistarelli, J. Bigun & E. Grosso (Eds.), *Advanced Studies in Biometrics* (pp. 19–42). Berlin: Springer.

Bussmann, J. B. J., Veltink, P. H., Koelma, F. K., van Lummel, R. C., & Stam, H. J. (1995). Ambulatory monitoring of mobility related activities: The initial phase of development of an activity monitor. *Eur J Phys Med Rehabil, 5*, 2–7.

Cappozzo, A., Della Croce, U., Leardini, A., & Chiari, L. (2005). Human movement analysis using stereophotogrammetry. Part 1: Theoretical background. *Gait Posture, 21*(2), 186–196.

Chastin, S. F., Dall, P. M., Tigbe, W. W., Grant, M. P., Ryan, C. G., Rafferty, D., et al. (2009). Compliance with physical activity guidelines in a group of UK-based postal workers using an objective monitoring technique. *Eur J Appl Physiol, 106*(6), 893–899.

Chiari, L., Della Croce, U., Leardini, A., & Cappozzo, A. (2005). Human movement analysis using stereophotogrammetry. Part 2: Instrumental errors. *Gait Posture, 21*(2), 197–211.

Crouter, S. E., Schneider, P. L., & Bassett, D. R., Jr. (2005). Spring-levered versus piezo-electric pedometer accuracy in overweight and obese adults. *Med Sci Sports Exerc, 37*(10), 1673–1679.

Crouter, S. E., Schneider, P. L., Karabulut, M., & Bassett, D. R., Jr. (2003). Validity of 10 electronic pedometers for measuring steps, distance, and energy cost. *Med Sci Sports Exerc, 35*(8), 1455–1460.

Cyarto, E. V., Myers, A. M., & Tudor-Locke, C. (2004). Pedometer accuracy in nursing home and community-dwelling older adults. *Med Sci Sports Exerc, 36*(2), 205–209.

Della Croce, U., Leardini, A., Chiari, L., & Cappozzo, A. (2005). Human movement analysis using stereophotogrammetry. Part 4: Assessment of anatomical landmark misplacement and its effects on joint kinematics. *Gait Posture, 21*(2), 226–237.

Department of Health. (2004). *At Least Five a Week: Evidence on the Impact of Physical Activity and Its Relationship to Health.* London: Department of Health.

Diggory, P., Gorman, M., Schwartz, J., & Helme, R. (1994). An automatic device to measure time spent upright. *Clin Rehabil, 8*, 353–357.

Ehrig, R. M., Taylor, W. R., Duda, G. N., & Heller, M. O. (2006). A survey of formal methods for determining the centre of rotation of ball joints. *J Biomech, 39*(15), 2798–2809.

Ehrig, R. M., Taylor, W. R., Duda, G. N., & Heller, M. O. (2007). A survey of formal methods for determining functional joint axes. *J Biomech, 40*(10), 2150–2157.

Eldridge, B., Galea, M., McCoy, A., Wolfe, R., & Graham, H. K. (2003). Uptime normative values in children aged 8 to 15 years. *Dev Med Child Neurol, 45*(3), 189–193.
Eldridge, B., Kimber, C., Wolfe, R., Galea, M., & Hutson, J. (2003). Uptime as a measure of recovery in children postappendectomy. *J Pediatr Surg, 38*(12), 1822–1825.
Eldridge, B., McCoy, A., Galea, M., Wolfe, R., & Graham, H. K. (2003). Variability in the measurement of uptime in children: A preliminary study. *Clin Rehabil, 17*(5), 499–503.
Follick, M. J., Ahern, D. K., Laser-Wolston, N., Adams, A. E., & Molloy, A. J. (1985). Chronic pain: Electromechanical recording device for measuring patients' activity patterns. *Arch Phys Med Rehabil, 66*(2), 75–79.
Gage, J. R., Schwartz, M. H., Koop, S. E., & Novacheck, T. F. (2009). *The Identification and Treatment of Gait Problems in Cerebral Palsy*. London: Mac Keith Press.
Galna, B., Peters, A., Murphy, A. T., & Morris, M. E. (2009). Obstacle crossing deficits in older adults: A systematic review. *Gait Posture, 30*(3), 270–275.
Gardener, E. A., Huppert, F. A., Guralnik, J. M., & Melzer, D. (2006). Middle-aged and mobility-limited: Prevalence of disability and symptom attributions in a national survey. *J Gen Intern Med, 21*(10), 1091–1096.
Glaister, B. C., Bernatz, G. C., Klute, G. K., & Orendurff, M. S. (2007). Video task analysis of turning during activities of daily living. *Gait Posture, 25*(2), 289–294.
Grant, P. M., Ryan, C. G., Tigbe, W. W., & Granat, M. H. (2006). The validation of a novel activity monitor in the measurement of posture and motion during everyday activities. *Br J Sports Med, 40*(12), 992–997.
Grood, E., & Suntay, W. (1983). A joint coordinate system for the clinical description of three-dimensional motions: Application to the knee. *Trans ASME J Biomech Eng, 105*, 136–143.
Halliday, S. E., Winter, D. A., Frank, J. S., Patla, A. E., & Prince, F. (1998). The initiation of gait in young, elderly, and Parkinson's disease subjects. *Gait Posture, 8*(1), 8–14.
Huddleston, J., Alaiti, A., Goldvasser, D., Scarborough, D., Freiberg, A., Rubash, H., et al. (2006). Ambulatory measurement of knee motion and physical activity: Preliminary evaluation of a smart activity monitor. *J Neuroeng Rehabil, 3*, 21.
Huxham, F., Baker, R., Morris, M. E., & Iansek, R. (2008a). Footstep adjustments used to turn during walking in Parkinson's disease. *Mov Disord, 23*(6), 817–823.
Huxham, F., Baker, R., Morris, M. E., & Iansek, R. (2008b). Head and trunk rotation during walking turns in Parkinson's disease. *Mov Disord, 23*(10), 1391–1397.
Huxham, F., Gong, J., Baker, R., Morris, M., & Iansek, R. (2006). Defining spatial parameters for non-linear walking. *Gait Posture, 23*(2), 159–163.
Iezzoni, L. I., McCarthy, E. P., Davis, R. B., & Siebens, H. (2001). Mobility difficulties are not only a problem of old age. *J Gen Intern Med, 16*, 235–243.
Institute of Medicine. (2002). *Dietary Reference Intakes for Energy, Carbohydrates, Fiber, Fat, Protein and Amino Acids (Macro-nutrients)*. Washington, DC: Institute of Medicine.
Jaeger, R. J., & Vanitchatchavan, P. (1992). Ground reaction forces during termination of human gait. *J Biomech, 25*(10), 1233–1236.
Jago, R., Watson, K., Baranowski, T., Zakeri, I., Yoo, S., Baranowski, J., et al. (2006). Pedometer reliability, validity and daily activity targets among 10- to 15-year-old boys. *J Sports Sci, 24*(3), 241–251.
Kadaba, M. P., Ramakrishnan, H. K., & Wootten, M. E. (1990). Measurement of lower extremity kinematics during level walking. *J Orthop Res, 8*(3), 383–392.

Kadaba, M. P., Ramakrishnan, H. K., Wootten, M. E., Gainey, J., Gorton, G., & Cochran, G. V. (1989). Repeatability of kinematic, kinetic, and electromyographic data in normal adult gait. *J Orthop Res, 7*(6), 849–860.

Kalman, R. E. (1960). A new approach to linear filtering and prediction problems. *Trans ASME J Basic Eng (D), 82*, 35–45.

Kuo, Y. L., Culhane, K. M., Thomason, P., Tirosh, O., & Baker, R. (2009). Measuring distance walked and step count in children with cerebral palsy: An evaluation of two portable activity monitors. *Gait Posture, 29*(2), 304–310.

Lee, I. M., & Buchner, D. M. (2008). The importance of walking to public health. *Med Sci Sports Exerc, 40*(7 Suppl), S512–S518.

Lord, S. R., Sherrington, C., Menz, H. B., & Close, J. C. T. (2007). *Falls in Older People: Risk Factors and Strategies for Prevention* (2nd ed.). Cambridge: Cambridge University Press.

Lu, T. W., & O'Connor, J. J. (1999). Bone position estimation from skin marker coordinates using global optimisation with joint constraints. *J Biomech, 32*(2), 129–134.

Luinge, H. J., & Veltink, P. H. (2004). Inclination measurement of human movement using a 3-D accelerometer with autocalibration. *IEEE Trans Neural Syst Rehabil Eng, 12*(1), 112–121.

Maddison, R., & Ni Mhurchu, C. (2009). Global positioning system: A new opportunity in physical activity measurement. *Int J Behav Nutr Phys Act, 6*, 73.

Melzer, D., Gardener, E., & Guralnik, J. M. (2005). Mobility disability in the middle-aged: Cross-sectional associations in the English Longitudinal Study of Ageing. *Age Ageing, 34*(6), 594–602.

Morris, M. E., Baker, R., Dobson, F., McGinley, J., Bilney, B., Dodd, K., et al. (2005). Clinical gait analysis in neurology. In J. Hausdorff & N. Alexander (Eds.), *Evaluation and Management of Gait Disorders* (pp. 247–271). Boca Raton, FL: Taylor & Francis.

Orendurff, M. S., Bernatz, G. C., Schoen, J. A., & Klute, G. K. (2008). Kinetic mechanisms to alter walking speed. *Gait Posture, 27*(4), 603–610.

Orendurff, M. S., Schoen, J. A., Bernatz, G. C., Segal, A. D., & Klute, G. K. (2008). How humans walk: Bout duration, steps per bout, and rest duration. *J Rehabil Res Dev, 45*(7), 1077–1089.

Ounpuu, O., Davis, R., & Deluca, P. (1996). Joint kinetics: Methods, interpretation and treatment decision-making in children with cerebral palsy and myelomeningocele. *Gait Posture, 4*, 62–78.

Pate, R. R., Pratt, M., Blair, S. N., Haskell, W. L., Macera, C. A., Bouchard, C., et al. (1995). Physical activity and public health: A recommendation from the Centers for Disease Control and Prevention and the American College of Sports Medicine. *JAMA, 273*(5), 402–407.

Peters, A., Baker, R., & Sangeux, M. (2009). Quantification of soft tissue artefact in lower limb human movement analysis—a systematic review. *Gait Posture*, in press.

Pirpiris, M. (2002). Single event multi-level surgery in spastic diplegia: Comprehensive outcome analysis. PhD thesis, University of Melbourne, Melbourne, Australia.

Pirpiris, M., & Graham, H. K. (2004). Uptime in children with cerebral palsy. *J Pediatr Orthop, 24*(5), 521–528.

Preece, S. J., Goulermas, J. Y., Kenney, L. P., & Howard, D. (2009). A comparison of feature extraction methods for the classification of dynamic activities from accelerometer data. *IEEE Trans Biomed Eng, 56*(3), 871–879.

Reinbolt, J. A., Schutte, J. F., Fregly, B. J., Koh, B. I., Haftka, R. T., George, A. D., et al. (2005). Determination of patient-specific multi-joint kinematic models through two-level optimization. *J Biomech, 38*(3), 621–626.

Rodriguez, D. A., Brown, A. L., & Troped, P. J. (2005). Portable global positioning units to complement accelerometry-based physical activity monitors. *Med Sci Sports Exerc, 37*(11 Suppl), S572–S581.

Roetenberg, D., Luinge, H. J., Baten, C. T., & Veltink, P. H. (2005). Compensation of magnetic disturbances improves inertial and magnetic sensing of human body segment orientation. *IEEE Trans Neural Syst Rehabil Eng, 13*(3), 395–405.

Roetenberg, D., Slycke, P. J., & Veltink, P. H. (2007). Ambulatory position and orientation tracking fusing magnetic and inertial sensing. *IEEE Trans Biomed Eng, 54*(5), 883–890.

Rosenbaum, P., Paneth, N., Leviton, A., Goldstein, M., & Bax, M. (2007). A report: The Definition and Classification of Cerebral Palsy April 2006. *Dev Med Child Neurol, 49*(Suppl 2), 8–14.

Rueterbories, J., Spaich, E. G., Larsen, B., & Andersen, O. K. (2010). Methods for gait event detection and analysis in ambulatory systems. *Med Eng Phys, 32,* 545–552.

Ryan, C. G., Grant, P. M., Tigbe, W. W., & Granat, M. H. (2006). The validity and reliability of a novel activity monitor as a measure of walking. *Br J Sports Med, 40*(9), 779–784.

Said, C. M., Goldie, P. A., Patla, A. E., & Sparrow, W. A. (2001). Effect of stroke on step characteristics of obstacle crossing. *Arch Phys Med Rehabil, 82*(12), 1712–1719.

Said, C. M., Goldie, P. A., Patla, A. E., Sparrow, W. A., & Martin, K. E. (1999). Obstacle crossing in subjects with stroke. *Arch Phys Med Rehabil, 80*(9), 1054–1059.

Sanders, S. H. (1980). Toward a practical instrument system for the automatic measurement of "up-time" in chronic pain patients. *Pain, 9*(1), 103–109.

Sanders, S. H. (1983). Automated versus self-monitoring of 'up-time' in chronic low-back pain patients: A comparative study. *Pain, 15*(4), 399–405.

Schepers, H. M., Roetenberg, D., & Veltink, P. H. (2010). Ambulatory human motion tracking by fusion of inertial and magnetic sensing with adaptive actuation. *Med Biol Eng Comput, 48*(1), 27–37.

Schepers, H. M., van Asseldonk, E. H., Buurke, J. H., & Veltink, P. H. (2009). Ambulatory estimation of center of mass displacement during walking. *IEEE Trans Biomed Eng, 56*(4), 1189–1195.

Schneider, P. L., Crouter, S. E., & Bassett, D. R. (2004). Pedometer measures of free-living physical activity: Comparison of 13 models. *Med Sci Sports Exerc, 36*(2), 331–335.

Schneider, P. L., Crouter, S. E., Lukajic, O., & Bassett, D. R., Jr. (2003). Accuracy and reliability of 10 pedometers for measuring steps over a 400-m walk. *Med Sci Sports Exerc, 35*(10), 1779–1784.

Schwartz, M. H., & Rozumalski, A. (2005). A new method for estimating joint parameters from motion data. *J Biomech, 38*(1), 107–116.

Sparrow, W. A., & Tirosh, O. (2005). Gait termination: A review of experimental methods and the effects of ageing and gait pathologies. *Gait Posture, 22*(4), 362–371.

Sproston, K., & Primatesta, P. (2002). *Health Survey for England 2002: The Health of Children and Young People.* London: The Stationery Office.

Tirosh, O., & Sparrow, W. A. (2005). Age and walking speed effects on muscle recruitment in gait termination. *Gait Posture, 21*(3), 279–288.

Tudor-Locke, C. (2003). *Manpo-Kei: The Art and Science of Step Counting*. Victoria, BC, Canada: Trafford.

Tudor-Locke, C., Hatano, Y., Pangrazi, R. P., & Kang, M. (2008). Revisiting "how many steps are enough?" *Med Sci Sports Exerc, 40*(7 Suppl), S537–S543.

US Department of Health and Human Services. (1996). *Physical Activity and Health: A Report of the Surgeon General*. Atlanta, GA: US Department of Health and Human Services, Centres for Disease Control and Prevention, National Centre for Chronic Disease Prevention and Health Promotion.

White, J., & Strong, J. (1992). Measurement of activity levels in patients with chronic pain. *Occup Ther J Res, 32*, 217–227.

Wild, S., Roglic, G., Green, A., Sicree, R., & King, H. (2004). Global prevalence of diabetes: Estimates for the year 2000 and projections for 2030. *Diabetes Care, 27*(5), 1047–1053.

World Health Organisation. (1980). *International Classification of Impairments, Disabilities and Handicaps*. Geneva: World Health Organisation.

World Health Organisation. (2001). *International Classification of Functioning, Disability and Health*. Geneva: World Health Organisation.

Wu, G., Siegler, S., Allard, P., Kirtley, C., Leardini, A., Rosenbaum, D., et al. (2002). ISB recommendation on definitions of joint coordinate system of various joints for the reporting of human joint motion—Part I: Ankle, hip, and spine. International Society of Biomechanics. *J Biomech, 35*(4), 543–548.

Wu, G., Siegler, S., Allard, P., Kirtley, C., Leardini, A., Rosenbaum, D., et al. (2003). ISB recommendation on definition of joint coordinate systems for the reporting of human joint motion. Part I: Ankle, hip and spine. *J Biomech, 36*(2), 303–304.

Zhang, K., Werner, P., Sun, M., Pi-Sunyer, F. X., & Boozer, C. N. (2003). Measurement of human daily physical activity. *Obes Res, 11*(1), 33–40.

Zijlstra, W., & Aminian, K. (2007). Mobility assessment in older people: New possibilities and challenges. *Eur J Ageing, 4*, 3–12.

Zijlstra, W., & Hof, A. L. (2003). Assessment of spatio-temporal gait parameters from trunk accelerations during human walking. *Gait Posture, 18*(2), 1–10.

# 8

# *Motion Sensors in Osteoarthritis: Prospects and Issues*

**Tim V. Wrigley**

**CONTENTS**

Introduction ........................................................................................................ 183
Movement Analysis in OA.................................................................................... 184
Laboratory-Based Motion Measures in OA............................................................ 185
Net External Knee Adduction Moment ............................................................... 185
Impact Forces ........................................................................................................ 187
Kinematic Differences........................................................................................... 188
Prospective Motion Sensor Technologies for Knee OA............................................ 189
Motion Sensors ...................................................................................................... 191
Precedents and Prospects for the Use of Motion Sensors in OA................................ 198
Field-Based Motion Measures Not Derived from Laboratory Measures.... 205
Conclusions............................................................................................................ 208
References.............................................................................................................. 210

## Introduction

Osteoarthritis (OA) in the joints of the lower limb is a disease that affects mobility. Its incidence and progression are thought to be related to joint loading during movement, particularly walking. Therefore, motion sensors have a potential role to play in assessment and management of the disease. OA is a degenerative disease that is most often thought of as primarily affecting the articular cartilage within the affected joints; however, it is in fact associated with a range of pathological changes in joints and a range of related impairments in function. It commonly affects certain joints (knee, hip, spine, hands) but others much less often (ankle, elbow, wrist). OA is a major cause of disability amongst older adults (Zhang and Jordan 2008). The exact cause of OA is unknown, and there is no cure. Treatment is directed towards management of symptoms, the key one of which is pain (although not everyone with *structural* OA evident on x-ray will have pain). For severe disease, surgical joint replacement may become necessary.

The relevance of motion sensors to OA mainly concerns knee and hip OA. Risk factors for OA development in these joints include ageing, female gender, obesity, a history of joint injury, some physically demanding occupations, uncertain inherited factors, bony malalignment and muscle weakness (Bierma-Zeinstra and Koes 2007; Felson 2004; Doherty 2001). Some of these factors may also affect how rapidly the disease progresses (Doherty 2001). Many of the preceding factors are associated with increased/abnormal joint loading. Hence there has been a large focus in the OA research literature on understanding the role of joint loading in OA.

To this end, gait analysis has been a key element in investigating the nature of routine joint loading in OA of the knee and hip joints, in particular. Knee joint loading has been directly associated with the progression of OA (Miyazaki et al. 2002) and possibly also its development (Amin et al. 2004). As there is no cure for OA, understanding how loading may be altered with conservative treatments, to potentially slow OA progression and improve symptoms, is particularly important. While joints can be surgically replaced when the disease becomes severe, not only is such surgery costly, but replaced joints also have a finite lifespan. Although this is improving all the time, surgical revision of a worn or malfunctioning replacement joint is technically difficult. Therefore, delaying the need for the initial joint replacement as long as possible is important.

This chapter focuses on the prospects for the use of motion sensors in assessing joint motion and load-related parameters, particularly during walking, in patients with OA, especially knee OA. However, many of the issues that remain challenging in the application of motion sensors for OA also apply to the use of such sensors in other musculoskeletal and neurological conditions. While there are specific movement parameters that are of key interest in OA, which might be measurable by sensors, similar parameters are important in other pathological conditions that affect human movement. Therefore, the chapter is a critical assessment of the prospects for useful movement measurement by discrete sensors—initially within the laboratory but also potentially in the clinic, home or even outdoors—that could be applied for patients with a range of medical conditions.

## Movement Analysis in OA

Traditionally, analysis of walking gait in patients with OA has been conducted either in the clinic, simply by the clinician observing the patient's walking gait, or in a gait laboratory, requiring complex and expensive 3D motion analysis systems. These are usually reflective marker-based, optical camera-based motion capture systems (based on stereophotogrammetry).

Clinical observation of gait is limited greatly by what the clinician can see and interpret visually, and 3D motion analysis of gait is not widely available. Therefore, 3D gait analysis has not been routinely used *clinically* in OA; it is primarily a research tool.

However, even with complex laboratory-based 3D gait analysis, it is not possible to fully understand how the knee and hip move or are loaded in a patient's daily activities (Bergmann et al. 2001; Foucher et al. 2010; Kutzner et al. 2010; Mundermann et al. 2008) based on recording a small number of walking trials under standardized conditions in a gait laboratory. It is the more typical cumulative loading (Maly 2008), over hours or days that can only be measured by long-term monitoring. *It is therefore in field-based motion monitoring that the major scope for motion sensors in OA lies.* However, there are still significant challenges to be overcome if this is to be achieved. Current work is still largely focused on establishing the utility of motion sensors for short-term measurements in relatively controlled laboratory and clinic environments. If this is successful, motion sensors may be able to move into more realistic environments, for long-term monitoring of joint health.

The progress which has already been made in the area of wearable motion sensors falls into two categories:

1. Sensors which measure the same parameters as those measured by the complex lab-based systems but in simpler form (e.g., measurement of a single parameter instead of the many parameters typically measured in the gait laboratory)
2. Sensors which measure novel parameters that are not measured in the laboratory, because they are only of relevance when recorded in more realistic settings, over long periods of time, that are representative of the range of movements that a patient makes in his or her daily life

We begin the chapter with a discussion of what has typically been measured in the 3D gait laboratory in patients with OA, to set the scene for the consideration of discrete motion sensors. That review considers the potential of such sensors for use in motion measurement in OA. Then, existing motion sensor developments for OA are reviewed. Finally, at the end of the chapter, we discuss sensor measures that by their nature can be assessed only outside the laboratory.

## Laboratory-Based Motion Measures in OA

### *Net External Knee Adduction Moment*

As most of the literature on movement analysis in OA concerns the knee, this is the focus of much of this chapter. A significant number of studies have shown differences between OA patients and healthy controls for basic *spatio-temporal* measures of walking gait, such as stride length and velocity; OA patients walk more slowly and have shorter stride lengths (Ornetti et al. 2010), and these effects are more evident with increasing OA disease severity (Astephen et al. 2008). However, laboratory gait analysis in knee OA research has focused on a *kinetic* parameter known as the *net external knee adduction moment* (KAM) during the stance phase of gait (Block and Shakoor 2010). This is due to its close association with the loading of the medial compartment of the knee joint (Schipplein and Andriacchi 1991; Zhao, Banks, Mitchell, et al. 2007), where most knee OA occurs (Ledingham et al. 1993). The KAM is a moment (torque) in the frontal plane that tends to push the knee into adduction (i.e., towards a more "bow-legged" position) during stance. It is largely due to the ground reaction force (GRF) vector—which the leg experiences when the foot is on the ground—passing medial to the knee joint in the frontal plane (Hunt et al. 2006; Figure 8.1). This external moment is in a net *adduction* direction throughout almost all of the stance phase in most individuals, whether healthy or with knee OA. As such, the balance of forces across the medial and lateral compartments of the knee is biased such that more compressive force is borne by the medial compartment (Andriacchi 1994; Schipplein and Andriacchi 1991).

The *peak* KAM is the knee loading parameter which has been related to OA disease *progression*; a higher peak KAM increases the risk for disease progression (Miyazaki et al. 2002). The peak KAM increases with OA severity (Sharma et al. 1998), varus malalignment (Specogna et al. 2007; Wada et al. 2001) and faster walking velocity (Mundermann et al. 2004), and it is also associated with pain (Thorp et al. 2007) and surgical outcomes following high tibial osteotomy surgery (Prodromos, Andriacchi, and Galante 1985). Whether or not the peak KAM is related to the *initiation* of OA disease is less certain; however, there is some evidence suggesting this (Amin et al. 2004). While the KAM is closely related to medial compartment loading, as a net external moment it is obviously not a direct measure of the compressive forces between the bone surfaces. Unfortunately, such forces can only be measured by an instrumented knee prosthesis; these have actually been implanted in a very small number of patients undergoing knee replacement for knee OA (D'Lima et al. 2008; Kutzner et al. 2010; Mundermann et al. 2008; Zhao, Banks, D'Lima, et al. 2007). The KAM is closely associated with such direct measures of compressive joint force (Zhao, Banks, Mitchell, et al. 2007). The compressive joint forces can also be *estimated* by complex musculoskeletal mathematical models; however, there are many disparate modelling approaches that are still evolving and have yet to be fully validated (Buchanan et al. 2004, 2005; Correa et al. 2010; Lin et al. 2010; Shelburne, Torry, and Pandy 2006; Winby et al. 2009).

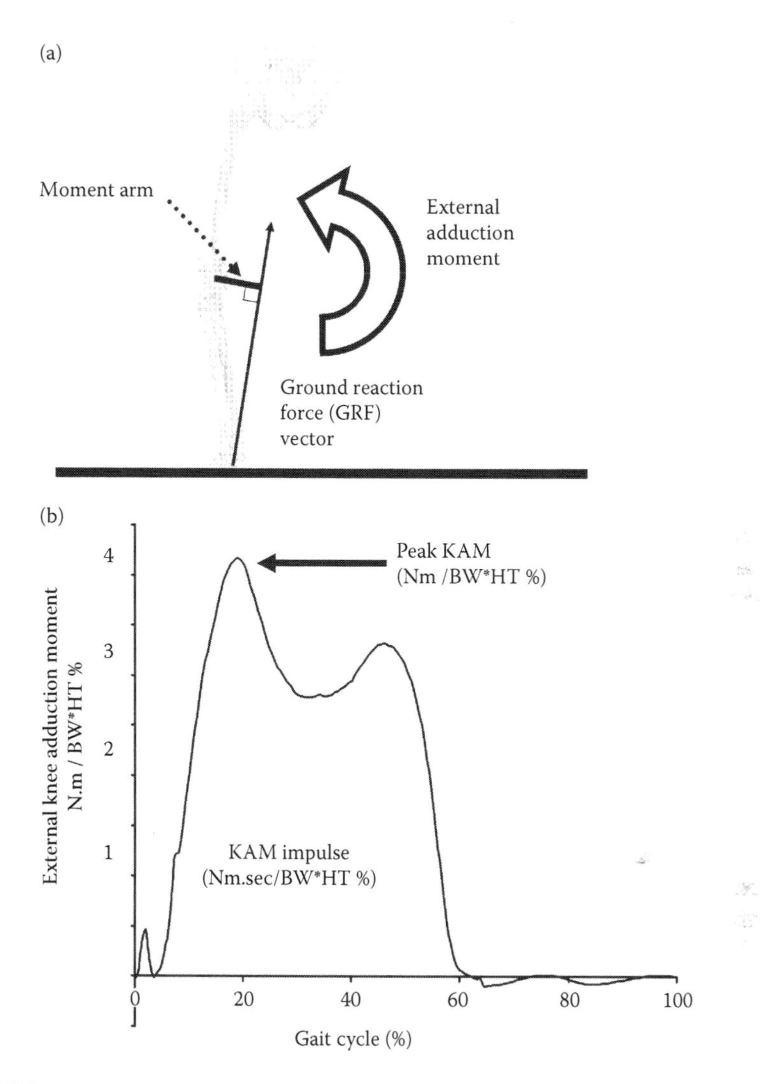


**FIGURE 8.1**
(a) Main factors involved in the net external knee adduction moment (KAM): frontal plane ground reaction force (GRF) and its moment arm; (b) Graph of the KAM, showing peak KAM and KAM impulse (area under curve).

### *Impact Forces*

Impact forces associated with the initial contact of the foot with the ground at the beginning of stance have also been of interest in OA, although they are not part of routine gait assessments (Collins and Whittle 1989; Gill and O'Connor 2003a, 2003b; Henriksen et al. 2006; Hunt, Hinman, et al. 2010; Liikavainio et al. 2007; Mikesky, Meyer, and Thompson 2000; Riskowski et al.

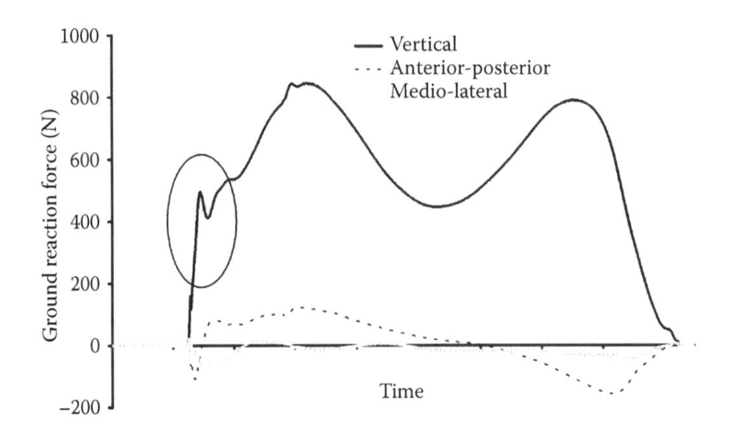


**FIGURE 8.2**
Ground reaction forces (GRF), showing heel strike transient (circled) in the vertical GRF just after foot–ground contact.

2005; Whittle 1999). Impact forces have been associated with OA development in animals (Radin et al. 1984) and knee pain in humans (Radin et al. 1991). However, their significance for human OA is not entirely clear. In a small group of OA patients and healthy control subjects, Henriksen et al. (2006) found no differences in impact forces between the groups. However, the actual impact forces within the human knee joint cannot be measured directly; therefore, they have mostly been inferred from the GRFs measured under the feet by laboratory force platforms or by an accelerometer mounted on the lower leg (Folman et al. 1986; Henriksen et al. 2006). For example, the nature of impact loading has been quantified by the heel strike transient (HST) in the vertical GRF at the very beginning of stance, exhibited by 20–30% of individuals with and without OA (Figure 8.2; Hunt, Hinman, et al. 2010; Radin, Whittle, and Yang 1986). The initial slope of the GRF has also been quantified to assess the rate of loading of the lower limb during initial stance. Impact forces appear to be greater in males than females, and they may increase with age (Henriksen et al. 2008; Robbins, Waked, and Krouglicof 2001).

## Kinematic Differences

In addition to the load-related (i.e., *kinetic*) movement parameters that have been of interest in OA such as the KAM and impact forces, patients with OA also demonstrate some changes in knee joint kinematics, i.e., the position and angular orientation of the knee joint during gait. These include changes in the sagittal plane knee joint flexion angle at the point of foot–ground contact and later during stance (Astephen et al. 2008; Kaufman et al. 2001; Maly, Costigan, and Olney 2006; Mundermann, Dyrby, and Andriacchi 2005; Rudolph, Schmitt, and Lewek 2007). In the frontal plane, the most common OA of the medial knee compartment is often associated with static varus malalignment (bow legs) (Andriacchi 1994; Brouwer et al. 2007; Cicuttini et al. 2004; Hunter et al. 2007; Sharma et al. 2001). This may also be associated with *dynamic* changes in frontal plane knee kinematics during gait. Patients with knee OA sometimes exhibit a varus thrust, where the knee undergoes a sudden varus (lateral) movement in the early part of the stance phase after initial contact with the ground, which has been defined as "the dynamic worsening or abrupt onset of varus alignment as the limb accepted weight, with a return to less varus alignment during lift-off and the swing phase of gait" (Chang et al. 2010, p. 1405). It occurs in 20–40% of patients with OA. The origins of the thrust are not clearly understood, but it is possibly associated with static varus malalignment, lateral joint laxity and limitations in muscle control. The varus thrust has been linked to faster progression of OA disease (Chang et al. 2004). Clinicians generally judge the presence of a varus thrust based on simple subjective observation of the patient walking (Hunt, Schache, et al. 2010). Varus–valgus angular motion of the knee has also been measured from frontal plane knee kinematics using conventional optical motion capture (in patients with or without varus thrust), although the significance of this measure is still under investigation (Foroughi et al. 2010; Hunt et al. 2008; Hunt, Schache, et al. 2010; van der Esch et al. 2008). As discussed later, accelerometry has also been used to assess the varus thrust (Ogata, Yasunaga, and Nomiyama 1997).

## Prospective Motion Sensor Technologies for Knee OA

As reviewed in the preceding section, the KAM, impact forces and certain sagittal and frontal plane kinematic measures are parameters of interest in knee OA. We now review the prospects for their measurement with discrete sensors, both within and outside the laboratory environment. The subsequent section reviews actual precedents for sensor use in OA, albeit limited.

In its simplest form, the KAM is the product of the frontal plane GRF magnitude and its frontal plane moment arm, that is, the distance from this force vector to the knee joint (Hunt et al. 2006). The higher the GRF and the further the knee is from the GRF vector in the frontal plane, the greater the KAM will be. The magnitude of the GRF is largely determined by body weight and walking velocity (Andriacchi, Ogle, and Galante 1977; Browning and Kram 2007; Messier et al. 1996). The position of the knee in relation to the GRF vector in the frontal plane is related to the knee's frontal plane alignment, particularly the extent of the commonly seen varus malalignment (bow legs). These determinants of the KAM are reflected in the way that absolute KAM values are normalized for body size differences, in order to allow comparison between patients of different size: this is most commonly done by dividing the *absolute* KAM (Newton metres, N·m) by body weight (Newtons, N) multiplied by a linear body dimension (metres, m), i.e., body weight × height (N·m) (Moisio et al. 2003).

While the camera-based motion capture systems use the more complex mechanical computation process of inverse dynamics to derive joint moments such as the KAM (Robertson 2004), the determination of the KAM, at its simplest, requires measures of the GRF vector (origin, orientation, magnitude) and the position of the knee joint centre. While these are readily measured in a 3D gait laboratory by floor-embedded force platforms and motion capture systems, moving such measurements out of the gait laboratory is problematic. While accurate force-instrumented shoes have been described, these are currently not practical or affordable for routine use (Liedtke et al. 2007; Liu, Inoue, and Shibata 2010; Schepers, Koopman, and Veltink 2007; Veltink et al. 2005). Foot pressure–measuring insoles are available, but they are still quite expensive and not very compact or durable; they also encumber the subjects with processing electronics and cables and may have accuracy issues when used over long periods of time. Furthermore, they generally measure only pressure/force *normal* (perpendicular) to the insole; this does not allow the magnitude and orientation of the important horizontal GRF vectors to be determined directly. However, Forner-Cordero et al. (2004, 2006) have shown that it is possible to estimate the GRF vectors from such insoles, within limits and in conjunction with a motion capture system to track the position of the feet/insoles, for the purpose of inverse dynamics analysis. Although the horizontal GRFs cannot be measured, reasonably successful attempts have been made to predict them (Fong et al. 2008; Rouhani et al. 2010; Savelberg and de Lange 1999). As such, perhaps the remaining barrier to the more widespread use of such insoles is the high expense of commercial pressure/force-sensing insole products, although the basic sensing technology is available more cheaply. If a smaller number of sensors than is typically used in the pressure-sensing insole systems (Bamberg et al. 2008; Morris and Paradiso 2002) were shown to be adequate for GRF estimation, then this problem would be greatly simplified.

Even if the GRF vector were measurable in the field during free walking, measuring knee joint *position* outside the laboratory with any discrete sensor is also currently problematic, as can be seen in the following discussion of sensor capabilities. Therefore, estimation of the KAM outside the gait laboratory is currently precluded by the absence of suitable GRF- and knee position–sensing technologies. As such, the best prospects for assessing knee joint loading in the field currently lie with proxy or surrogate measures that *indirectly* reflect knee joint loading, rather than measuring it more directly.

Where a patient can initially undergo lab-based gait analysis to determine his or her KAM at typical walking speed, this can be combined in a simple approach with field-based measures of step count (i.e., by pedometer) to estimate overall knee load exposure during typical daily activities. The peak KAM could be used for this, or measures such as the KAM impulse—the integrated area under the KAM-time curve (Maly 2008; Robbins et al. 2009); see Figure 8.1. Such cumulative load measures are still experimental, and, while promising, they have yet to be widely investigated. Some existing precedents for the monitoring of steps taken and activity levels in patients with OA are discussed in the final section of this chapter.

Therefore, the current prospects for useful field-based motion sensing in knee OA are in the area of kinematic measures that are known to differ in knee OA and also could feasibly be used to *infer* joint load exposure. Furthermore, these measures should be made over long periods, as patients go about their activities of daily living, to assess cumulative joint loading.

## Motion Sensors

The major discrete motion-sensing/tracking technologies that have found use in human motion measurement can be divided into a number of categories:

- Mechanical (e.g., electrogoniometers)
- Ultrasonic
- Electromagnetic field
- Inertial (accelerometers and gyroscopes)
- Hybrid (i.e., combinations of two or more of the preceding)

At a minimum, a fully useful sensor-based motion analysis system for walking gait should be able to measure anatomical landmark positions and body segment/joint angles. While the discrete sensors described in the following offer a range of motion parameters including acceleration and velocity, and sometimes measure or derive angle and position estimates, such parameters are not necessarily sufficient to measure body position and joint angles with similar fidelity to what can be achieved in the gait laboratory with camera-based motion analysis systems. The practical application of these sensors to human movement analysis is also often associated with issues related to factors such as fixation to the body, skin movement artefact, sensitivity to signal-distortion sources in the environment, uncertain relationship to *anatomical* motion, measurement of relative orientation *between* sensors, etc. While some of these issues also affect the camera-based motion analysis systems commonly used in gait laboratories (e.g., skin movement artefact; Cappozzo 2008; Della Croce et al. 2005), it seems that discrete sensors currently suffer from a greater number of such issues. We now consider each of the preceding technologies in turn.

Mechanical motion-sensing systems such as *electrogoniometers* consist of a potentiometer or transducer technology that measures the joint *angle* directly when attached across a joint on the body, such as the knee. They are often quite encumbering to the patient (Morlock et al. 2001; Myles et al. 2002), are difficult to maintain in correct anatomical alignment and have some issues in relation to accuracy (Legnani et al. 2000; Shiratsu and Coury 2003). A knee brace can be instrumented with a potentiometer to measure knee angle (Riskowski et al. 2009); however, braces are not a viable long-term option for patients (although braces are worn by a small proportion of knee OA patients).

Discrete *ultrasonic sensors* can measure the distance between a transmitter and a microphone, based on pulse transit time. 3D *position* can be calculated by triangulation of the distances of a transmitter from more than one microphone in a structured arrangement. Transmission is, however, limited by distance and the need for line of sight (like camera-based systems); sampling frequency may also be somewhat limited (Kindratenko 2001). While not a commonly used technology in human motion analysis, commercial systems are available (Zebris GmbH, Isny, Germany). Three transmitters per body segment are needed to derive 3D segment orientation. In small-volume, fixed environments, ultrasonic systems have been implemented with fixed microphone arrays in combination with inertial sensing, exploiting the position-sensing capability of the former, and the latter for sensor orientation (Intersense Inc, Billerica, Massachusetts, USA).

*AC or DC electromagnetic field sensors* can measure absolute *position* and *orientation* in relation to a local magnetic source a short distance away, but that restricts their practical range (Nixon et al. 1998). These sensors found popularity in human animation because of their real-time capabilities, which predated those of the camera-based motion capture systems (Ascension Technology Corporation, Milton, Vermont, USA; Polhemus, Vermont, USA). This capability facilitates immediate assessment of the measured motion. Magnetic field sensors are not dependent on line of sight, as the fields pass through human bodies. However, these sensors are subject to distortion due to metal in the environment (LaScalza, Arico, and Hughes 2003), which, if permanently fixed in place, can be mapped and compensated for in routine capture environments. Interference is also due to mains power wiring and computer monitors, mainly with DC sensors (Foxlin 2002). These sensors have been used for human gait analysis, for example, walking on a treadmill (Mills et al. 2007). But perhaps because of the small operating range, they have found greater use for measurement of stationary activities, for example, mounted on the back for analysis of spinal motion (Hindle et al. 1990; Pearcy and Hindle 1989), as well as measurement of 6DOF (degrees of freedom) joint motion in vitro (Amis et al. 2008).

Despite their limitations, these sensors have an advantage over the increasingly popular inertial sensors (accelerometers and gyroscopes) to be discussed next, in their ability to measure *position* directly and accurately. Together with the ability to measure sensor angular orientation, they are the closest of all the discrete sensors to the type of data output produced by the conventional optical motion capture systems most often used in gait analysis. Indeed, while optical motion capture systems require at least three non-colinear, small, spherical, reflective markers to estimate each 3D body segment's orientation, magnetic field sensors provide this information from one sensor (albeit a cabled sensor subject to soft tissue artefact and difficulties in consistently defining its exact position and orientation in relation to the underlying body segment anatomy). The position-sensing capability also allows the use of a handheld sensor-instrumented probe during the patient calibration phase, to digitize anatomical landmark positions in relation to each segment's sensor coordinate system for subsequent gait analysis (Mills et al. 2007). The global location of those virtual anatomical landmark positions is then calculated during movement in relation to their segment's measured, mounted-sensor position and orientation and then used to estimate body segment kinematics in the same fashion as if the landmarks had been measured by an optical motion capture system.

*Inertial sensors*—that is, accelerometers and gyroscopes—are becoming the most commonly used discrete sensors for human motion analysis research, due to their small size, reasonable cost and ready availability. However, they have some limitations that make them difficult to use. They remain largely research tools, having yet to find their way into many production human motion-measuring devices. Because of their increasing popularity and availability, we cover them in some depth here.

Most *accelerometers* are sensitive to both gravitational and translational/rotational acceleration about each axis of the sensor (uni-, bi- and triaxial devices are available). Distinguishing between these different acceleration sources can complicate their use. Piezoelectric accelerometers are not sensitive to gravitational acceleration, having an effective low-frequency limit of approximately 0.5 Hz. However, this is not always an advantage, as the ability to measure the gravitational field (alone) when an accelerometer is stationary is often exploited to estimate sensor angle in relation to the vertical (gravitational) axis (but not *around* a vertical axis) (Tuck 2007). This ability can be of use in the measurement of static body postures (Wong and Wong 2008). However, during movement, separation of the gravitational acceleration from the translational and rotational acceleration associated with the movement requires additional information, complicating the measurement. Either multiple accelerometers on a single body segment can be used (Willemsen, Frigo, and Boom 1991; Willemsen, van Alste, and Boom 1990), or a single accelerometer can be used in conjunction with another sensor, such as a gyroscope. Accelerometer output can theoretically be double-integrated to yield *position*, but in practice, drift and the difficulty in separating the different components of acceleration have meant that this has rarely been pursued successfully. Despite their sensitivity to different acceleration sources, direct measurement of acceleration from discrete sensors—on its own—is sometimes the focus of human movement analysis

(Kavanagh and Menz 2008; Mathie et al. 2004). It may have application in an OA context in relation to the impacts between the foot and the ground at the beginning of the stance phase, as discussed earlier. In other cases, such impacts may represent an artefact that needs to be removed from the signal prior to its use for other purposes, such as segment angle estimation.

In simple terms, the accelerometer signal is due to a combination of gravity and movement of the sensor. In more detail, the signal (s) from an accelerometer attached to a moving body segment—at position **r** and with its sensitive axis having orientation vector **u** in the *segment* coordinate system—represents the projection of the acceleration of that point **r** along the vector **u** and can be expressed as (Giansanti et al. 2003; van den Bogert, Read, and Nigg 1996)

$$s = \mathbf{u} \bullet (R^{-1}(a - g) + \dot{\omega} \times \mathbf{r} + \omega \times (\omega \times \mathbf{r})),$$

where $R$ is the orientation matrix of the body segment with respect to the global coordinate system, $a$ is the linear acceleration of the segment centre of mass (COM), $g$ is the gravitational acceleration ($[0\ 0\ {-g}]^T$), and $\omega$ is the segment angular velocity in the segment coordinate system (and $\dot{\omega}$ its angular acceleration).

As signified by the presence in the preceding equation of the local position vector **r** of the accelerometer in the segment's coordinate system, the accelerometer signal depends on where the accelerometer is placed on the body segment. While indicating the complex nature of the accelerometer signal, the preceding equation does identify some potential for simplification. In the simplest possible case, when an accelerometer that is physically aligned with the body segment's coordinate system is stationary (or moving at constant velocity), all acceleration terms except gravity disappear, and the signal reduces to a simpler form from which the sensor's/segment's orientation can be derived by simple trigonometry (Tuck 2007):

$$s = R^{-1} \begin{bmatrix} 0 \\ 0 \\ -g \end{bmatrix}.$$

When the accelerometer is not stationary, and where segment orientation can be derived by additional means available in the laboratory (e.g., optical motion capture), elements of the signal that are irrelevant to a given research question can be removed from the accelerometer signal, such as that due to the gravitational acceleration (Henriksen et al. 2006) and angular acceleration (Lafortune and Hennig 1991). (This is also the basis of strapdown navigation algorithms, using accelerometers combined with gyroscopes, which are discussed later.) If the gravitational acceleration can be removed, then the linear velocity of the sensor may be derived by integration. Another integration step then yields displacement (position). However, integration is associated with drift artefact which can severely limit the accuracy of such estimates. The integration steps are

$$v = \int a.dt + v_0,$$

where $v$ is the linear velocity (m/s), $a$ is the (nongravitational) acceleration (m/s$^2$), and $v_0$ is the initial velocity (m/s), and

$$p = p_0 + \int v.dt,$$

where $p$ is the position (m), $v$ is the linear velocity (m/s), and $p_0$ is the initial position (m).

A completely different application of accelerometry—that is, activity monitoring—is discussed in the last section of this chapter. Interestingly, accelerometers have another potentially powerful application in human movement analysis. One of the more problematic aspects of camera-based motion analysis, utilizing inverse dynamics to estimate net joint moments (such as the KAM), is the need for body segment acceleration estimates. In marker-based motion analysis systems, these must be derived by double differentiation of segment position, an inherently noisy process. Therefore, accelerometers, which derive this quantity directly, *should* have a ready advantage for inverse dynamics. However, while the necessary procedures have been published (van den Bogert, Read, and Nigg 1996), this capacity has other practical difficulties (such as the need for multiple accelerometers per body segment) and therefore has rarely been exploited.

*Gyroscopes* measure angular velocity about one or more axes of the sensor (as for accelerometers, uni-, bi- and triaxial devices are available). Angular velocity is rarely of direct interest in human movement analysis (Damiano et al. 2006). Generally it is the absolute angle of a body segment in space, from which its angle in relation to another body segment can be derived (e.g., knee joint angle, being the angle between the shank and thigh), which is of primary kinematic interest in 3D gait analysis. When the angular velocity signal from a gyroscope is integrated, this provides the *change* in angle; if the *absolute* angle is desired, the initial angle of the device must be known, which is then added to the angle change from the integration. Any integration is subject to drift error in the angle estimate, due to sensor DC bias (i.e., gyroscopes do not measure exactly zero when the gyroscope is stationary), and this error accumulates over time. Unlike the accelerometer, the gyroscope is not influenced by the gravitational field. Also, the position of the gyroscope on the limb does not affect the measured angular velocity (Tong and Granat 1999). Integration of the gyroscope angular velocity to yield angle (θ) is given by

$$\theta = \theta_0 + \int \omega . dt,$$

where θ is the angle (°), ω is the angular velocity (°/s), and $\theta_0$ is the initial angle (°).

The outputs of accelerometers and gyroscopes can be combined to make up for the limitations of one sensor type by compensating with information from another sensor type, using fusion algorithms, sometimes based on a Kalman filter (Luinge and Veltink 2005; Welch 2009). When the gyroscope output is integrated to provide angular orientation, drift can be corrected when the sensor is detected as being stationary or moving at constant velocity (zero acceleration), by using an accelerometer to estimate orientation in relation to the vertical gravitational field (Williamson and Andrews 2001). For example, times of foot contact with the ground can be used to reset the velocity estimates of foot-mounted sensors (Sabatini et al. 2005; Veltink et al. 2003), and the foot angle can be derived from an accelerometer sensing its inclination angle from gravitational acceleration.

Combinations of accelerometers and gyroscopes to estimate *position* are often based on algorithms from strapdown navigation systems (Foxlin 2002; Sabatini 2006; Welch and Foxlin 2002). In this scheme, the gyroscope signal is first integrated to determine sensor orientation; this is then used to *estimate* the gravitational acceleration, which is a component of the accelerometer signal. This component is then removed from the accelerometer signal, allowing the double integration of the remaining nongravitational acceleration to derive position (Figure 8.3). However, this is much more easily said than done. As Foxlin (2002) pointed out, these algorithms were designed for aeronautical navigation, where large, expensive, highly accurate sensor systems require an accuracy only of the order of hundreds of metres or more to do their job well. Their application to human movement measurement, using

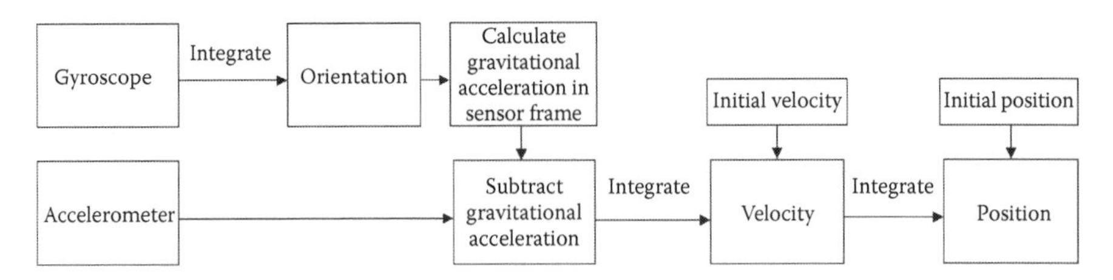


**FIGURE 8.3**
Strapdown navigation algorithm for the estimation of position from an inertial sensor incorporating an accelerometer and a gyroscope.

much smaller and cheaper sensors, requires an accuracy of the order of millimetres. As such, it is not surprising that Foxlin (2002) speculated that 6DOF human motion tracking using inertial sensors alone may never be viable. So while inertial sensors are becoming the most popular discrete sensors for human motion analysis research, even their proponents concede that *position* measurement, in particular, is not feasible with combined inertial sensors (gyroscopes and accelerometers; Favre et al. 2010), or accelerometers alone (Giansanti et al. 2003), to the level of accuracy achievable with conventional gait analysis systems.

*Hybrid* systems have been created that add an additional sensor, such as a magnetometer, to compensate for remaining deficiencies in the inertial sensors. A magnetometer is essentially an electronic compass, measuring heading (i.e., sensor direction in the horizontal plane) in relation to the earth's magnetic field (which is not provided by the other two sensor types). However, magnetometers are not without their own drawbacks, which can be severe, mainly due to their sensitivity to other ferromagnetic influences in their surrounding environment (de Vries et al. 2009; Roetenberg, Baten, and Veltink 2007).

Current commercially available inertial sensors that have found use for human motion analysis span a broad range but fall largely into three main categories:

1. Relatively inexpensive separate inertial sensing devices, i.e., single- or multi-axis accelerometers and gyroscopes (e.g., Analog Devices, Norwood, Massachusetts, USA; Freescale Semiconductor, Austin, Texas, USA), that must be physically and electronically integrated by the user if combined operation is desired.
2. Relatively inexpensive integrated sensors incorporating a single- or multi-axis accelerometer and gyroscope and, sometimes, a magnetometer (e.g., SparkFun Electronics, Boulder, Colorado, USA; Phidgets Inc., Calgary, Alberta, Canada). These sensors output the raw signals from each of their component sensors but provide no capability for combining the measures to provide the most commonly desired linear and angular kinematics, in particular, position and angular orientation.
3. More expensive sensor packages similar to those in (2) but often targeted specifically at human motion analysis and with the addition of proprietary, onboard, data fusion algorithms to yield more useful derived measurements, such as angular orientation (Euler/Cardan angles) and even position (e.g., Xsens Technologies BV, Enschede, Netherlands; Microstrain Inc, Williston, Vermont, USA). These systems may also incorporate proprietary compensation algorithms to improve system performance in situations where it may be affected by distortion, for example, that due to ferrous metal

in the environment on magnetometers. Such sensors may also be available packaged as a complete system of multiple sensors for full-body movement tracking, sometimes incorporating wireless signal transmission (e.g., Xsens Technologies BV, Enschede, Netherlands; Animazoo, West Sussex, UK).

## Precedents and Prospects for the Use of Motion Sensors in OA

Having discussed the theory associated with various discrete motion sensor technologies, we now review the extant literature that has moved beyond basic motion sensor research and development to their application for human movement analysis, in particular for OA. The literature on inertial sensors, in particular, for human motion analysis is large, which reflects the ease with which such devices can now be obtained. Furthermore, this literature reflects a large array of rather diverse approaches proposed for the use of such sensors in human motion analysis, which makes distilling useful information on usable solutions somewhat difficult. Indeed, in contrast to the large *volume* of literature, the literature on truly viable solutions for general human gait analysis is currently comparatively small. This reflects the difficulties in dealing with the limitations of the current generation of this technology. Furthermore, there have been relatively few studies thus far that have investigated the use of motion sensors *specifically* for OA. Therefore, the coverage here considers both applications *for* OA and those which provide suitable motion information that *could* be used for measuring the parameters of interest in knee OA, which were outlined earlier in the chapter.

We have already conceded that measurement of the KAM with discrete sensors is not feasible. That leaves kinematic and other measures. Amongst these, there are several studies that have investigated the occurrence of the varus (lateral) thrust in knee OA using sensors. As already noted, this is a sudden movement of the knee joint in a lateral direction in early stance just after foot–ground contact. Ogata, Yasunaga, and Nomiyama (1997) used a uniaxial accelerometer mounted at the tibial tuberosity to assess medio-lateral acceleration in patients with knee OA, as well as a strain gauge foot transducer to signal foot contact. All knees with medial OA showed the varus thrust pattern, as well as 50% of normal knees. Note that this is substantially higher than the reported incidence of varus thrust as determined by visual observation (Chang et al. 2010), the method used clinically. A similar varus thrust has been measured by the same accelerometer technique in younger knees with anterior cruciate ligament deficiency (Yoshimura et al. 2003), which is not uncommon in older knees with OA (Hill et al. 2005).

Impact forces in OA have been investigated with accelerometers (Henriksen et al. 2006; Voloshin and Wosk 1983; Voloshin, Wosk, and Brull 1981). However, as noted earlier, the significance of such impacts for human OA is somewhat uncertain. A knee brace instrumented with an accelerometer and potentiometer has been used to train subjects to modulate knee angle and acceleration prior to foot–ground contact, to reduce loading rate after contact (Riskowski 2010; Riskowski et al. 2009). But it has yet to be trialled in OA patients.

The most complex use of sensors specifically for OA is that of Turcot and colleagues (2008), aimed at resolving accelerations at the knee joint surface in knee OA patients. This is *not* a measure that is able to be made with conventional camera-based gait analysis systems, although some laboratories are increasing the number of markers on a single body segment to enhance the ability to estimate movements at the joint surface (Andriacchi and Dyrby 2005). However, the importance of joint accelerations in knee OA is not yet clear. But joint instability is a problem for some OA patients (Felson et al. 2007; Fitzgerald, Piva, and Irrgang 2004), and abnormal motions at the joint surfaces have been suggested to be implicated in the aetiology of OA (Andriacchi and Mundermann 2006; Andriacchi et al. 2004). Turcot and colleagues (2008) described a system made up of two inertial sensors, each consisting of a dual-axis accelerometer and gyroscope, with the signals recorded to a portable data logger. The devices were attached to a knee exoskeleton made up of thigh and tibial frames linked across the knee, originally designed to minimize soft tissue artefact in optical motion capture of knee joint kinematics (Sudhoff et al. 2007). While the exoskeleton provides a more rigid attachment for the sensors than direct placement on the skin, it does represent a significant encumbrance for the patient that would preclude its unsupervised use outside the laboratory; however, this device appears aimed at laboratory use only. It does, however, highlight one of the difficulties in the use of motion sensors, which would otherwise be attached to the skin and thus subject to potential soft tissue movement artefact (Forner-Cordero et al. 2008). Further complexities of this sensor technique include the need for simultaneous use of an optical motion capture system (for tracking reflective markers mounted on the exoskeleton to determine the position and orientation of each accelerometer's coordinate system) and a knee x-ray (to determine the position of the knee joint surface in relation to the tibial coordinate system). After estimating the gravitational and angular acceleration components (using measurements from the optical motion capture system), the internal linear acceleration of the tibia and femur at the joint surfaces was finally derived. A total of 18 parameters were extracted from the acceleration records, covering anterior-posterior (A-P), medio-lateral (M-L) and proximo-distal (P-D) linear acceleration maximums, minimums and ranges during the loading phase (2–12% gait cycle) and from pre-swing to mid-swing (55–90% gait cycle). Eight of these were found to be different between the nine OA patients and nine healthy controls studied, mainly in the M-L and A-P directions. It is possible that the M-L acceleration parameters were related to the varus thrust discussed earlier. Only two parameters of the untranslated, external acceleration at the sensor were different between the two groups. Note that Henriksen et al. (2006) found no difference in *longitudinal* tibial peak acceleration between asymptomatic and OA knees; however, they did not look at the M-L and A-P accelerations and did not resolve the accelerations at the joint surface, as done by Turcot et al. (2008).

As there has been limited work done on sensor use for knee or hip OA, here we highlight the work of a major research group that has been particularly active in the area of inertial sensors for lower-limb movement in general. As such, this is not intended as an exhaustive review of the use of inertial sensors for human motion analysis. However, it satisfies our aim to illustrate the evolution that has occurred in approaches over the last decade or so and the point that has now been reached. This group at the Laboratory of Movement Analysis and Measurement (LMAM) at the Ecole Polytechnique Federale de Lausanne (EPFL) in Switzerland has been active in this area over an extended period. The diverse range of approaches employed over a decade of research is illustrative of the difficulties in identifying an optimal approach to the use of inertial sensors for human movement analysis; these approaches are summarized in the following. Some of these have been applied to patients with OA (Aminian et al. 2004).

Aminian et al. (2004) described a system of four gyroscopes attached bilaterally to the shank and thigh, connected to a data logger, for use with patients with hip OA (both pre- and post-joint replacement surgery). Joint angles were estimated by integrating the angular velocity signals from the gyroscopes; the total ranges of thigh and shank rotation were calculated. Spatio-temporal gait parameters (gait cycle duration, stance time, double support time, stride length) were also estimated from the gyroscope signals (Aminian et al. 2002). The system was compared to a camera-based motion capture system. While there was good agreement between the two systems, this total angular range measure for the *separate* shank and thigh (rather than the hip or knee joint angle) is not usually a measure of interest in gait analysis; but the use of such a segmental angular difference measure rather than an *absolute* angular measure avoided the difficulty of estimating absolute start angles in gyroscope-based systems. However, these measures did show the ability to discriminate between hip OA patients with different levels of functional impairment.

A subsequent paper from this group (Favre et al. 2006) described an extension of this approach involving derivation of the *change* in *knee* joint angle (range of motion) by integrating the difference in the two signals from the thigh and shank gyroscopes. They later described an algorithm for deriving *absolute* uniplanar joint angles, more akin to those typically assessed in gait analysis (albeit 2D), based on a dual-axis accelerometer and uniaxial gyroscope (Dejnabadi, Jolles, and Aminian 2005). This integrated device was placed on the body segments above and below the knee, i.e., the thigh and shank. When placed in a semi-arbitrary position on a body segment, a sensor returns data in its own coordinate system, akin to the technical frame (coordinate system) of three non-colinear reflective markers used in conventional camera-based motion capture systems (Cappozzo et al. 2005). To convert this arbitrary coordinate system into an anatomically meaningful one, a transformation to an anatomical frame must be derived. In optical motion capture, this is usually done by a process of anatomical calibration, where particular anatomical landmarks are digitized by a pointer or a reflective marker, allowing the transformation between technical and anatomical frames to be calculated. In the described inertial sensor system, anatomical calibration to the patient required a digital photo of the patient in several fixed joint positions in relation to a calibration structure, from which the position and orientation of the sensors in relation to anatomical landmarks on the limb were measured. The photo was necessary to measure sensor positions, since the inertial sensors cannot measure their positions accurately. The algorithm used to derive joint angles required that the sensor positions be known accurately but did not require the problematic integration of the gyroscope signal. A subsequent paper extended that approach to include separate shank and thigh segment orientations (Dejnabadi et al. 2006).

A modified instrumentation approach and different processing algorithm using a triaxial accelerometer and gyroscope were next described by Favre et al. (2008). This system measured shank and thigh rotation in relation to a common global reference frame, as in conventional laboratory-based gait analysis systems. The difficult problem of mathematically aligning the two sensors' coordinate systems on the thigh and shank segments was addressed by having the patient perform a hip abduction–adduction movement, with the knee held firmly in full extension; as such, the angular velocities of the two segments in the common reference frame should be equal. The quaternion-based sensor fusion algorithm for sensor angle derivation was again based on gyroscope integration, supplemented by static orientation measurement from the accelerometer when the sensor was stationary (Favre et al. 1996). Drift error was estimated by a low-pass digital filter (0.2 Hz) and removed.

This system represented a significant advance in the use of purely inertial sensors, providing absolute angle data in a form similar to that of the conventional motion capture systems used in gait laboratories. However, a number of elements of the sensor-based approach, in particular, the alignment movement, would probably limit its use to the laboratory. The overall level of accuracy achieved was not dissimilar to that attainable with conventional motion capture systems available at the time. Indeed, one of the difficulties in attempting to validate inertial sensor systems is that there are also errors in the conventional motion capture systems used as the reference for comparison. While there have been technical improvements in the resolution of those systems since that time, other factors such as skin marker placement do place some limits on their accuracy and reliability (Cappozzo 2008; McGinley et al. 2009). Favre et al. (2009) described an enhancement to the preceding approach, involving mounting the sensors on a knee exoskeleton and perfoming functional calibration to align sensor data with segment rotation axes involving pure passive flexion/extension and rotation movements of the shank in relation to the thigh.

Alternative approaches for the determination of the segment anatomical coordinate system in relation to a sensor coordinate system and a fixed global reference coordinate system have been described for systems incorporating not only accelerometers and gyroscopes but also a magnetometer. These approaches are based on the performance of standardized active/passive movements (O'Donovan et al. 2007), combined with known, measured, static joint angles (Cutti et al. 2010; Ferrari et al. 2010), or a sensor-instrumented pointer apparatus (Picerno, Cereatti, and Cappozzo 2008). These techniques are all based on custom software. The vendor-provided sensor software does implement a more basic procedure to reset the sensor coordinate system to that of the segment it is attached to, while the segment is in a known orientation (to the extent that this is able to be achieved). For example, Bergmann et al. (2009) approximated alignment of the sensors with the sagittal plane and the vertical during standing, using the vendor's method, whereby once the sensor is attached to the body segment, the vertical axis is reset to the sensed vertical gravitational axis, while the forward axis remains in the same plane as it was when the sensor was initially placed on the segment, aligned with the sagittal plane (but again orthogonal to the vertical axis to form a right-hand coordinate system). This approach is based on the fact that, when accelerometers and magnetometers are used together, a reference coordinate system/frame using the earth's gravitational and magnetic field axes can be determined, and this coordinate system exists everywhere; i.e., it is not confined to a laboratory. For example, such a frame may have its x-axis pointing to magnetic North (as long as this is consistently sensed and not distorted), its z-axis pointing up (gravitational field) and its y-axis perpendicular to the other two, i.e., pointing West.

The Swiss LMAM group's most recent contribution to the evolution of the inertial sensors work on body segment orientation is Favre et al.'s (2010) description of a *hybrid* system, incorporating the addition of a conventional electromagnetic motion capture system, with each magnetic sensor co-located on the body with the previously described inertial sensor system. The electromagnetic system was not used just for validation as previously but now played a role in the overall measurement approach: the position of anatomical landmarks on the body was digitized using a pointer instrumented with an electromagnetic sensor (remember that these sensors are capable of accurate position measurement, unlike inertial sensors). Anatomically based segment coordinate systems were then derived from these landmark positions, expressed in the common technical coordinate systems of each co-located magnetic/inertial sensor pair. The segmental angular velocities measured by the gyroscopes were projected onto their segment's anatomical coordinate system. Joint angles were determined by gyroscope integration via the fusion algorithm, as previously. The system was assessed with healthy normal subjects as well as patients with ankle OA. The results were comparable to those of conventional motion capture systems (Della Croce et al. 2005).

This hybrid system perhaps reflects a realization that inertial sensors—that is, accelerometers and gyroscopes *on their own*—are not (yet?) capable of providing a comprehensive motion capture system for the lower limbs, even in the laboratory. Other hybrid approaches have also been described, using a body-worn electromagnetic source (compared to the fixed source in most electromagnetic tracking systems), and include attempts to derive accurate position estimates in addition to orientation (Roetenberg, Baten, and Veltink 2007; Roetenberg, Slycke, and Veltink 2007; Schepers, Roetenberg, and Veltink 2010).

Mariani et al. (2010) from LMAM described a new, custom wireless inertial sensing system (Van de Molengraft et al. 2009) aimed more at estimation of *spatio-temporal* parameters of gait and incorporating a triaxial accelerometer and gyroscope attached to the back of the shoes. This system was designed to measure the 3D position of the heels for estimation of total distance travelled and other spatio-temporal parameters. The system provides a clear exposition of a strapdown-style algorithm for measurement of position from inertial sensors in human motion analysis, with periodic drift resetting by exploiting the regular occurrence of a zero-velocity event for the body segment concerned, in this case the foot when it is flat on the ground during stance phase. Foot-flat events were detected from the gyroscope angular velocity (Salarian et al. 2004). The 3D orientation of the foot was determined when the foot was stationary at these times from the gravitational vector as sensed on each axis of the accelerometer. This was used as the initial orientation with the reset, integrated gyroscope signal for foot-orientation estimation during the subsequent swing phase of the gait cycle. This orientation was used to transform the acceleration to the global frame, from which the gravitational vector was then removed. Integration of this gravity-free acceleration then yielded foot velocity, and then another integration yielded foot position. To counter the remaining position drift, a novel approach of subtracting a sigmoid-like curve was used. The standard spatio-temporal parameters of stride length (SL) and stride velocity (SV) were calculated, as well as the more novel parameters of foot clearance and turning angle. Against a camera-based system, the sensor system showed reasonable accuracy (e.g., $1.5 \pm 6.8$ cm for stride length) in tracking a range of gait tasks, including straight walking, turns, initiations and terminations.

While commercial systems employing inertial sensors for human motion analysis have been on the market for a number of years, these systems often also use magnetometers in addition to accelerometers/gyroscopes. As noted earlier, the limitations of magnetometers are well known; however, the limitations previously for inertial sensors operating *alone* could mean that magnetometers may currently be *necessary* for human motion analysis, albeit in restricted environments free of magnetic distortions. But such hybrid systems have not been comprehensively validated for lower-body gait analysis against conventional motion capture systems. Existing approaches described in the research literature for the use of these commercial sensors in such contexts still require special procedures and software to derive joint kinematics (Cutti et al. 2010; Ferrari et al. 2010; O'Donovan et al. 2007; Picerno, Cereatti, and Cappozzo 2008). Vendors continue to refine proprietary algorithms associated with inertial and hybrid systems, particularly aimed at minimizing the effect of magnetic distortions (Roetenberg, Baten, and Veltink 2007), enhancing the accuracy of fusion algorithms and adding other aiding motion-sensing technologies to improve accuracy.

However, these inertial/magnetometer-based systems have found successful application in human motion capture for computer animation, where perhaps accuracy is less important than sufficiently realistic visualization of human or animated nonhuman creature movement. Calibration of the system to a human subject for animation is generally based on simply entering the body segment lengths of the subjects and then having them assume standard, standing calibration, neutral and squatting poses. The positions of known anatomical landmarks are not used to derive anatomically based segment coordinate systems. In order to move the animated body through space, such systems derive segment orientation estimates, but they also estimate the positions of segment end-point joints by use of a hierarchical, linked model of known-length body segments. This process may be accurate enough for animation, but it is unlikely to be sufficiently accurate for gait analysis.

While the criterion against which sensors must be assessed is the conventional, usually camera-based, 3D gait analysis system, it is worth noting that until relatively recently, discrete motion sensors actually had one significant advantage over such camera-based laboratory motion capture systems: this is the fact that they are inherently capable of *real-time* motion measurement. In contrast, the camera-based systems required extensive offline processing before measurements were produced. While the first camera-based systems for real-time 3D reconstruction of retro-reflective marker positions on the body were actually developed well over a decade ago, in the early 1990s (Furnee 1998), it is only in recent years that the major commercial camera-based motion capture systems have offered this capability. The latest state-of-the-art systems from the major manufacturers of camera-based motion capture systems do now generally provide real-time marker position tracking, and, with suitable processing software (which may need to be custom developed), they can derive kinematic and kinetic measures in real time with minimal latency. However, this capability is often not fully exploited, and considerable postprocessing is commonly still required to yield final measurements from camera-based systems. As such, there may be applications where the real-time capability of discrete sensor systems could still give them an advantage for real-time applications in the laboratory.

So, in summary, in late 2010 inertial sensors are the most popular discrete sensors for human motion measurement. But while the basic inertial sensing components are now relatively cheap, their integration into a usable motion-sensing system that produces useful outputs of body segment and joint positions is not sufficiently well proven (or cheap enough at the current level of utility) to warrant widespread use with the expectation of equivalent accuracy and data types to conventional camera-based gait analysis systems. Furthermore, it is not currently clear which will be the most profitable directions for use of inertial sensors. Will these combinations of accelerometers and gyroscopes finally emerge as robust systems that can be used on their own, possibly in the field with minimal supervision? Alternatively, will the integrated use of inertial sensors and magnetometers—currently requiring closed, vendor-specific, proprietary algorithms—emerge with new algorithms to make them more accurate and viable in unrestricted environments? Or are there new technologies, hopefully providing more direct position and orientation estimates, that may emerge and take over from the admittedly difficult array of current sensor technologies?

It took camera-based motion capture several decades of evolution in hardware and software to move from the ready availability of the basic hardware (i.e., video cameras) to the powerful and generally accurate systems in use for 3D gait analysis today (albeit with some remaining issues, such as soft tissue artefact). Proponents of inertial sensors have suggested that such a period of further development and validation still awaits inertial sensor use in human gait analysis (Favre et al. 2010).

## Field-Based Motion Measures Not Derived from Laboratory Measures

In addition to attempts to use discrete motion sensors to replicate laboratory measures, there are other movement- and loading-related parameters that can be measured in field environments, such as home and outdoors. These are more general measures that are relevant to both healthy and impaired individuals, not just those with OA. Such measures are directed at how much movement an individual makes, and at what intensity. This is important to understanding what OA patients do in their daily activities. To this end, we have already discussed the possibility of combining a laboratory-based estimate of knee joint loading with a pedometer-based measure of step count, to estimate cumulative knee load exposure during typical daily activities (Maly 2008; Robbins et al. 2009). However, a laboratory assessment will not often be available, but the use of pedometers and related sensors described in this section may still yield useful data on patients' walking and other activities.

A range of generic activity monitors and pedometers have been employed in lower-limb pathologies, including OA. These technologies have been in wide use for research assessing activity levels across all age groups, from healthy children and adults to those with various medical conditions. Pedometers range from the very cheap and simple to more complex devices based on more robust footfall-counting technologies. Devices based on accelerometers and proprietary algorithms for multiple-activity-level discrimination are considerably more expensive. Note that measures derived from the global positioning system (GPS) are not considered here, as GPS is restricted to outdoor use; however, suitable implementations may provide sufficiently accurate data on distances travelled outdoors (but it would be necessary to distinguish between self-propelled motion and motion in a motor vehicle, for example).

Pedometers are typically attached to the body at some distance from the point of foot–ground contact (e.g., the waist), although ankle- and shoe-mounted devices are also available. They vary in their ability to accurately count foot contacts. Not everyone contacts the ground with high enough movement transients to trigger the step counter, especially if they are walking slowly (Tudor-Locke et al. 2002). When accurate pedometers do give a gross indication of the overall exposure to gait-related loading in individuals with OA. While this walking activity may be self-chosen or related to occupational activity, pedometer recordings can assess the extent to which the OA limits the individual's overall mobility. It is well known that OA limits physical activity levels, with the result that OA patients are less active than recommended (Centers for Disease Control 1997; Dunlop et al. 2001; Hootman et al. 2003). Pedometry also has a role to play as a treatment modality, encouraging patients to increase their daily step counts (Bravata et al. 2007; Merom et al. 2007; Stovitz et al. 2005; Talbot et al. 2003; Tudor-Locke and Lutes 2009).

Activity monitors typically employ accelerometers and use more complex proprietary algorithms to convert the body-mounted accelerometer signal into indices of movement intensity. But they may also have basic pedometry capabilities. While often described generically as accelerometry, such devices utilize accelerometers and their algorithms for the very specific purpose of judging the intensity of daily activities, which is very different to the applications of accelerometers mentioned earlier in this chapter. Activity-monitoring accelerometers do not attempt to measure the characteristics of movements at single joints like the sensors mentioned earlier. However, that may suit the purpose of many investigations. But because people with OA tend to be in older age groups, who are not usually highly active, there may not be a great need to employ activity monitors that are sensitive to a wide range of activity intensities. Pedometry may be sufficient for this group, as walking represents their major form of physical activity (Colbert et al. 2010).

In relation to pedometer use, it is interesting to consider the nature of typical walking patterns, as these will dictate the necessary capabilities of the measurement system. Orendurff et al. (2008) used an activity monitor to count the number of steps typically taken by healthy adults in each bout of normal walking, finding that most walking bouts (60%) lasted 30 s or less, with 40% of all bouts consisting of fewer than 12 steps and 75% of fewer than 40 steps. However, rest intervals between walking bouts were typically very short, generally lasting 20 s or less. Therefore, pedometers and activity monitors should ideally be able to sense footfalls for a diverse range of gait phenomena, including stops and starts, accelerations and decelerations. Of course, pedometers can only distinguish the *number* of steps; they cannot measure step length or distance travelled, for example (while *estimates* of these are sometimes made, they are based on assumptions such a constant step length).

Because accelerometers are sensitive to gravity, they have also been used to estimate body orientation, to determine the daily time spent in various gross body postures. The extent to which the gravitational vector is sensed on each axis of a dual-axis accelerometer can be used to determine the overall orientation of the body segments to which it is attached, i.e., whether the trunk and lower-limb orientation indicate that the wearer is upright, sitting or lying (Zijlstra and Aminian 2007). This is not generally a capability of generic activity monitors based on accelerometers. It requires devices specifically aimed at this application or custom devices.

Amongst the applications in OA, Farr et al. (2008) used an accelerometer-based activity monitor to study patients with early knee OA. Only 30% of patients reached recommended daily activity levels, with women being significantly less active than men. In those with late-stage hip and knee OA, de Groot et al. (2008) found a similar impact on activity limitation for the two joints measured by the activity monitor, but the activity level was not as low compared to healthy controls as was expected and not as low as the patients themselves reported their activity to be. Brandes et al. (2008) studied hip and knee patients prior to joint replacement surgery, using an accelerometer-based system. They found that walking accounted for 10.5 ± 5% of the daily activity (from early morning until bedtime), standing 32.6 ± 12.3%, sitting 43.9 ± 15.1% and lying 12.5 ± 11.1%.

A number of studies have looked at activity levels of OA patients *after* joint replacement surgery (Naal and Impellizzeri 2010). Walker et al. (2002) found a large increase in activity 6 months after knee replacement surgery compared to pre-surgery, using an accelerometer-based monitor. Also, self-reported activity did not correlate with measured activity. Morlock et al. (2001) used an electrogoniometer- and inclinometer-based system for activity monitoring in hip replacement patients. They found that patients divided their time between sitting (44%), standing (25%), walking (10%), lying down (6%) and stair climbing (0.4%), similar to the pre-surgery patients studied by Brandes et al. (2008). Schmalzried et al. (1998) used a pedometer to determine steps taken by patients at least 6 months after hip or knee replacement surgery. They found a huge range, from 395 to 17,718 steps/day, an approximately 45-fold difference. The same group also reported a 15-fold difference in steps per day between patients with the same score on the UCLA Activity Rating Scale (Zahiri et al. 1998), indicating that such self-reported scales may lack resolution and accuracy (Wagenmakers et al. 2008). Kinkel et al. (2009) recently used an ankle-mounted activity monitor to study steps per day in patients after hip replacement surgery, finding 6-fold differences within the same age group. Lower step counts were found in older patients, females and those with higher body mass indices. The same device has been found to report a substantially greater number of steps than a pedometer in patients after hip replacement, which was interpreted as underreporting by the pedometer (Silva et al. 2002).

In summary, pedometry/activity monitoring is a well-established and more mature technology than the use of discrete sensors for deriving conventional gait analysis measures. While there is a large range of technologies available—with varying accuracy that may be activity/population-specific to some extent—many devices have been shown to provide useful information on important aspects of habitual patient movement patterns, which are not necessarily revealed by patient self-report measures.

## Conclusions

Welch and Foxlin (2002) subtitled their review of motion sensors "No Silver Bullet, but a Respectable Arsenal". While they were looking at a different area of application (tracking head movements for head-mounted displays), and progress in motion sensors has undoubtedly been made in the intervening years, the phrase "no silver bullet" remains an apt description of the absence of an ideal discrete sensor device for human gait analysis, either in the laboratory or in the field, for OA or any other pathology. The existing devices may work well in controlled environments but not outside them. They may work well for particular human movements but not others. They may measure some conventional gait analysis parameters but not others. They may utilize complex proprietary algorithms that perform well or not so well under particular circumstances (which may be poorly defined or difficult to predict), or they may come with no processing algorithms at all, requiring the users to develop the custom software that best suits their intended application. In light of this, it is interesting to speculate on what a silver-bullet human motion/gait analysis discrete sensor device would look like:

- Attaches to the body by simple but robust means
- Is very small and unobtrusive
- Requires no special interaction by the wearer to set up or maintain accurate measurements
- Is operable without continuous specialist supervision
- Utilizes robust algorithms to relate the measured data to movement of the underlying anatomy
- Provides accurate, real-time anatomical *position* and *orientation* measures, not subject to drift or other errors, under any normal or pathological movement conditions, consistent with what is available from state-of-the-art motion capture systems utilizing other technologies (e.g., optical)
- Does not require any cables, utilizing wireless transmission for data logging or real-time analysis locally or remotely
- Operates for at least a day—ideally longer—without the need for battery replacement or recharging
- Works equally well in all environments, indoors and outdoors

It remains to be seen if the preceding requirements are realistic. As it stands, estimation of joint loading in OA is not feasible except in a laboratory equipped with a full motion capture system, including force platform(s). In the meantime, pedometry/activity monitoring allows one to *infer* exposure to joint loading activities in patients with OA. If accurate pedometry can be combined with a laboratory-derived measure of joint loading such as the KAM or KAM impulse, good estimates of joint load exposure should be achievable. These should enhance our ability to understand the importance of joint loading in OA, in relation to disease incidence and progression (deterioration in joint structure and symptoms), and more so than can be achieved based on a typical single session of laboratory gait analysis.

There is no prospect that direct measurement of intra-articular joint loads in practical form—that is, not requiring joint surgery—is on the horizon with any known technology. Even based on musculoskeletal modelling, joint force estimation requires inverse dynamics or similar estimates of joint moments. So whether joint loading is inferred from the KAM or estimated from modelling, GRFs and joint position and orientation measures are required. To this end, it is to be hoped that progress is made on practical force-measuring shoes and technologies for the measurement of joint position. When this is achieved, better estimates of joint loading in the field may be within reach.

## References

Amin, S., N. Luepongsak, C. A. McGibbon, M. P. LaValley, D. E. Krebs, and D. T. Felson. 2004. Knee adduction moment and development of chronic knee pain in elders. *Arthritis Rheum* 51 (3):371–76.

Aminian, K., B. Najafi, C. Bula, P. F. Leyvraz, and P. Robert. 2002. Spatio-temporal parameters of gait measured by an ambulatory system using miniature gyroscopes. *J Biomech* 35 (5):689–99.

Aminian, K., C. Trevisan, B. Najafi, H. Dejnabadi, C. Frigo, E. Pavan, A. Telonio, F. Cerati, E. C. Marinoni, P. Robert, and P. F. Leyvraz. 2004. Evaluation of an ambulatory system for gait analysis in hip osteoarthritis and after total hip replacement. *Gait Posture* 20 (1):102–7.

Amis, A. A., P. Cuomo, R.B.S. Rama, F. Giron, A. M. J. Bull, R. Thomas, and P. Aglietti. 2008. Measurement of knee laxity and pivot-shift kinematics with magnetic sensors. *Oper Techn Orthop* 18:196–203.

Andriacchi, T. P. 1994. Dynamics of knee malalignment. *Orthop Clin North Am* 25 (3):395–403.

Andriacchi, T. P., and C. O. Dyrby. 2005. Interactions between kinematics and loading during walking for the normal and ACL deficient knee. *J Biomech* 38 (2):293–98.

Andriacchi, T. P., and A. Mundermann. 2006. The role of ambulatory mechanics in the initiation and progression of knee osteoarthritis. *Curr Opin Rheumatol* 18 (5):514–18.

Andriacchi, T. P., A. Mundermann, R. L. Smith, E. J. Alexander, C. O. Dyrby, and S. Koo. 2004. A framework for the in vivo pathomechanics of osteoarthritis at the knee. *Ann Biomed Eng* 32 (3):447–57.

Andriacchi, T. P., J. A. Ogle, and J. O. Galante. 1977. Walking speed as a basis for normal and abnormal gait measurements. *J Biomech* 10 (4):261–68.

Astephen, J. L., K. J. Deluzio, G. E. Caldwell, and M. J. Dunbar. 2008. Biomechanical changes at the hip, knee, and ankle joints during gait are associated with knee osteoarthritis severity. *J Orthop Res* 26 (3):332–41.

Bamberg, S. J., A. Y. Benbasat, D. M. Scarborough, D. E. Krebs, and J. A. Paradiso. 2008. Gait analysis using a shoe-integrated wireless sensor system. *IEEE Trans Inf Technol Biomed* 12 (4):413–23.

Bergmann, G., G. Deuretzbacher, M. Heller, F. Graichen, A. Rohlmann, J. Strauss, and G. N. Duda. 2001. Hip contact forces and gait patterns from routine activities. *J Biomech* 34 (7):859–71.

Bergmann, J. H., R. E. Mayagoitia, and I. C. Smith. 2009. A portable system for collecting anatomical joint angles during stair ascent: A comparison with an optical tracking device. *Dyn Med* 8:3.

Bierma-Zeinstra, S. M., and B. W. Koes. 2007. Risk factors and prognostic factors of hip and knee osteoarthritis. *Nat Clin Pract Rheumatol* 3 (2):78–85.

Block, J. A., and N. Shakoor. 2010. Lower limb osteoarthritis: Biomechanical alterations and implications for therapy. *Curr Opin Rheumatol* 22 (5):544–50.

Brandes, M., R. Schomaker, G. Mollenhoff, and D. Rosenbaum. 2008. Quantity versus quality of gait and quality of life in patients with osteoarthritis. *Gait Posture* 28 (1):74–9.

Bravata, D. M., C. Smith-Spangler, V. Sundaram, A. L. Gienger, N. Lin, R. Lewis, C. D. Stave, I. Olkin, and J. R. Sirard. 2007. Using pedometers to increase physical activity and improve health: A systematic review. *JAMA* 298 (19):2296–304.

Brouwer, G. M., A. W. van Tol, A. P. Bergink, J. N. Belo, R. M. Bernsen, M. Reijman, H. A. Pols, and S. M. Bierma-Zeinstra. 2007. Association between valgus and varus alignment and the development and progression of radiographic osteoarthritis of the knee. *Arthritis Rheum* 56 (4):1204–11.

Browning, R. C., and R. Kram. 2007. Effects of obesity on the biomechanics of walking at different speeds. *Med Sci Sports Exerc* 39 (9):1632–41.

Buchanan, T. S., D. G. Lloyd, K. Manal, and T. F. Besier. 2004. Neuromusculoskeletal modeling: Estimation of muscle forces and joint moments and movements from measurements of neural command. *J Appl Biomech* 20 (4):367–95.

Buchanan, T. S., D. G. Lloyd, K. Manal, and T. F. Besier. 2005. Estimation of muscle forces and joint moments using a forward-inverse dynamics model. *Med Sci Sports Exerc* 37 (11):1911–16.

Cappozzo, A. 2008. The observation of human joint movement. In *IFMBE Proceedings 22*, edited by J. Vander Sloten, P. Verdonck, M. Nyssen and J. Haueisen. Berling: Springer-Verlag. pp. 126–129.

Cappozzo, A., U. Della Croce, A. Leardini, and L. Chiari. 2005. Human movement analysis using stereophotogrammetry. Part 1: Theoretical background. *Gait Posture* 21 (2):186–96.

Centers for Disease Control. 1997. Prevalence of leisure-time physical activity among persons with arthritis and other rheumatic conditions—United States, 1990–1991. *MMWR Morb Mortal Wkly Rep* 46 (18):389–93.

Chang, A., K. Hayes, D. Dunlop, D. Hurwitz, J. Song, S. Cahue, R. Genge, and L. Sharma. 2004. Thrust during ambulation and the progression of knee osteoarthritis. *Arthritis Rheum* 50 (12):3897–903.

Chang, A., M. Hochberg, J. Song, D. Dunlop, J. S. Chmiel, M. Nevitt, K. Hayes, C. Eaton, J. Bathon, R. Jackson, C. K. Kwoh, and L. Sharma. 2010. Frequency of varus and valgus thrust and factors associated with thrust presence in persons with or at higher risk of developing knee osteoarthritis. *Arthritis Rheum* 62 (5):1403–11.

Cicuttini, F., A. Wluka, J. Hankin, and Y. Wang. 2004. Longitudinal study of the relationship between knee angle and tibiofemoral cartilage volume in subjects with knee osteoarthritis. *Rheumatology (Oxford)* 43 (3):321–24.

Colbert, L. H., C. E. Matthews, T. C. Havighurst, K. Kim, and D. A. Schoeller. 2010. Comparative validity of physical activity measures in older adults. *Med Sci Sports Exerc* 43 (5): 867–76.

Collins, J. J., and M. W. Whittle. 1989. Impulsive forces during walking and their clinical implications. *Clin Biomech* 4 (3):179–87.

Correa, T. A., K. M. Crossley, H. J. Kim, and M. G. Pandy. 2010. Contributions of individual muscles to hip joint contact force in normal walking. *J Biomech* 43 (8):1618–22.

Cutti, A. G., A. Ferrari, P. Garofalo, M. Raggi, and A. Cappello. 2010. "Outwalk": A protocol for clinical gait analysis based on inertial and magnetic sensors. *Med Biol Eng Comput* 48 (1):17–25.

Damiano, D. L., E. Laws, D. V. Carmines, and M. F. Abel. 2006. Relationship of spasticity to knee angular velocity and motion during gait in cerebral palsy. *Gait Posture* 23 (1):1–8.

de Groot, I. B., J. B. Bussmann, H. J. Stam, and J. A. Verhaar. 2008. Actual everyday physical activity in patients with end-stage hip or knee osteoarthritis compared with healthy controls. *Osteoarthr Cartilage* 16 (4):436–42.

Dejnabadi, H., B. M. Jolles, and K. Aminian. 2005. A new approach to accurate measurement of uniaxial joint angles based on a combination of accelerometers and gyroscopes. *IEEE Trans Biomed Eng* 52 (8):1478–84.

Dejnabadi, H., B. M. Jolles, E. Casanova, P. Fua, and K. Aminian. 2006. Estimation and visualization of sagittal kinematics of lower limbs orientation using body-fixed sensors. *IEEE Trans Biomed Eng* 53 (7):1385–93.

Della Croce, U., A. Leardini, L. Chiari, and A. Cappozzo. 2005. Human movement analysis using stereophotogrammetry. Part 4: Assessment of anatomical landmark misplacement and its effects on joint kinematics. *Gait Posture* 21 (2):226–37.

de Vries, W. H., H. E. Veeger, C. T. Baten, and F. C. van der Helm. 2009. Magnetic distortion in motion labs, implications for validating inertial magnetic sensors. *Gait Posture* 29 (4):535–41.

D'Lima, D. D., N. Steklov, B. J. Fregly, S. A. Banks, and C. W. Colwell, Jr. 2008. In vivo contact stresses during activities of daily living after knee arthroplasty. *J Orthop Res* 26 (12):1549–55.

Doherty, M. 2001. Risk factors for progression of knee osteoarthritis. *Lancet* 358 (9284):775–76.

Dunlop, D. D., L. M. Manheim, J. Song, and R. W. Chang. 2001. Arthritis prevalence and activity limitations in older adults. *Arthritis Rheum* 44 (1):212–21.

Farr, J. N., S. B. Going, T. G. Lohman, L. Rankin, S. Kasle, M. Cornett, and E. Cussler. 2008. Physical activity levels in patients with early knee osteoarthritis measured by accelerometry. *Arthritis Rheum* 59 (9):1229–36.

Favre, J., R. Aissaoui, B. M. Jolles, J. A. de Guise, and K. Aminian. 2009. Functional calibration procedure for 3D knee joint angle description using inertial sensors. *J Biomech* 42 (14):2330–35.

Favre, J., X. Crevoisier, B. M. Jolles, and K. Aminian. 2010. Evaluation of a mixed approach combining stationary and wearable systems to monitor gait over long distance. *J Biomech* 43 (11):2196–202.

Favre, J., B. M. Jolles, R. Aissaoui, and K. Aminian. 2008. Ambulatory measurement of 3D knee joint angle. *J Biomech* 41 (5):1029–35.

Favre, J., B. M. Jolles, O. Siegrist, and K. Aminian. 1996. Quaternion-based fusion of gyroscopes and accelerometers to improve 3D angle measurement. *Electron Lett* 42 (11):612–14.

Favre, J., F. Luthi, B. M. Jolles, O. Siegrist, B. Najafi, and K. Aminian. 2006. A new ambulatory system for comparative evaluation of the three-dimensional knee kinematics, applied to anterior cruciate ligament injuries. *Knee Surg Sports Traumatol Arthrosc* 14 (7):592–604.

Felson, D. T. 2004. An update on the pathogenesis and epidemiology of osteoarthritis. *Radiol Clin North Am* 42 (1):1–9, v.

Felson, D. T., J. Niu, C. McClennan, B. Sack, P. Aliabadi, D. J. Hunter, A. Guermazi, and M. Englund. 2007. Knee buckling: Prevalence, risk factors, and associated limitations in function. *Ann Intern Med* 147 (8):534–40.

Ferrari, A., A. G. Cutti, P. Garofalo, M. Raggi, M. Heijboer, A. Cappello, and A. Davalli. 2010. First in vivo assessment of "Outwalk": A novel protocol for clinical gait analysis based on inertial and magnetic sensors. *Med Biol Eng Comput* 48 (1):1–15.

Fitzgerald, G. K., S. R. Piva, and J. J. Irrgang. 2004. Reports of joint instability in knee osteoarthritis: Its prevalence and relationship to physical function. *Arthritis Rheum* 51 (6):941–46.

Folman, Y., J. Wosk, A. Voloshin, and S. Liberty. 1986. Cyclic impacts on heel strike: A possible biomechanical factor in the etiology of degenerative disease of the human locomotor system. *Arch Orthop Trauma Surg* 104 (6):363–65.

Fong, D. T., Y. Y. Chan, Y. Hong, P. S. Yung, K. Y. Fung, and K. M. Chan. 2008. Estimating the complete ground reaction forces with pressure insoles in walking. *J Biomech* 41 (11):2597–601.

Forner-Cordero, A., H. J. Koopman, and F. C. van der Helm. 2006. Inverse dynamics calculations during gait with restricted ground reaction force information from pressure insoles. *Gait Posture* 23 (2):189–99.

Forner-Cordero, A., M. Mateu-Arce, I. Forner-Cordero, E. Alcantara, J. C. Moreno, and J. L. Pons. 2008. Study of the motion artefacts of skin-mounted inertial sensors under different attachment conditions. *Physiol Meas* 29 (4):N21–N31.

Forner-Cordero, A., H. J. Koopman, and F. C. van der Helm. 2004. Use of pressure insoles to calculate the complete ground reaction forces. *J Biomech* 37 (9):1427–32.

Foroughi, N., R. M. Smith, A. K. Lange, M. K. Baker, M. A. Fiatarone Singh, and B. Vanwanseele. 2010. Dynamic alignment and its association with knee adduction moment in medial knee osteoarthritis. *Knee* 17 (3):210–16.

Foucher, K. C., L. E. Thorp, D. Orozco, M. Hildebrand, and M. A. Wimmer. 2010. Differences in preferred walking speeds in a gait laboratory compared with the real world after total hip replacement. *Arch Phys Med Rehabil* 91 (9):1390–95.

Foxlin, E. 2002. Motion tracking requirements and technologies. In *Handbook of Virtual Environment Technology*, edited by K. Stanney. Mahwah, NJ: Lawrence Erlbaum. pp. 163–210.

Furnee, H. 1998. Real-time motion capture systems. In *Three-Dimensional Analysis of Human Locomotion*, edited by P. Allard, A. Cappozzo, A. Lundberg and C. Vaughan. New York. Wiley. pp. 85–108.

Giansanti, D., V. Macellari, G. Maccioni, and A. Cappozzo. 2003. Is it feasible to reconstruct body segment 3-D position and orientation using accelerometric data? *IEEE Trans Biomed Eng* 50 (4):476–83.

Gill, H. S., and J. J. O'Connor. 2003a. Heelstrike and the pathomechanics of osteoarthrosis: A pilot gait study. *J Biomech* 36 (11):1625–31.

Gill, H. S., and J. J. O'Connor. 2003b. Heelstrike and the pathomechanics of osteoarthrosis: A simulation study. *J Biomech* 36 (11):1617–24.

Henriksen, M., R. Christensen, T. Alkjaer, H. Lund, E. B. Simonsen, and H. Bliddal. 2008. Influence of pain and gender on impact loading during walking: A randomised trial. *Clin Biomech (Bristol, Avon)* 23 (2):221–30.

Henriksen, M., E. B. Simonsen, T. Graven-Nielsen, H. Lund, B. Danneskiold-Samsoe, and H. Bliddal. 2006. Impulse-forces during walking are not increased in patients with knee osteoarthritis. *Acta Orthop* 77 (4):650–56.

Hill, C. L., G. S. Seo, D. Gale, S. Totterman, M. E. Gale, and D. T. Felson. 2005. Cruciate ligament integrity in osteoarthritis of the knee. *Arthritis Rheum* 52 (3):794–99.

Hindle, R. J., M. J. Pearcy, A. T. Cross, and D. H. T. Miller. 1990. Three-dimensional kinematics of the human back. *Clin Biomech* 5 (4):218–28.

Hootman, J. M., C. A. Macera, S. A. Ham, C. G. Helmick, and J. E. Sniezek. 2003. Physical activity levels among the general US adult population and in adults with and without arthritis. *Arthritis Rheum* 49 (1):129–35.

Hunt, M. A., T. B. Birmingham, J. R. Giffin, and T. R. Jenkyn. 2006. Associations among knee adduction moment, frontal plane ground reaction force, and lever arm during walking in patients with knee osteoarthritis. *J Biomech* 39 (12):2213–20.

Hunt, M. A., T. B. Birmingham, T. R. Jenkyn, J. R. Giffin, and I. C. Jones. 2008. Measures of frontal plane lower limb alignment obtained from static radiographs and dynamic gait analysis. *Gait Posture* 27 (4):635–40.

Hunt, M. A., R. S. Hinman, B. R. Metcalf, B. W. Lim, T. V. Wrigley, K. A. Bowles, G. Kemp, and K. L. Bennell. 2010. Quadriceps strength is not related to gait impact loading in knee osteoarthritis. *Knee* 17 (4):296–302.

Hunt, M. A., A. G. Schache, R. S. Hinman, and K. M. Crossley. 2010. Varus thrust in medial knee osteoarthritis: Quantification and effects of different gait-related interventions using a single case study. *Arthritis Care Res (Hoboken).*

Hunter, D. J., J. Niu, D. T. Felson, W. F. Harvey, K. D. Gross, P. McCree, P. Aliabadi, B. Sack, and Y. Zhang. 2007. Knee alignment does not predict incident osteoarthritis: The Framingham osteoarthritis study. *Arthritis Rheum* 56 (4):1212–18.

Kaufman, K. R., C. Hughes, B. F. Morrey, M. Morrey, and K. N. An. 2001. Gait characteristics of patients with knee osteoarthritis. *J Biomech* 34 (7):907–15.

Kavanagh, J. J., and H. B. Menz. 2008. Accelerometry: A technique for quantifying movement patterns during walking. *Gait Posture* 28 (1):1–15.

Kindratenko, V. 2001. A comparison of the accuracy of an electromagnetic and a hybrid ultrasound-inertia position tracking system. *Presence-Teleop Virt Environ* 10 (6):657–63.

Kinkel, S., N. Wollmerstedt, J. A. Kleinhans, C. Hendrich, and C. Heisel. 2009. Patient activity after total hip arthroplasty declines with advancing age. *Clin Orthop Relat Res* 467 (8):2053–58.

Kutzner, I., B. Heinlein, F. Graichen, A. Bender, A. Rohlmann, A. Halder, A. Beier, and G. Bergmann. 2010. Loading of the knee joint during activities of daily living measured in vivo in five subjects. *J Biomech* 43 (11):2164–73.

Lafortune, M. A., and E. M. Hennig. 1991. Contribution of angular motion and gravity to tibial acceleration. *Med Sci Sports Exerc* 23 (3):360–63.

LaScalza, S., J. Arico, and R. Hughes. 2003. Effect of metal and sampling rate on accuracy of Flock of Birds electromagnetic tracking system. *J Biomech* 36 (1):141–44.

Ledingham, J., M. Regan, A. Jones, and M. Doherty. 1993. Radiographic patterns and associations of osteoarthritis of the knee in patients referred to hospital. *Ann Rheum Dis* 52 (7):520–26.

Legnani, G., B. Zappa, F. Casolo, R. Adamini, and P. L. Magnani. 2000. A model of an electro-goniometer and its calibration for biomechanical applications. *Med Eng Phys* 22 (10):711–22.

Liedtke, C., S. A. Fokkenrood, J. T. Menger, H. van der Kooij, and P. H. Veltink. 2007. Evaluation of instrumented shoes for ambulatory assessment of ground reaction forces. *Gait Posture* 26 (1):39–47.

Liikavainio, T., J. Isolehto, H. J. Helminen, J. Perttunen, V. Lepola, I. Kiviranta, J. P. Arokoski, and P. V. Komi. 2007. Loading and gait symmetry during level and stair walking in asymptomatic subjects with knee osteoarthritis: Importance of quadriceps femoris in reducing impact force during heel strike? *Knee* 14 (3):231–38.

Lin, Y. C., J. P. Walter, S. A. Banks, M. G. Pandy, and B. J. Fregly. 2010. Simultaneous prediction of muscle and contact forces in the knee during gait. *J Biomech* 43 (5):945–52.

Liu, T., Y. Inoue, and K. Shibata. 2010. A wearable force plate system to successively measure multi–axial ground reaction force for gait analysis. *2009 IEEE International Conference on Robotics and Biomimetics (ROBIO)* 19–23 Dec. 2009. Piscataway, NJ: IEEE. pp. 836–841.

Luinge, H. J., and P. H. Veltink. 2005. Measuring orientation of human body segments using miniature gyroscopes and accelerometers. *Med Biol Eng Comput* 43 (2):273–82.

Maly, M. R. 2008. Abnormal and cumulative loading in knee osteoarthritis. *Curr Opin Rheumatol* 20 (5):547–52.

Maly, M. R., P. A. Costigan, and S. J. Olney. 2006. Role of knee kinematics and kinetics on performance and disability in people with medial compartment knee osteoarthritis. *Clin Biomech (Bristol, Avon)* 21 (10):1051–59.

Mariani, B., C. Hoskovec, S. Rochat, C. Bula, J. Penders, and K. Aminian. 2010. 3D gait assessment in young and elderly subjects using foot-worn inertial sensors. *J Biomech* 43 (15):2999–3006.

Mathie, M. J., A. C. Coster, N. H. Lovell, and B. G. Celler. 2004. Accelerometry: Providing an integrated, practical method for long-term, ambulatory monitoring of human movement. *Physiol Meas* 25 (2):R1–R20.

McGinley, J. L., R. Baker, R. Wolfe, and M. E. Morris. 2009. The reliability of three-dimensional kinematic gait measurements: A systematic review. *Gait Posture* 29 (3):360–69.

Merom, D., C. Rissel, P. Phongsavan, B. J. Smith, C. Van Kemenade, W. J. Brown, and A. E. Bauman. 2007. Promoting walking with pedometers in the community: The step-by-step trial. *Am J Prev Med* 32 (4):290–97.

Messier, S. P., W. H. Ettinger, T. E. Doyle, T. Morgan, M. K. James, M. L. O'Toole, and R. Burns. 1996. Obesity: Effects on gait in an osteoarthritic population. *J Appl Biomech* 12:161–72.

Mikesky, A. E., A. Meyer, and K. L. Thompson. 2000. Relationship between quadriceps strength and rate of loading during gait in women. *J Orthop Res* 18 (2):171–75.

Mills, P. M., S. Morrison, D. G. Lloyd, and R. S. Barrett. 2007. Repeatability of 3D gait kinematics obtained from an electromagnetic tracking system during treadmill locomotion. *J Biomech* 40 (7):1504–11.

Miyazaki, T., M. Wada, H. Kawahara, M. Sato, H. Baba, and S. Shimada. 2002. Dynamic load at baseline can predict radiographic disease progression in medial compartment knee osteoarthritis. *Ann Rheum Dis* 61 (7):617–22.

Moisio, K. C., D. R. Sumner, S. Shott, and D. E. Hurwitz. 2003. Normalization of joint moments during gait: A comparison of two techniques. *J Biomech* 36 (4):599–603.

Morlock, M., E. Schneider, A. Bluhm, M. Vollmer, G. Bergmann, V. Muller, and M. Honl. 2001. Duration and frequency of every day activities in total hip patients. *J Biomech* 34 (7):873–81.

Morris, S. J., and J. A. Paradiso. 2002. A compact wearable sensor package for clinical gait monitoring. *Offspring* 1 (1):7–15.

Mundermann, A., C. O. Dyrby, and T. P. Andriacchi. 2005. Secondary gait changes in patients with medial compartment knee osteoarthritis: Increased load at the ankle, knee, and hip during walking. *Arthritis Rheum* 52 (9):2835–44.

Mundermann, A., C. O. Dyrby, D. D. D'Lima, C. W. Colwell, Jr., and T. P. Andriacchi. 2008. In vivo knee loading characteristics during activities of daily living as measured by an instrumented total knee replacement. *J Orthop Res* 26 (9):1167–72.

Mundermann, A., C. O. Dyrby, D. E. Hurwitz, L. Sharma, and T. P. Andriacchi. 2004. Potential strategies to reduce medial compartment loading in patients with knee osteoarthritis of varying severity: Reduced walking speed. *Arthritis Rheum* 50 (4):1172–78.

Myles, C. M., P. J. Rowe, C. R. Walker, and R. W. Nutton. 2002. Knee joint functional range of movement prior to and following total knee arthroplasty measured using flexible electrogoniometry. *Gait Posture* 16 (1):46–54.

Naal, F. D., and F. M. Impellizzeri. 2010. How active are patients undergoing total joint arthroplasty? A systematic review. *Clin Orthop Relat Res* 468 (7):1891–1904.

Nixon, M. A., B. C. McCallum, W. R. Fright, and N. B. Price. 1998. The effect of metals and interfering fields on electromagnetic trackers. *Presence* 7 (2):204–18.

O'Donovan, K. J., R. Kamnik, D. T. O'Keeffe, and G. M. Lyons. 2007. An inertial and magnetic sensor based technique for joint angle measurement. *J Biomech* 40 (12):2604–11.

Ogata, K., M. Yasunaga, and H. Nomiyama. 1997. The effect of wedged insoles on the thrust of osteoarthritic knees. *Int Orthop* 21 (5):308–12.

Orendurff, M. S., J. A. Schoen, G. C. Bernatz, A. D. Segal, and G. K. Klute. 2008. How humans walk: Bout duration, steps per bout, and rest duration. *J Rehabil Res Dev* 45 (7):1077–89.

Ornetti, P., J. F. Maillefert, D. Laroche, C. Morisset, M. Dougados, and L. Gossec. 2010. Gait analysis as a quantifiable outcome measure in hip or knee osteoarthritis: A systematic review. *Joint Bone Spine*. 77 (5):421–5.

Pearcy, M. J., and R. J. Hindle. 1989. New method for the non-invasive three-dimensional measurement of human back movement. *Clin Biomech* 4 (2):73–79.

Picerno, P., A. Cereatti, and A. Cappozzo. 2008. Joint kinematics estimate using wearable inertial and magnetic sensing modules. *Gait Posture* 28 (4):588–95.

Prodromos, C. C., T. P. Andriacchi, and J. O. Galante. 1985. A relationship between gait and clinical changes following high tibial osteotomy. *J Bone Joint Surg Am* 67 (8):1188–94.

Radin, E. L., R. B. Martin, D. B. Burr, B. Caterson, R. D. Boyd, and C. Goodwin. 1984. Effects of mechanical loading on the tissues of the rabbit knee. *J Orthop Res* 2 (3):221–34.

Radin, E. L., K. H. Yang, C. Riegger, V. L. Kish, and J. J. O'Connor. 1991. Relationship between lower limb dynamics and knee joint pain. *J Orthop Res* 9 (3):398–405.

Radin, E.L., M. Whittle, and K. H. Yang. 1986. The heelstrike transient, its relationship with the angular velocity of the shank, and effects of quadriceps paralysis. In *Advances in Bioengineering*, edited by S. Lantz. New York: American Society of Mechanical Engineering. pp. 121–3.

Riskowski, J. L. 2010. Gait and neuromuscular adaptations after using a feedback-based gait monitoring knee brace. *Gait Posture* 32 (2):242–47.
Riskowski, J. L., A. E. Mikesky, R. E. Bahamonde, T. V. Alvey III, and D. B. Burr. 2005. Proprioception, gait kinematics, and rate of loading during walking: Are they related? *J Musculoskelet Neuronal Interact* 5 (4):379–87.
Riskowski, J. L., A. E. Mikesky, R. E. Bahamonde, and D. B. Burr. 2009. Design and validation of a knee brace with feedback to reduce the rate of loading. *J Biomech Eng* 131 (8):084503.
Robbins, S. M., T. B. Birmingham, G. R. Jones, J. P. Callaghan, and M. R. Maly. 2009. Developing an estimate of daily cumulative loading for the knee: Examining test-retest reliability. *Gait Posture* 30 (4):497–501.
Robbins, S., E. Waked, and N. Krouglicof. 2001. Vertical impact increase in middle age may explain idiopathic weight-bearing joint osteoarthritis. *Arch Phys Med Rehabil* 82 (12):1673–77.
Robertson, D. G. E. 2004. *Research Methods in Biomechanics*. Champaign, Illinois: Human Kinetics.
Roetenberg, D., C. T. Baten, and P. H. Veltink. 2007. Estimating body segment orientation by applying inertial and magnetic sensing near ferromagnetic materials. *IEEE Trans Neural Syst Rehabil Eng* 15 (3):469–71.
Roetenberg, D., P. J. Slycke, and P. H. Veltink. 2007. Ambulatory position and orientation tracking fusing magnetic and inertial sensing. *IEEE Trans Biomed Eng* 54 (5):883–90.
Rouhani, H., J. Favre, X. Crevoisier, and K. Aminian. 2010. Ambulatory assessment of 3D ground reaction force using plantar pressure distribution. *Gait Posture* 32 (3):311–16.
Rudolph, K. S., L. C. Schmitt, and M. D. Lewek. 2007. Age-related changes in strength, joint laxity, and walking patterns: Are they related to knee osteoarthritis? *Phys Ther* 87 (11):1422–32.
Sabatini, A. M. 2006. Inertial sensing in biomechanics: A survey of computational techniques bridging motion analysis and personal navigation. In *Computational Intelligence for Movement Sciences: Neural Networks, Support Vector Machines and Other Emerging Techniques*, edited by R. Begg, M. Palaniswami. Hershey, PA: Idea Group. (pp. 70–100).
Sabatini, A. M., C. Martelloni, S. Scapellato, and F. Cavallo. 2005. Assessment of walking features from foot inertial sensing. *IEEE Trans Biomed Eng* 52 (3):486–94.
Salarian, A., H. Russmann, F. J. Vingerhoets, C. Dehollain, Y. Blanc, P. R. Burkhard, and K. Aminian. 2004. Gait assessment in Parkinson's disease: Toward an ambulatory system for long-term monitoring. *IEEE Trans Biomed Eng* 51 (8):1434–43.
Savelberg, H. H., and A. L. de Lange. 1999. Novel Award Third Prize Paper. Assessment of the horizontal, fore-aft component of the ground reaction force from insole pressure patterns by using artificial neural networks. *Clin Biomech (Bristol, Avon)* 14 (8):585–92.
Schepers, H. M., H. F. Koopman, and P. H. Veltink. 2007. Ambulatory assessment of ankle and foot dynamics. *IEEE Trans Biomed Eng* 54 (5):895–902.
Schepers, H. M., D. Roetenberg, and P. H. Veltink. 2010. Ambulatory human motion tracking by fusion of inertial and magnetic sensing with adaptive actuation. *Med Biol Eng Comput* 48 (1):27–37.
Schipplein, O. D., and T. P. Andriacchi. 1991. Interaction between active and passive knee stabilizers during level walking. *J Orthop Res* 9 (1):113–19.

Schmalzried, T. P., E. S. Szuszczewicz, M. R. Northfield, K. H. Akizuki, R. E. Frankel, G. Belcher, and H. C. Amstutz. 1998. Quantitative assessment of walking activity after total hip or knee replacement. *J Bone Joint Surg Am* 80 (1):54–59.

Sharma, L., D. E. Hurwitz, E. J. Thonar, J. A. Sum, M. E. Lenz, D. D. Dunlop, T. J. Schnitzer, G. Kirwan-Mellis, and T. P. Andriacchi. 1998. Knee adduction moment, serum hyaluronan level, and disease severity in medial tibiofemoral osteoarthritis. *Arthritis Rheum* 41 (7):1233–40.

Sharma, L., J. Song, D. T. Felson, S. Cahue, E. Shamiyeh, and D. D. Dunlop. 2001. The role of knee alignment in disease progression and functional decline in knee osteoarthritis. *JAMA* 286 (2):188–95.

Shelburne, K. B., M. R. Torry, and M. G. Pandy. 2006. Contributions of muscles, ligaments, and the ground-reaction force to tibiofemoral joint loading during normal gait. *J Orthop Res* 24 (10):1983–90.

Shiratsu, A., and H. J. Coury. 2003. Reliability and accuracy of different sensors of a flexible electrogoniometer. *Clin Biomech (Bristol, Avon)* 18 (7):682–84.

Silva, M, E. F. Shepherd, W. O. Jackson, F. J. Dorey, and T. P. Schmalzried. 2002. Average patient walking activity approaches 2 million cycles per year. *J Arthroplasty* 17 (6):693–97.

Specogna, A. V., T. B. Birmingham, M. A. Hunt, I. C. Jones, T. R. Jenkyn, P. J. Fowler, and J. R. Giffin. 2007. Radiographic measures of knee alignment in patients with varus gonarthrosis: Effect of weightbearing status and associations with dynamic joint load. *Am J Sports Med* 35 (1):65–70.

Stovitz, S. D., J. J. VanWormer, B. A. Center, and K. L. Bremer. 2005. Pedometers as a means to increase ambulatory activity for patients seen at a family medicine clinic. *J Am Board Fam Pract* 18 (5):335–43.

Sudhoff, I., S. Van Driessche, S. Laporte, J. A. de Guise, and W. Skalli. 2007. Comparing three attachment systems used to determine knee kinematics during gait. *Gait Posture* 25 (4):533–43.

Talbot, L. A., J. M. Gaines, T. N. Huynh, and E. J. Metter. 2003. A home-based pedometer-driven walking program to increase physical activity in older adults with osteoarthritis of the knee: A preliminary study. *J Am Geriatr Soc* 51 (3):387–92.

Thorp, L. E., D. R. Sumner, M. A. Wimmer, and J. A. Block. 2007. Relationship between pain and medial knee joint loading in mild radiographic knee osteoarthritis. *Arthritis Rheum* 57 (7):1254–60.

Tong, K., and M. H. Granat. 1999. A practical gait analysis system using gyroscopes. *Med Eng Phys* 21 (2):87–94.

Tuck, K. 2007. Tilt sensing using linear accelerometers. Application note AN3461. Austin, Texas: Freescale Semiconductor.

Tudor-Locke, C., and L. Lutes. 2009. Why do pedometers work? A reflection upon the factors related to successfully increasing physical activity. *Sports Med* 39 (12):981–93.

Tudor-Locke, C., J. E. Williams, J. P. Reis, and D. Pluto. 2002. Utility of pedometers for assessing physical activity: Convergent validity. *Sports Med* 32 (12):795–808.

Turcot, K., R. Aissaoui, K. Boivin, M. Pelletier, N. Hagemeister, and J. A. de Guise. 2008. New accelerometric method to discriminate between asymptomatic subjects and patients with medial knee osteoarthritis during 3-d gait. *IEEE Trans Biomed Eng* 55 (4):1415–22.

Van de Molengraft, J., S. Nimmala, B. Mariani, K. Aminian, and J. Penders. 2009. Wireless 6D inertial measurement platform for ambulatory gait monitoring (abstract). In *6th International Workshop on Wearable, Micro and Nanosystems for Personalised Health (pHealth 2009) Programme.* http://cired.no/project/pHealth2009/pHealth%202009%20Programme_final.pdf (accessed 30/10/2010). Oslo, Norway: SINTEF.

van den Bogert, A., L. Read, and B. M. Nigg. 1996. A method for inverse dynamic analysis using accelerometry. *J Biomech* 29 (7):949–54.

van der Esch, M., M. Steultjens, J. Harlaar, N. Wolterbeek, D. L. Knol, and J. Dekker. 2008. Knee varus-valgus motion during gait—A measure of joint stability in patients with osteoarthritis? *Osteoarthr Cartilage* 16 (4):522–25.

Veltink, P. H., C. Liedtke, E. Droog, and H. van der Kooij. 2005. Ambulatory measurement of ground reaction forces. *IEEE Trans Neural Syst Rehabil Eng* 13 (3):423–27.

Veltink, P. H., P. Slycke, J. Hemssems, R. Buschman, G. Bultstra, and H. Hermens. 2003. Three dimensional inertial sensing of foot movements for automatic tuning of a two-channel implantable drop-foot stimulator. *Med Eng Phys* 25 (1):21–28.

Voloshin, A. S., and J. Wosk. 1983. Shock absorption of meniscectomized and painful knees: A comparative in vivo study. *J Biomed Eng* 5 (2):157–61.

Voloshin, A., J. Wosk, and M. Brull. 1981. Force wave transmission through the human locomotor system. *J Biomech Eng* 103 (1):48–50.

Wada, M., Y. Maezawa, H. Baba, S. Shimada, S. Sasaki, and Y. Nose. 2001. Relationships among bone mineral densities, static alignment and dynamic load in patients with medial compartment knee osteoarthritis. *Rheumatology (Oxford)* 40 (5):499–505.

Wagenmakers, R., M. Stevens, W. Zijlstra, M. L. Jacobs, I. van den Akker-Scheek, J. W. Groothoff, and S. K. Bulstra. 2008. Habitual physical activity behavior of patients after primary total hip arthroplasty. *Phys Ther* 88 (9):1039–48.

Walker, D. J., P. S. Heslop, C. Chandler, and I. M. Pinder. 2002. Measured ambulation and self-reported health status following total joint replacement for the osteoarthritic knee. *Rheumatology (Oxford)* 41 (7):755–58.

Welch, G. F. 2009. History: The use of the Kalman filter for human motion tracking in virtual reality. *Presence-Teleop Virt Environ* 18 (1):72–91.

Welch, G. F., and E. Foxlin. 2002. Motion tracking: No silver bullet, but a respectable arsenal. *IEEE Comput Graph* 22 (6):24–38.

Whittle, M. W. 1999. Generation and attenuation of transient impulsive forces beneath the foot: A review. *Gait Posture* 10 (3):264–75.

Willemsen, A. T., C. Frigo, and H. B. Boom. 1991. Lower extremity angle measurement with accelerometers—Error and sensitivity analysis. *IEEE Trans Biomed Eng* 38 (12):1186–93.

Willemsen, A. T., J. A. van Alste, and H. B. Boom. 1990. Real-time gait assessment utilizing a new way of accelerometry. *J Biomech* 23 (8):859–63.

Williamson, R., and B. J. Andrews. 2001. Detecting absolute human knee angle and angular velocity using accelerometers and rate gyroscopes. *Med Biol Eng Comput* 39 (3):294–302.

Winby, C. R., D. G. Lloyd, T. F. Besier, and T. B. Kirk. 2009. Muscle and external load contribution to knee joint contact loads during normal gait. *J Biomech* 42 (14):2294–2300.

Wong, W. Y., and M. S. Wong. 2008. Detecting spinal posture change in sitting positions with tri-axial accelerometers. *Gait Posture* 27 (1):168–71.

Yoshimura, I., M. Naito, M. Hara, and J. Zhang. 2003. The effect of wedged insoles on the lateral thrust of anterior cruciate ligament-insufficient knees. *Am J Sports Med* 31 (6):999–1002.

Zahiri, C. A., T. P. Schmalzried, E. S. Szuszczewicz, and H. C. Amstutz. 1998. Assessing activity in joint replacement patients. *J Arthroplasty* 13 (8):890–95.

Zhang, Y., and J. M. Jordan. 2008. Epidemiology of osteoarthritis. *Rheum Dis Clin North Am* 34 (3):515–29.

Zhao, D., S. A. Banks, D. D. D'Lima, C. W. Colwell Jr., and B. J. Fregly. 2007. In vivo medial and lateral tibial loads during dynamic and high flexion activities. *J Orthop Res* 25 (5):593–602.

Zhao, D., S. A. Banks, K. H. Mitchell, D. D. D'Lima, C. W. Colwell Jr., and B. J. Fregly. 2007. Correlation between the knee adduction torque and medial contact force for a variety of gait patterns. *J Orthop Res* 25 (6):789–97.

Zijlstra, W., and K. Aminian. 2007. Mobility assessment in older people: New possibilities and challenges. *Eur J Ageing* 4 (1):3–12.

# 9

# The Challenges of Monitoring Physical Activity in Children with Wearable Sensor Technologies

**Gita Pendhakar, Daniel T.H. Lai, Alistair Shilton, and Remco Polman**

**CONTENTS**

Introduction ........ 222
Why Sensor Monitoring for PA in Children? ........ 224
Challenges in Monitoring ........ 224
  Hardware and Software Challenges ........ 225
  Data Interpretation Challenges ........ 226
  Behavioural Challenges ........ 228
Case Study on Ambulatory Gait Monitoring in Idiopathic Toe Walking Children ........ 230
  Introduction ........ 230
  Gait Monitoring in ITW Children ........ 231
    What Is ITW? ........ 231
    Causes of ITW ........ 232
    Consequences of ITW ........ 232
    Gait Assessment in ITW Children ........ 232
    Lack of an Objective Method for Ambulatory Monitoring in ITW ........ 233
    Challenges in Ambulatory Monitoring of the Gait in ITW Children ........ 233
    Differences in Gait Features in Toe Walking and Normal Stride ........ 237
  Experiments, Algorithm Development and Statistical Analysis ........ 238
    Acceleration Measurement Methods ........ 238
    Algorithm Development for Identifying Strides ........ 239
  Development of a Miniature System Using Sensors to Monitor and Assess the Gait in ITW Children Remotely ........ 239
    Hardware Description ........ 241
    Analogue Output Connector ........ 241
    Battery Charger ........ 242
    External On/Off Switch ........ 242
    Serial RS232 Interface ........ 242
  Case Study Conclusion ........ 242
Conclusion ........ 243
References ........ 243

## Introduction

Physical activity (PA) monitoring in children remains an important challenge for epidemiologists, exercise scientists, clinicians and behavioural researchers as they attempt to utilize PA in combating child-related disorders such as obesity in children and young adults. Work has primarily focused on the development of improved metrics for measuring PA intensity. These metrics have been composed of direct observations, subjective surveys and, recently, portable activity monitors based on inertial sensors (Freedson, Pober, and Janz 2005). Portable monitors using accelerometer sensors have been used since the early 1990s, with development having focused on miniaturization and cost minimization. The challenges in interpreting sensor data as well as calibration and reliability issues have, however, been the focus of research in the last decade (post-2000). Although work is still continuing in the development of models to interpret PA from sensors, field trials have already begun. During 2003–2004, the National Health and Nutrition Examination Survey (NHANES) released survey results of 7000 participants above 6 years of age who wore an accelerometer during the day for 7 days. This large survey concluded that accelerometers could objectively capture physical activities performed by their users; however, inconsistencies with calibration, correlation and validation of sensor measured PA still remain.

The major impediment to research so far has been the difficulty of accurately measuring and quantifiying PA, mainly arising from the lack of a reliable long-term monitoring system which can be worn for long durations. Current accurate monitoring systems for measuring motion have been confined to gait laboratories equipped with expensive video systems, e.g., Vicon (Inition UK) and Optotrak Certus (NDI Canada). Recently, however, several commercial systems have been developed using microelectromechanical machines (MEMS) sensors, which measure body accelerations and cleverly interpret them as PAs or, more aptly, body postures, e.g., sitting, standing, running and lying down. Simple portable devices such as pedometers measure step counts but offer limited activity information and primitive energy expenditure (EE) estimates; they work only when the device is correctly worn. This, however, has only served to increase the proliferation of activity monitors on the market, which range from basic units such as activePAL and Dynastream AMP 331 to more complex devices such as the Actigraph and Ossur Prosthetic Activity Monitor™ (Godfrey, Conway, Meagher, and ÓLaighin 2008). The Intelligent Device for Energy Expenditure and Activity (IDEEA) is one of the more successful activity monitors. It requires five sensor attachments and is capable of detecting up to 32 different activities. The monitor provides descriptive data such as activity duration, speed and estimation of EE (Zhang, Werner, Sun, Pi-Sunyer, and Boozer 2003). Zhang et al. (2003)

found the IDEEA to be highly accurate for detecting most activities, except cycling and jumping. The majority of these monitors employ a range of accelerometer sensors, which makes their accuracy highly dependent on the body location they are programmed for. For example, an accelerometer positioned at the hip would not accurately measure motion involving only the arms, e.g., washing dishes. In addition, these devices assume that all motion is due to the user, and hence they are inaccurate when the user is stationary in a moving environment, e.g., driving a car, sitting in a playground swing or using a treadmill (Leenders, Nelson, and Sherman 2003). It has been recently reported that these devices do not perform well if the user has pathological or atypical gait and that they are inaccurate at measuring kinematics such as distance or direction travelled (Kuo, Culhane, Thomason, Tirosh, and Baker in press). These devices are restricted to providing information correlating PA to personal health under ideal movement conditions. Furthermore, currently accepted PA measures such as the Borg exertion rating and talk test cannot be calculated by these devices, while measures like the metabolic equivalent of task (MET) and target cardiac rates do not directly relate acceleration data to cardiac measurements, i.e., electrocardiograms (ECG) (Ann V. Rowlands and Roger G. Eston 2007).

It is evident that further improvements to PA monitors require investigations into sensor technology and detection algorithms to circumvent the limitations that have been mentioned whilst remaining robust and acceptable for target users, e.g., children. In this chapter, we first motivate the need for PA monitoring in children and provide a brief survey of the current commercially available activity monitors. We then describe the challenges from a design perspective for monitors incorporating multiple-sensor technology with the latest developments in computational intelligence and signal-processing models for monitoring PA. While previous activity monitors relied on single-axis acceleration analysis, new devices are integrating multiple-axis sensor data (sensor fusion of accelerometer and gyroscope data) with physiological measurements, e.g., muscle activity (electromyography [EMG]), whilst maintaining existing features, e.g., portability, cost-effectiveness and subject privacy. The inclusion of physiological measurements would potentially provide improved verification of user participation in a perceived activity, but it presents added problems in terms of data interpretation. We present a case study utilizing inertial sensor technology to monitor the severity of idiopathic toe walking (ITW) found in children. In addition, our research results will be widely applicable in the areas of exercise, sports performance, nutrition and studies on disorders such as obesity and diabetes. Future research can potentially apply our results to studies of different user populations, for example, for monitoring child development and studying healthy lifestyles in young adults for disorder prevention in the later years.

## Why Sensor Monitoring for PA in Children?

PA monitoring in children is particularly important for understanding the factors involved in child development and growth. These factors will facilitate the design of interventions for addressing child disorders and, more recently, for assessing rehabilitation progress in children after corrective surgery. The primary challenge faced by new PA-monitoring methodologies is to develop determinants of PA in children which are useful for quantifying the amount of PA in addition to being highly correlated with human physiology. This will enable the development of more effective programmes to deal with lifestyle-related problems in children, for example, obesity and sedentary lifestyles. New PA-monitoring techniques are also required to improve the accuracy of PA measurements. Children, for example, have been shown to be unreliable in using self-reports to effectively record the PA they were engaged in. Furthermore, poor compliance in reporting frequently contributed to errors too high for proper PA quantification. In addition, new PA monitors are required in assessment applications, particularly in rehabilitation progress in post-surgery situations, e.g., after gait-corrective surgery on children with cerebral palsy.

## Challenges in Monitoring

A number of PA monitors are available commercially with a variety of features. These range from simple, cheap pedometer-type devices to more sophisticated and expensive uni- and triaxial accelerometers and even global positioning system (GPS) devices. All have different specifications with respect to tolerance, sampling rate, frequency range, onboard memory and interfacing requirements, making the selection of an appropriate device for a given application a nontrivial undertaking.

The most basic type of automatic PA monitors is pedometers (Chen, Anton, and Jelal 2009; Schneider, Crouter, and Bassett 2004), which in their most basic form are simple electromechanical devices that count sudden acceleration events presumed to be caused by the foot hitting the ground in a step when the subject is either walking or running. While popular, such devices are of limited utility in applications such as child activity monitoring, as they may fail to capture nonstructured but potentially strenuous activities such as climbing and digging in a sandpit. For this reason we do not consider these in detail here. A good summary of commercially available pedometer-type PA monitors is given by Chen et al. (2009).

More advanced PA monitors such as the ActiGraph, the Actical, the Actiwatch, the RT3 Triaxial Research Tracker and the Caltrac (Trost, McIver, and Pate 2005) measure and record acceleration in one, two or three directions. This is typically achieved through MEMS devices using a cantilevered beam and a piezoelectric crystal. Measurements taken in this way may then be thresholded and used to count sudden acceleration events in the same way as a pedometer, or more advanced processing may be used to estimate EE (Trost et al. 2005). The advantage of such devices over simple pedometers is that they have the potential to capture a wide range of physical activities and hence give a more accurate picture of a subject's PA, including nonstructured activities such as digging in a sandpit. A number of accelerometer-based PA monitors of this type are summarized by Trost et al. (2005).

Finally, GPS-based monitors have been used to measure the distance covered by a subject over a given period of time and hence, it is assumed, their total EE (Chen et al. 2009). Like the pedometer, however, such devices are unlikely to be capable of capturing nonstructured "play" activities in children. Furthermore, GPS tracking is viable only in outdoor environments and requires significant power to operate, thus making it impractical for our purposes.

## Hardware and Software Challenges

Some key considerations when selecting PA monitors are

1. Price: the cost of purchase, maintenance, interfacing and application
2. Precision: the desired accuracy
3. Number of sensors: how many sensors are to be placed on the body
4. Sensor placement: where sensors are placed on the body
5. Epoch length: the period of time over which accelerations are integrated between readings (McClain, Abraham, Brusseau, and Tudor-Locke 2008)
6. Monitoring period: how long the activity needs to be monitored

Clearly, these considerations are interrelated. For example, while uniaxial accelerometers may be cheaper in per-unit terms, they may not be able to provide an accurate reflection of PA unless multiple sensors are placed on various parts of the body, whereas if more expensive triaxial accelerometers are used, fewer may be required to obtain the same accuracy. Moreover, while the use of many sensors may be desirable to achieve high precision, this may be expensive and interfere with the activities being measured, leading paradoxically to less representative results and potential noncompliance on behalf of the participant. The number of sensors and their placement are two key considerations when designing a PA monitor. If too many sensors are used or their placement is poorly chosen, then participant compliance may be a problem. It has been suggested that small improvements in accuracy which result from the use of more than one sensor do not warrant the increased burden on the participant (Trost et al. 2005). However, it should be noted that this is based on a very simple model for estimating EE, and hence this result may not reflect the potential of using multiple sensors.

## Data Interpretation Challenges

A major challenge in deploying portable sensors for monitoring PA in children is the interpretation of sensor data, in particular correlating the data to a physiological measurement which describes biological changes in the body or even to changes in behavioural patterns. This is not particularly easy, as can be seen in past research on metrics for quantifying PA based on a host of physiological measurements such as MET, oxygen consumption and doubly labelled water.

MET is a measure of the human metabolic rate in terms of body heat production. One MET is equivalent to the heat produced by a human in a relaxed position, usually taken to be 58 $W/m^2$ (Ainsworth et al. 2000) or 1 $kcal{\cdot}kg^{-1}{\cdot}h^{-1}$. Under the assumption that the mean body surface area (Du-Bois area) is approximately 1.8 $m^2$, the total heat of 1 MET would be 104 W. MET values can range from as low as 0.9 (sleeping) to as high as 18 (running at 17.5 km/h). The limitation with MET is that the values vary between persons performing the same activity. The baseline of 1 MET is dependent on body mass, age and other physiological factors, and hence the equivalence of 1 MET to 1 $kcal{\cdot}kg^{-1}{\cdot}h^{-1}$ may hold true only for healthy persons (40-year-old 70 kg man) but vary significantly between individuals.

Oxygen consumption can be measured using several techniques. Maximal or peak oxygen consumption, i.e., $VO_2$ max, is defined as the maximum capacity of a human to use oxygen during incremental exercise. This method requires strenuous PA (usually on a treadmill or cycle ergometer) in graded exercise to fully stress the aerobic energy system. $VO_2$ max is achieved when oxygen consumption remains steady even though the activity intensity or workload is increased. Fick's equation provides the definition of $VO_2$ max, but measuring it can be difficult as the procedure places undue stress on the cardiovascular system, requires sophisticated laboratory equipment and expertise and has a number of additional limitations (Midgley, McNaughton, Polman, and Marchant 2007). However, field methods such as the Cooper test have been developed to estimate this quantity (Freedson et al. 2005).

The doubly labelled water method is used to measure the average metabolic rate, also known as the *field metabolic rate*. The method involves using heavier isotopes of hydrogen (H-2, also known as deuterium) and oxygen (O-18) to form deuterium oxide ($D_2O^{18}$), which is heavy water. The average elimination of deuterium and O-18 is then measured over a period of time. It turns out that O-18 in the body is lost through carbon dioxide and body water (urine, sweat), while deuterium is lost only through body water. This concept can be used to infer the metabolic rate by computing the ratio of oxygen used in metabolism to carbon dioxide eliminated. Measurements involve taking two body water samples (at the beginning and the end of the measurement period) and carbon dioxide measurements at fixed intervals during the measurement period.

New portable PA systems sporting inertial sensing technology are faced with the challenge of correlating kinematic measurements with physiological measurements. For example, accelerometers measure acceleration due to body movement, while physiological measures, e.g., MET, measure body heat dissipation. In addition to the differences in measurement quantities, physiological associations are not constant during child development. For example, the cost of energy in movement per mass decreases with child maturation (Freedson et al. 2005). Nevertheless, several researchers have attempted to develop regression models to correlate accelerometer measures with physiological measures. Most work so far has focused on activity counts, while newer work has attempted to correlate the raw acceleration signal with some physiological measure.

The major drive has been to calibrate accelerometer readings against physiological measures. Early work attempted to calibrate activity counts against the duration of time when a subject's heart rate was above a threshold value but achieved low correlations between 0.5 and 0.7 (Janz 1994). Sallis et al. calibrated oxygen consumption against a Caltrac accelerometer, reporting good correlations (0.89) between activity counts and EE (Sallis, Buono, Roby, Carlson, and Nelson 1990). Linear regression equations were also derived to link MET values to accelerometer activity counts, for example, Freedson's equation (Freedson et al. 2005):

$$\begin{aligned}\text{METs} = {} & 2.757 + (0.0015 \times \text{counts per minute}) \\ & - (0.08957 \times \text{age (yr)}) \\ & - (0.000038 \times \text{counts per minute} \times \text{age (yr)}).\end{aligned}$$

Equations relating oxygen consumption to activity were also devised, e.g. (Freedson et al. 2005),

$$VO_2\ (\text{mL}\cdot\text{kg}^{-1}\cdot\text{min}^{-1}) = 7.7104 + 0.002631974\ (\text{counts per minute}).$$

These studies were conducted using commercially available PA monitors such as the Caltrac, Actigraph and RT3 Triaxial Research Tracker. The limitation of using raw accelerometer data (accelerometer worn on the ankle for 1 week) was that the activity counts were not associated with total EE as assessed by doubly labelled water. Poor correlation between these quantities was surmised to be caused by prolonged monitoring and the diversity of physical activities engaged in by the participants. Most of these measurements were made using a uniaxial accelerometer, and few studies have considered a triaxial accelerometer or a diversity of other inertial sensors. One study even suggested that there was little or no difference between using a uniaxial accelerometer and using sensors with more degrees of sensing

(Freedson et al. 2005). It is evident that further research is required with multiaxial sensors to fully utilize the information they offer to detect the type of PA and its intensity. A major problem with the definition of acceleration metrics calibrated against physiological measures is that the resulting regression models tend to be population-specific; i.e., they represent only the trend in the population groups under investigation and do not work well for groups outside the study. This poor generalization is a prevailing challenge and could be addressed by using more advanced function estimation methods capable of generating more generalized models across a larger variety of populations. Attempts to correlate sensor data with observational analysis of PA have also been made, finding greater correlation for the CSA-7164 accelerometer monitor than for the Actiwatch (Kelly, Reilly, Grant, and Paton 2004). It should be noted that a recent study of 277 individuals that attempted to simultaneously correlate both MET and $VO_2$ with accelerometer data has also concluded that linear regression models are inadequate for capturing relationships between sensor data and physiological data (Kozey, Lyden, Howe, Staudenmayer, and Freedson 2010).

Some recent work has examined the use of pattern recognition and Hidden Markov Models (HMMs) in a more refined classification of PA. Also, multiaxial sensor data have been combined using a semi-active Markov model to perform activity detection in swimmers (Thomas et al. 2010). Neural network (NN) models were also recently used to identify five different activities (sitting, standing, stair climbing, walking and cycling) for 49 young adult participants who wore Actigraph units on their hips and ankles (de Vries, Garre, Engbers, Hildebrandt, and van Buuren 2010). It was found that combined information from two sensor units led to better classification of the activities (83%, compared to 80.4% for the hip alone or 77.7% for the ankle alone).

## Behavioural Challenges

To date, a number of methods have been used to assess PA patterns in children and adolescents. Recently, eight physical assessment approaches were identified (Dollman et al. 2009), a number of which we have already referred to in this chapter. More objective methods, making use of electronic monitoring technologies, include heart rate monitoring, accelerometry and pedometry. Methods which are more subjective in nature include direct observation, self-report, parent report, teacher report and diaries or logs. Within these different assessment approaches, there are also a number of options for the researcher when making the choice of the most appropriate methodology or instrument for a complex task. In the same overview, Dollman et al. also highlighted a number of factors which should be taken into consideration when selecting a physical assessment instrument. These include the type of population (e.g., urban/rural, gender, migration status) and in particular their age or developmental status, the sample size, the burden on the participants and researcher, the delivery mode, the time frame in which PA is being assessed (day, week, semester, year), the information required (e.g., accuracy), data management (e.g., computer interface and download options, data storage, memory capacity), measurement error, validity and reliability, the available budget and the nature of the study (intervention, programme evaluation, surveillance).

The use of questionnaires to assess PA levels has been shown to be challenging. Problems are associated with the notion that these instruments often assess only one domain-specific form of PA which accounts for only a fraction of total daily EE (Tremblay, Esliger, and Tremblay 2007). In addition, the "human factor" also results in a number of problems. Participants have a tendency to overreport socially desirable health behaviours including PA (Esliger and Tremblay 2007). On the other hand, there is a tendency to underreport routine, intermittent, moderately intense activities like walking (nonpurposeful activities). This problem is augmented in young children. Their accumulation of daily PA is generally acquired through informal play and games. Retrospective assessment of such behaviour via self-report instruments is subject to significant human error. Questionnaires are limited in their scope and cannot address questions which can be addressed with activity monitors. For example, issues related to the frequency, intensity and duration of PA and its beneficial effects on health are more accurately addressed by activity monitors. Growing evidence indicates that the use of more objective monitoring devices provides a more comprehensive and stronger association with health outcomes in comparison to questionnaires which assess PA patterns (Janz 2006). Although subjective assessment of PA is problematic, it should be noted that the use of activity monitors is not without its limitations. In particular, the misinterpretation of data due to the differing methodologies is a potential problem.

Studying the PA behaviour of children or adolescents brings a number of additional complications. These include the accurate detection of typical PA patterns in children, which have been characterized by short and sporadic bouts of activity; by the notion that children mature physically, cognitively and socially at different rates; and by the notion that the use of technology might influence subsequent PA behaviour (McClain and Tudor-Locke 2009).

Children of different ages differ in terms of their anatomy, physiology and cognitive and psychosocial capabilities. It is therefore not surprising that typical behaviours or PA patterns differ greatly between young children, adolescents and adults. Biological maturation and skill development allow children to become more active as they mature. For example, very young children mainly sleep and eat, and younger children exhibit only unstructured and incidental PA behaviours and engage in active play rather than formal exercise, whereas adolescent PA patterns resemble those of adults more (Dwyer, Baur, and Hardy 2009). These developmental differences have important implications in terms of actual assessment as well as in terms of the determinants, mediators or outcomes of these behaviours. For example, sampling of data has generally taken place in 1 min epochs. Because of the unstructured and sporadic nature of younger children's activity behaviour, reducing this period would be necessary to obtain valid and reliable data (Corder, Brage, and Ekelund 2007).

The use of technologies could potentially reduce the burden on the participants in the assessment of PA. However, some devices require participants' attention and are therefore less suitable for use with younger individuals where their use might result in a reduction in adherence or missing data. There are also a number of practical issues which need to be considered when assessing PA in children. These include the size and weight of the device and its placement on the body, as well as its tamper-proofness. Children are curious by nature and as such accessible on/off or reset buttons could be problematic. To this end, researchers have used sealed recording devices to overcome this potential difficulty (e.g., Hohepa, Schofield, Kolt, Scragg, and Garrett 2008).

Another issue which needs consideration is that children often engage in locomotor activities which are accompanied by other activities. This could include bouncing a ball or jumping. Assessment of such activities would most likely result in an underestimation of EE. Hence, movement speeds would be lower during the execution of such movements (Duncan, Badland, and Schofield 2009).

A problem is that most devices cannot provide information on the mode or context of PA (e.g., home, school or community) or the reasons why certain behaviours are undertaken. In addition, most devices still have difficulty in registering some movement modes including cycling, climbing and swimming. Therefore, there appears to be a need for additional qualitative information from participants, but this is subject to memory lapses, over- and underreporting of PA behaviours by participants or social desirability bias in parental reporting (Dollman et al. 2009).

Finally, most monitoring devices have been designed to assess PA levels rather than sedentary behaviours. However, there is accumulating evidence that physical inactivity is independently associated with health (Ekelund et al. 2006). As such, there is a need to consider both sedentary and activity behaviour. Both might, again, have different antecedents, mediators and outcomes, which may vary with age or developmental status.

## Case Study on Ambulatory Gait Monitoring in Idiopathic Toe Walking Children

### Introduction

A case study on gait monitoring in idiopathic toe walking children is detailed in the following. Although the problem of gait monitoring sounds fairly simple, it is quite complex in ITW children due to their habitual toe walking behaviour. The habitual toe walking in ITW children poses a behavioural challenge, as these children often have a tendency to modify their gait if they want to and especially when they are being monitored in a gait laboratory. Hence, an accurate assessment of their gait when being observed in a gait laboratory could be an issue. In this case, an innovative system using sensors was developed for ambulatory monitoring; however, there were hardware and software challenges while designing and developing the system, as mentioned in the following.

Considering some of the previously mentioned challenges, in order to accurately assess and determine the gait in ITW children, the following method was adopted. An initial assessment based on observational gait analysis was carried out at the clinic to determine the severity of toe walking. The calf flexibility for these children was also evaluated using goniometers. The ITW children were then treated as per the severity of toe walking. These children were also assessed using sensors. The accuracy of the sensor-based assessment method was compared with the accuracy obtained from observational analysis and parental questionnaires, which is the traditional method followed at the clinic. Because the sensor data were more accurate than the observational assessment, a miniature system was developed which was embedded in the heel of the children's boot and used to assess the gait in ITW children.

## Gait Monitoring in ITW Children

### *What Is ITW?*

Toe walking is a condition in children whereby the child has an abnormal gait and adopts a strong plantar flexed position with the foot and walks on the head of the metatarsals rather than with a heel-toe gait. Toe walking normally occurs in children with a variety of neurological conditions and can be associated with several diagnoses (Shulman 1997). Children who have spastic cerebral palsy, muscular dystrophy, spinal injury or acute myopathy often toe walk as they adopt a plantar flexed posture during walking.

ITW children are children who do not have any medical conditions; however, they still walk with their feet strongly plantar flexed (Caselli, Rzonca, and Lue 1988). Idiopathic toe walkers are diagnosed by excluding all known causes of toe walking including neuromuscular and orthopaedic disorders and physical injuries caused by accidents. A thorough medical history, gait evaluation, musculoskeletal examination and neurological examination are necessary to distinguish an ITW child from a toe walking child with any of the previously mentioned conditions (Caselli et al. 1988). The severity of toe walking varies from a gait where the children walk with their heels lifted just off the ground and appear to "bounce" as they walk to a gait where they balance on the tips of their toes as they walk and the heels are never placed on the ground even when they are standing still. They are often able to place the heels on the ground if they want to, but, if left unsupervised, they re-adopt a plantar flexed posture during walking. Toe walking in ITW children invariably is symmetrical in involvement (Patrick, Ugo, and Kerrigan 2001).

### *Causes of ITW*

There are various theories to explain toe walking in ITW children. First, it could be an inherited factor. There is strong evidence to suggest that this condition is hereditary, and it does occur far more often in some families than others (Hirsch and Wagner 2004; Sala, Shulman, Kennedy, Grant, and Chu 1999; Sobel, Caselli, and Velez 1997). Second, there is evidence through EMG studies that indicates that in some children the ITW gait is the direct result of an abnormal order in which muscles of the lower leg are activated during walking (Kalen, Adler, and Bleck 1986).

### *Consequences of ITW*

ITW develops into a habitual toe walking practice in ITW children, which in turn leads to a fixed contracture of the Achilles tendon. Hence ITW children develop shortened calf muscles (Sobel et al. 1997b), which may lead to a number of problems. Another important effect of toe walking is that it results in an improper balance of the body because the centre of gravity is shifted slightly forward, and hence the children have a tendency to fall often (Caselli et al. 1988). Apart from these problems, the ITW children sometimes experience physical disability, accompanied by a sense of social isolation and ostracism within their peer group. This often leads to social problems.

### *Gait Assessment in ITW Children*

As ITW is relatively common and not a life-threatening condition, the need for treating this condition is not clearly defined in the literature (Patrick et al. 2001). The treatment for ITW varies depending on the severity of toe walking. ITW management options extend from observation alone to surgery and could vary depending on the treating physician's opinion. The most important diagnostic procedure for ITW children is physical examination of their gait pattern by an experienced clinician. Parental observation is another important tool used to assess the severity of toe walking in ITW children. The parents are often asked to complete a questionnaire, which provides the required information for treating ITW children (Caselli et al. 1988). Nonoperative treatments are preferred for less severe cases of toe walking, and these include the use of botulinum toxin (BOTOX®), casting, orthotics and calf exercises. Surgery is normally chosen if any of these medical treatments does not work or if the ITW child has severely shortened calf muscles. However, it is observed that most of the ITW children deliberately walk with a heel-toe gait when their gait is being assessed at the clinic in order to give the clinician the impression that they have a normal gait. Hence, in order to assess the gait of ITW children accurately, it must be assessed and monitored in a way such that the children are not aware of being monitored and assessed.

### Lack of an Objective Method for Ambulatory Monitoring in ITW

Currently, in most clinics, the severity of toe walking in ITW children is assessed by a trained clinician visually observing their gait in a clinic or by parental observation, which is purely subjective. Ankle flexibility is often measured at the clinic. However, ankle flexibility cannot be directly correlated to the severity of toe walking. It is very expensive to get the gait assessed in a gait laboratory. The problem with ITW children is that they can modify their gait and walk with a heel-toe gait if they want to. Hence, there is a need for objective methods of gait assessment such as portable low-cost devices which can provide reliable and repeatable results and can be used for such cases of gait assessment and monitoring.

### Challenges in Ambulatory Monitoring of the Gait in ITW Children

Some of the behavioural, technological and data interpretation challenges in monitoring the gait in ITW children are discussed in the following.

#### *Behavioural Issues*

The toe walking gait in ITW children is habitual in nature. Hence these children are capable of walking normally with a heel-toe gait if they want to or if they are asked to. However, when they are in their own comfort zone, they walk on their toes. Whenever these children are being assessed in the clinic, they tend to pretend that they usually walk normally in order to avoid any further treatments. Hence, the assessment of their gait in a clinic or a gait laboratory is not useful and accurate. This behavioural issue is quite frustrating to the parents and the clinician as their real gait cannot be assessed and treated.

Ambulatory gait monitoring is a new method of gait monitoring which can be carried out at any remote location using sensors. Due to the availability of various miniature sensors which can be used for gait assessment, the technology for ambulatory gait monitoring is developing rapidly (Aminian et al. 1998). Ambulatory monitoring using accelerometers has proved accurate and beneficial and have a lot of applications in healthcare areas, and it is a useful tool to assess the gait in ITW children.

#### *Sensor Selection and Placement Issues*

For ambulatory monitoring and assessment, selection of appropriate sensors is important, as the accuracy in distinguishing the toe walking gait from normal gait would depend on the sensors and extracting the appropriate gait features from the sensor data. The sensors should be small and unobtrusive, so that they can be belted, attached or embedded in a suitable manner and used for long-term gait monitoring (Mathie, Coster, Lovell, and Celler 2004). Positioning of the sensors is another important issue, as the accuracy of the gait parameter sensed by the sensor would depend on the position where it is placed and its orientation.

Taking into consideration all the preceding requirements, accelerometers have been considered the best choice of sensors for distinguishing toe walking gait from normal gait, and it was decided that embedding the accelerometers in the heel of the children's boots would provide the required gait features.

In order to evaluate the accuracy of using accelerometers to distinguish toe walking stride from normal stride, a number of experiments have been conducted on ITW children and normal children. The miniature accelerometer evaluation board was embedded in the heel of standard lace-up boots as shown in Figure 9.1 for gait measurements in ITW children. The compact 5 V batteries used to power the accelerometer evaluation board were placed on the side of the boot along with a switch to turn the power on and off as required.

There are a few advantages in embedding the accelerometer in the heel of the boots. First, the heel of the boots provided a proper housing for placing the compact accelerometer board as compared to possible force sensors in the ball of the boots, which flexes much more. Second, as the ITW children rarely have a heel strike, damage to an accelerometer board embedded in the heel of the boots is unlikely. Third, this position gives a peak signal (increase in magnitude) when the heel touches the ground, making a heel strike easy to detect. The other gait parameters such as stance and swing

**FIGURE 9.1**
Boots with an accelerometer board embedded in the heel.

phases can be accurately determined by placing the accelerometer in the heel of the boot. Two different sizes of boots were embedded with two different accelerometer boards to enable acceleration measurements for younger (5–8 years) and older (9–12 years) ITW children during the experiments at the medical centre.

*Sensor Data Interpretation*

Based on the position of the accelerometer, the accelerometer data have to be interpreted accurately so as to determine the gait features that distinguish normal and toe walking gait. The following discusses the theory of gait and the accelerometer data that are useful for gait analysis.

One gait stride can be divided into stance (60%) and swing (40%) phases. The stance phase in a gait cycle consists of early, mid- and late stance. Acceleration caused during walking is the total acceleration in the direction of each axis of the transducer in the heel during gait (walking), i.e., the sum of static (determined by gravity and the angle of tilt of the relevant axis with respect to gravity) and dynamic (due to motion when the foot moves from one position to another) accelerations. Acceleration during the mid-stance phase in a normal gait cycle corresponds only to static acceleration as the heel is resting on the ground and not moving. Figure 9.2 shows the boot (with the accelerometer board embedded) during mid-stance of normal walking, when the foot is flat on the ground. The dual-axis accelerometer is placed in the heel of the boot with the x-axis of the accelerometer being parallel to the foot on the horizontal surface, *h*, and the y-axis being perpendicular to the foot, or parallel to the vertical axis, *v*. In a normal heel-toe gait, during mid-stance, the angle made by the foot with respect to ground is zero.

However, if the boots are inclined to the ground or the horizontal surface *h* at an angle θ, taken as positive toe down, then, as shown in Figure 9.3, the equations for the accelerometer x-axis and y-axis ($a_x$ and $a_y$) signals

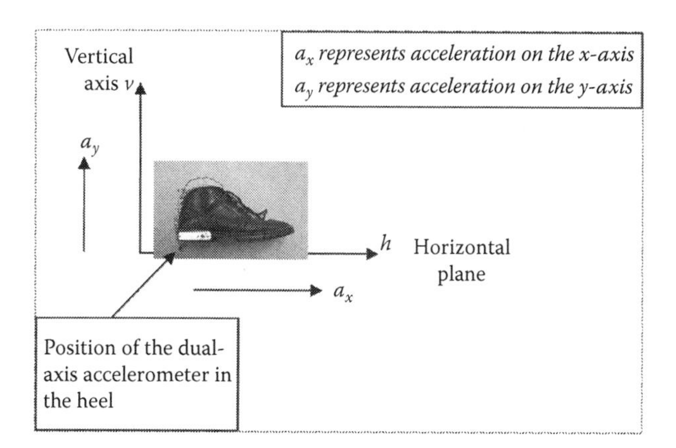


**FIGURE 9.2**
Orientation and reference axes for the boot on the ground.

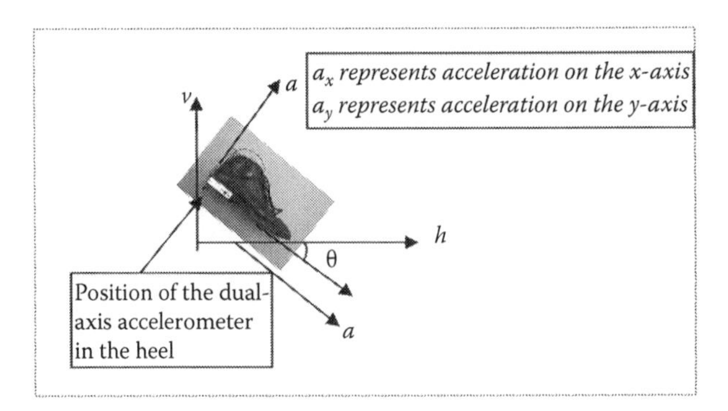


**FIGURE 9.3**
Boot inclined to the ground.

can be written in terms of horizontal and vertical accelerations as shown in Equations 9.1 and 9.2, as discussed by Svensson and Holmberg (2005):

$$a_x \approx g\ \sin\theta + a_v \sin\theta + a_h \cos\theta \tag{9.1}$$

$$a_y \approx -g\ \cos\theta + a_v \cos\theta + a_h \sin\theta \tag{9.2}$$

where $a_x$ is the acceleration on the x-axis, $a_y$ is the acceleration on the y-axis, $a_v$ is the vertical acceleration, $a_h$ is the horizontal acceleration, $g$ is the gravitational acceleration and θ is the angle made by the foot with respect to the ground for toe-down position. Equations 9.1 and 9.2 can be simplified for early, mid- and late stance in normal gait and toe walking gait as explained in the following subsection.

However, in the case of ITW children, the early stance is initiated by a toe strike, where the heel is up, and hence the foot is at a positive angle θ, as shown in Figure 9.3. Any part of the foot in touch with the ground has zero dynamic acceleration. In early stance of a normal stride, the dynamic acceleration of the heel will be very nearly zero, and even in a toe walking stride, dynamic acceleration of the heel will be small, especially along the x-axis. Ignoring these small dynamic accelerations during stance, we are left with just the static accelerations shown in Equations 9.3 and 9.4 and mentioned by Pendharkar, Morgan, and Percival (2005b):

$$a_x \approx g\ \sin\theta \tag{9.3}$$

$$a_y \approx -g\ \cos\theta. \tag{9.4}$$

In Figure 9.3, $h$ represents the horizontal surface of the ground, and the angle θ with the arrowhead is the angle of the ground with respect to the foot.

Equations 9.3 and 9.4 can be further solved to obtain the angle θ made by the foot with respect to the ground:

$$\theta = \tan^{-1}(-a_x/a_y). \tag{9.5}$$

In a normal stride, the foot remains flat (stationary) on the ground during mid-stance. This phase is identified by zero dynamic acceleration as the foot is stationary. As the angle of inclination of the foot with respect to the ground, θ, is zero, the upwards vertical-axis static acceleration component ($a_y$) is a constant equal to minus gravity, and the horizontal-axis static acceleration component ($a_x$) is zero. The equations for mid-stance of a normal stride are shown in Equations 9.6 and 9.7:

$$a_x \approx 0 \tag{9.6}$$

$$a_y \approx -g. \tag{9.7}$$

However, in toe walking stride, during mid-stance, the heel is still off the ground and hence is at a positive angle θ with respect to the ground, as shown in Figure 9.3. Hence, considering the dynamic accelerations to be negligible, the equations for mid-stance for ITW children would be the same as Equations 9.3 and 9.4. The foot angle might reasonably be expected to be more nearly constant during mid-stance than early stance, so that the dynamic accelerations will also be smaller; that is, the approximation will be more accurate.

In a normal stride, late stance is characterized by the heel lifting off the ground. This causes changes in the static components as in Equations 9.3 and 9.4 but also in some dynamic components. The heel essentially moves in a circle around the toes. The dynamic y-axis acceleration represents the heel accelerating as it leaves the ground and increases its angular velocity. The dynamic x-axis acceleration represents the centripetal acceleration of the heel about the centre of rotation near the toes. In ITW children, because the foot is at a positive angle with respect to the ground, there is no distinct change from mid-stance to late stance, though the dynamic components may increase, as described for normal gait. Hence, for normal and ITW strides, the acceleration signal that characterizes the late stance would be approximated by Equations 9.3 and 9.4, as the dynamic components are small.

### *Differences in Gait Features in Toe Walking and Normal Stride*

Based on the preceding discussion, it can be concluded that mid-stance is one of the gait features which can be used to distinguish a normal from a toe walking stride. A period of $a_x = 0$ and $a_y = -g$ is observed during mid-stance in normal gait only. In toe walking children, during mid-stance the values of the acceleration signals are $a_x \neq 0$ and $a_y \neq -g$. The other distinguishing feature between the two is the heel strike, which is absent in a toe walking stride. During heel strike, there is a rapid increase in vertical, and hence primarily y-axis, acceleration, which reaches a peak positive amplitude (Menz and Lord 2003). However, in toe walking children, the heel strike is missing.

Hence, these two features, heel strike and flat-footed stance, are likely candidates to distinguish a normal stride from an ITW stride.

## Experiments, Algorithm Development and Statistical Analysis

### *Acceleration Measurement Methods*

#### *Sensor Calibration*

The calibration of the sensors was carried out by keeping the foot flat on the ground with the boots on, as shown in Figure 9.2. The voltage output of the acceleration signals $a_x$ and $a_y$ when the boots were flat on the ground was recorded as

$$a_x = 1.94 \text{ V}$$

$$a_y = 2.08 \text{ V}.$$

These values of $a_x$ and $a_y$ when the boot is flat on the ground with zero angle of inclination represent the mid-stance in a normal stride and were used as reference values for the algorithms designed. For mid-stance, when the angle made by the foot with respect to the ground is zero, the equivalent voltage for 1 g can be calculated as

$$-g = a_y = 2.08 \text{ V}. \tag{9.8}$$

#### *Gait Acceleration Measurement*

For this study, the gait acceleration signals of the right heel were measured and assessed as the subject walked on a treadmill for a certain amount of time. The accelerometer evaluation board was embedded only in the heel of the right boot. ITW gait is normally symmetric, and it is assumed that if the right foot had a heel strike, then the left foot would also have a heel strike or vice versa. However, if the gait measurements were carried out on subjects with asymmetrical gait (for e.g., Cerebral Palsy of Acquired Brain Injury subjects), two separate accelerometer evaluation boards would have to be embedded, one in each boot (right and left), in order to get a correct assessment of the gait.

The experiments to measure the gait acceleration while walking were conducted at a clinic in Melbourne, Australia. Two groups of 10 children were recruited for the experiments. One group consisted of ITW children, and the second group consisted of normal children who were used as the control group. The children in both groups ranged from 5 to 12 years old and included both boys and girls. All the children weighed between 19 and 40 kg. Both groups of children walked on a treadmill for 2 min wearing the boots embedded with the accelerometer evaluation board and wired to the Fluke handheld oscilloscope. The ITW children were asked to walk in a manner that they found comfortable, which resulted in a toe walking gait. Figure 9.4

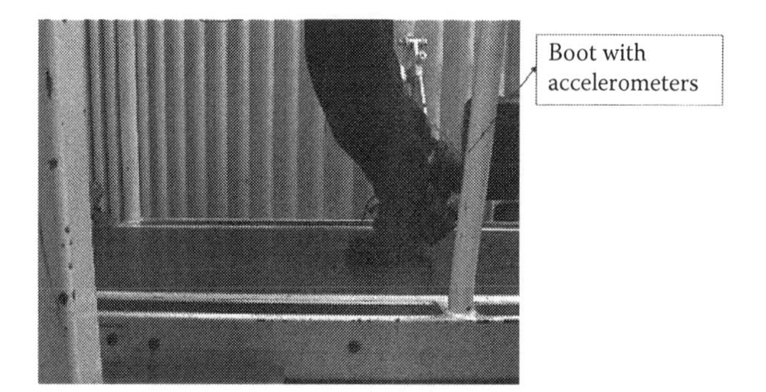


**FIGURE 9.4**
Idiopathic toe walking child walking on a treadmill with the boots with an accelerometer board embedded in the heel.

shows an ITW child walking on his toes with the boots (embedded with the accelerometer) on the treadmill at the clinic in Melbourne. The handheld oscilloscope is not seen in the picture.

### *Algorithm Development for Identifying Strides*

Heel acceleration data were recorded for all the children in the two groups and downloaded onto a PC for further analysis. First of all, the data collected from the raw acceleration signals $a_x$ and $a_y$ from the experiments were plotted with respect to time and examined to identify individual strides. Based on the differences observed (mid-stance) between a normal and a toe walking stride, an innovative algorithm was developed in Igor Pro to process the signals in order to identify and distinguish a toe walking stride from a normal stride.

#### *Graphical Representation of the Heel Acceleration Signals*

Figure 9.5 shows the plot locating the mid-stance and shows the final $a_x$ and $a_y$ signals for a normal gait pattern. In the group of children with a normal gait, $a_x$ represents zero acceleration during mid-stance as the foot is stationary, and the angle of inclination of the heel with respect to the ground is zero.

Figure 9.6 shows an example of the toe walking gait in ITW children. In severe ITW (children who walk on the tip of their toes), during stance, the static component of the acceleration $a_y$, due to the foot being angled down, is not equal to $-g$, as shown in Figure 9.6 (Pendharkar, Morgan, and Percival 2005a). In less severe case of ITW, the children may occasionally have a heel strike marked by a positive peak that is much smaller than a normal heel strike peak.

### Development of a Miniature System Using Sensors to Monitor and Assess the Gait in ITW Children Remotely

Once it was realized that using accelerometers and the algorithm developed, the toe walking strides could be differentiated from normal strides

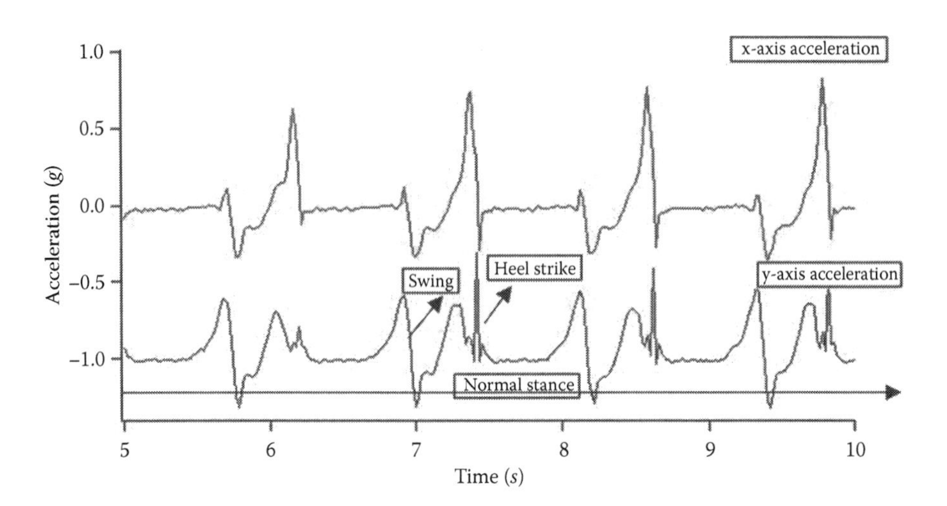


**FIGURE 9.5**
The acceleration signals $a_x$ (x-axis acceleration) and $a_y$ (y-axis acceleration) in units of $g$ with respect to time in seconds for normal gait.

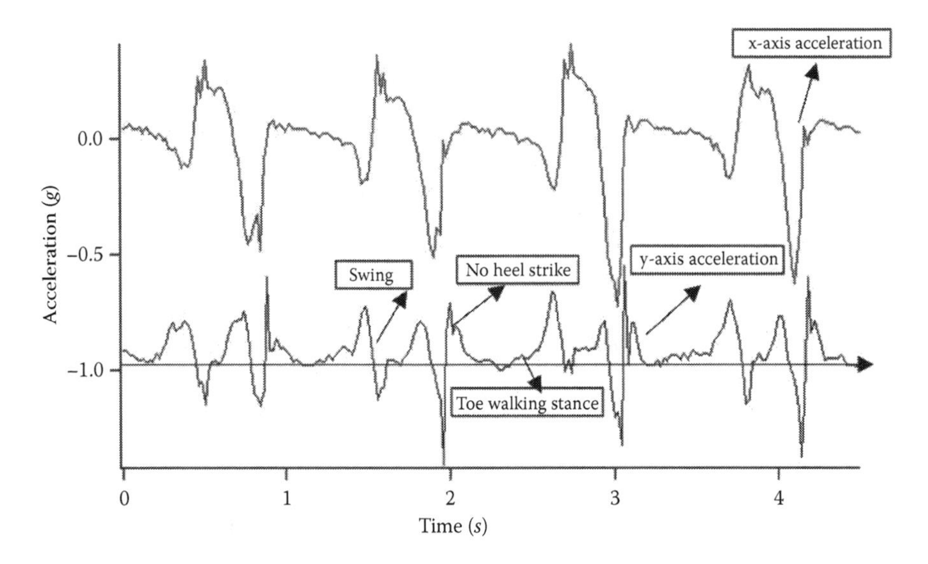


**FIGURE 9.6**
The acceleration signals $a_x$ (x-axis acceleration) and $a_y$ (y-axis acceleration) in units of $g$ with respect to time in seconds for severe toe walkers.

accurately, using mid-stance gait features, we designed and developed a compact miniature system which could be embedded in the heel of the boots and used remotely to assess the gait in ITW children. The printed circuit board (PCB) of the motherboard and the daughterboard were placed in a compact aluminium case, as shown in Figure 9.7.

**FIGURE 9.7**
The miniature system.

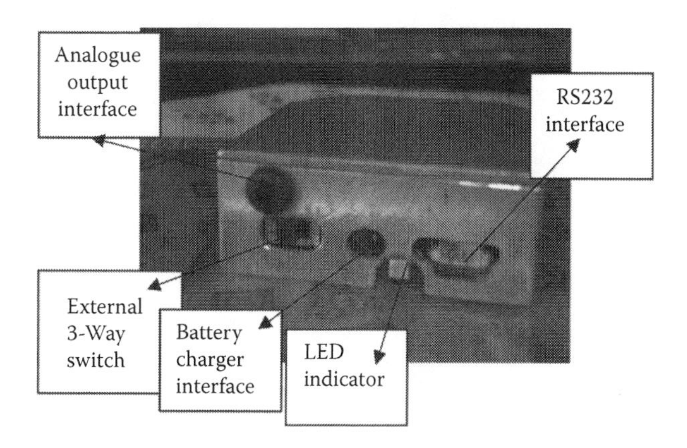


**FIGURE 9.8**
External connections for the miniature system.

### *Hardware Description*

Figure 9.8 shows a picture of the aluminium case fitted with the motherboard and the daughterboard developed for the miniature system. A NiMH (1.2 V) battery is small and compact, and hence five 1.2 V batteries were used in series so that they could be fitted easily in the aluminium case. The aluminium case was embedded in the heel of standard lace-up boots as shown in Figure 9.9. The lace-up boots were similar to the ones used at Monash Medical Centre (MMC). Assuming that ITW gait is symmetric (Sobel, Caselli, and Velez 1997), the miniature system was embedded in only one boot.

### *Analogue Output Connector*

As shown in Figure 9.9, a connector is provided to interface the two analogue outputs (x-axis and y-axis) of the ADXL 202 accelerometer to an oscilloscope whenever required for validation purposes.

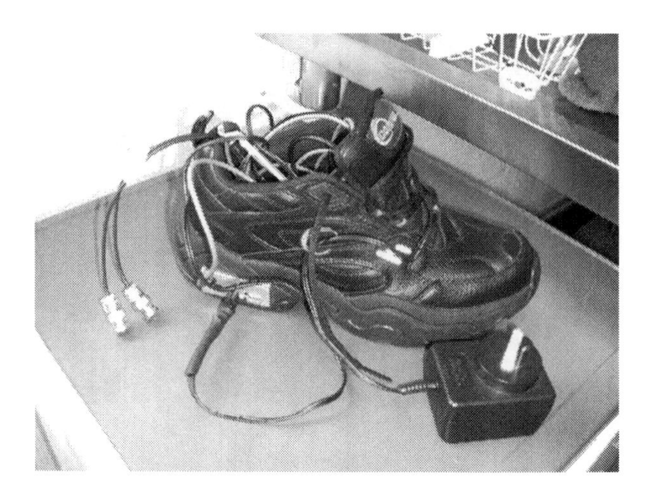

**FIGURE 9.9**
Boot with the miniature system embedded in the heel.

### *Battery Charger*

The power supply batteries used for this system can be charged by connecting a charger to one of the connections on the side of the boot, as shown in Figure 9.9.

### *External On/Off Switch*

A three-way switch was used to indicate if the system is on, off or charging. When the switch is in the extreme left position, it indicates that it is in charging mode. If the switch is in the extreme right position, it indicates that the power to the miniature system is on, and the centre position indicates that the power is switched off. This switch is located on the side of the boot, as shown in Figure 9.9.

### *Serial RS232 Interface*

The serial RS232 interface, shown in Figure 9.9, is used to serially connect the miniature system to the PC in order to download the information onto the PC.

## Case Study Conclusion

The miniature system that was designed and developed to monitor the gait activity was compact and could be easily embedded in the heel of the boots and could be used remotely. This addresses the behavioural challenge posed by the ITW children modifying their gait when being observed. A single dual-axis accelerometer was used in the miniature system, which was useful in distinguishing toe walking strides from normal strides. Using an innovative algorithm and computer intelligence, the accelerometer data could be analysed and used to distinguish toe walking strides from normal strides. The algorithm developed using mid-stance gait features had an accuracy of 98.5%. This algorithm was implemented in the microcontroller, and walking experiments were carried out, which were validated using video recordings. However, when the algorithm was implemented in the microcontroller and tested for gait monitoring, the accuracy of distinguishing the toe walking strides from normal strides was reduced to 83%. This indicated that there were hardware and software limitations. The compact miniature system can be embedded in the boots or attached to the boots, or it can be belted on a body part so as to be used for any other application in gait analysis involving foot movement.

## Conclusion

It is very important to monitor PA in children with disorders such as obesity. Monitoring PA in children using wearable sensors provides an objective means of assessment which could significantly improve the accuracy of assessment as compared to subjective methods of assessment. Although there are a number of portable systems available on the market to monitor PA, many of these are able to provide accurate information only under ideal movement conditions. This is particularly problematic for child monitoring as a significant proportion of PA in children is unplanned, unstructured and unpredictable. Due to the availability of various low-cost sensors, it is possible to design and develop systems to monitor specific activities. However, there are several challenges involved in implementing such systems in the real world. These include both hardware and software challenges, data interpretation challenges and behavioural challenges, making the design and development of such systems difficult. The case study is an example of a gait activity-monitoring solution which provides some ideas on how the various challenges can be solved to obtain an activity-monitoring system with high accuracy for monitoring specific activities. It is evident that further research is required to develop more efficient activity-monitoring systems.

## References

Ainsworth, B. E., Haskell, W. L., Whitt, M. C., Irwin, M. L., Swartz, A. M., Strath, S. J., et al. (2000). Compendium of Physical Activities: An update of activity codes and MET intensities. *Medicine & Science in Sports & Exercise*, 32(9), S498–S516.

Aminian, K., Robert, P., Buchser, E. E., Rutschmann, B., Hayoz, D., & Depairon, M. (1998). Physical activity monitoring based on accelerometry: Validation and comparison with video observation. *Medical and Biological Engineering and Computing, 37*, 304–308.

Ann V. Rowlands and Roger G. Eston. (2007). The measurement and interpretation of children's physical activity. *Journal of Sports Science and Medicine, 6*, 270–276.

Caselli, M. A., Rzonca, E. C., & Lue, B. Y. (1988). Habitual toe-walking: Evaluation and approach to treatment. *Clinics in Podiatric Medicine & Surgery, 5*(3), 547–559.

Chen, C., Anton, S., & Jelal, A. (2009). *A Brief Survey of Physical Activity Monitoring Devices* (No. MPCL-08-09): Mobile Pervasive Computing Research, University of Florida. Document Number).

Corder, K., Brage, S., & Ekelund, U. (2007). Accelerometers and pedometers: Methodology and clinical application. *Current Opinion in Clinical Nutrition, 10*, 597–603.

de Vries, S. I., Garre, F. G., Engbers, L. H., Hildebrandt, V. H., & van Buuren, S. (2010). Evaluation of neural networks to identify types of activity using accelerometers. *Medicine & Science in Sports & Exercise,* Publish Ahead of Print, 10.1249/MSS.1240b1013e3181e5797d.

Dollman, J., Okely, A. D., Hardy, L., Timperio, A., Salmon, J., & Hills, A. P. (2009). A hitchhiker's guide to assessing young people's physical activity: Deciding what method to use. *Journal of Science and Medicine in Sport, 12*, 518–525.

Duncan, J. S., Badland, H. M., & Schofield, G. (2009). Combining GPS with heart rate monitoring to measure physical activity in children: A feasibility study. *Journal of Science and Medicine in Sport, 12*, 583–585.

Dwyer, G. M., Baur, L. A., & Hardy, L. L. (2009). The challenge of understanding and assessing physical activity in preschool-age children: Thinking beyond the framework of intensity, duration and frequency of activity. *Journal of Science and Medicine in Sport, 12*, 534–536.

Ekelund, U., Brage, S., Froberg, K., Harro, M., Anderssen, S. A., Sardinha, L. B., et al. (2006). TV viewing and physical activity are independently associated with metabolic risk in children: The European Youth Heath Study. *PLoS Medicine, 3*, e488.

Esliger, D. W., & Tremblay, M. S. (2007). Physical activity and inactivity profiling: The next generation. *Applied Physiology, Nutrition and Metabolism, 32*, S195–S207.

Freedson, P., Pober, D., & Janz, K. F. (2005). Calibration of accelerometer output for children. *Medicine & Science in Sports & Exercise, 37*(11), S523–S530.

Godfrey, A., Conway, R., Meagher, D., & ÓLaighin, G. (2008). Direct measurement of human movement by accelerometry. *Medical Engineering & Physics 30*, 1364–1386.

Hirsch, G., & Wagner, B. (2004). The natural history of idiopathic toe walking: A long term follow-up of fourteen conservatively treated children. *Journal of Acta Paediatrica, 93*, 196–199.

Hohepa, M., Schofield, G., Kolt, G. S., Scragg, R., & Garrett, N. (2008). Pedometer-determined physical activity levels of adolescents: Differences by age, sex, time of week and transportation mode to school. *Journal of Physical Activity & Health, 5*, S140–S152.

Janz, K. F. (1994). Validation of the CSA accelerometer for assessing children's physical activity. *Medicine & Science in Sports & Exercise, 26*, 369–375.

Janz, K. F. (2006). Physical activity in epidemiology: Moving from questionnaire to objective measurement. *British Journal of Sports Medicine, 40*, 191–192.

Kalen, V., Adler, N., & Bleck, E. E. (1986). Electromyography of idiopathic toe walking. *Journal of Pediatric Orthopedics, 6*(1), 31–33.

Kelly, L. A., Reilly, J. J., Grant, S., & Paton, J. Y. (2004). Objective measurement of physical activity in pre-school children: Comparison of two accelerometers against direct observation. *Medicine & Science in Sports & Exercise, 36*(5), S329.

Kozey, S. L., Lyden, K., Howe, C. A., Staudenmayer, J. W., & Freedson, P. S. (2010). Accelerometer output and MET values of common physical activities. *Medicine & Science in Sports & Exercise, 42*(9), 1776–1784 1710.1249/MSS.1770b1013e3181d1479f1772.

Kuo, Y.-L., Culhane, K. M., Thomason, P., Tirosh, O., & Baker, R. (2009). Measuring distance walked and step count in children with cerebral palsy: An evaluation of two portable activity monitors. *Gait & Posture, 29*(2), 304–310.

Leenders, N., Nelson, T., & Sherman, W. (2003). Ability of different physical activity monitors to detect movement during treadmill walking. *International Journal of Sports Medicine, 24*(1), 43–50.

Mathie, M., Coster, A., Lovell, N., & Celler, B. (2004). Accelerometry: Providing an integrated, practical method for long-term, ambulatory moitoring of human movement. *Physiological Measurements, 25*, R1–R20.

McClain, J. J., Abraham, T. L., Brusseau, T. A. J., & Tudor-Locke, C. (2008). Epoch length and accelerometer outputs in children: Comparison to direct observation. *Medicine & Science in Sports & Exercise, 40*(12), 2080–2087 2010.1249/MSS.2080b2013e3181824d3181898.

McClain, J. J., & Tudor-Locke, C. (2009). Objective monitoring of physical activity in children: Considerations for instrument selection. *Journal of Science and Medicine in Sport, 12*, 526–533.

Menz, H. B., & Lord, S. R. (2003). Acceleration patterns of the head and pelvis when walking on level and irregular surfaces. *Gait & Posture, 18*, 35–46.

Midgley, A., McNaughton, L. R., Polman, R., & Marchant, D. (2007). Criteria for determination of the maximal oxygen uptake: A brief critique and recommendations for future research. *Sports Medicine, 37*(12), 1019–1028.

Patrick, O. R., Ugo, D. C., & Kerrigan, D. C. (2001). Propulsive adaptation to changing gait speed. *Journal of Biomechanics, 34*, 197–202.

Pendharkar, G., Morgan, D., & Percival, P. (2005a). Distinguishing toe-walking gait using accelerometer in idiopathic toe-walking (ITW) children. Paper presented at the 3rd International Conference on Biomechanics of the Lower Limb in Health Disease and Rehabilitation, Manchester, UK.

Pendharkar, G., Morgan, D., & Percival, P. (2005b, 3–5 December). Processing accelerometry data to detect stance phase in gait and differentiate toe-walking steps from normal steps in idiopathic toe-walking children. Paper presented at the Intelligent Sensors Sensor Network and Information Processing, Melbourne, Australia.

Sala, D. A., Shulman, L. S., Kennedy, R. F., Grant, A. D., & Chu, M. L. Y. (1999). Idiopathic toe-walking: A review. *Developmental Medicine and Child Neurology, 41*, 846–848.

Sallis, J. F., Buono, M. J., Roby, J., Carlson, D., & Nelson, J. A. (1990). The Caltrac accelerometer as a physical activity monitor for school-age children. *Medicine & Science in Sports & Exercise, 22*, 698–703.

Schneider, P. L., Crouter, S. E., & Bassett, D. R. (2004). Pedemeter measures of free-living physical activity: Comparison of 13 models. *Medicine & Science in Sports & Exercise, 36*(2), 331–335.

Shulman, L. H. (1997). Developmental implications of idiopathic toe walking. *Journal of Paediatrics, 130*(4), 541–546.

Sobel, E., Caselli, M. A., & Velez, Z. (1997a). Effect of persistent toe walking on ankle equinus: Analysis of 60 idiopathic toe walkers. *Journal of the American Podiatric Medical Association, 87*(1), 17–22.

Sobel, E., Caselli, M. A., & Velez, Z. (1997b). Effect of persistent toe walking on ankle equinus. Analysis of 60 idiopathic toe walkers. *Journal of the American Podiatric Medical Association, 87*(1), 17–22.

Svensson, W., & Holmberg, U. (2005). Foot and ground measurements using portable sensors. Paper presented at the International Conference on Rehabilitation Robotics, Chicago, IL, USA.

Thomas, O., Sunehag, P., Dror, G., Yun, S., Kim, S., Robards, M., et al. (2010). Wearable sensor activity analysis using semi-Markov models with a grammar. *Pervasive and Mobile Computing, 6*(3), 342–350.

Tremblay, M. S., Esliger, D. W., & Tremblay, A. (2007). Incidental movement, lifestyle-embedded activity and sleep: New frontiers in physical activity assessment. *Applied Physiology, Nutrition and Metabolism, 32*, S2.

Trost, S. G., McIver, K. L., & Pate, R. R. (2005). Conducting accelerometer-based activity assessments in field-based research. *Medicine & Science in Sports & Exercise, 37*(11), 531–543.

Zhang, K., Werner, P., Sun, M., Pi-Sunyer, F. X., & Boozer, C. N. (2003). Measurement of human daily physical activity. *Obesity Research, 11*(1), 33–40.

# 10

## *Ambulatory and Remote Monitoring of Parkinson's Disease Motor Symptoms*

**Joseph P. Giuffrida and Edward J. Rapp**

**CONTENTS**

Introduction to Parkinson's Disease .... 248
Clinically Driven Design Input Specifications .... 253
  Patient Characteristics .... 253
  Clinician Characteristics .... 256
Technology Development .... 258
  Finger-Worn Motion Sensor Unit .... 259
  Wrist-Worn Command Module .... 261
  Wireless Data Telemetry .... 263
  Software Development .... 264
  Human Factors .... 265
  Test Engineering .... 265
System Validation to Clinical Standards .... 266
  Automated Tremor Assessment Compared to the Clinical Standard .... 267
  Automated Bradykinesia Assessment Compared to the Clinical Standard .... 272
  Quantitative and Independent Bradykinesia Feature Extraction .... 274
  Patient Acceptance .... 275
Challenges to Widespread Clinical Use .... 276
Support and Acknowledgements .... 279
References .... 279

The translation of existing medical technologies to mainstream clinical use can significantly impact healthcare access, costs and outcomes, including telehealth technologies for remote diagnosis and mobile, portable diagnostics for home-based monitoring. Parkinson's disease (PD) is a neurodegenerative movement disorder with no cure that currently affects 6.3 million people worldwide (Davis, Edin, and Allen 2010). However, unlike the cardiac market, in which Holter monitors are a clinically accepted standard for assessing arrhythmias, quantitative and automated systems for ambulatory assessment of Parkinson's disease motor symptoms are currently not in widespread clinical use. This limits the opportunity for continuous remote monitoring of motor symptoms to capture complex fluctuation patterns and optimize treatment protocols, stifles utilization of sensitive motion-sensing technology to detect subtle motor changes for new pharmaceutical interventions and limits access to treatment optimization for socio-economically and geographically disparate patient populations not in close proximity to movement disorder specialists. The development of Kinesia™, a wireless system of biokinetic sensors for quantitative and more continuous assessment of motor symptoms, addresses this need for telehealth technologies in the Parkinson's disease market. The patient-worn technology utilizes a sensor network consisting of microelectromechanical accelerometers and gyroscopes for motion sensing integrated with wireless transmission of the biokinetic data to a computer for analysis. The patient puts the system on the finger and wrist and then follows on-screen video instructions to complete upper-extremity motor tasks typically performed in the clinic. Software and algorithms automatically process the sensor data into reports to document symptom severity. Utilizing this technology, clinicians can capture clinically relevant parameters of movement disorders typically addressed during an office visit but can also do so in the patient's home environment on a more continuous basis with a standardized and quantitative assessment platform.

This chapter discusses the detailed development of this sensor network technology for assessing Parkinson's motor symptoms and how several challenges were resolved by engineering teams to produce a clinically deployable system. First, it was important to understand the clinical implications of Parkinson's disease from both the patient's and the clinician's perspective, such that an appropriate set of input specifications could be developed addressing both clinical and technological requirements and constraints. This included development of human interfaces tailored to patients with movement disorder characteristics, as well as providing clinicians with relevant symptom information instead of engineering-based "squiggly lines". Next, hardware, firmware, software and mechanical engineering development was leveraged with biomedical research to produce a quantitative standardized platform for assessing Parkinson's motor symptoms with report outputs correlated to traditional clinical standards. Finally, several additional tasks successfully commercialized the system, including regulatory clearance and reimbursement strategies; while these are non-engineering-based tasks, they are integral challenges to producing a wireless sensor technology that will be accepted for widespread clinical use.

## Introduction to Parkinson's Disease

Imagine trying to move your fingers to turn the pages of this book, but instead of being able to complete that relatively simple task, your fingers suddenly get stuck and refuse to respond to your brain's command to move.

After several seconds of hesitation, your fingers unlock and you can turn the page; however, this muscle "freezing" happens periodically during activities of daily living such as trying to tie your shoes, eat or walk. This freezing and hesitation of movement is known as *bradykinesia*, a motor symptom that is a major source of functional impairment for Parkinson's disease patients (Guyton and Hall 1996). As if it is not enough that you periodically get stuck moving, the disease now decides to cause your hands to shake uncontrollably while you are trying to write, type or simply relax and watch television. This uncontrollable shaking, or tremor, is often the most visible sign of Parkinson's disease (O'Suilleabhain and Dewey 2001). As you continue your day, you now notice some joints are stiffer than normal, and it is difficult to move them. This is referred to as *rigidity*, another motor symptom of Parkinson's disease, which occurs when antagonistic muscles controlling a joint (such as the biceps and triceps at the elbow) are both receiving high levels of involuntary activation (Rhoades and Pflanzer 1996). Patients affected by this movement disorder may encounter different levels of tremor, bradykinesia and rigidity throughout the day, as they fluctuate between motor symptoms of too much involuntary movement and limited control over desired movements.

Parkinson's disease, a progressive neurodegenerative disorder, affects the motor system and is characterized by cardinal symptoms in the upper and lower extremities of tremor, bradykinesia and rigidity of musculature. While many treatment interventions focus on the cardinal motor symptoms, Parkinson's disease can also affect gait and balance, speech, olfaction, sleep and cognition. Currently, in the United States, there are over one million people living with Parkinson's disease, and 50,000 new cases are reported each year (Davis, Edin, and Allen 2010). The disease onset most frequently occurs between 50 and 65 years of age (Davis, Edin, and Allen 2010). As the world's population continues to increase and people live longer, with no cure or existing technology to accurately diagnose Parkinson's, the incidence should continue to increase, creating a significant and growing market for movement disorder treatments and diagnostic systems.

Parkinson's disease is caused by a loss of dopamine-producing neurons in the substantia nigra region of the brain. While the exact reason for this neuron death remains unknown, significant strides have been made in the treatment of motor symptoms. Pharmaceutical interventions such as levodopa are typically prescribed to treat motor symptoms when patients are first diagnosed (Lozano 2001). However, over time, *dyskinesias*, or wild involuntary movements as a side effect of drug therapy, can develop as a motor complication. In addition to pharmaceuticals targeted to treat motor symptoms, new drug development is now focusing on neuroprotective strategies to slow or stop the progression of the disease. Another treatment option, deep brain stimulation, improves motor symptoms and reduces motor fluctuations and dyskinesia in patients whose symptoms cannot be managed with medications (Benabid et al. 2006). As of 2006, over 35,000 deep brain stimulation systems have been implanted worldwide (Benabid et al. 2006). Numerous studies show the benefits of electrically stimulating the subthalamic nucleus and the globus pallidus internus regions of the brain in Parkinson's patients (Kumar 2002; Moro et al. 2006).

The efficacy of these treatment interventions is often judged by alleviation of patient symptoms and improved quality of life. The current clinical standard in evaluating symptoms is the Unified Parkinson's Disease Rating Scale (UPDRS), a qualitative ranking system developed by a panel of movement disorder experts (Goetz et al. 2007). The UPDRS includes multiple subsections to monitor several areas of disease impact, including motor and non-motor experiences of daily living, a motor examination and motor complications. The motor examination section includes several movements the patient completes to elicit motor symptoms while a clinician qualitatively assesses the symptoms through visual examination and assigns a score from 0 to 4 (Table 10.1). The UPDRS examination is normally completed during an office visit in the presence of a trained clinician. In addition to the UPDRS, clinicians may ask patients to keep home journals to record symptom severities and when they took medication at various times during the day. Capturing this information is important because motor symptoms and side effects fluctuate during the day based on the timing and dose of medications.

While the UPDRS has been shown to be an effective tool for in-clinic motor symptom assessment, only obtaining a snapshot of patient motor symptoms in the office may not provide the temporal resolution required to see how symptoms fluctuate during the day in response to treatment protocols. For example, because patients are going to be out in public for their clinician appointments, they may have timed their medication such that it has maximum benefit during the time of the examination to minimize symptoms in public. Therefore, the UPDRS motor scores captured in the office may be low at the time but may not indicate the true pattern of symptom severity occurring in the context of the patients' everyday life. Similarly, relying on patient journals for increased temporal resolution of symptom information does not provide an objective or standardized method of treatment. Clinicians currently lack effective, affordable and quantitative technologies that can be easily self-administered by patients for monitoring symptoms on a more continuous basis. Furthermore, while expert clinicians are often very good at using the UPDRS to rate the severity of motor features, the assessment remains qualitative, which introduces inter- and intra-rater reliability issues. For example, different clinicians may have different visual interpretations of motor features and score the same patient differently. Additionally, it is possible that the same clinician could score the same motor feature differently on different days due to the subjective nature of the examination. Providing a standardized, objective assessment of motor symptoms would help to normalize evaluations across clinician assessments. Finally, the resolution of the UPDRS scale ranges from 0 to 4 in increments of whole numbers. While this may be acceptable as a gross measure of motor features, increased resolution

**TABLE 10.1**

Excerpt from Section III of the Movement Disorder Society (MDS)-Unified Parkinson's Disease Rating Scale (UPDRS) Exam

| **UPDRS Part III: Motor Examination** |
|---|
| 3.4 FINGER TAPPING |
| Instruct the patient to tap the index finger on the thumb 10 times as quickly AND as big as possible. Rate each side separately, evaluating speed, amplitude, hesitations, halts and decrementing amplitude. |
| 0: Normal: No problems. |
| 1: Slight: Any of the following: a) the regular rhythm is broken with one or two interruptions or hesitations of the tapping movement; b) slight slowing; c) the amplitude decrements near the end of the 10 taps. |
| 2: Mild: Any of the following: a) 3 to 5 interruptions during tapping; b) mild slowing; c) the amplitude decrements midway in the 10-tap sequence. |
| 3: Moderate: Any of the following: a) more than 5 interruptions during tapping or at least one longer arrest (freeze) in ongoing movement; b) moderate slowing; c) the amplitude decrements starting after the 1st tap. |
| 4: Severe: Cannot or can only barely perform the task because of slowing, interruptions or decrements. |
| 3.5 HAND MOVEMENTS |
| Instruct the patient to make a tight fist with the arm bent at the elbow so that the palm faces the examiner. Have the patient open the hand 10 times as fully AND as quickly as possible. If the patient fails to make a tight fist or to open the hand fully, remind him/her to do so. Rate each side separately, evaluating speed, amplitude, hesitations, halts and decrementing amplitude. |
| 0: Normal: No problem. |
| 1: Slight: Any of the following: a) the regular rhythm is broken with one or two interruptions or hesitations of the movement; b) slight slowing; c) the amplitude decrements near the end of the task. |
| 2: Mild: Any of the following: a) 3 to 5 interruptions during the movements; b) mild slowing; c) the amplitude decrements midway in the task. |
| 3: Moderate: Any of the following: a) more than 5 interruptions during the movement or at least one longer arrest (freeze) in ongoing movement; b) moderate slowing; c) the amplitude decrements starting after the 1st open-and-close sequence. |
| 4: Severe: Cannot or can only barely perform the task because of slowing, interruptions or decrements. |
| 3.6 PRONATION-SUPINATION MOVEMENTS OF HANDS |
| Instruct the patient to extend the arm out in front of his/her body with the palms down; then to turn the palm up and down alternately 10 times as fast and as fully as possible. Rate each side separately, evaluating speed, amplitude, hesitations, halts and decrementing amplitude. |
| 0: Normal: No problems. |
| 1: Slight: Any of the following: a) the regular rhythm is broken with one or two interruptions or hesitations of the movement; b) slight slowing; c) the amplitude decrements near the end of the sequence. |
| 2: Mild: Any of the following: a) 3 to 5 interruptions during the movements; b) mild slowing; c) the amplitude decrements midway in the sequence. |
| 3: Moderate: Any of the following: a) more than 5 interruptions during the movement or at least one longer arrest (freeze) in ongoing movement; b) moderate slowing c) the amplitude decrements starting after the 1st supination-pronation sequence. |
| 4: Severe: Cannot or can only barely perform the task because of slowing, interruptions or decrements. |

*(Continued)*

**TABLE 10.1 (CONTINUED)**

Excerpt from Section III of the Movement Disorder Society (MDS)-Unified Parkinson's Disease Rating Scale (UPDRS) Exam

| UPDRS Part III: Motor Examination |
|---|
| 3.15 POSTURAL TREMOR OF THE HANDS |
| Rate the highest amplitude seen. Instruct the patient to stretch the arms out in front of the body with palms down. The wrist should be straight and the fingers comfortably separated so that they do not touch each other. Observe this posture for 10 seconds. |
| 0: Normal: No tremor. |
| 1: Slight: Tremor is present but less than 1 cm in amplitude. |
| 2: Mild: Tremor is at least 1 but less than 3 cm in amplitude. |
| 3: Moderate: Tremor is at least 3 but less than 10 cm in amplitude. |
| 4: Severe: Tremor is at least 10 cm in amplitude. |
| 3.16 KINETIC TREMOR OF THE HANDS |
| This is tested by the finger-to-nose maneuver. With the arm starting from the outstretched position, have the patient perform at least three finger-to-nose maneuvers with each hand reaching as far as possible to touch the examiner's finger. The finger-to-nose maneuver should be performed slowly enough not to hide any tremor that could occur with very fast arm movements. Repeat with the other hand, rating each hand separately. The tremor can be present throughout the movement or as the tremor reaches either target (nose or finger). Rate the highest amplitude seen. |
| 0: Normal: No tremor. |
| 1: Slight: Tremor is present but less than 1 cm in amplitude. |
| 2: Mild: Tremor is at least 1 but less than 3 cm in amplitude. |
| 3: Moderate: Tremor is at least 3 but less than 10 cm in amplitude. |
| 4: Severe: Tremor is at least 10 cm in amplitude. |
| 3.17 REST TREMOR AMPLITUDE |
| As part of this rating, the patient should sit quietly in a chair with the hands placed on the arms of the chair (not in the lap) and the feet comfortably supported on the floor for 10 seconds with no other directives. Rest tremor is assessed separately for all four limbs and also for the lip/jaw. Rate only the maximum amplitude that is seen at any time as the final rating. |
| Extremity ratings |
| 0: Normal: No tremor. |
| 1: Slight: < 1 cm in maximal amplitude. |
| 2: Mild: > 1 cm but < 3 cm in maximal amplitude. |
| 3: Moderate: 3 – 10 cm in maximal amplitude. |
| 4: Severe: > 10 cm in maximal amplitude. |

*Source:* Goetz, CG et al., *Mov Disord.*, 23, 2129–2170, 2008. http://www.ncbi.nlm.nih.gov/pubmed/19025984

might be important for several applications such as pharmaceutical development, especially with respect to neuroprotective drugs. This class of drug development aims to slow or stop disease progression and, if effective, may lead to very subtle changes or "slow downs" in motor symptom progression compared to control subjects. These slight changes in motor symptoms may fall between the integers used for the UPDRS evaluation. Objective motion sensors quantifying feature parameters may provide increased sensitivity for detecting subtle changes with far greater resolution than the human eye. Therefore, while the UPDRS remains an important clinical tool and the clinical gold standard for assessing the complete picture of the Parkinson's disease patient, the motor aspect may be greatly augmented through the implementation of home-based, wireless motion-sensing technologies to capture motor symptom fluctuations with greater resolution in the context of the patient's own environment.

## Clinically Driven Design Input Specifications

The development of a wireless sensor network to monitor the severity of Parkinson's disease motor symptoms begins with the challenge of carefully translating the clinically based needs and requirements to a set of design specifications (Giuffrida et al. 2009). These specifications provide the context and framework for a team of engineers to design, develop, test and clinically validate the new technology platform. Often, and especially in the case of Kinesia commercialization, the development of novel medical technologies requires several different areas of engineering expertise, including electrical, mechanical, firmware and software. However, it is the *biomedical* engineer who executes the critical aspect of transferring the clinical requirements and context of the intended market to engineering-driven specifications. Carefully analysing and summarizing the intended patient characteristics, symptoms to be monitored and clinician requirements, as well as interpreting between clinical jargon and standard engineering notations to form a coherent set of input specifications, can make or break a new medical device technology. Therefore, before one dives into the selection of microelectromechanical systems (MEMS)–based sensors or the radio frequency (RF) band to utilize in a wireless data protocol, it is first important to define the intended patient and clinician populations that will be interfacing with the technology.

### Patient Characteristics

As described earlier, the Parkinson's disease patient end user will range from approximately 50 years of age through the later years of life, namely, middle-aged to elderly. This has several important implications for system development. First, while computer technology has grown tremendously over the last two decades, at the current time of this publication, much of the intended Parkinson's disease patient population may not rely on computer-based technology as much as younger generations. Therefore, the patient computer interface must be incredibly user-friendly and simple with no assumptions about computer experience. Additionally, the elderly Parkinson's disease population may have other health complications such as decreased hearing or vision, which has implications for system feedback modalities. In addition, the system should fit into the constraints of existing treatments and therapies for Parkinson's disease. For example, some patients who no longer respond to medication may be implanted with deep brain stimulation systems to treat motor symptoms or may use walkers and canes to assist with gait and balance. The technology should be designed to not interfere with these existing treatment strategies but to work with them to improve overall quality of life. This means that the wearable human interface design should not prohibit someone from using a cane or walker, and any wireless data transmission networks should not interfere with any implanted deep brain stimulation systems.

Finally, while it may seem obvious, it must be remembered that the Parkinson's patient population has movement disorder symptoms that, at times, cause too much involuntary movement (tremors or dyskinesias) and, at other times, compromise control of voluntary movements (bradykinesia or rigidity). These symptoms fluctuate back and forth throughout the day in response to activities and treatment protocols. This basic consideration of movement flows into several areas of the design inputs before engineers can even start using that motion to develop algorithms that quantify symptom severity. First, the patient motions due to Parkinson's disease and medication side effects occur in three-dimensional space in both linear translation and rotation; therefore, the sensing mechanism should be capable of detecting these types of motions (Giuffrida et al. 2009). Second, since the goal is to measure motion to capture quantitative features of the movement symptoms, the system being used to measure those symptoms should not interfere with or "dampen" those features. This requires the sensing portion of the system to be as small and lightweight as possible. Otherwise, in cases of extremely low tremor, the sensitivity of the device would be compromised since the weight of the patient-worn unit could dampen very slight tremor. Reducing the device sensitivity would detract from a major market need, which is to improve the resolution of symptom severity ratings compared to the UPDRS. Furthermore, the movement symptoms experienced by a patient can be very violent at times, especially in the case of dyskinesias. Therefore, minimizing the number of wires and untethering the patient from the computer portion of the system would maximize patient safety and further minimize interference with the measurement of motion. Finally, the fact that the intended users have movement symptoms impacts the ergonomic human interface design. It must be simple and easy for the user to put on and take off regardless of the symptoms they are currently experiencing.

In addition to understanding the implications movement symptoms have at a gross level, system development must also consider the unique features of the symptoms themselves. Kinesia development initially focused on two major symptoms: tremor and bradykinesia. While Parkinson's patients deal with many other motor and non-motor symptoms, the focus had to begin somewhere, and these are two of the most prevalent symptoms often targeted by therapies. Tremor, often targeted by research and treatments (O'Suilleabhain and Dewey 2001), is an involuntary movement characterized under three conditions: at rest, while maintaining a specific posture and during movement. Rest tremors usually occur at a frequency of approximately 4–7 Hz, while action or postural tremor is higher, usually around 9–11 Hz (O'Suilleabhain and Dewey 2001). The peak frequency of tremor is fairly stationary, and the maximum amplitude is typically under 10 g and 1200°/s. The amplitude and frequency bandwidth characteristics of the movement disorder motor symptoms provide important constraints for sensor selections, analogue and digital filtering, and data acquisition sampling rates. The standard clinical method for analysing tremor is qualitative assessment by a clinician and assignment of a score (0–4) based on the UPDRS.

While tremor is often the most common Parkinson's disease symptom, bradykinesia is often most troubling to the patient (Guyton and Hall 1996). Bradykinesia refers to delays or hesitations in initiating, and slowness in executing, movements (Dunnewold, Jacobi, and van Hilten 1997; Poluha, Teulings, and Brookshire 1998). The standard clinical method for analysing bradykinesia is a qualitative assessment by a clinician and a score assignment (0–4) based on the UPDRS, assigned while the subject completes repetitive finger-tapping, repetitive hand opening-closing and pronation/supination. Objective assessment of movements associated with Parkinson's disease can be challenging for movement disorder specialists (Poluha, Teulings, and Brookshire 1998), as they are to take into account several different features of the movement such as speed, amplitude and rhythm, as well as changes in those features over time. A key feature of bradykinesia that differentiates it from tremor is that it needs to be evaluated within the context of a known task. This is why a clinician instructs a patient through repetitive finger-tapping, hand grasps and pronation/supination of the forearm to evaluate bradykinesia (Goetz et al. 2003). During the task, clinicians are watching the speed and amplitude of the movement in addition to any changes in those features during the repetitive movement, as well as hesitations that occur. Therefore, in the context of the ambulatory environment, without knowing whether a patient is actually trying to complete a task, it is difficult to determine if he or she is simply relaxing watching television or getting stuck while trying to turn the pages of a book. Therefore, this places the constraint on the system that the patient should be receiving real-time task-based instructions, and the recorded motion data should be synchronized to those instructions.

Considering the preceding features of Parkinson's patients and motor symptoms, it is clear that some mechanism of three-dimensional motion capture is required. The motion capture mechanism must be in a small, compact package that minimizes components worn by the patient but still provides real-time feedback. The donning and doffing method must be simple and able to be performed irrespective of movement disorder symptoms. The interactive software interface should require minimal instruction and utilize clear images and/or sounds that will guide a patient through a motor evaluation.

### Clinician Characteristics

The input specifications for a new medical technology should focus not only on the patient end users but also on the clinicians who will ultimately prescribe the system for clinical evaluation and interpret the results. The purpose of the system for clinicians is to send it home with patients for them to use over the course of several days, review the results of motor symptom fluctuations in response to treatment and then make changes to that treatment protocol to optimize symptomatic benefit. Therefore, the system should provide easy and clinically relevant set-up and review stages for clinicians to utilize in clinical practice. Engineering development firms oftentimes have a tendency to speak and provide outputs in engineering jargon and "squiggly lines" which may not translate well into clinical outcome measures or fit into the existing expectations for standard of care. For example, while the frequency of the peak amplitude in the power spectrum for a bradykinesia finger-tapping task might be of great interest to clinical engineering researchers, the practicing clinician requires an outcome measure that quickly relates to patient outcomes. Several important issues related to clinician acceptance of the new sensor technology were addressed during Kinesia development.

First, an obvious but challenging aspect to gaining acceptance of a new medical technology in clinical practice is that it must provide benefits to clinicians and patients compared to available existing technologies. For the Parkinson's disease market, the traditional standard of care to evaluate motor symptoms is the UPDRS, which must be completed in the presence of a trained clinician and only affords a snapshot in time of patient symptoms (Goetz et al. 2003). A major benefit to clinicians would be to capture patient motor symptoms at home over an extended period of time to evaluate how symptoms fluctuate during the day in response to treatments. The clinician's need for up to a day's worth of patient monitoring drives the battery and memory requirements, and ultimately impacts the size and weight of the patient-worn system. Another important benefit of Kinesia compared to the current subjective standard is standardized, objective outcome measures to reduce any inter- and intra-rater reliability issues associated with the UPDRS (Giuffrida et al. 2009). This requires that manufacturing the Kinesia system includes a strict calibration routine to minimize any variability between sensor units. Clinicians are often constrained for time due to the large number of patients in clinical practice; therefore, a large benefit of Kinesia is that it allows them to obtain a more detailed assessment of the patient in less time. Finally, there are often patients with limited access to movement disorder experts due to socio-economic conditions or geography (Hubble et al. 1993). Therefore, providing a low-cost system that can be utilized in remote and ambulatory environments allows clinicians to optimize treatment protocols for these underserved populations.

While the Kinesia system demonstrates clinical benefits, it is important to tailor the outcome measures of the system within the context of existing standard-of-care measures. For Kinesia development, the existing standard of care is the UPDRS. Therefore, while the motion sensors are capable of providing detailed information about angular velocities, peak frequencies and average linear accelerations over time, these values do not have clinical significance for a physician unless they are correlated and scaled to the existing outcome measure. This drives two requirements: (1) the output measures should be reported during tasks similar to those completed during the motor section of a UPDRS examination, and (2) algorithm development should process detailed kinematic sensor network features into measures that can be normalized and correlated to the 0 (no symptoms) to 4 (severe symptoms) values clinicians are accustomed to utilizing (Goetz et al. 2003). Furthermore, while the symptoms should be on the same scale, a clinical benefit compared to the UPDRS is to take advantage of the increased resolution provided by the quantitative sensor data and provide 0 to 4 severity measures at a greater-than-integer resolution. Finally, there are several variables that are important for clinicians in assessing patient outcomes in addition to motor symptoms alone. These include documentation of when medications were taken, subjective assessments from patients concerning how they feel and subjective assessments from patients concerning their ability to complete activities of daily living. This drives a requirement that the system not only measures patient motion features in the home setting but also provides an interactive patient journal that will help drive clinically relevant measures. Integrating all these features into a concise clinical report allows clinicians to make treatment decisions based on a full spectrum of patient-centred measures.

A final design constraint includes the environments in which the system will be prescribed for use by clinicians. The environment includes the settings of hospital and home, as well as the particular geographic regions. Based on several of the mentioned patient and clinician requirements for minimizing interference with motion, maximizing patient safety and providing real-time feedback, an engineering solution is to implement a wireless sensor network to transfer motion data from the patient to a computer. Therefore, it is important to understand these settings and geographic regions as they place great constraints on the selection of the RF transmission band. A transmission frequency band must not interfere with existing systems that may be in the home or clinical environment. For example, the system must not interfere with other types of monitoring equipment in constant use in the hospital setting. Furthermore, it must not be interfered with by common technologies in the home such as cordless phones. In addition to determining acceptable frequency bands and power levels based on the application setting, the radio requirements and available communication bands differ across international markets.

## Technology Development

Based on the clinical and market-driven input specifications determined in the preceding, integration of hardware, firmware, software and mechanical development produces a clinically deployable system capable of measuring Parkinson's disease movement disorder symptoms. In general, the patient-worn hardware developed for the Kinesia system includes two components: a finger-worn sensor unit and a wrist-worn command module, connected by a thin, flexible wire (Figure 10.1). The user-worn system captures three-dimensional motion and wirelessly transmits the sensor data to a computer (Tables 10.2 and 10.3). While the sensor network provides the foundation for the system, software development is critical

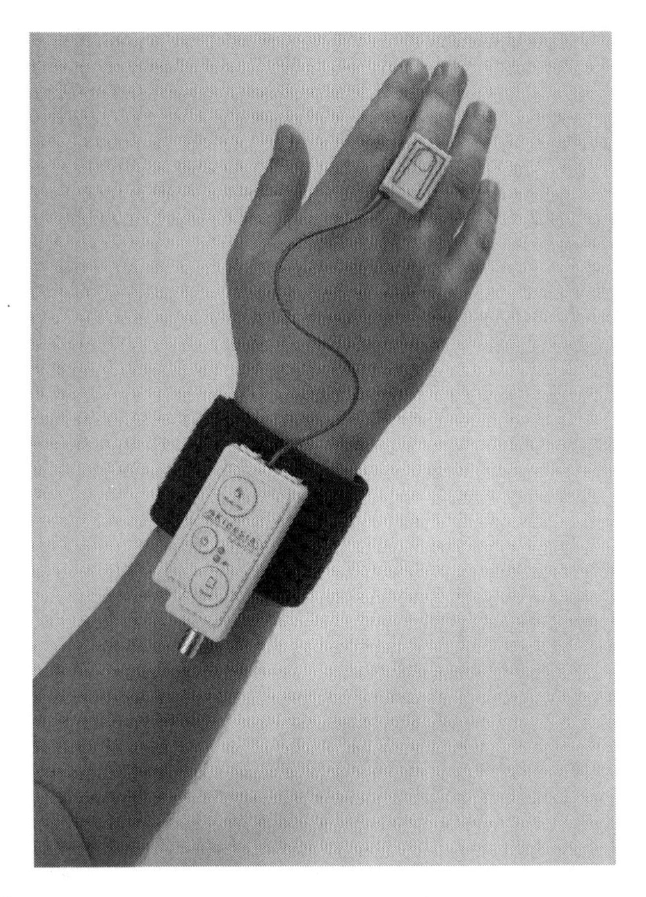

**FIGURE 10.1**
The Kinesia patient-worn unit includes a finger-worn component that contains motion sensors and a wrist-worn component that integrates data acquisition, wireless data telemetry and a rechargeable battery. (From Giuffrida, J. P., et al., *Movement Disorders*, 24, 723–730, 2009.)

**TABLE 10.2**

Kinesia System Specifications

| | |
|---|---|
| Sensor module size | 0.75 × 0.86 × 1.09 in. |
| Sensor module weight | 0.4 oz. |
| Command module size | 3.15 × 1.66 × 0.815 in. |
| Command module weight | 3.0 oz |
| Wireless link | Bluetooth™ Radio |
| Data transmission range | ~100 feet (line of sight) |
| Data rate | 57.6 kbps, 4800 bytes/s throughput |
| Transmission frequency | 2.400–2.483 GHz |
| Transmission bandwidth | 1 MHz |
| Batteries | Rechargeable lithium polymer 850 mA·h |
| Battery life | 8 h |

**TABLE 10.3**

Kinesia Sensor Unit Specifications

| | Angular Rate | Acceleration |
|---|---|---|
| Sensor type | MEMS gyroscopes | MEMS accelerometers |
| Number of channels | 3 orthogonal channels | 3 orthogonal channels |
| Input range | ±1200°/s | ±5 g |
| Minimum frequency (3 dB) | 0 Hz | 0 Hz |
| Maximum frequency (3 dB) | 15 Hz | 15 Hz |
| Input noise | <1.6°/s RMS | <2 mg RMS |
| CMRR | N/A | N/A |
| A/D resolution | 12-bit | 12-bit |
| Sampling rates | 128 sps | 128 sps |
| DC offset rejection | N/A | N/A |
| Input impedance | N/A | N/A |

*Note:* CMRR, common mode rejection ratio; MEMS, microelectromechanical systems; RMS, root mean square; SPS, samples per second.

because it provides additional interface layers for both the patient and the clinician. The software provides a visual display to guide patients through an evaluation, records and processes data synchronized to specific motor symptom evaluation tasks and generates symptom severity reports related to the traditional clinical standards. This section details the technology development and system integration required for ambulatory monitoring of Parkinson's disease.

## Finger-Worn Motion Sensor Unit

To fully describe the movement symptoms, three independent translational degrees of freedom (position) and three independent rotational degrees of freedom (orientation) need to be quantified. A human arm is considered to have seven degrees of freedom (DOF). The shoulder gives pitch, yaw and roll; the elbow allows for pitch; and the wrist allows for pitch, yaw and roll. Only three of these DOF are necessary to move the hand to any point in space, but the additional DOF allow people to grasp things from different angles. The fingers on the hand add an additional two redundant DOF. Although it is not a requirement to describe which joints are used to move an index finger in space, it is a requirement to fully describe the motion of that finger. This requirement dictates a minimum of three orthogonal accelerometers and three orthogonal angular rate gyroscopes. As previously described, the amplitude and frequency bandwidth characteristics of the movement disorder motor symptoms provide important constraints for sensor selections, filtering, data acquisition sampling rates, memory requirements and wireless communication data rates.

The first design decision is whether to build the motion-sensing unit or integrate an off-the-shelf solution. Is it more cost-effective to use an existing 6 DOF inertial measurement unit (IMU) or design a custom sensor unit including gyroscopes and accelerometers? While using an existing IMU may decrease design costs, a custom sensor board allows for a much more compact design and has lower per-unit costs. Furthermore, using an existing IMU still requires a new printed circuit board design and enclosure and so may not save much over a custom design.

Several accelerometer options were evaluated before the Analog Devices ADXL325 was selected. This MEMS accelerometer is a small, low-power, three-axis ±5 g accelerometer with signal-conditioned voltage outputs. It is able to measure the static acceleration of gravity, as well as dynamic acceleration resulting from motion, shock or vibration. Another advantage this accelerometer provides is that the bandwidth of the frequency response can be limited using external capacitors. The ADXL325 is available in a small, low-profile, 4 × 4 × 1.45 mm package, which satisfies the requirement for small size.

Likewise, multiple gyroscopes were evaluated before the Standard Microsystems LPR5150AL was selected. This part uses MEMS technology and provides dual-axis pitch and roll gyroscopes. The companion chip, LY5150ALH, provides a yaw gyroscope. Both devices measure ±1500°/s with analogue outputs and are available in small, low-profile, 5 × 5 × 1.5 mm packages. The gyroscopes detect up to 140 Hz with the bandwidth limited by external capacitors. However, the strongest advantage is that both chips are mounted in the same plane. A single plane simplifies the mechanical design needed to meet the challenge of a compact enclosure for patient wear.

During the circuit component analysis, a curious difference between the accelerometers and the gyroscopes can be noted with regard to power consumption and battery requirements. The accelerometers require considerably lower power than the gyroscopes. A MEMS gyroscope is the combination of one actuator and one accelerometer integrated into a single micromachined structure. It includes a sensing element composed of a single driving mass, kept in continuous oscillating movement and able to react when an angular rate is applied based on the Coriolis principle. Since the mass is kept in continuous motion, the gyroscope requires higher power consumption than the accelerometers. This operational power requirement of the gyroscope produces longer start-up times and does not allow the opportunity for low-power mode between samples. Therefore, the gyroscopes dominate the power requirements of the sensor unit.

The sensor unit is designed with the data acquisition sampling built in to minimize the number of wires that need to connect to the command module. Additionally, sending a digital serial data stream from the sensor unit to the command module reduces artefact noise that can typically occur on analogue data lines. The Analog Devices ADUC7022 was selected as the microcontroller primarily because it incorporates high-performance, multichannel, 12-bit data acquisition, Flash/EE (electrically erasable programmable) and RAM (random access) memories in a single 6 $mm^2$ package. The analogue motion sensor signals are high-pass filtered at 15 Hz to remove artefacts and sampled at 128 Hz to meet the requirement of recording movement disorder motor symptoms, which typically have a maximum frequency component of 15 Hz. The sampled data are sent to the wrist module over an RS485 bidirectional asynchronous communication link that reduces the interconnection cable to four wires: power, ground and two differential data. The bootloader feature of the microcontroller allows the firmware to be reprogrammed over the communication link, an important feature for field upgradability.

Once each sensor unit is completely assembled, the final production step is calibration. The calibration allows for the firmware to compensate for any zero offset and gain variation in the individual sensors. Also, calibration compensates for any minor misalignment of the MEMS sensor chips during manufacturing. A rotating fixture was developed to spin the sensor unit inside of a six-sided cube. The calibration procedure drives the sensor along all six degrees of measurement and records the sensor data, and then the calibration software computes the calibration "constants". The calibration constants are loaded into the microcontroller's nonvolatile memory. This allows the sensor firmware to mathematically correct the sensor data before transmitting them to the command module.

## Wrist-Worn Command Module

The input specifications require minimizing the size and weight of the motion-sensing units to avoid damping movement symptoms and decreasing sensitivity. Therefore, much of the required electronics and power supply are packaged into a wrist-worn command module connected to the sensor unit through a thin, flexible cable (Figure 10.1). The command module supplies power to the sensor unit, provides memory for long-term data storage and transmits motion sensor data to a computer via a wireless link.

The command module provides an accurate time (within a millisecond) and date/time stamp on each data packet. Data files saved to the computer over the real-time link or downloaded from the onboard memory are synchronized with the computer clock. The command module uses nonvolatile memory for storing patient data when the wireless link is not used and as a backup for data transmitted over the wireless link.

The microcontroller selected for the command module is an Atmel AT32UC3B0256 microcontroller. This microcontroller is a high-performance, 32-bit, RISC microprocessor yet is designed for low power consumption. The microcontroller incorporates on-chip Flash memory (256 KB) for program storage and static RAM memory (32 KB) for temporary data storage. A nonremovable micro SD card is used for nonvolatile storage of the patient data. The Peripheral Direct Memory Access controller, built into the microcontroller, enables data transfers between the SD card and the static RAM memory without processor involvement, reducing power and processing requirements.

To further minimize power consumption, the command module uses the microcontroller's low-power mode and the on-chip brown-out detector to monitor the power supply. The real-time clock and associated timer keep track of the time of day, even during system shutdown. One of the universal synchronous/asynchronous receiver/transmitters (USART) communicates with the radio module (described further later), and a second USART communicates using RS485 with the sensor module. The built-in, full-speed USB 2.0 controller interfaces to the computer to transfer patient data from the SD card.

The firmware program development, debugging and testing are completed using an in-circuit emulator and the Atmel AVR32 programming tools, including a high-level "C" compiler. The firmware controls all of the functions of the command module including packetizing the serial data stream received from the sensor unit into a wireless data transmission stream that is transmitted to the computer, controlling the power supplies and low-power mode and maintaining the system clock. A packet counter is included with each piece of data transmitted over the wireless link to ensure that no data packets are dropped during the transmission, and this also allows the data stored on the SD card to be compared with data transmitted over the wireless link. The USB bootloader also allows the firmware to be updated in the field.

One of the challenges in making the device low power is managing the multiple internal power supplies and clock sources. The different power supplies include the digital I/O power, the sensor module power and the microprocessor core power supply. In addition to having individual control over the different power sources, the microcontroller goes into a low-power (sleep) mode when not collecting data. Only the slow oscillator and real-time clock are functional during sleep mode. When collecting data, the microcontroller comes out of sleep mode, runs on the high-speed clock to collect the data from the sensor module and writes the data to the data buffer. When the data buffer is full, a packet of data is written to the SD card and transmitted over the wireless link to the computer. With careful management of power, a single small lithium-ion battery provides up to 8 h of continuous use.

A medical-grade Lemo connector allows the command module to be docked to a computer using a USB cable. When the command module is connected to the computer, the SD card appears as a removable flash drive. Power from the USB port is used to recharge the lithium-ion battery. Finally, the command module integrates a membrane switch label with LED indicators for power status and battery-charging states.

## Wireless Data Telemetry

Several of the clinically driven input requirements led to the implementation of a wireless network for transmitting motion data from the patient-worn unit to the computer. These include minimizing interference with or damping of the motion features, maximizing patient safety and comfort, providing real-time feedback and synchronizing motion data with computer-based video task instructions. Therefore, a 2.4 GHz Bluetooth™ radio embedded into the patient-worn command module wirelessly transmits data to a computer. The 2.4 GHz radio module was selected for several reasons. First, in general, the chip allows wireless data communications to satisfy the requirement that the patient can be untethered from the recording system yet still receive real-time feedback while using the system. Next, several RF bands can be selected. However, the 2.4 GHz band is selected due to the fact that this band is unlicensed in most international markets. Additionally, this band allows transmission of the data up to 50 feet in line of sight at acceptable power levels.

Bluetooth was selected because the technology is mature, it transmits at a high data rate, and there are multiple manufacturers of small, low-cost, complete radio modules. The radio modules provide a simple interface to a microcontroller, receiving data from the microcontroller USART using hardware flow control. The microprocessor in the radio module handles the "over-the-air" protocols that provide for error detection and retransmission and for radio control such as RF power and frequency control as established by the US Federal Communication Commission and other international requirements. Also, on the computer end of the RF link, there are a multitude of low-cost Bluetooth dongles available. Many desktop computers and laptops already have Bluetooth radios embedded in their systems. However, for anyone who has ever tried to pair a Bluetooth device with a Windows-based computer, the connection procedure can be somewhat complex and frustrating. Completing the wireless link with a small, lightweight CleveMed base station unit eliminates that cumbersome task. By building a layer of separation between the Bluetooth radio module and the computer, the software and the USB port driver can be supplied with the Kinesia system. This software can now control the Bluetooth discovery and pairing operations and remove the complication of radio set-up from the clinician or the patient.

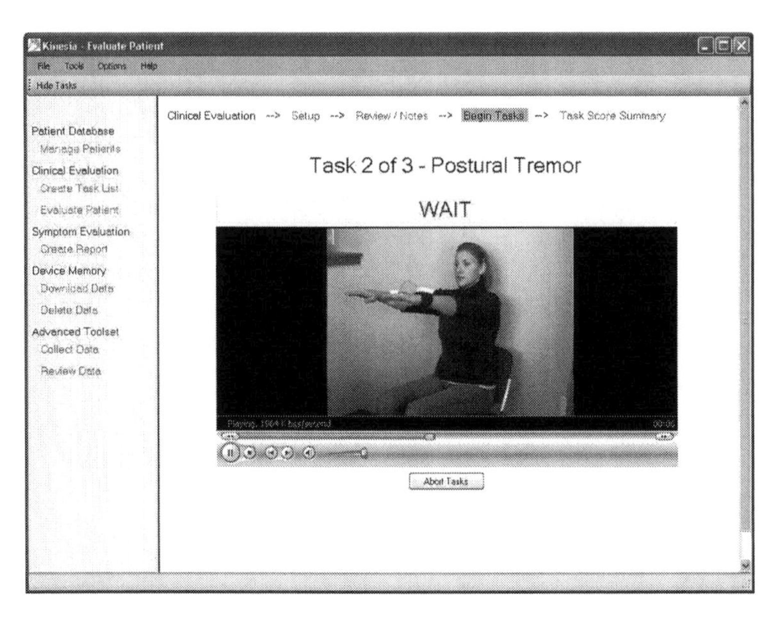


**FIGURE 10.2**
The Kinesia software interface provides an audiovisual interface that automatically guides subjects through a series of upper-extremity motor tasks normally completed during a Unified Parkinson's Disease Rating Scale (UPDRS) evaluation.

## Software Development

The clinical interface software (Figure 10.2) integrates numerous features including a patient database, real-time display and signal processing, quantitative data review features, instructional clinical videos to automatically guide patients through an evaluation and report generation. The software interface displays, records and processes data in real time over the wireless link. The patient database manages recorded data files and patient information such as stored user names, subject IDs, demographic information and any study notes. The real-time display quantitatively analyses incoming motion sensor data in both the time and frequency domains. Each channel of data can be filtered with customizable parameters. Additionally, the real-time streaming data can be saved to a file in ASCII data format for easy import into third-party compatible programs for custom analysis.

Placing the software into evaluation mode (Figure 10.2) puts the system into automated motor assessment. Once the system is placed in this mode, pressing any button on the computer will begin a series of video-guided tasks. This mode simplifies computer interaction for the patient as no computer navigation is required to begin, complete and end the task evaluations. At the start of each task, verbal and video instructions explain to the patient exactly what upper-extremity task he or she should complete, including the subset of motor tasks described in Table 10.1. The patient then receives verbal and visual cues to begin the task, and a timer counts down to provide the subject with feedback on the time remaining. The time of each task ranges from 15 to 30 s. When the timer stops, the patient receives verbal and visual cues to relax. When all seven tasks have been completed, the software returns to the main menu, awaiting the push of another button to begin the process again. Clinician reports are then automatically generated to document symptom severity features for each task, utilizing the algorithms discussed in detail later in this chapter.

### Human Factors

The greatest technological innovations can be developed for the medical field, and impressive algorithms can be created and validated to rate the severity of the most complex disease states in the ambulatory environment, but if the patient is unable to easily put the system on and take it off, the new technology will be destined to fail with regard to commercial acceptance and clinical utility. This presents an extremely challenging case for the Kinesia system design since these patients are often elderly and may be experiencing movement disorder symptoms due to Parkinson's disease while trying to put on, take off or wear the patient unit.

First, the donning and doffing of the system is simplified by integrating the components into finger- and wrist-worn components since most users would likely be comfortable wearing a ring and a watch. The system is designed to be put on using only one hand since the other hand will actually be wearing the system. An adjustable Velcro sleeve is first positioned around the wrist and pulled snug with one hand. The command module can then be slipped over the sleeve with a belt-type clip to hold it in place. Finally, the finger module can easily be slipped over the finger and tightened by pulling on a Velcro strap. The compact and wireless design further improves cosmetic appearance as well as ergonomics since the user does not have bulky wires leading from his or her body to a computer. Finally, placing the sensor unit on a patient's finger provides another important benefit besides comfortable wear. First, this particular location affords the ability to accurately assess many of the motor tasks typically used to evaluate for tremor and bradykinesia as part of the standard UPDRS examination. The tasks of resting, holding arms outstretched, touching the nose, finger-tapping, opening and closing the hand, and pronating and supinating the forearm can all be captured with the sensor in this single location.

### Test Engineering

The final Kinesia system design is verified for effectiveness and safety before clinical trials. First, test protocols are developed and executed to confirm that the Kinesia system meets or exceeds all of the input specifications. For example, test protocols verify that size and weight goals are met, battery life under different use cases is measured, and the wireless transmission range and data accuracy are confirmed. Perhaps the most challenging testing is to confirm the performance of the accelerometers and the gyroscopes in the sensor unit. Not only is accuracy over the dynamic range a requirement, but frequency response is equally important. A rotating fixture is used to spin the Kinesia system, as being untethered and wireless provide yet another boon for testing purposes. The challenge is to build a test fixture capable of better dynamic performance than the sensor unit can measure. The frequency response of the accelerometers is evaluated using the motor shaft horizontal. The motor speed in revolutions per second (rps) produces a sinusoidal gravitational vector at that frequency in hertz. This demonstrates accurate calibration and ensures proper performance of the accelerometer and the low-pass filter. Repeating the test on each surface of a "calibration cube" verifies each axis independently.

Testing the frequency response of the gyroscopes proves to be a more difficult challenge. The rotation speed of the motor axis vertical is varied in a sinusoidal pattern at the different test frequencies. Once the performance of the motor is confirmed, the amplitude of the output of the gyroscope is plotted against frequency to demonstrate accurate calibration and to ensure proper performance of the gyroscope and the low-pass filter. As with the accelerometers, repeating the test on each surface of our calibration cube verifies each axis independently.

Finally, all Kinesia system components are assembled and tested as a complete system. The system-level test protocol combines the complete system operation with the software as dictated by the input specifications and described in the user's manual. As with any medical device, there is a plethora of applicable standards and guidance documents. These documents guide the design and validation to ensure that the Kinesia system is safe for the clinician and patient to use, does not generate electromagnetic interference for other equipment and medical devices and is not susceptible to electromagnetic interference from other sources. Although using a precertified Bluetooth radio module greatly simplifies the radio link testing, test protocols verify data integrity and range of transmission. Electromagnetic interference (EMI) levels are measured to ensure compliance with the existing standards. Examples of some test standards include (1) IEC CAN\CSA 60601-1-1, class I, Type BF, Medical Electrical Equipment, Part 1: General Requirements for Safety, (2) IEC 60601-1-2:2000, Medical electrical equipment _General Requirements for Safety. Electromagnetic compatibility and (3) UL 2601 – 1 Standard for Safety for Medical Electrical Equipment Part 1, General Requirements for Safety.

## System Validation to Clinical Standards

Once engineering teams design, manufacture and verify the input specifications of a clinically deployable system, the next challenge is clinical study validation. While system *verification* typically implies that the input specifications have been achieved, *validation* of a system typically means it meets the requirements of the end user's needs and environment. Therefore, the Kinesia sensor system was utilized in a clinical study to validate that the algorithm outputs correlate to the existing UPDRS standard, as well as to validate patient acceptance. The clinical validation studies described in the following were completed in collaboration with the Movement Disorders Center at University Hospitals of Cleveland (Cleveland, Ohio), the University of Cincinnati Movement Disorders Center (Cincinnati, Ohio) and the Clinical Neuroscience Center (Southfield, Michigan). As described earlier, several motor and non-motor features of Parkinson's that affect quality of life are required to accurately describe the disease. However, it creates an unreasonable focus to begin by tackling every symptom, and this greatly extends the time frame to deliver a useful technology that solves an unmet market need. Therefore, Kinesia development first focuses on assessing tremor and bradykinesia features, as those have been identified as two of the most common and troubling symptoms to patients and are the ones most often targeted by therapies. In the future, the technology base could be expanded to address additional symptoms.

## Automated Tremor Assessment Compared to the Clinical Standard

The first objective of the clinical validation study was to demonstrate the clinical utility of automated tremor assessment (Giuffrida et al. 2009). Kinesia was used to collect data from 60 Parkinson's subjects as they completed a subset of the UPDRS upper-extremity motor examination including rest, postural and kinetic tremor tasks (Table 10.1). Since Kinesia had not yet gained Food and Drug Administration (FDA) clearance to go on the market at the time of this study, all subjects provided informed consent prior to participation. For some subjects, each arm was tested once when symptoms were occurring, and, for others, the more affected arm was tested twice, with and without symptoms. In all, 87 data collection trials were performed with a variety of patient tremor symptoms.

Each subject was set up with the Kinesia sensor unit on the finger and the command module on the wrist (Figure 10.1). Next, the subject was seated in front of an interactive software interface (Figure 10.2) that automatically instructed him or her to perform the three tremor tasks including rest, postural and kinetic tremors. Rest tremor was evaluated for 30 s, while the subject remained still with hands resting in the lap. Postural tremor was evaluated for 20 s with arms extended in front of the body. Kinetic tremor was evaluated while the subject repeatedly extended the arm and touched the nose for 15 s. Data were wirelessly transmitted from the patient-worn unit to the computer while the subject completed the tasks. Each trial was videotaped and later randomized by subject and task. The randomized sequence of videos was then scored as per the UPDRS by two movement disorder neurologists. UPDRS scoring criteria were used to rate tremor types independently (Table 10.1).

The objective of clinical validation was to record motion data from subjects as they completed movement tasks to elicit tremor, process that data into quantitative variables describing tremor severity and use those variables as inputs to an algorithm to output tremor scores highly correlated to clinician UPDRS scores. First, quantitative variables describing tremor severity were computed for each trial. The six kinematic signals included three linear accelerations along and three angular velocities about orthogonal axes. Each kinematic signal was band-pass filtered from 3 to 10 Hz using a second-order Butterworth filter to focus the area of interest to the typical tremor range and filter out voluntary motions. A power spectrum was calculated using a fast Fourier transform. Several time- and frequency-based measures were computed for each kinematic channel including peak power, frequency of the peak power, root mean square (RMS) of angular velocity and RMS of angle.

A multiple linear regression model correlated quantitative features of the kinematic signals with average clinician UPDRS scores for each tremor task. To eliminate variability in clinician UPDRS scores, the average of the clinician scores was used as the "gold standard". The following linear model regressed average clinician scores against computed quantitative variables:

$$R = b_0 + \vec{B}_a \cdot \vec{P}_a + \vec{B}_g \cdot \vec{P}_g, \tag{10.1}$$

where $R$ is the clinician's rating and $\vec{B}_a$, $\vec{B}_g$, $\vec{P}_a$ and $\vec{P}_g$ are all three-dimensional vectors. $\vec{P}_a$ and $\vec{P}_g$ are the quantitative variables for the three accelerometers and three gyroscopes, respectively, and $\vec{B}_a$, $\vec{B}_g$ and $b_0$ are the regression coefficients. Regressions were performed using different sensor combinations including accelerometers only, gyroscopes only and all sensor data, as well as the different quantitative variables described in the preceding. Sensor and quantitative variable combinations that produced the largest correlations were determined.

Once quantitative features extracted from the sensors most highly correlated to UPDRS scores were determined for each tremor task, generalization to new data was evaluated using a "one-left-out" technique. This means a single regression was computed using all but one data point. The resulting regression model and coefficients were then used to compute an output score for the left-out data point. The analysis was repeated leaving each data point out once. The coefficient of determination ($r^2$) and RMS error between regression model outputs and average clinician scores were computed for all generalization data. This analysis was computed separately for rest, postural and kinetic tremors.

The Kinesia system successfully captured movement features from the sensor network that are related to movement disorder tremor severity. Figure 10.3 depicts data recorded from typical subjects with slight and moderate postural tremor. Data for subjects with slight tremor and an average UPDRS score of 1.25 (Figure 10.3a through c) show low-amplitude acceleration and angular velocity with low-amplitude power spectra. In contrast, a subject with moderate tremor and an average UPDRS score of 3.0 (Figure 10.3d through f) produces large-amplitude oscillations on all accelerometer and gyroscope channels. These large-amplitude, time-domain oscillations result in sharp power spectra peaks centred on the tremor frequency.

Several quantitative sensor variables correlated well with average clinician UPDRS scores. The coefficient of determination ($r^2$) values from each regression are summarized in Table 10.4. The logarithm of peaks in the power spectra of all six kinematic channels produced the largest correlation with clinician scores for rest and postural tremors, while the RMS amplitude of all sensors

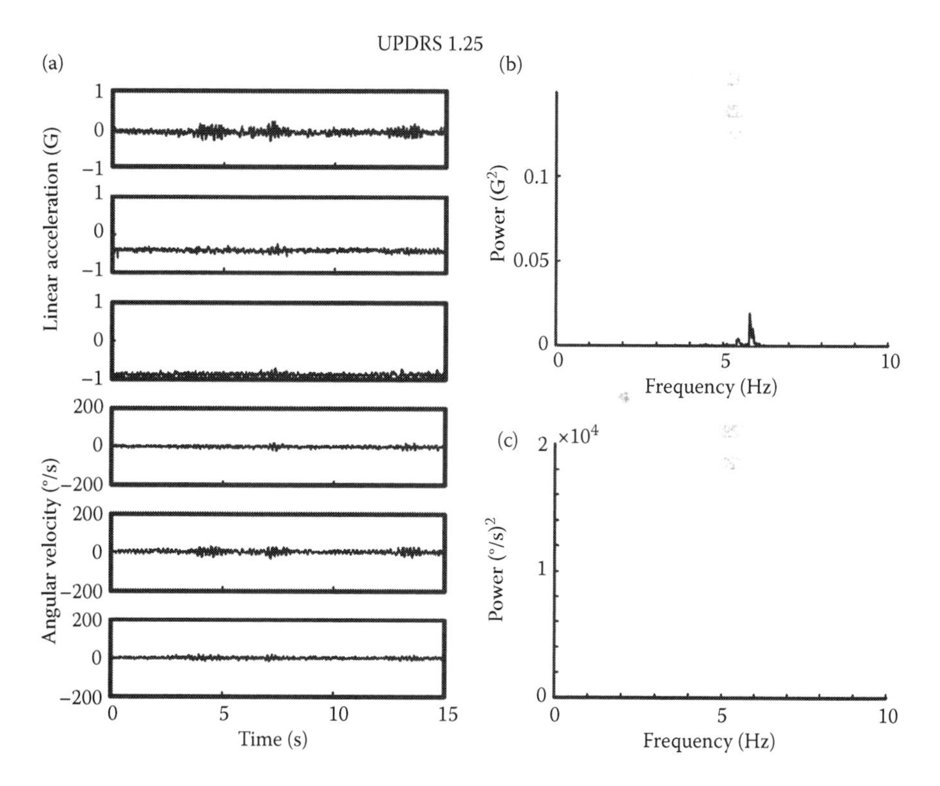


**FIGURE 10.3**
(a) Time-domain signals recorded from the three accelerometers and three gyroscopes for a subject with slight postural tremor (average Unified Parkinson's Disease Rating Scale [UPDRS] score 1.25); (b) power spectrum of the accelerometer channel with the largest amplitude; (c) power spectrum of the gyroscopic channel with the largest amplitude; (d) time-domain signals recorded from the three accelerometers and three gyroscopes for a patient with moderate postural tremor (average UPDRS score 3.0); (e) power spectrum of the accelerometer channel with the largest amplitude; (f) power spectrum of the gyroscopic channel with the largest amplitude. (From Giuffrida, J. P., et al., *Movement Disorders*, 24, 723–730, 2009).

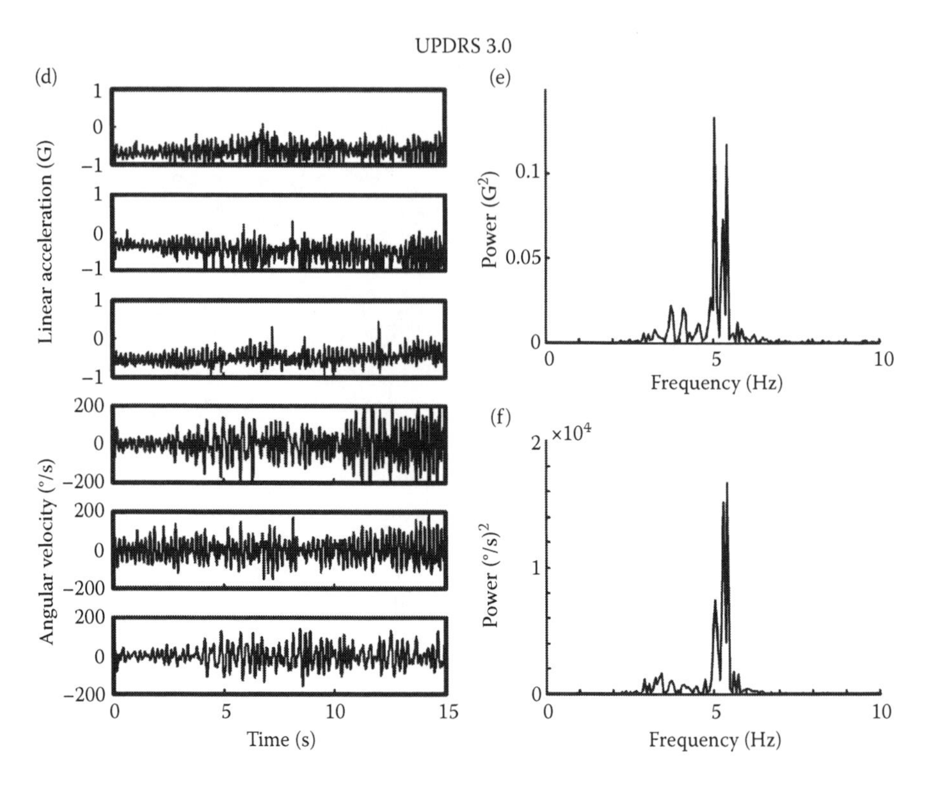


**FIGURE 10.3 (CONTINUED)**
(a) Time-domain signals recorded from the three accelerometers and three gyroscopes for a subject with slight postural tremor (average Unified Parkinson's Disease Rating Scale [UPDRS] score 1.25); (b) power spectrum of the accelerometer channel with the largest amplitude; (c) power spectrum of the gyroscopic channel with the largest amplitude; (d) time-domain signals recorded from the three accelerometers and three gyroscopes for a patient with moderate postural tremor (average UPDRS score 3.0); (e) power spectrum of the accelerometer channel with the largest amplitude; (f) power spectrum of the gyroscopic channel with the largest amplitude. (From Giuffrida, J. P., et al., *Movement Disorders*, 24, 723–730, 2009).

produced the largest correlation with kinetic tremor. The natural logarithm of peak power yielded the largest correlations for both accelerometers and gyroscopes in the rest and postural tremor tasks. The RMS value of the signal produced the largest correlations for both accelerometers and gyroscopes in the kinetic tremor task. The largest correlations were achieved when both accelerometer and gyroscope data were combined in the analysis.

The one-left-out analysis demonstrated that the developed regression model generalizes to new data. Good correlations and low errors between model outputs and clinician scores are achieved for the one-left-out analysis. Figure 10.4 compares model scores for new data points to average clinician UPDRS scores for each tremor task. The RMS error between model and UPDRS scores is low for each task, while the $r^2$ value is high.

**TABLE 10.4**

Kinematic Variable Correlations to Clinical Scores

| Quantitative Variables | Rest | Postural | Kinetic |
|---|---|---|---|
| Log(accelerometer power) | 0.86 | 0.89 | 0.41 |
| Log(gyroscope power) | 0.87 | 0.88 | 0.52 |
| Log(all sensors power) | 0.89 | 0.90 | 0.54 |
| Log(accelerometer RMS) | 0.81 | 0.86 | 0.38 |
| Log(gyroscope RMS) | 0.88 | 0.87 | 0.49 |
| Log(all sensors RMS) | 0.89 | 0.89 | 0.55 |
| Accelerometer RMS | 0.60 | 0.62 | 0.60 |
| Gyroscope RMS | 0.76 | 0.64 | 0.67 |
| All sensors RMS | 0.80 | 0.65 | 0.69 |

*Note:* RMS, root mean square. Quantitative variables were correlated with average clinician Unified Parkinson's Disease Rating Scale (UPDRS) scores for three tremor tasks. The $r^2$ values for each regression are shown for each variable and task, with the highest correlation highlighted in grey.

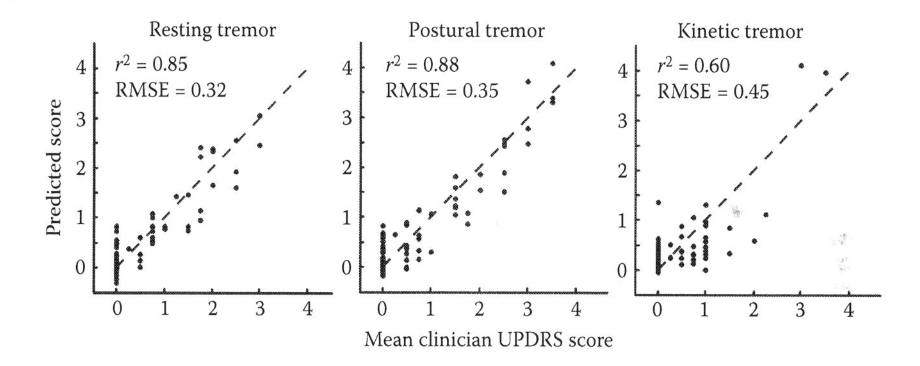


**FIGURE 10.4**

Predicted scores from the "one-left-out" model versus mean clinician Unified Parkinson's Disease Rating Scale (UPDRS) score. Dotted lines correspond to a perfect fit. The coefficient of determination ($r^2$) and root-mean-square errors (RMSE) between predicted and actual scores are displayed. (From Giuffrida, J. P., et al., *Movement Disorders*, 24, 723–730, 2009).

Correlations obtained using Kinesia are extremely high for rest and postural tremors and good for kinetic tremor. Results are consistent with, and in some cases better than, previous studies. One potential reason for superior results is that due to the wireless sensor design and real-time feedback, the Kinesia system knew the specific task a subject was performing during the evaluation, in contrast with previous studies during which the tasks a subject was completing during the tremor evaluation were not controlled. The Kinesia tremor evaluation could be completed without the presence of a clinician, as the interactive interface guides subjects to perform the same tremor tasks normally performed in the clinic and assesses data during those tasks. This allows different algorithms to be applied to different tremor tasks to optimize results across the different types of tremor that may be present in Parkinson's disease.

## Automated Bradykinesia Assessment Compared to the Clinical Standard

While tremor is the most recognized Parkinson's motor symptom, bradykinesia is often most troubling to patients since performing simple movements can become extremely difficult. Therefore, in addition to automatically rating the severity of tremor, the sensor network was utilized in the same clinical study at the Movement Disorders Center of University Hospitals of Cleveland to examine bradykinesia. In the literature, the term *bradykinesia* refers to delays or hesitations in initiating movements and slowness in executing movements; it is often used synonymously with the terms *akinesia* and *hypokinesia*. Technically, *bradykinesia* refers only to slowness of movement; however, lack of spontaneous movement (akinesia) and smaller-than-desired movements (hypokinesia) are often grouped together as bradykinesia, as was done in this clinical validation study (Espay et al. 2009).

Bradykinesia was evaluated by three tasks as per the UPDRS, including 15 s of finger-taps, 15 s of hand movements and 15 s of rapid alternating movements (pronation/supination of the forearm). Each trial was videotaped, and clips were edited and randomized by subject and bradykinesia task. The randomized videos were then presented to two clinicians for scoring (0–4) as per the UPDRS (Table 10.1). The average value of the scores was computed for each trial to minimize scoring variability. According to the subset of the UPDRS tasks that can elicit bradykinesia in the upper extremity, ratings should reflect speed, amplitude, hesitations, fatiguing, arrests in movement and changes in these variables over time. Figure 10.5 shows two examples of data recorded from gyroscope sensors during the hand grasps task, for subjects with mild (Figure 10.5a) and severe bradykinesia (Figure 10.5b). The average hand grasps UPDRS score for mild bradykinesia was 0.5. Signals appear sinusoidal, with very regular frequency and only a slight decrease in amplitude near the end. However, as evident from the plot, severe bradykinesia (mean UPDRS score was 3.5) exhibits decreased amplitude, varied frequency, numerous hesitations and severe fatiguing.

Quantitative variables describing bradykinesia severity were computed for each trial. Each kinematic sensor signal was band-pass filtered from 0.3 to 3.5 Hz using a second-order Butterworth filter to minimize any tremor artefacts that existed. Multiple quantitative variables were computed for the kinematic channels, including RMS acceleration, RMS angular velocity, RMS displacement, RMS rotation, frequency of movement, peaks in the power spectra of the acceleration and angular velocity signals, and the standard deviation of a 1 s

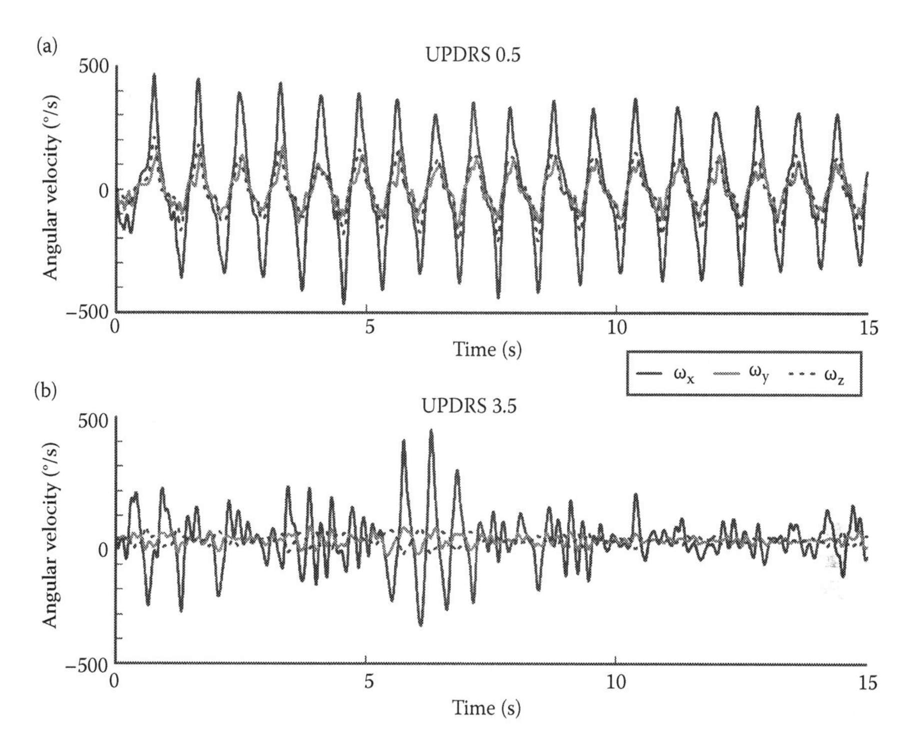


**FIGURE 10.5**
Gyroscope signals recorded during the hand grasps task are shown for mild bradykinesia (a) and severe bradykinesia (b).

sliding window of the RMS amplitude of angular velocity around the axis of motion divided by the mean, known as the *coefficient of variation*.

Bradykinesia proved more difficult to quantify with respect to direct correlation to UPDRS scores. Various multiple linear regression models attempted to correlate quantitative features of the kinematic signals with average clinician UPDRS scores for each bradykinesia task. However, quantitative features were much less correlated with clinician UPDRS scores during bradykinesia tasks than tremor tasks. RMS amplitude produced the best correlations to clinical bradykinesia UPDRS scores, with $r^2$ values of 0.28, 0.34 and 0.56 for finger-tapping, hand grasps and pronation/supination, respectively. The low correlations may be due to the fact that differences in bradykinesia are clearly visible in extreme cases, while rating moderate characteristics is more difficult. Additionally, the ambiguity of rating various bradykinesia manifestations via the UPDRS bradykinesia tasks may have led to inconsistent ratings among clinicians. For finger-tapping, hand grasps and pronation/supination, the RMS error between the two clinician raters was, on average, over a full point on the 0 to 4 scale of the UPDRS. This variability may indicate that too many variables are currently involved in the existing rating scale for bradykinesia, and each clinician may weigh the importance of each variable separately when interpreting a score. This further indicates the need for quantitative sensor technologies in medical devices that can detect subtle differences and extract independent feature characteristics with greater resolution than the naked eye. However, this also represents a challenge for clinical acceptance of new technology because, while rating features of bradykinesia independently may provide more sensitive measures of movement disorder symptoms, it is not what clinicians are accustomed to doing and represents a new methodology.

## Quantitative and Independent Bradykinesia Feature Extraction

Recently, a Modified Bradykinesia Rating Scale (MBRS) was developed for rating the bradykinesia manifestations of speed, amplitude and rhythm separately (Espay et al. 2009; Kishore et al. 2007). This scale may provide a better method for evaluating the individual components of bradykinesia. By rating speed, amplitude and rhythm separately, many of the ambiguities present in the UPDRS could be eliminated. Furthermore, since it is unclear if the underlying neural mechanisms causing the different bradykinesia manifestations are the same, rather than targeting bradykinesia as a whole, it is possible that novel pharmaceutical or deep brain stimulation therapies could be developed for targeting specific bradykinesia manifestations. Regardless of the rating scale employed, quantitative, objective assessments are necessary for capturing the complex motor fluctuations of Parkinson's disease in response to therapy.

A small pilot study was completed with the Movement Disorders Center at the University of Cincinnati that utilized Kinesia to extract independent features of bradykinesia, rather than assess it as a whole, and correlate those measures to the new MBRS. Seven subjects were asked to complete finger-taps, hand grasps and rapid alternating movements with the sensor unit placed on the index finger. Each trial was videotaped, and clips were edited and randomized by subject and bradykinesia task. The randomized videos were then presented to four clinicians for scoring. Subjects were rated separately for speed, amplitude and rhythm per the modified bradykinesia rating scale (MBRS); 0 corresponds to normal and 4 corresponds to severe (Espay et al. 2009). The average value of the scores was computed for each trial to minimize scoring variability.

Quantitative variables were processed from the motion signals and compared to the MBRS scores. The variables included RMS position, RMS velocity, peak power, peak frequency and the standard deviation of the RMS computed in 1 s sliding windows divided by the mean, known as the coefficient of variation. These variables exhibited a linear relationship with MBRS scores in the finger-tapping task when using only the signal recorded from the gyroscope measuring angular velocity around the axis of movement (Figure 10.6). Similar results were obtained using other signals for the hand grasps and alternating movement tasks. While the pilot data were collected with only seven subjects, correlations between quantitative assessment

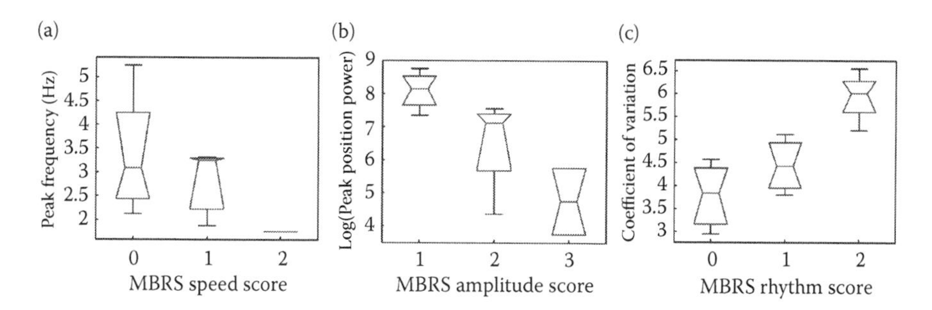


**FIGURE 10.6**
Several quantitative features exhibited a linear relationship with Modified Bradykinesia Rating Scale (MBRS) ratings. The analysis used the x-axis gyroscope measuring angular velocity on the index finger during the finger-tapping task. (a) Peak frequency decreases with increasing MBRS speed score; (b) the logarithm of the peak power decreases with increasing MBRS amplitude score; (c) the coefficient of variation increases with increasing MBRS rhythm score.

and independent scoring of different bradykinesia features were already improved compared to the previous study analysis, as exhibited by the linear relationships (Figure 10.6). This indicates that quantitative motion-sensing technology may be incredibly useful for distinguishing key features of clinical disorders compared to traditional subjective rating scales. It is likely that identifying specific features that objectively quantify speed, amplitude and rhythm will aid in the development of deep brain stimulation (DBS) and other therapies that can better target specific manifestations of bradykinesia (Hauser 2006). However, challenges remain to gaining clinical acceptance of breaking out the traditional scoring into separate measures, and this can be achieved only through greater research and peer-reviewed publications that demonstrate the benefits of treating each feature independently.

## Patient Acceptance

After clinical data collection, each subject was asked to complete a short questionnaire that evaluated comfort and ease of use. The questionnaire focused on cosmetic acceptability, obtrusiveness and ability to follow system instructions. Table 10.5 shows the results from 40 Parkinson's disease subjects who completed the questionnaire. All subjects found the device comfortable, lightweight and unobtrusive. Nearly all participants indicated that they would be capable of using Kinesia at home, and most would wear it in public. Finally, all subjects could successfully use the interactive software interface to follow on-screen instructions and complete the upper-extremity tasks. While further unsupervised home-based studies are required to verify these results, the initial clinical validation study strongly indicates that the intended market will be able to effectively utilize the system in their own homes for ambulatory monitoring of Parkinson's disease motor symptoms (Giuffrida et al. 2009).

**TABLE 10.5**

Questionnaire Results Indicating Clinical Patient Acceptance

| Questions | Yes (%) |
|---|---|
| Was the device comfortable to wear? | 100 |
| Did the device feel heavy? | 0 |
| Did the device restrict your arm movement? | 0 |
| Did the device restrict your hand movement? | 0 |
| Would you wear the device at home? | 94 |
| Would you perform arm movements at home similar to those during an office exam? | 88 |
| Would you have trouble putting the device on at home? | 3 |
| Would you wear the device in public? | 55 |

## Challenges to Widespread Clinical Use

The Kinesia technology provides a standardized, portable platform for assessing quantitative movement disorder features on a more continuous basis. While previous studies have shown good correlations to clinical scoring of movement disorder motor symptoms using prototype laboratory systems (Lukhanina, Kapoustina, and Karaban 2000; Norman, Edwards, and Beuter 1999; Riviere, Reich, and Thakor 1997), this work advances the concept into a clinically deployable system through two particular technological innovations. First, wireless data transmission is a key technology feature to meet several challenges indicated by the patient's and clinician's requirements. It maximizes cosmetic acceptability and subject safety, as well as allowing unimpeded motion capture. Furthermore, the wireless link provides real-time synchronization between video-instructed tasks and data analysis. Since subjects follow the video interface guiding them through different arm posture tasks to evaluate symptoms while real-time data are transmitted to the computer, the system knows which task a subject is performing and can apply appropriate analytical algorithms for a particular task. The other technological innovation that allows most of the challenges to be met is the availability of MEMS technologies for sensing three-dimensional motion. The miniature sensor network that can be easily integrated into the engineering design allows the system to be developed in a small, lightweight form factor with minimal current consumption requirements for long-term use. These innovations allow a technological advance by moving sensing systems from laboratory-based equipment into a clinical package with high patient acceptance that replicates several features normally evaluated in a routine, real-world clinical examination.

Developing a wireless sensor system for remote, portable diagnosis and home-based monitoring of Parkinson's disease motor systems requires a tremendous amount of engineering effort, detailed in the preceding. Furthermore, several non-engineering-based efforts are also required for commercial success. These include obtaining regulatory approval for medical devices, securing reimbursement and often changing the culture of traditional clinical care. Unless these challenges are met, even the most advanced and innovative medical technologies will remain in the research and development laboratory and never reach the intended patient population to improve quality of life.

In order to legally market and sell the Kinesia system in the United States as a medical device, it was important to obtain FDA clearance to market the system through a regulatory approval process. There are typically two regulatory pathways available to gain clearance from the FDA to market a new medical device: the pre-market approval (PMA) or the 510 (k). The 510 (k) pathway to regulatory approval requires that a company demonstrates the safety and efficacy of the device and shows substantial equivalence in both technology and intended use to previously cleared-to-market devices. While there were no systems on the market exactly similar to Kinesia, several previous FDA cleared-to-market systems were available that encompassed the various technological and intended-use components of Kinesia. Therefore, Kinesia was determined to be a Class II non-exempt device that could be cleared to market by showing substantial equivalence to these predicates. Data from bench-top verification and field-testing, as well as the clinical study data, were also critical to collaboratively proving safety and efficacy. Utilizing this documentation as part of CleveMed's quality control system, Kinesia was FDA cleared to market in 2007 through an FDA 510(k) application.

Obtaining FDA clearance allows a medical device manufacturer to market and sell the system in the United States; however, expanding the system to international markets is also important. This process of meeting medical device regulations and obtaining certifications in all international markets can be a daunting task, as each region or country often has its own policies. For example, at the time of this publication, much of the market in Southeast Asia accepts FDA clearance as satisfactory for the import and sale of medical devices, while Europe requires the CE mark and compliance to the medical device directive. Still other countries require independent registration or ISO certification to market any medical devices in their country. The guidelines are different for different areas, so it is important to understand these regulations early in the development process so as not to limit the market potential of the system because a particular technological characteristic or paperwork trail was ignored early in development.

In addition to obtaining regulatory approval, true commercialization of a new medical technology in the United States is achieved only once reimbursement is available for the new technology. Clinicians are most likely to utilize a new technology if it demonstrates benefits to the patient and they can get paid for their efforts in utilizing the technology. In the United States, Medicare and third-party-payer healthcare insurance companies most often provide that reimbursement. Therefore, the manufacturers of new medical devices have two options, either integrating their technology into existing reimbursement categories and codes or, often, in the case of new technologies, creating new reimbursement codes and procedures since there often is no existing code that applies to the new technology. This is often a long and challenging process which begins with petitioning the American Medical Association to obtain a temporary category III CPT code for the new device or procedure. A category III code traditionally does not provide reimbursement and is often denied when submitted as a claim by providers, but it is mainly intended to track usages and billing by clinicians so that a permanent category I code with a payment schedule can be established. However, this creates a catch-22 and huge challenge for medical device companies since a large number of documented clinical uses are required to transition from a category III to a category I, but, as already explained, clinicians are not likely to even attempt to use a new technology unless they will be reimbursed for it in the first place. This creates a long process for manufacturers of new medical technologies to work with clinicians to utilize the technology and demonstrate efficacy, need and reduced healthcare costs.

Reducing healthcare costs while improving patient care is where the ambulatory sensor networks described throughout this text can potentially have a large impact. Monitoring patients in the home environment should improve care since disease states can be monitored in the context of a person's own environment. It should also reduce healthcare costs through fewer visits to the doctor's office and advanced algorithms for detecting and diagnosing particular conditions. These are important points to insurance companies; however, these types of technologies can be seen as threatening to practicing clinicians in terms of both taking the human element out of patient care and reducing the doctors' revenue streams and livelihood. Herein lies one of the last big challenges for the practical implementation of these types of ambulatory sensor networks: changing the traditional culture of patient care. While these types of technologies can greatly enhance the clinical diagnosis and treatment of medical conditions, clinicians play a much larger role in overall patient care and should never be completely taken out of the loop. The intent of these types of sensor networks is to provide clinicians with a more detailed view of symptoms and disease states than can be obtained in a single office visit. By utilizing this information in addition to traditional assessments, clinicians should be able to more effectively optimize treatment protocols to improve patient care. By optimizing the treatment protocols early on, side effects and complications leading to long-term disability and in-hospital care costs could be minimized to reduce overall healthcare costs. For example, in the case of the Parkinson's disease market, the UPDRS should still be considered an important tool that allows a clinician to evaluate the patient as a whole, including motor and non-motor symptoms and activities of daily living. While the UPDRS can be utilized in the clinic, the Kinesia system can be used at home to provide more detailed information on symptom fluctuations to optimize treatments and reduce motor fluctuations. Clinicians will still be responsible for reviewing and interpreting the data collected by the ambulatory sensor networks. By reducing motor fluctuations through optimized medication titration, events such as trips and falls may be avoided. More quickly optimizing medication protocols and minimizing adverse events caused by motor symptoms may reduce overall healthcare costs.

The Parkinson's disease patient population presents a unique market in which ambulatory monitoring of motor symptoms with sensor networks can provide important insight into disease fluctuations to improve patient care. Ultimately, the goal is to provide an ambulatory system for automated motor symptom assessment for use in clinics or in the homes of patients with movement disorder motor symptoms such as those attributed to Parkinson's disease. In-clinic use would provide a standardized, objective platform to score the severity of patient tremor symptoms with increased amplitude resolution. Sending the system home with a patient would allow a clinician to capture complex fluctuating patterns of Parkinson's symptoms in response to medication timing and dose, which would increase the temporal resolution of patient evaluation enormously compared to sporadic office visits (Couzin 2008; Samii et al. 2006). In addition to clinician use, the system may also show utility for pharmaceutical studies, providing a standardized platform to evaluate drug efficacy and potentially bring drugs to market faster through improved sensitivity of symptom monitoring.

## Support and Acknowledgements

This work was supported by several Small Business Innovative Research Grants from the National Institute of Health, National Institute of Neurological Disorders and Stroke (2R44NS043816-04) and National Institute on Aging (5R44AG033520-03), and a Commercialization Grant from the State of Ohio. Several key collaborators also made this work possible including David Riley, MD, Brian Maddux, MD, PhD, Alberto Espay, MD, Peter LeWitt, MD, and Dustin Heldman, PhD.

## References

Benabid, A. L., Deuschl, G., Lang, A. E., Lyons, K. E., & Rezai, A. R. (2006). Deep brain stimulation for Parkinson's disease. *Movement Disorders,* 21 Suppl 14, S168–S170.

Couzin, J. (2008). Parkinson's disease: Streamlined clinical trials, from a home computer. *Science*, 320, 1143.

Davis, K. L., Edin, H. M., & Allen, J. K. (2010). Prevalence and cost of medication nonadherence in Parkinson's disease: Evidence from administrative claims data. *Movement Disorders*, 25, 474–480.

Dunnewold, R. J., Jacobi, C. E., & van Hilten, J. J. (1997). Quantitative assessment of bradykinesia in patients with Parkinson's disease. *Journal of Neuroscience Methods*, 74, 107–112.

Espay, A., Giuffrida, J., Chen, R., Vaughan, J., Duker, A., & Heldman, D. (2009). Differential response of bradykinesia and hypokinesia to levodopa in Parkinson's disease. *Parkinsonism & Related Disorders*, 15, S126–S134.

Giuffrida, J. P., Riley, D. E., Maddux, B. N., & Heldman, D. A. (2009). Clinically deployable Kinesia technology for automated tremor assessment. *Movement Disorders*, 24, 723–730.

Goetz, C. G., Fahn, S., Martinez-Martin, P., Poewe, W., Sampaio, C., Stebbins, G. T., ... LaPelle, N. (2007). Movement Disorder Society-Sponsored Revision of the Unified Parkinson's Disease Rating Scale (MDS-UPDRS): Process, format, and clinimetric testing plan. *Movement Disorders*, 22, 41–47.

Goetz CG, Tilley BC, Shaftman SR, Stebbins GT, Fahn S, Martinez-Martin P, Poewe W, ... LaPelle N. (2008). Movement Disorder Society-sponsored revision of the Unified Parkinson's Disease Rating Scale (MDS-UPDRS): scale presentation and clinimetric testing results. *Mov Disord*. 23:2129–2170.

Goetz, C. G., Poewe, W., Rascol, O., Sampaio, C., Stebbins, G. T., Fahn, S., ... Van Hilten, B. J. (2003). The Unified Parkinson's Disease Rating Scale (UPDRS): Status and recommendations. *Movement Disorders*, 18, 738–750.

Guyton, A. C., & Hall, J. E. (1996). *Textbook of Medical Physiology*, 9th ed. Philadelphia: W. B. Saunders.

Hauser, R. A. (2006). Long-term care of Parkinson's disease: Strategies for managing "wearing off" symptom re-emergence and dyskinesias. *Geriatrics*, 61, 14–20.

Hubble, J. P., Pahwa, R., Michalek, D. K., Thomas, C., & Koller, W. C. (1993). Interactive video conferencing: A means of providing interim care to Parkinson's disease patients. *Movement Disorders*, 8, 380–382.

Kishore, A., Espay, A. J., Marras, C., Al-Khairalla, T., Arenovich, T., Asante, A., ... Lang, A. E. (2007). Unilateral versus bilateral tasks in early asymmetric Parkinson's disease: Differential effects on bradykinesia. *Movement Disorders*, 22, 328–333.

Kumar, R. (2002). Methods for programming and patient management with deep brain stimulation of the globus pallidus for the treatment of advanced Parkinson's disease and dystonia. *Movement Disorders*, 17 (Suppl 3), S198–S207.

Lozano, A. M. (2001). Deep brain stimulation for Parkinson's disease. *Parkinsonism & Related Disorders*, 7, 199–203.

Lukhanina, E. P., Kapoustina, M. T., & Karaban, I. N. (2000). A quantitative surface electromyogram analysis for diagnosis and therapy control in Parkinson's disease. *Parkinsonism & Related Disorders*, 6, 77–86.

Moro, E., Poon, Y. Y., Lozano, A. M., Saint-Cyr, J. A., & Lang, A. E. (2006). Subthalamic nucleus stimulation: Improvements in outcome with reprogramming. *Archives of Neurology*, 63, 1266–1272.

Norman, K. E., Edwards, R., & Beuter, A. (1999). The measurement of tremor using a velocity transducer: Comparison to simultaneous recordings using transducers of displacement, acceleration and muscle activity. *Journal of Neuroscience Methods*, 92, 41–54.

O'Suilleabhain, P. E., & Dewey, R. B., Jr. (2001). Validation for tremor quantification of an electromagnetic tracking device. *Movement Disorders*, 16, 265–271.

Poluha, P. C., Teulings, H. L., & Brookshire, R. H. (1998). Handwriting and speech changes across the levodopa cycle in Parkinson's disease. *Acta Psychologica (Amst)*, 100, 71–84.

Rhoades, R., & Pflanzer, R. G. (1996). *Human Physiology*, 3rd ed. Fort Worth: Saunders College Publishers.

Riviere, C. N., Reich, S. G., & Thakor, N. V. (1997). Adaptive Fourier modeling for quantification of tremor. *Journal of Neuroscience Methods*, 74, 77–87.

Samii, A., Ryan-Dykes, P., Tsukuda, R. A., Zink, C., Franks, R., & Nichol, W. P. (2006). Telemedicine for delivery of health care in Parkinson's disease. *Journal of Telemedicine & Telecare*, 12, 16–18.

# 11

# *Nocturnal Sensing and Intervention for Assisted Living of People with Dementia*

**Paul J. McCullagh, William M.A. Carswell, Maurice D. Mulvenna, Juan C. Augusto, Huiru Zheng, and W. Paul Jeffers**

**CONTENTS**

Introduction .......... 283
Overview of Sensing in Healthcare .......... 284
Peculiarities of Nocturnal Sensing and Interventions .......... 288
NOCTURNAL Sensing/Intervention Platform .......... 291
Addressing User Acceptance .......... 295
Conclusions .......... 297
Acknowledgements .......... 297
References .......... 298

## Introduction

Research on ambient assistive living (AAL) technologies is capturing significant attention from the scientific community and attracting investment from relevant areas of public services like the healthcare sector and government. Much of that research has focused on the idea of the smart home as an extension of social and healthcare services that can supplement the attention given by healthcare professionals, family and friends. Smart home systems are based on the deployment of technology within a house to increase safety and health by deploying sensors and actuators which allow the system to detect when help is required and provide some assistance.

Sensors allow us to measure some aspect of reality. There are a variety of sensors to be considered like those that detect temperature or levels of an environmental substance (e.g., carbon monoxide), movement, pressure, biometric information, etc. A smart home system can detect that help is required because there is an explicit request on behalf of an occupant or because the information gathered from the sensors suggests that. The system may then decide to act on this using *actuators*, i.e., devices that act upon the environment. Examples of actuators can be speakers, monitors, a printed diagnosis, an SMS or a voice call to a carer.

Most of the contributions reported in the assistive technology literature focus on the most active period of the day (daylight time) because that is their interest or because they assume that the night-time will be well covered by the same technology and associated procedures. Our project, NOCTURNAL (Night Optimised Care Technology for UseRs Needing Assisted Lifestyles), assumes that the night-time and daylight periods of the day are different enough to require separate analysis (McCullagh et al. 2009).

This chapter explains the technological infrastructure of NOCTURNAL within the context of sensing and actuation used in AAL, highlights the peculiarities of assisting people in the hours of darkness, reflects on the consequences to society of the AAL-associated technology and explains how those elements have been taken into account to define the infrastructure that is being developed in our project.

## Overview of Sensing in Healthcare

Ambient intelligent (AmI) systems for healthcare are being developed as part of the fundamental shift from hospital-centred to home-centred models of care within the health service. This development is timely, given the need for a new healthcare system that can cope with an increasingly ageing society (World Health Organisation 2005).

Augusto and McCullagh (2007, p. 4) defined an AmI system as *"a digital environment that proactively, but sensibly, supports people in their daily lives"*. AmI systems are supported by "smart environments", physical environments enriched with sensing and actuating devices. This combination provides a platform to develop the autonomous semi-intelligent behaviour exhibited by systems like smart homes (Cook et al. 2009).

A significant catalyst which has prompted AmI-supported healthcare has been the rapid development and expansion of the sensor market combined with the affordability of technology and gadgets. A full range of small, reliable and nonintrusive sensory equipment is available and can be easily installed into any environment. The challenge is to develop software which utilizes data from these sensory devices and actuates a help response within the sensory environment. This constitutes a specialized application of the traditional architecture of intelligent systems being developed in the last decades (see Figure 11.1).

In essence, an enriched environment of sensors and other electronic devices networked together can act as a "personal aid". The personal aid can respond according to each scenario, either based on real-time events alone or by analysing previously collected data sets, and actuate a response which benefits the users within their environment. Increasingly, these systems are

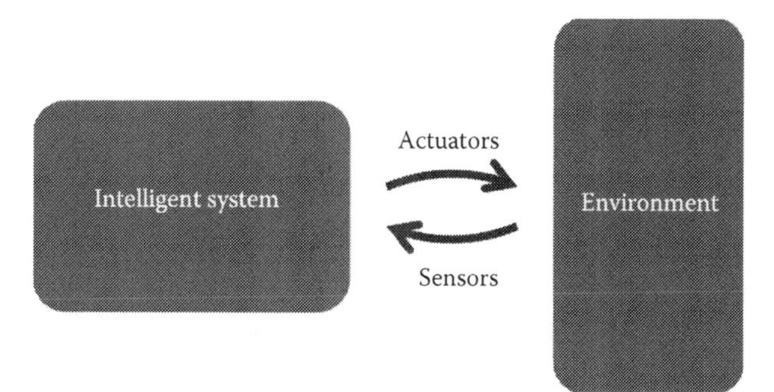


**FIGURE 11.1**
Diagram to illustrate the passing of event information to the intelligent system, which actuates an appropriate action back into the environment.

developed through multiagent systems (MAS), autonomous programs that perceive changes in the environment (for example, a smart home) and act upon it. There are three main types of sensors used to measure such conditions, illustrated in Table 11.1. Of these types, there are also many versions which use different sensing principles and/or may operate within predetermined ranges.

Many sensors exist to aid the sensitive issue of tracking. These include radio frequency identification (RFID) detectors, global positioning system (GPS) modules, ultrasonic transducers, passive infrared detectors (PID) and pressure mats. RFID tags have been widely used in research projects for tracking participants (Altus et al. 2000; Arcelus et al. 2007; Miura et al. 2008). The tags work by transmitting a unique identification on a radio frequency which is picked up by strategically positioned readers. Within healthcare, active RFID tags which contain a battery and transmit signals autonomously predominate, to ensure continuous 24-hour monitoring. The use of RFID tags has been mainly limited to the hospital and nursing home environments. A more advanced version of tracking location is achieved by GPS satellites, which broadcast signals from medium-earth orbit (18,000 km above the earth) that GPS receivers use to provide latitudinal, longitudinal and altitudinal coordinates. By the process of triangulation, using three satellites, the location of any person carrying a GPS device can be pinpointed to an accuracy of approximately 10 m. This type of tracking has been used for people who suffer from illnesses such as amnesia or dementia but have the physical ability to lead an active life in the community (Shoval et al. 2008). This form of tracking is reliant on the client being in possession of a GPS receiver, either a handheld device or a chip attached to a piece of clothing (Lin et al. 2006).

PIR sensors form the basis for most home-based tracking systems. They work by detecting the movement of warm objects (in most cases people) in an unobstructed field of vision. Originally they were used for intruder

**TABLE 11.1**

Sensor Types and Characteristics

| Sensor Types | Examples | Description | Uses | Special Settings |
|---|---|---|---|---|
| Signal emission | Magnetic switches/exit sensors | Used on doors | Tracking movement from room to room | Size and distance range |
| | I-Button [ib] | Information tag | Gaining access to computers, buildings, etc. | |
| | Ultrasound | Echoes radio/sound waves against objects | Measuring distance and location, detecting movement | |
| | Light-dependent resistor (LDR) | Voltage-driven circuit measure | Measuring light levels in an area | |
| | Radio frequency identification tags (RFID) | A traceable signal | Tracking local movement | |
| | Passive infrared sensor (PIR) | Detects movements | Detecting movement | |
| | Wireless sensors | Wire-free, run on batteries | Detecting movement | |
| | Electricity power supply | Measures current | Monitoring appliances | |
| | Global positioning satellite systems (GPS) | Satellite picks up location | Tracking and guidance | |
| Pressure | Bed occupancy sensor | Measures presence, movement and pressure | Analysing sleeping quality and patterns | Numerical range and weight distribution |
| | Pressure mats | Mats with a pressure sensor to detect presence | Detecting movement | |
| | Piezoresistive pressure sensor | Measures pressure | | |
| | Digital air pressure sensor, digital barometric pressure sensor | Detect air changes | Detecting leaking gas, etc. | |
| | Thermostats | Measure temperature | Detecting changes in room temperature | |
| Motion (based on natural forces) | Mercury tilt switch | Measuring stability levels | Ensuring devices such as wheelchairs remain flat on the ground | |
| | Accelerometers | Measure speed and directional movement | Detecting loss/deterioration in balance and falls | |

alarms and for outdoor security lights, but they have now been integrated into the healthcare system as a method of detecting movement activities and inferring well-being for vulnerable elderly people in hospitals (Noury et al. 2008) or living at home (Lee et al. 2007).

Motion detection is a distinct area within tracking. The purpose of motion detection is not always to determine where people are going but rather physical attributes such as presence. Motion detectors include some of the previously mentioned sensors as well as pressure sensors, light sensors, accelerometers, cameras and tilt switches. Pressure mats have found application in beds to monitor presence (Chong and Kumar 2003) and on floors to monitor aspects of movement, such as wandering activity or balance (Srinivasan et al. 2005). Most bed sensors comprise a pressure strip that monitors pressure and provides a Boolean response, i.e., in/out of bed. Floor mats can contain many different pressure points, each measuring differences in the force exerted. The output can be interpreted as simple presence or provide more information, such as direction, walking patterns or size of pressure area to indicate whether someone is walking stably or has fallen (Redfern et al. 1997).

An accelerometer is an electromechanical device that measures acceleration forces, useful for detecting instances of trips, falls or sudden actions. Accelerometers have been used to monitor people with a propensity to fall, although other healthcare options include measurement of heart activity (Haskell et al. 1993) and epileptic seizures (Tormans et al. 2007). Cameras and microphones have been used in more controlled experimentation, mainly due to the ethical issues related to privacy with imagery devices. Visual representation from cameras can take different modalities. Tracking may be facilitated by using cameras which can identify pattern markers placed in the environment or by using heat sensing cameras to detect body temperature (Zhou and Hu 2004) and hence provide additional functions compared to PIDs.

A mercury tilt switch is a switch which allows or interrupts the flow of electric current in an electrical circuit depending on the switch's physical position or alignment relative to Earth's gravity or other inertial force. Tilt switches have been used to retrieve information such as posture, i.e., how long a person spends upright, sitting or lying down (Digorry et al. 1994; Lewis et al. 2001), or to monitor mobility devices such as wheelchairs to detect incidents where the mobility device was in an unnatural position, indicating the likelihood of an accident or potential need for help (Galligan et al. 2003).

RFID tags have not been limited to the field of tracking; they can also be used as a method for identification. Nurses using a personal digital assistant (PDA) equipped with a portable RFID reader could check patient identification (ID) emitted from the RFID tag against a database before administering medication (Bacheldor 2007). Another device used as an ID tag is the I-button, which is almost like a key card, almost like a key card for people to use to access computers, lifts, buildings, etc. The information contained in the button could actuate an appropriate response from an associated device after being scanned. Even though it is relatively small, like a pendant, it can prove troublesome as it needs to be carried around; short-range radio detection means the I-button must be placed close to its associated sensor (Davidson et al. 2003).

Other types of sensors are used to monitor physical situations such as heart conditions. Photodetectors are sensors of light energy. Light-dependent resistors (LDR) work using a resistor which changes in resistance according to light intensity. Within healthcare they have been used to measure heart activity by receiving light from a light-emitting diode (LED) clipped onto an earlobe. The LED converts blood vessel pressure into an electronic signal (Livreri et al. 2008). Smart garments have been created with the capability to monitor vital signs and human posture by using electro-active polymer actuators (De Rossi et al. 2003). Indeed, the development of actuating fabrics for detecting vital signs and motion has been previously investigated (De Rossi et al. 2002; Mazzoldi et al. 2002).

In comparison to fixed positions sensors, wireless sensors provide greater mobility and facilitate better area coverage. They are cheaper to install and reduce the danger of trips and falls (Goldberg and Wickramasinghe 2003). An added advantage is the ability to utilize additional devices without needing to physically disturb an environment (Wickramasinghe and Mishra 2004).

As previously stated, sensors combine to form a network in which multiple readings form the basis for actuating an appropriate response. There are many examples of multifunctional sensor inputs which combine well together to create a smart environment, such as Vigil (Holmes et al. 2007), which proves the viability and employability of such systems.

## Peculiarities of Nocturnal Sensing and Interventions

There is a significant body of research that has described the needs of people with dementia and has produced solutions to address those needs using information and communication technologies (ICT). Key examples of such research include the ENABLE and COGKNOW projects. ENABLE (Enabling Technologies for People with Dementia) assessed the effect of assistive devices to support memory, to provide pleasure and comfort and to facilitate communication (calendar, medicine reminder, lamp, locator, gas cooker monitor and picture phone) among people with dementia. Issues with regard to their quality of life, and the burden on their carers, were addressed (Adlam et al. 2003). In COGKNOW, the aim was to develop a system to support people with dementia in different aspects of their daily life including remembering, maintaining social contacts, performing daily activities and feeling safe (Davies et al. 2009). These projects focused on the needs of people with dementia but primarily during the daytime hours. This is a natural area of need for people with dementia, with an emphasis on supporting people as they interact and are most active in daily life.

The needs of people with dementia at night-time have not received a similar degree of attention in terms of ICT research involving telecare and telehealth services. Certainly their needs can be very different to daytime needs, but they are equally, if not more, important (Carswell et al. 2009). First, the environment at night is very different. The person with dementia will often be in an environment where disorientation is more likely due to low light conditions. Second, the person with dementia is more likely to be confused and disorientated as he or she awakens, either naturally as he or she awakes from sleep or if he or she is exhibiting "sundowning", in which the behaviour of people with dementia changes as the evening and night fall (Scarmeas et al. 2007). Older people generally experience changes in their sleeping behaviours. This includes going to sleep early but awakening earlier as well, having more fragmented sleep patterns and suffering from insomnia and from sleep apnoea, which is "increasingly seen among older people and is significantly associated with cardio- and cerebrovascular disease as well as cognitive impairment" (Wolkove et al. 2007, p. 1299). People with dementia may also be likely to "wander", causing distress to themselves and their carers when they "escape" from their room and building, sometimes requiring the intervention of emergency services (Hermans et al. 2007). Third, people with dementia are more likely to be alone or to not have as immediate access to carers as during the daytime. This can cause any stress and anxiety experienced as they awake in the dark to be significant and to increase unchecked and more rapidly than would otherwise be the case if a carer was with them.

These three issues of the environment, the situation of the person with dementia and the access to caregivers set the background for enumeration of the needs of people with dementia and their carers during the night hours. The use and role of sensors and actuators in supporting people with dementia and their carers during the night hours is different from during daylight hours for a number of interrelated reasons.

There are fewer activities of daily living experienced during night-time, and most of the activity is in the bedroom and bathroom. Therefore, the complexity of the system needed to recognize the basic behaviours from the data is somewhat decreased. However, there may be a paucity of data as activities at night-time may be sparse, and the underlying purpose of the actions that generate the sensor data may be difficult to infer.

There is an important role in supporting the carer of a person with dementia during night-time. This can be achieved by providing an intervention that alleviates the burden of care on the carers, enabling them to achieve uninterrupted sleep, for example (McKibbin et al. 2005), or assisting the carer in supporting the person with dementia by automating some tasks.

People with dementia may have a much more pronounced and immediate need for help and support as they awake in an unknown, dark environment. While this is not an emergency, the sensors and actuators must be capable of quickly creating an intervention that moderates the anxiety of the person with dementia.

Therefore, at night-time, the sensors conduct their activities as they would during the day, but fewer data are available for comprehension compared to daytime sensing. During night-time the actuators have a different role to fulfil to during daytime in managing the lighting for the person with dementia, in particular providing illumination and lighted guidance to the bathroom and back. In addition to light, sound can play a more important role during night-time. In particular, the use of music as a therapeutic intervention has been shown to reduce the anxiety of people with dementia (Aldridge 2000). Light and music together can provide a powerful intervention capability, and it has been shown that there is significant reduction in restlessness immediately after people with dementia in care homes experienced multisensory environments through the use of lighting, tactile surfaces, meditative music and the odour of essential oils (Robinson et al. 2006). Better lighting can also play an important role in reducing the risk of falls, in particular when people with dementia seek to go to the bathroom and back. Night-time falls may be reduced by proper lighting, reduction of nocturia, avoidance of bedtime sedatives and use of a bedside commode, if walking to the bathroom is considered unsafe (Moylan and Binder 2007). The Joseph Rowntree Foundation report (Kerr et al. 2008) made recommendations for home management of older people. These recommendations were to ensure that, where appropriate, relevant technology should be used, for example, guidance around noise, light, safety and silent call systems, as well as to ensure that systems are in place for night staff to have all the equipment and facilities required to provide good night-time care.

In summary, important requirements for nocturnal care of people with dementia are (a) more specialized algorithms to infer behaviour from sparse data; (b) specially designed interventions, perhaps including music and video, that provide therapeutic support to people to reduce anxiety; and (c) sophisticated guidance, perhaps through the use of lightning.

The NOCTURNAL decision-making process is summarized in Figure 11.2. The system understands the environment through a variety of sensors. These data can be partially processed and acted upon through a local decision support system (DSS). The information is uploaded to the NOCTURNAL server, which stores the information in a database. More complex intelligent data analysis can be conducted there, and the result of that analysis can trigger actions by the system, such as contacting healthcare professionals, carers and the house occupants (including the person with dementia). Healthcare professionals and researchers can interrogate the database for more complex data retrieval. For more details, see Zheng et al. (2009).

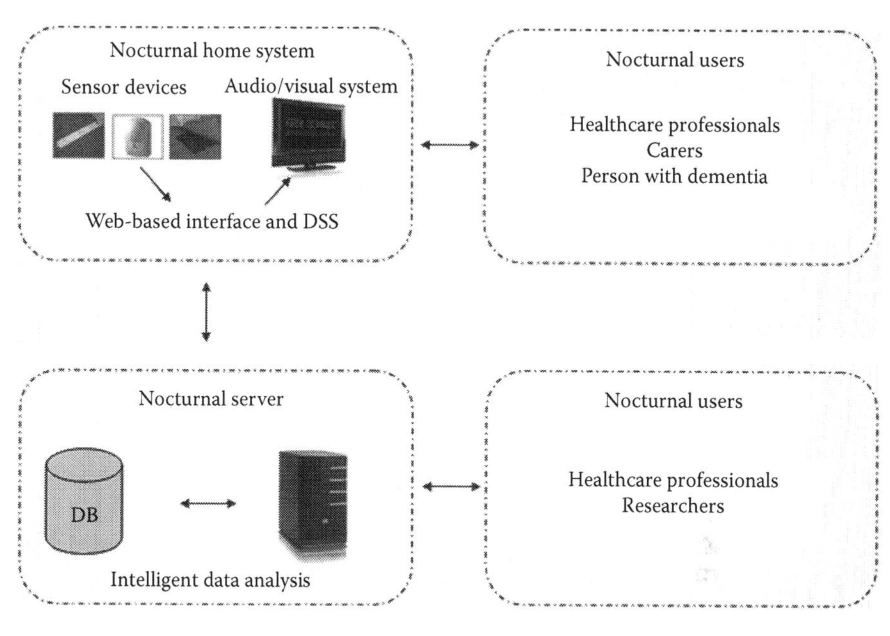


**FIGURE 11.2**
Gathering of data and operational flow in NOCTURNAL. DSS, decision support system.

## NOCTURNAL Sensing/Intervention Platform

NOCTURNAL can be set up to monitor every room within the home of the person with dementia. At present, the sensor range consists of PID detectors, pressure strips, door contact switches and voltage monitors. This range of sensors is provided by Tunstall.* Tracking the clients within their homes is one of the most important goals of the project. Receiving data from any of the sensory equipment helps to pinpoint the location of a client, which in turn will enable the system to serve the client better.

One PID (Figure 11.3a) is located in a corner of each room, positioned to ensure that it will not detect the client's movements in any other room. Pressure strips, approximately 1 m long by 5 cm wide (Figure 11.3b), are used as a bed sensor. There are two strips per bed, placed at different intervals and overlapping in the middle (Figure 11.3c). Door contact sensors (Figure 11.3d) are used in a number of ways. The majority of door contact sensors are located in the kitchen area. At present, door contacts are set to measure the opening and closing of the fridge, food cupboard and front and back doors. Voltage sensor adaptors (Figure 11.3e) are connected to plug-in appliances and are used to measure frequency of usage. Appliances currently being

* http://www.tunstall.co.uk

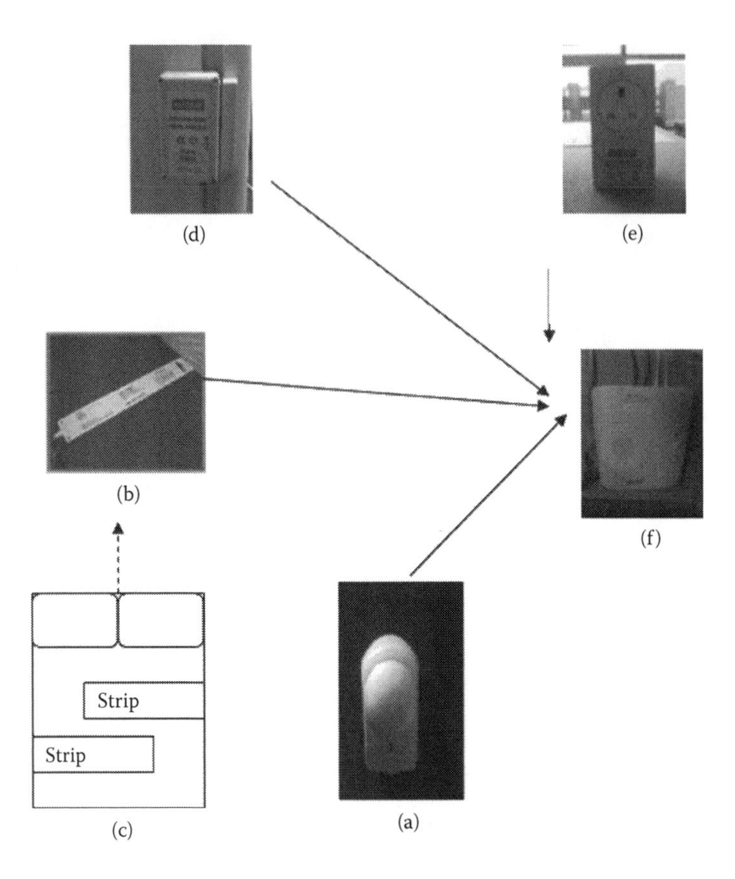


**FIGURE 11.3**
Data collection to support algorithms providing intelligence throughout the home. (a) Passive infrared detector; (b) pressure strips; (c) strip layout; (d) door contact sensor; (e) voltage sensor; (f) remote call box.

monitored are the kettle and microwave. All sensory equipment is attached to a remote call box (Figure 11.3f) which collects the data (appliance identification, time and duration of use) in comma separated variable (CSV) format and sends it off to a secure server. Whenever an emergency is triggered, the call box puts out or receives a call from the healthcare team at Fold Telecare,* who provide current home healthcare to elderly people in Northern Ireland.

The aim of the completed network is to have a sensor-enriched environment which will aid the client by introducing ambient control mechanisms throughout the house (e.g., using the X10 communication/automation protocol to control lighting). Whatever the client decides to do at night will be aided in a number of ways. The system builds upon the current set-up, where

* http://www.foldgroup.co.uk/telecarehousing.php

the healthcare team (from FOLD Telecare) remotely monitors the status of sheltered dwellings and can respond to an emergency request or alerts (e.g., fall pendant alarm pressed) by telephone. Figure 11.4 illustrates the overall NOCTURNAL architecture. A home hub can be used to provide feedback and can be deemed to be a "looking glass" for the clients, both the carer and the person with dementia.

Some of the additional features focus around providing greater guidance. Smart sensing will help to automatically control the lights in a number of

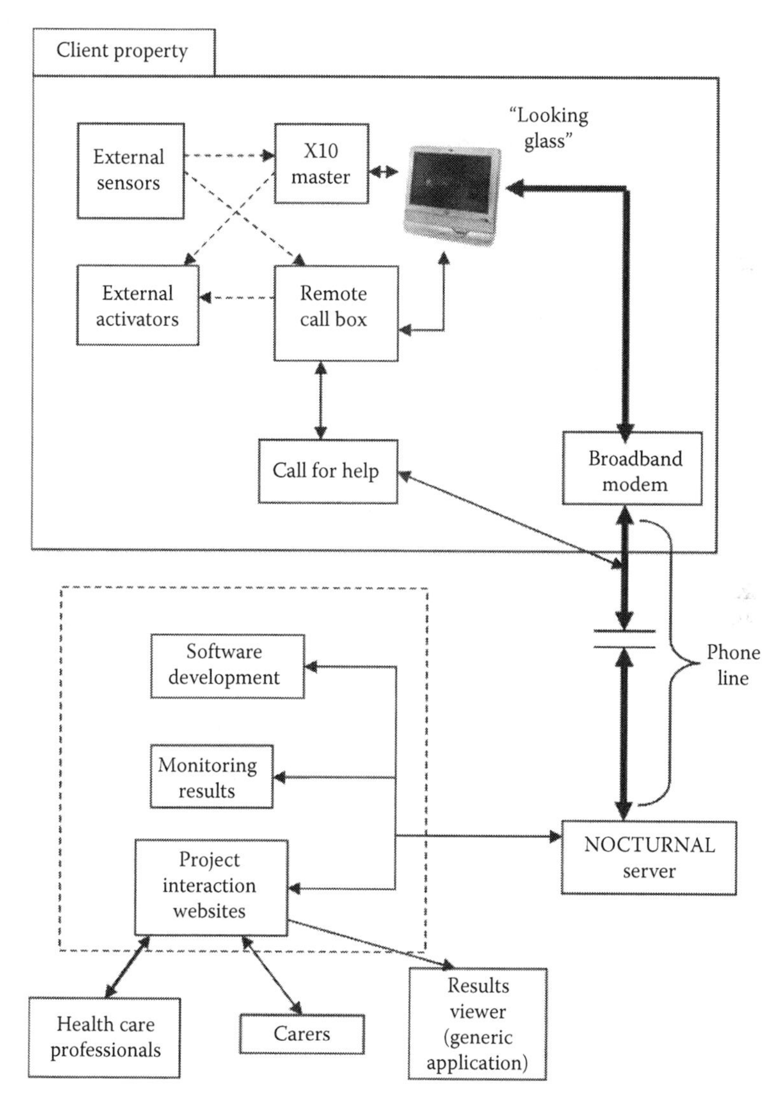


**FIGURE 11.4**
Complete NOCTURNAL architecture.

beneficial ways, subject to conditions. First, it will ensure that any areas the client wanders into light up. Not only will the new room light up, but it will also maintain an illuminated pathway back to the safety of where the client ought to be. More complicated lighting guidance algorithms will also be developed to encourage clients to follow a logical path when needed, for example, lighting the way to the bathroom to ensure that the client is less likely to wander off to another location instead.

A primary goal of NOCTURNAL is to encourage good sleeping habits. One of the AmI features that NOCTURNAL assists with is encouraging the client to experience a regular day/night pattern. A combination of bed and PIR sensors will detect restless sleep, and NOCTURNAL will actuate combative measures such as raising light levels and/or playing low-level music. Studies have shown varying degrees of success when applying these techniques (Sixsmith and Gibson 2007; Ziv et al. 2007). The assessment of each situation combined with personal client knowledge will help actuate the correct solution for each individual, enabling him or her to return to sleep.

An important, multifunctional feature for each dwelling will be an audiovisual unit. This unit will form a type of avatar with extra input received from the array of sensors located throughout the house. On the back end, the audiovisual unit will be able to collect all the in-house sensor information and form a decision support system (DSS) using intelligent agents to actuate the appropriate response. All the information collected can be exported to a database for knowledge management (KM), which facilitates the development of future services for individual clients. Service updates can be remotely uploaded into the local audiovisual units in each home.

One of the functions the unit aims to address is the encouragement of better cognitive ability. The unit will display pictures and play music for reminiscence therapy, which has also been shown to alleviate anger, confusion and tension (Wang 2007; Woods et al. 2005). Wireless speakers linked to the audiovisual unit will be spread throughout the house to help facilitate better communication with the client. This will enable audio support, for example verbal prompting if the fridge door is inadvertently left open. Prompts can be used to support task completion (e.g., washing hands after toileting).

Whilst the system will react with real-time solutions based on real-time sensor input and short-term recorded pattern deviations, a tailored healthcare plan from professionals can be developed as a consequence of the sensory information gathered over much longer periods. These data will allow the healthcare professional to analyse information gathered with regard to the client's habits/rituals over varying time scales. From this they will be able to provide structured therapy sessions throughout the day, which can be brought to the client through the NOCTURNAL visual display unit, and also provide intelligent agents with information which would enable them to provide better responses to the real-time sensory data they receive within the client's home.

## Addressing User Acceptance

NOCTURNAL provides assistive technology which aims at improving the quality of life of people with dementia whilst at the same time preserving privacy. NOCTURNAL could pose a threat to personal dignity as data on activities of daily living, e.g., duration of sleep, wandering and toileting, are sensed and evaluated with respect to what is "normal". An "abnormal" activity will require the system to respond by alerting a carer or possibly by the direct use of technology (lighting or audio prompt). Many see this as an unacceptable "big brother"* or *The Truman Show*† approach to healthcare. However, technology change is inexorable and can be used for benefit. A report to the Information Commissioner (Wood 2006, p. 2) provides the following commentary: "We live in a surveillance society. It is pointless to talk about surveillance society in the future tense. In all the rich countries of the world everyday life is suffused with surveillance encounters, not merely from dawn to dusk but 24/7."

NOCTURNAL addresses ethical concerns by adhering to best-practice ethics guidelines (Kluge 2003) in the gathering of information from people and about people in order to identify which is the best way to provide assistance. As personal data will be collected, the assisted person and their carer are informed and their consent sought. The assisted person can have his or her information removed at any time, in accordance with Data Protection Act. Persistent data stored on computer systems are anonymous and encrypted, protecting the personal data.

When asked about quality-of-life issues, older people's first preference is for "comfortable and secure homes" (Clark and Crissel 2006). Whilst accepting the potential for intrusion, the ethical dilemma is to weigh up the pros and cons of such an intervention. A 2009 report from the Alzheimer's Society, Counting the Cost addressed healthcare in hospitals. A press release (Alzheimer's Society 2009) stated: "Keeping people with dementia at home for longer, is normally the preferred mode for the person with dementia and their carer. Poor hospital care also had a negative impact on the people's dementia and physical health. The majority of people with dementia leave hospital worse than when they arrive with a third having to enter a care home as they are unable to return home." It argued the economic

* A fictional character in George Orwell's novel *Nineteen Eighty-Four.*

† *The Truman Show* is a 1998 movie where the house and living environment monitors the occupant's every move and broadcasts it to a receptive audience for entertainment purposes.

perspective: "At least £80 million a year and probably hundreds of millions could be saved if people with dementia are enabled to leave hospital one week earlier. Hospitals must commit to reducing the length of stay if we are to stop people with dementia deteriorating in hospital and lessen the chance of people being discharged to a care home."

The National Health Service Confederation estimated that the cost of helping older people to cope with daily routines of living in their own homes will likely to treble to about £30 million by 2026. The current health and social care system will not be able to sustain this burden, and there is increasing political pressure to provide social insurance to fund elderly care.

The South Eastern Trust (Northern Ireland Health and Social Care) is spending at least £60 Million on institutional care in 2009/2010, and thus there is an obvious need to contain such expenditures.

Thus, from a societal perspective, people with dementia are better off at home. As demographic trends indicate that the population will include a larger proportion of older people in the near future, society needs pervasive technology. Telemedicine solutions based on sensors and ICT, such as NOCTURNAL, can assist the person with dementia and his or her carer to sustain home-based care for as long as possible.

However, there is also a danger that by increasing the amount of technology we could contribute to social isolation, by reducing the need for interaction with carers and healthcare professionals. NOCTURNAL assistive technology and telecare can orientate the person that it is daytime or night-time, switch on the lights automatically if the person gets up at night and alert a carer or monitoring centre that the person needs assistance. This can potentially free carers to spend time with the person, and it may enable a carer to get a good night's rest, knowing that if the person gets up at night, the carer will be alerted. Here, technology can benefit the individual. However, the Trent Dementia Services Development Centre (AT Dementia, 2009, p. 2) stated: "If assistive technology does not meet the individual needs and preferences of the person it may be ineffective or may even cause additional confusion or distress. For example, assistive technology and telecare may not be the answer if the person switches off or unplugs the equipment, the person is confused or distressed by any alarm sounds or recorded messages, or if there are insufficient carers or care workers to respond to an alert."

Assistive technology on its own cannot provide human contact and personal care. Many older people experience loneliness and social isolation. Ideally, technology should be provided as an addition to contact and care, not as a replacement. All technology has the potential to work for good or bad, and as such we should not be surprised that assistive technology obeys the "law of unintended consequences" (Merton 1936). This can be ameliorated by involving users (people with dementia, carers and healthcare professionals) in the design process, an approach in keeping with the TRAIL lab philosophy (Galbraith et al. 2008): "TRAIL is focused on supporting this diverse set of stakeholders as we develop new technologies, research perspectives, processes and integrated service solutions that deliver real value to our users, the ageing people in the North of Ireland and further afield across Europe" (TRAIL 2009). This approach has been adopted by NOCTURNAL to support user acceptance.

## Conclusions

Recent developments in the assistive technology industry have made available technology which can be used to deploy new applications closely related to the needs of people's daily lives. Policymakers in the healthcare system have also been gradually moving the focus of healthcare from the traditional hospital focus to the home. At the same time, our society has become accepting of technology (mobile phones, GPS technology in the car, computers and the Internet at home) that can be used to assist our daily lives. This confluence of factors has paved the way for the use of technology that can provide assistance at home to increase independence. This is particularly meaningful for older people, especially if they are affected by chronic conditions such as dementia.

This article reports on a sensor-based system, NOCTURNAL, which is developing an innovative market solution to provide assistance during the hours of darkness to this segment of the population. A key part of the system relates to the accurate sensing and "understanding" of the activities engaged in by the people our system assists during the night. The sensing component is based on off-the-shelf solutions. The "understanding" component is crucial to beneficial use and is linked to concepts of data analysis, data visualization, data mining and AmI. One other important element of the system is the guidance offered in the way of warnings and alarms to the call centre but also in the form of reassurance and help through multimodal interaction (audiovisual reminders, advice, orientation, etc.) that helps the house occupant to stay safe at home. In developing the system, it has been necessary to weigh the benefits against the potential intrusion of such technology and always to ensure that the technology strives to act in support of the needs of the users (people with dementia and carers).

## Acknowledgements

NOCTURNAL is jointly funded by the Assisted Living Innovation Program of the Technology Strategy Board and EPSRC (UK).

## References

Adlam, T., R. Faulkner, R. Orpwood, K. Jones, J. Macijauskiene, and A. Budraitiene. 2004. The installation and support of internationally distributed equipment for people with dementia. *IEEE Transactions on Information Technology in Biomedicine* 8(3): 253–57.

Aldridge, D. 2000. *Music Therapy in Dementia Care*. Jessica Kingsley, England.

Altus, D.E., R.M. Mathews, P.K. Xaverius, K.K. Engelman, and B.A. Nolan. 2000. Evaluating an electronic monitoring system for people who wander. *American Journal of Alzheimer's Disease and Other Dementias* 15(2): 121.

Alzheimer's Society. 2009. *Counting the Cost: Caring for People with Dementia on Hospital Wards.* London: Alzheimer's Society. http://www.alzheimers.org.uk/site/scripts/news_article.php?newsID = 579 (accessed November 2009).

Arcelus, A., M.H. Jones, R. Goubran, and F. Knoefel. 2007. Integration of smart home technologies in a health monitoring system for the elderly. In *Proceedings of the 21st International Conference on Advanced Information Networking and Applications Workshops (AINAW'07)* IEEE Xplore, pp. 820–25.

AT Dementia. 2009. The benefits and limitations of assistive technology. http://www.atdementia.org.uk/content_files/files/The_benefits_and_limitations_of_assistive_technology.pdf (accessed May 2011).

Augusto, J.C., and P.J. McCullagh. 2007. Ambient intelligence: Concepts and applications. *International Journal on Computer Science and Information Systems* 4(1): 1–28.

Bacheldor, B. 2007. RFID-enabled handheld helps nurses verify meds. *RFID Journal*, July 2007. http://www.rfidjournal.com/article/articleview/3470/

Carswell, W., P.J. McCullagh, J.C. Augusto, S. Martin, M.D. Mulvenna, H. Zheng, H.Y. Wang, J.G. Wallace, K. McSorley, B. Taylor, and W.P. Jeffers. 2009. A review of the role of assistive technology for people with dementia in the hours of darkness. *Technology and Healthcare* 17(4): 281–304.

Chong, C., and S.P. Kumar. 2003. Sensor networks: Evolution, opportunities, and challenges. *Proceedings of the IEEE* 91(8): 1247–56.

Clark, H., and J. Crissel. 2006. *Addressing What Older People Want?* Policy context, Evidence submitted to the Older People's Inquiry into 'That Bit of Help', Edited by Norma Raynes, Heather Clark and Jennifer Beecham, Joseph Rowntree Foundation, New York. http://www.infj.ulst.ac.uk/~ccjg23/files/ThatBitofHelp.pdf (accessed May 2011).

Cook, D.J., J.C. Augusto, and V.R. Jakkula. 2009. Ambient intelligence: Technologies, applications, and opportunities. *Journal of Pervasive and Mobile Computing* 5: 277–98.

Davidson, A.J., F. Aujard, B. London, M. Menake, and G.D. Block. 2003. Thermochron iButtons: An inexpensive method for long-term recording of core body temperature in untethered animals. *Journal of Biological Rhythms* 18(5): 430–32.

Davies, R.J., C.D. Nugent, M. Donnelly, M. Hettinga, F. Meiland, F. Moelaert, M.D. Mulvenna, J.E. Bengtsson, D. Craig, and R.M. Droes. 2009. A user-driven approach to develop a cognitive prosthetic to address unmet needs of people with mild dementia. *Journal of Pervasive and Mobile Computing* 5(3): 253–67.

De Rossi, D., F. Carpi., F. Lorussi, A. Mazzoldi, E.P. Scilingo, and A. Tognetti. 2002. Electroactive fabric for distributed, conformable and interactive systems. Paper presented at IEEE SENSORS 2002 Conference (Hyatt Orlando, Kissimmee, Florida, 11–14 June 2002).

De Rossi, D., F. Carpi, F. Lorussi, A. Mazzoldi, R. Paradiso, E.P. Scilingo, and A. Tognetti. 2003. Electroactive fabrics and wearable biomonitoring devices. *AUTEX Research Journal* 3(4): 180–85.

Digorry, P., M. Gorman, and R. Helme. 1994. An automatic device to measure time spent upright. *Clinical Rehabilitation* 8: 353–57.

Galbraith, B., M.D. Mulvenna, S. Martin, and E. McGloin. Living Labs: Helping to meet the needs of ageing people? In Mann, W.C., (ed), *Aging, Disability and Independence—Selected Papers from the 4th International Conference on Aging, Disability and Independence (2008)*, IOS Press, Assistive Technology Research Series, Vol. 22, pp. 105–118, Amsterdam.

Galligan, C., G. Morose, and J. Giordani. 2003. *An Investigation of Alternatives to Mercury Containing Products*. Lowell, MA: Lowell Center for Sustainable Development.

Goldberg, S., and N. Wickramasinghe. 2003. Paper presented at 21st century healthcare—The wireless panacea. The 36th Hawaii International Conference on System Sciences, January 6–9, Big Island, Hawaii.

Haskell, W.L., M.C. Yee, A. Evans, and P.J. Irby. 1993. Simultaneous measurement of heart rate and body motion to quantitative physical activity. *Medicine and Science in Sports and Exercise* 25: 109–15.

Hermans D.G., U.H. Htay, and R. McShane. 2007. Non-pharmacological interventions for wandering of people with dementia in the domestic setting. Cochrane Database of *Systematic Reviews* 1: CD005994.

Holmes, D., J.A. Teresi, M. Ramirez, J. Ellis, J. Eimicke, J. Kong, L. Orzechowska, and S. Silver. 2007. An evaluation of a monitoring system intervention: Falls, injuries, and affect in nursing homes. *Clinical Nursing Research* 16: 317–35.

Kerr, D., H. Wilkinson, and C. Cunningham. 2008. *Supporting Older People in Care Homes at Night*. Joseph Rowntree Foundation, New York.

Kluge, E.-H. 2003. *A Handbook of Ethics for Health Informatics Professionals*. British Computer Society Health Informatics Committee. Endorsed by International Medical Informatics Association. British Computer Society, England.

Lee, S.W., G.S. Lee, and Y.J. Kim, B.O. Cho, and N.H. Lee. 2007. A healthcare system for elders living alone using low-power wireless sensors. In *Proceedings of Fourth Int'l Conference on Information Technology and Applications*, Harbin, China, January, 2007.

Lewis, M.J., B. Reilly, L.A. Houghton, and P.J. Whorwell. 2001. Ambulatory abdominal inductance plethysmography: Towards objective assessment of abdominal distension in irritable bowel syndrome. *Gut* 48: 216–20.

Lin, C.C., M.J. Chiu, C.C Hsiao, R.G. Lee, and Y.S. Tsai. 2006. Wireless health care service system for elderly with dementia. *IEEE Transactions on Information Technology in Biomedicine* 10: 696–704.

Livreri, P., G.M. Di Blasi, and G. Terrazzino. 2008. IrDa-based heart frequency monitoring for mobile healthcare applications, sensors and microsystems. In Mignani, A.G., R. Falciai, C. Di Natale, and A.D'Amico (eds) *Proceedings of 10th Italian Conference* Firenze, Italy, 15–17 February 2005 (pp. 105–10) World Scientific, Singapore.

Mazzoldi, A., D. De Rossi, F. Lorussi, E.P. Scilingo, and R. Paradiso, R. 2002. Smart textiles for wearable motion capture systems. *Autex Research Journal* 2(4): 199–203.

McCullagh, P.J., W. Carswell, J.C. Augusto, S. Martin, M.D. Mulvenna, H. Zheng, H.Y. Wang, J.G. Wallace, K. McSorley, B. Taylor, and W.P. Jeffers. 2009. State of the art on night-time care of people with dementia. In *Proceedings of the Conference on Assisted Living 2009,* published by Institution of Engineering and Technology, England, London, 24–25 of March, 2009.

McKibbin, C.L., S. Ancoli-Israel, J. Dimsdale, C. Archuleta, R. von Kanel, P. Mills, T.L. Patterson, and I. Grant. 2005. Sleep in spousal caregivers of people with Alzheimer's disease. *Journal of Sleep Research* 14(2): 177–85.

Merton, R.K. 1936. The unanticipated consequences of purposive social action. *American Sociological Review* 1(6): 894–904.

Miura, M., S. Ito, R. Takatsuka, and S. Kunifuji. 2008. Aware group home enhanced by RFID technology. In *Proceedings of 12th International Conference on Knowledge-Based and Intelligent Information & Engineering Systems,* Croatia (pp. 847–54).

Moylan, K., and E. Binder. 2007. Falls in older adults: Risk assessment, management and prevention. *American Journal of Medicine* 120(6): 493.e1–e6.

Noury, N., T. Hadidi, M. Laila, A. Fleury, C. Villemazet, V. Rialle, and A. Franco. 2008. Level of activity, night and day alternation, and well being measured in a smart hospital suite. In *Proc. 30th Annual Int. Conference of the Institute of Electrical and Electronic Engineering -Engineering in Medicine and Biology Society 2008* (pp. 3328–31).

Redfern, M.S., P.L. Moore, and C.M. Yarsky. 1997. The influence of flooring on standing balance among older persons. *Human Factors,* 39(3): 445–55.

Robinson, L., D. Hutchings, L. Corner, F. Beyer, H. Dickinson, A. Vanoli, T. Finch, J. Hughes, C. Ballard, C. May, and J. Bond. 2006. A systematic literature review of the effectiveness of non-pharmacological interventions to prevent wandering in dementia and evaluation of the ethical implications and acceptability of their use. *Health Technology Assessment* 10(26): 1–124.

Scarmeas N., J. Brandt, D. Blacker, M. Albert, G. Hadjigeorgiou, B. Dubois, D. Devanand, and L. Honig. 2007. Disruptive behavior as a predictor in Alzheimer disease. *Archives of Neurology* 64(12): 1755–61.

Shoval, N., G.K. Auslander, T. Freytag, R. Landau., F. Oswald, U. Seidl, H. Wahl, S. Werner, and J. Heinik. 2008. The use of advanced tracking technologies for the analysis of mobility in Alzheimer's disease and related cognitive diseases. *BMC Geriatrics* 8: 7, http://www.biomedcentral.com/1471-2318/8/7 (accessed May 2011).

Sixsmith, A., and G. Gibson. 2007. Music and the wellbeing of people with dementia. *Ageing and Society* 27: 127–45.

Srinivasan, P., D. Birchfield, G. Qian, and A. Kidane. 2005. Design of a pressure sensitive floor for multimodal sensing. In *Ninth International Conference on Information Visualisation* (pp. 41–46). Proceeding IV '05 Proceedings of the Ninth International Conference on Information Visualisation IEEE Computer Society Washington, DC, USA 2005.

Tormans, D., T. Swinnen, K. Cuppens, S. Omloop, B. Geraets, P. Colleman, V. Claes, B. Ceulemans, and B. Vanrumste. 2007. Nocturnal monitoring of pediatric patients with epilepsy based on accelerometers. In *Conference Proceedings of the 18th ProRISC Annual Workshop on Circuits, Systems and Signal Processing.* (ProRISC 2007) 29–30 November 2007, the Netherlands; Technology Foundation, Utrecht, 2007.

TRAIL. 2009. *Technologies for Rurality, Ageing and Independent Living*. http://trail.ulster.ac.uk/ (accessed May 2011).

Wang, J.J. 2007. Group reminiscence therapy for cognitive and affective function of demented elderly in Taiwan. *International Journal of Geriatric Psychiatry* 22(12): 1235–40.

Wickramasinghe, N., and S.K. Mishra. 2004. A wireless trust model for healthcare. *International Journal of Electronic Healthcare* 1(1): 60–77.

Wolkove, N., O. Elkholy, M. Baltzam, and M. Palayew. 2007. Sleep and aging: 1. Sleep disorders commonly found in older people. *Canadian Medical Association Journal* 176(9): 1299–304.

Wood, D.M. 2006. *A Report on the Surveillance Society. For the Information Commissioner by the Surveillance Studies Network*. http://www.ico.gov.uk (accessed November 2009).

Woods, B., A. Spector, C. Jones, M. Orrell, and S. Davies. 2005. Reminiscence therapy for dementia. *Cochrane Database of Systematic Reviews* 2: CD001120.

World Health Organisation. 2005. Ageing and life course http://www.who.int/ageing/en (accessed May 2011).

Zheng, H., W. Carswell, S. Martin, M. Mulvenna, P. McCullagh, P. Jeffers, J. Augusto, J. Wallace, H. Wang, B. Taylor, and K. McSorley. 2009. NOCTURNAL: Night Optimised Care Technology for UseRs Needing Assisted Lifestyles. In *Proceedings of eChallenges e-2009 Conference*.

Zhou, H., and H. Hu. 2004. Human motion tracking for rehabilitation—A survey, *Biomedical Signal Processing and Control* 3(1): 1–18, http://www.sciencedirect.com/science/article/pii/S1746809407000778

Ziv, N., A. Granot, S. Hai, A. Dassa, and I. Haimov. 2007. The effect of background stimulative music on behavior in Alzheimer's patients. *Journal of Music Therapy* 44: 329–43.

# 12

# *Experiences in Developing a Wearable Gait Assistant for Parkinson's Disease Patients*

**Marc Bächlin, Daniel Roggen, Meir Plotnik, Jeffrey M. Hausdorff, and Gerhard Tröster**

**CONTENTS**

Introduction ............ 304
Freezing of Gait ............ 304
Limitations of Pharmacological FOG Treatment ............ 305
The State of the Art in FOG Treatment ............ 305
Personal Health Assistant ............ 306
Initial Insight and Our Contribution ............ 307
Technological Evaluation ............ 309
Identification of Potential Sensor Modalities ............ 309
Sensor Selection ............ 311
Laboratory Prototype ............ 312
The Wearable Computer ............ 313
Context Sensors and Annotation ............ 314
Prototyping of Context-Aware Applications ............ 315
Platform Adaptation for an FOG Assistant ............ 316
Sensor Selection, Configuration and Placement ............ 318
Online Context Recognition of Freeze ............ 318
User Feedback ............ 318
Controlled Clinical Proof-of-Concept Study ............ 318
Organizational Steps ............ 320
Participants ............ 321
Validation Protocol ............ 321
Technical Validation Results ............ 323
Subjective Validation Results ............ 324
Lessons Learned and Future Steps ............ 326
Future Steps ............ 331
Conclusion ............ 332
References ............ 334

## Introduction

Parkinson's disease (PD) is a common neurological disorder caused by the progressive loss of dopaminergic and other subcortical neurons (Braak et al. 2004). PD patients often suffer from impaired motor skills (Jankovic 2008). Besides a flexed posture, tremor at rest, rigidity, akinesia (or bradykinesia) and postural instability, motor blocks are a common negative effect of PD. Motor blocks (freezing) most commonly affect the patients' legs during walking and are generally referred to as *freezing of gait* (FOG). Clinical assessment of PD is largely based on subjective patient reports. The Hoehn and Yahr (H&Y) scale* is commonly used to describe symptoms of PD progress (Hoehn and Yahr 1967).

### Freezing of Gait

FOG typically manifests as a sudden and transient inability to move. PD patients who experience FOG frequently report that their feet are inexplicably glued to the ground during the FOG episodes (Schaafsma et al. 2003). FOG is difficult to measure as it is highly sensitive to environmental triggers, cognitive input and medication. For example, FOG occurs frequently at home and much less frequently in the doctor's office or in a gait laboratory (Nieuwboer et al. 1998). Evaluation of FOG conditions is usually done using an FOG questionnaire (FOG-Q; Giladi et al. 2000). Five subtypes of freezing have been described by Schaafsma et al. (2003): start hesitation, turn hesitation, hesitation in tight quarters, destination hesitation and open-space hesitation.

FOG usually appears during the late stages of PD. However, several studies observed FOG even at early stages of the disease (Lamberti et al. 1997; Giladi et al. 2001). About 50% of all PD patients regularly show FOG symptoms (Fahn 1995; Lamberti et al. 1997; Giladi 2001; Macht et al. 2007). Ten per cent of PD patients with mild symptoms and 80% of those severely affected regularly experience freezing. FOG occurs more frequently in men than in women and less frequently in patients whose main symptom is tremor (Giladi et al. 2001).

FOG has been associated mainly with PD but can be seen in other extrapyramidal movement disorders, such as progressive supranuclear palsy (PSP) and its subtype, the "pure freezing syndrome", and in vascular parkinsonism (Giladi et al. 1997; Factor et al. 2002). FOG entails grave mobility problems that have a severe impact on patients' health-related quality of life (Boer et al. 1996). Patients who suffer from frequent freezing episodes often report a

* The Hoehn and Yahr (H&Y) scale uses five stages to indicate the relative level of disability: Stage 1: Symptoms on one side of the body only. Stage 2: Symptoms on both sides of the body; no impairment of balance. Stage 3: Balance impairment; mild to moderate disease; physically independent. Stage 4: Severe disability but still able to walk or stand unassisted. Stage 5: Wheelchair-bound or bedridden unless assisted.

negative influence on their functional independence, including a reduction in mobility and social activities (Boer et al. 1996; Martinez-Martin 1998).

Due to their erratic and unpredictable nature, FOG episodes often lead to falls (Bloem et al. 2004), a major cause of morbidity and mortality in PD. Hip fractures are a severe consequence of falls, as these are associated with a high morbidity and mortality in PD (Coughlin and Templeton 1980). In fact, within 10 years after diagnosis of PD, about 25% of patients will experience hip fracture (Johnell et al. 1992). Therefore, the motivation to try to avert FOG episodes is high, and technological solutions such as the present device are particularly suitable, given the motor dynamics of the FOG episode, which can be identified automatically.

## Limitations of Pharmacological FOG Treatment

Pharmacological management of PD is difficult and often ineffective at relieving FOG. The most common form of treatment used to manage motor symptoms in PD patients is levodopa (LD). The effect of LD on parkinsonian symptoms wears off over time, and the effective periods varies between 2 and 6 h. In some patients, this wearing-off effect is expressed in a gradual deterioration in motor performance; in others, the deterioration is relatively sharp and unexpected. The latter patients are known to exhibit motor response fluctuations. For these patients, clear ON and OFF periods can be distinguished, where ON periods are times when the medication is effective, and OFF periods are times when it is no longer effective. As the disease progresses, the effective duration of each drug dose shortens, and more frequent LD administration is necessary (Giladi et al. 2001). In addition, the development of involuntary movements (i.e., dyskinesia) and the OFF/ON phenomenon further limit mobility and complicate dosing.

Although FOG episodes generally appear more frequently during the OFF state, gait deficits in PD patients are often resistant to pharmacological treatment (Bloem et al. 2004). Therefore, effective nonpharmacological treatments need to be developed as an adjunct therapy to relieve symptoms and improve mobility.

## The State of the Art in FOG Treatment

Various behavioural "tricks" have been developed by clinicians and patients to overcome freezing attacks. These tricks include marching to command, stepping over a walking stick or cracks in the floor, walking to music or a beat and shifting body weight. Such external cues are commonly considered effective in alleviating FOG symptoms in PD patients (Hashimoto 2006; Okuma 2006).

Lim et al. (2005) performed an extensive review of the effects of external rhythmical cueing on gait in PD patients and found strong evidence for improvements in walking speed with the help of auditory cues. Insufficient evidence was found for the effectiveness of visual and somatosensory cueing. Similarly, Nieuwboer et al. (2009) showed that auditory cueing is more advantageous than visual and somatosensory cueing.

Rhythmic auditory stimulation (RAS) was shown to be particularly effective at improving gait among PD patients. Regular metronome ticking sounds were applied as RAS with a rate of 110% compared to the natural walking rate of the tested patient. This served to enhance patients' gait speed and reduced gait variability (i.e., it improved gait stability; Hausdorff et al. 2007). Nevertheless, there was no relative advantage to using this method to improve gait in patients with PD who also suffer from FOG (PD + FOG) compared to PD patients who do not suffer from FOG (PD-FOG; Nieuwboer et al. 2009). Interestingly, a study in which PD + FOG patients used the metronome recordings for cueing at home showed no effect in reducing the freezing symptoms (Cubo et al. 2004).

## Personal Health Assistant

Driven by cost and quality issues, the health systems in developed countries will undergo a fundamental change in this decade (Tröster 2005). We believe that the physician-operated and hospital-centred health system will be extended by a consumer-operated personal prevention, early risk detection and wellness system. INDEPENDENT* also took the position that technological developments can be beneficial to the quality of life (Sixsmith et al. 2007).

A "personal health assistant" (PHA) comprises a wearable sensing and communicating system, seamlessly embedded in one's daily outfit. Several on-body sensors identify the biometric and contextual status of the wearer continuously. The embedded computer generates individualized feedback to the user and to the surroundings, affording an effective means of prevention, disease management and rehabilitation.

Examples are the healthwear developed by the Massachusetts Institute of Technology Media Lab.† The healthwear is a wearable system with sensors that can continuously monitor the user's vital signs, motor activity, social interactions, sleep patterns and other health indicators (Pentland 2004). Lorincz et al. (2004) developed CodeBlue, a common software infrastructure, to address the challenges of resource limitations in protocol design, application development and security models. CodeBlue integrates sensor nodes and other wireless devices into a disaster-response setting and provides facilities for ad hoc network formation, resource naming and discovery, security and in-network aggregation of sensor-produced data.

---

* INDEPENDENT was an interdisciplinary project to explore the potential of technology and design solutions to enhance the quality of life of people with dementia, to help them to live independently and to empower them without compromising their rights or privacy.

† http://hd.media.mit.edu

## Initial Insight and Our Contribution

The limitations of pharmacological approaches and the research results on nonpharmacological treatment show potential for alternative approaches. This was our initial insight and presented us the opportunity for a wearable gait assistant for PD patients which provides the appropriate support when required. Starting with such an initial insight the development of a concrete system consisted of several stages, shown in Figure 12.1.

In the following three sections, we present the first three stages done towards the following goals:

1. Stage: Technological evaluation
   a. Identification of potential sensor modalities and signal-processing approaches for context awareness
   b. Pruning of the solution tree, by including user aspects such as wearability, comfort, energy use
2. Stage: Laboratory prototype
   a. Estimation of the system's robustness, with respect to hardware, software and algorithms
   b. Assessment of the robustness of the context recognition
   c. Estimation of the operating conditions of the context-recognition algorithm (e.g., identify for which modes of locomotion the selected approach may not work)
   d. Identification of the key algorithmic parameters, in order to adjust the system characteristics (e.g., sensitivity, latency) to the patients
3. Stage: Clinical trial
   a. Assessment of the system's behaviour in realistic conditions, in particular context awareness
   b. Identification of the engineering problems revealed by real-world testing
   c. Collection of the real-world data to adjust the system's parameters
   d. Collection of the user feedback on the technology and the system's operations: potential benefits, drawbacks, user wishes, suggestions or criticism

In the section "Lessons Learned and Future Steps" we summarize the best practice and lessons learned from our work and explain the next development steps towards a commercial system. In the final section we conclude our work and give an outlook on the next development steps.

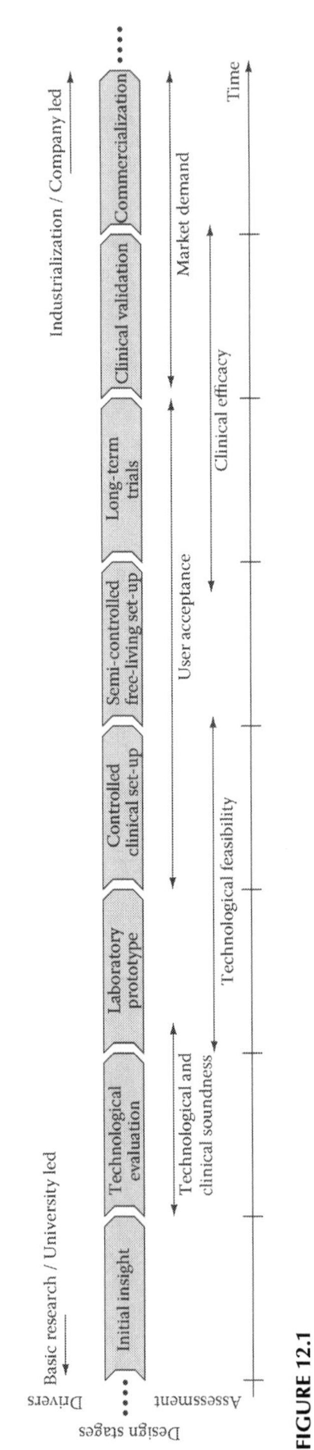


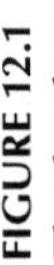

**FIGURE 12.1**
The development stages of a context-aware wearable assistant, highlighting the major assessment at each design stage.

## Technological Evaluation

As we have seen in the introduction, the identification of a need from the clinicians' and patients' side together with new scientific knowledge into a medical problem triggered an initial insight, where the concept of a context-aware health-assistant was drafted. From a product perspective, we envisioned the following assistant: a highly miniaturized, wearable system composed of a limited, optimized number of unobtrusive wireless on-body (or clothing-integrated) sensors and an audio feedback capability unobtrusively integrated, for instance, in an earplug. From an engineering perspective, a key question was the technical soundness of the concept. Since context awareness is the key enabler of the assistant, special attention was placed on the exploration of on-body sensing and pattern recognition for context awareness.

In order to help PD + FOG patients during daily life, we envisioned a device that can provide context-aware acoustic feedback to assist the patient. As freezing is linked to the user's context, such a wearable assistant ought to be context-aware. It must be capable of recognizing the situation of freeze. Furthermore, it may take advantage of contextual clues that indicate the onset of a freezing event to achieve a rapid reaction.

### Identification of Potential Sensor Modalities

FOG is a sudden and transient inability to move. Bonato et al. (2004) presented the first evidence that data mining and signal processing allow the presence and severity of motor dysfunctions in PD patients to be recognized. Hausdorff and colleagues (2003) examined the ground reaction force signal, measured with force-sensitive insoles in the shoes worn by PD patients who were walking normally or experiencing FOG episodes. Using time series and fractal analysis methods, they found that FOG is not a frozen akinetic state, nor is freezing a random, uncorrelated attempt to overcome motor blockades (Hausdorff et al. 2003). Instead, the measured force signals oscillated in an organized pattern. More recently, Moore et al. (2008) measured the vertical acceleration of the left shank of 11 PD patients and analysed the power spectra over 6 s signal intervals. They discovered that high-frequency components of leg movements were present during FOG in the 3-8Hz frequency band, and were not apparent during normal standing or walking. Moore introduced a freeze index (FI) to identify objectively FOG events offline. This FI is defined as the power in the "freeze" band (3–8 Hz) divided by the power in the "locomotor" band (0.5–3 Hz). FOG is detected using a "freeze" threshold. FI values above this threshold are identified as FOG events. Relevant sensors include force-sensitive insoles and acceleration sensors.

Furthermore, we identified the following context situations with other potential sensor modalities (Bächlin, Roggen, Plotnik, Inbar et al. 2009):

- Situational aspects
- Local aspects
- Cognitive-affective aspects
- Physiological aspects

In the following, we explain each of these aspects, together with the relevant sensor modalities, which might help to improve the context-aware feedback.

*Situational aspects:* Schaafsma et al. (2003) analysed the occurrence of FOG depending on the type of walking situation. This work led to a detailed characterization of the occurrence of freezing in the following conditions (the numbers in parentheses give the prevalence of freezing episodes for the corresponding situation): turns (63%), start of walking (23%), walking in narrow spaces (12%), upon reaching destination (9%) and walking in an open runway (3%). Further insights, although not quantified, show that freezing may also occur while crossing narrow spaces (e.g., entering an elevator) and that freezing is more common in crowded places (Bloem et al. 2004). With sensors to detect the walking situation, these numbers may be used as domain-specific knowledge. Together with the prior probabilities of freezing, it may allow adjusting the freezing-detection algorithm according to the situation of the user. Based upon user preferences, the acoustic feedback may even be automatically activated upon entering, e.g., narrow passages as a preventive measure. Relevant sensors include gyroscopes to detect turns and proximity sensors such as laser or ultrasound devices placed on the shank to scan around the user to detect tight quarters or an open runway. A dense presence of people can be sensed by the same proximity sensors (i.e., a naive detection of the number of obstacles around the user at close distance).

*Local aspects:* Patients also report that they often experience FOG at the same location in their everyday surroundings. Therefore, using the patient's current location and his location history in conjunction with his previous instances of FOG will be a good way to predict the conditions of further FOG instances. To determine the location of a person, there are several possibilities including global positioning system (GPS), Wi-Fi and cellphone tower signals, compass, ultrasound sensors, infrared proximity sensors, accelerometer sensors and cameras. GPS, a satellite-based positioning system, allows outdoor global localization but does not operate in difficult environments (e.g., indoors, urban canyons). Wi-Fi or cellphone tower signals allow localization indoors as well as outdoors (La Marca et al. 2005; Otsason et al. 2005), but they generally need a map of the radio beacon location. Accurate localization is possible in environments instrumented with localization beacons (e.g., radio, ultrasound). Absolute localization (e.g., satellite) combined with dead reckoning using body-worn sensors is a way to address this issue. Dead reckoning in wearable computing has been investigated in combination with a compass to estimate the heading direction and velocity-estimation methods, e.g., step counting (Lee and Mase 2002; Randell et al. 2003; Schindler et al. 2006). We introduced, characterized and tested a vision-based dead reckoning system for wearable computing that allows tracking of the user's trajectory in an unknown and non-instrumented environment by integrating the optical flow (Roggen et al. 2007).

*Cognitive-affective aspects:* Attention, which may be related in part to stress, and cognitive loading (e.g., dual tasking) are also known to increase the likelihood of freezing. This is often used in clinical studies to enhance artificially the number of observed freezes (Bloem et al. 2004). Further factors affecting freezing include stress, anxiety, depression and cognitively challenging situations (Giladi and Hausdorff 2006). While the recognition of stress or cognitive load remains challenging, especially outside of the laboratory, a number of modalities are promising to investigate as a way to assess the cognitive-affective state of the user and eventually adjust prior probabilities of freezing. Relevant sensors include galvanic skin response (GSR), Electrocardiography (ECG), blood pressure and Electrooculography (EOG) sensors. Stuyven et al. (2000) showed that eye movement and blink rate as measured by an EOG sensor are related to cognitive load. Our ongoing work towards wearable EOG goggles will eventually allow for long-term gaze tracking and blink recognition outside of the laboratory (Bulling et al. 2009). It is known that stress and fear are reflected in the physiology, such as galvanic skin response, blood pressure and heart rate. We showed that galvanic skin response related to the fight-or-flight reflex can be sensed despite moderate levels of physical activity, which is a promising step towards eventually detecting fear in natural conditions (Schumm et al. 2008). Other work showed some success in deriving stress from multimodal sensors (Benoit et al. 2009).

*Physiological aspects:* Offline analysis of the gait prior to freeze evidenced abnormal stride length and cadence during the three steps prior to freezing (Nieuwboer et al. 2001). The online detection of such anomalies in the gait may eventually be used to trigger the cueing shortly before the freeze occurs. Bartsch et al. (2007) exhaustively reviewed methods to characterize gait in PD patients, such as synchrony or fluctuation analysis. Finally, preliminary insights suggest an increase in heart rate at the onset of freezing compared to 10 s before the freeze (Meidan et al. 2009). Ongoing analyses aim to identify up to which time prior to freezing the heart rate increases. The detection of a heart rate increase not correlated with a physical activity increase during walking may thus suggest the onset of freeze.

## Sensor Selection

As the identification of potential sensor modalities has shown, there are multiple ways in which to achieve context awareness. During the early stages of the project, which tended to be exploratory, we had to choose the most promising kind and number of sensors. Choosing how to best recognize context and react to it is a trade-off between robust context recognition, effectiveness and wearability. For our first wearable context-aware health assistant, we decided on the sensor modality based on the following three factors

*Technological feasibility:* The sensing, context-recognition and feedback principles must be sound and realizable in the envisioned form factor. The system must be robust for real-world use. It must be low power enough for the envisioned operation time of about 2 h in a clinical setting and up to a maximum of 1 day in subsequent use at home.

*User acceptance:* The wearable system must be comfortable to wear, and the sensors should not influence the patient, either due to their physical dimension or due to their attachment method.

*Clinical efficacy:* The wearable assistant must have a demonstrable efficacy, and safety issues must be addressed.

During the initial insight stage, we set out to investigate the effectiveness of context-aware cueing for patients. Thus, we were most interested in detecting FOG occurrence. The system should detect the onset of freezes from the sensors and trigger RAS until the subject resumes walking. Sensing contextual factors indicative of potential upcoming freeze events would require obtrusive sensing modalities (e.g., heart rate) that would affect wearability. They were put aside until context-aware cueing based on simpler context recognition modalities had been evaluated in a clinical set-up. Thus, the wearable assistant will provide RAS only during an actual or impending FOG event. The device will act as a context-aware wearable assistant that activates only when necessary and remains transparent in other situations.

We selected motion sensors because they offer rich information and are unobtrusive. They directly measure limb-movement patterns related to gait freeze but also provide information on modes of locomotion (Veltink et al. 1996) and general physical activity. With additional processing, they may provide information about gait irregularity prior to freeze. We chose wireless sensors because they allow rapid comparative evaluation of different on-body sensing locations. At the cost of transparency or wearability, the platform should log raw sensor data and system behaviour for post hoc analyses.

## Laboratory Prototype

The exploratory nature of the early stages of development of a wearable assistant requires flexibility in the choice of sensors concerning their modalities, numbers and placement on the body. Further, enough computational power must be available to assess and develop context-recognition algorithms of various complexities. Nevertheless, the platform should remain small enough and offer sufficient battery lifetime to be unobtrusive when tested with patients.

In our laboratory development process, we typically use laptops or netbooks for preliminary investigations. In clinical and free-living trials, we favour a custom wearable computer that can be optimized for small size, long battery life and wearability. In the following three subsections, we first present the individual building blocks, and we then show how they were integrated, configured and programmed for the application of FOG assistance in the subsection "Platform Adaptation for an FOG Assistant". Within the laboratory prototype stage, we comparatively evaluated different sets of candidate sensors and processing methods in simulated conditions on colleagues and ourselves. Although no patients were involved, these recordings in simulated conditions benefited from real data. Our strategy to achieve smooth transitions through these platforms is (i) to have a set of sensors that can be interfaced seamlessly across platforms and (ii) to achieve a common code base for software and context-recognition algorithms. For these reasons, we chose wireless miniature sensor nodes based on Bluetooth's RFCOMM layer, a Linux operating system across the boards and an open-source context-recognition framework based on C/C++ code.

### The Wearable Computer

Within the DAPHNet project,* we developed a platform for combined physiological signal acquisition and context awareness. The platform is wearable, has extended battery life, offers enough computational power to run context-recognition algorithms and is flexible so that a variable number of sensors can be added to it. The core of the platform is a custom embedded wearable computer that interfaces to the sensors and to the user. It is based on an Intel XScale PXA270 processor at 400 MHz with 256 MB RAM and 64 MB FLASH for the kernel and the file system. It runs a Linux 2.6.20 operating system. It offers processing power comparable to an ultra-portable PC. The computing system is modular at the hardware level in order to realize different feedback and sensing modalities. The system offers by default USB and Bluetooth as extension interfaces, allowing connections to a diversity of physiological and nonphysiological sensors (Bächlin et al. 2007). The system can be extended by ZigBee or ANT wireless interfaces with USB dongles. In order to avoid protruding parts and unintended disconnection of these dongles, we provided an internal USB bay within the system's housing. Furthermore, there are space and interconnection possibilities for an

* DAPHNet stands for Dynamic Analysis of Physiological Networks and is a Future and Emerging Technologies Project supported by the European 6th Framework Program, under Grant 018474-2.

internal Printed Circuit Board (PCB) extension board. This extension board is interfaced to the CPU over a serial link and to a frontal 3.5 mm jack. It is meant to prototype various signal-acquisition and conditioning hardware (e.g., for ECG or galvanic skin response sensing) or to provide user feedback. An MMC memory slot is provided to store data logs.

Due to the general-purpose and energy-efficient processor, the system allows running online algorithms while the power consumption is between 1.5 and 2 W. On the 3.7 V, 3.3 Ah battery the system can run for more than 6 h. The packaged wearable computer is 132 × 82 × 30 mm in size and weighs 231 g. The complete miniaturized computing system can be integrated in the buckle of a belt (Amft et al. 2004). The platform interfaces to custom or commercial sensors, any number of which can be combined to acquire information about the user's context, in particular the user's physiological state and physical activity parameters.

## Context Sensors and Annotation

For high-quality acquisition of physiological parameters, we developed a driver for the Mobi system from TMSI* that meets our usability design goals concerning the physiological signal acquisition most closely. The Mobi system is a compact (114 × 98 × 37 mm, 165 g) sensing unit designed for the particular task of monitoring physiological signals. The system is medically certified, and it allows acquisition of eight analogue physiological signals. Four of the eight channels sampled with up to 2048 Hz, the others with 128 Hz. It includes, in particular, sensors for ExG (i.e., ECG, electromyography or EOG), $SpO_2$ (blood oxygen saturation), galvanic skin conductivity, respiration and skin temperature. Sensor data are streamed over a Bluetooth wireless link.

For physical context acquisition (manipulative gestures, modes of locomotion, physical activity), we developed a miniature motion sensor including a three-axis accelerometer (ADXL330) and a two-axis gyroscope (IDG650). The combination is commonly used for manipulative gesture recognition and gait analysis (Pappas et al. 2004; Bamberg et al. 2008). The motion sensors are 22 × 41 × 12 mm in size (with packaging: 25 × 44 × 17 mm) and weigh less than 22 g with a rechargeable 300 mAh Li-ion battery. The battery lifetime is 6 h (Bächlin et al. 2007).

Further, a custom ECG sensor for applications requiring custom on-body signal processing was developed. It may be used for online heart-rate variability analysis or online artefact detection directly in the node. The potential difference from two electrodes is picked up by a low-gain (10×) instrumentation amplifier and then low-pass filtered ($f_c = 35$ Hz) and amplified (380×) for reading by the ADC converter. In addition, a hardware eighth-order elliptic filter ($f_c = 35$ Hz) may be activated.

* Twente Medical Systems International BV, Netherlands.

In order to collect annotations* during studies, we developed a wireless keypad. It is typically used by the experimenter but may also be used by the subject to mark moments of interest (e.g., if the system does not operate as expected). It behaves as a virtual "sensor" by streaming key presses to the wearable computer. When the software is run on a standard computer or laptop, annotation is also supported with a regular keyboard.

All the custom sensor nodes share (i) a common architecture based on an Atmel ATmega324 microcontroller, where only the sensing part differs; (ii) a common firmware, where only node-specific aspects are configured; and (iii) a common data transmission protocol on top of the Bluetooth RFCOMM layer. Motion sensors and ECG sensors also share a common form factor, which allows them to share the same plastic case. All custom sensors and the keypad support data streaming at user-configurable speeds from 8 to 256 Hz. The data transmission protocol is a frame-based protocol including a start header, the sensor ID, a packet counter, the payload and a 16-bit Fletcher checksum. By streamlining the hardware and firmware, we simplify the development of additional sensor nodes that conform to the node architecture and communication protocol, and hence we can seamlessly acquire their data in our context-recognition framework.

All our custom wireless devices described here are available as "open hardware" and technical details are presented in Roggen et al., 2010.

## Prototyping of Context-Aware Applications

We proceeded in two steps to rapidly prototype context-aware applications. During the technological evaluation phase, we acquired data with laboratory prototypes. The data were used for a first offline evaluation of the context-recognition algorithms in a post hoc process using rapid analysis tools (e.g., MATLAB®). In the laboratory prototype stage, we implemented these algorithms to run online (real-time) on the wearable computer for the trials in a clinical set-up. In order to minimize the differences in the code base, we harnessed the CRN Toolbox (context-recognition toolbox), an open-source data acquisition and processing framework developed by the University of Passau (Bannach et al. 2008). It is based on a representation where data flow through interconnected processing blocks written in C++. In the first step, it allowed us to acquire data efficiently from multiple sensor modalities using data acquisition and storage blocks. In the second step, we used it to do online processing by developing the appropriate processing blocks.

In order to achieve a high degree of reliability for the overall system, we modified the data acquisition logic of the CRN Toolbox with an architecture that allows for sensor disconnection and automatic reconnection at

* Annotations or labels are used during the development of a context-aware system: they mark moments of interest, such as when the user performs a specific activity. Annotations together with the corresponding sensor data are then used with machine learning techniques to train classifiers that link sensor signal patterns to specific contexts.

runtime. Thus, individual sensors that fail, or disconnect due to a poor radio link, do not affect the rest of the system. Once the radio reception betters, the sensors are automatically reconnected. Discovery and interconnection of sensor nodes are also simplified: which devices are on or off, and the order in which they are activated, is not important, which minimizes human error during experiments.

When multiple subsystems are jointly recorded or processed online, it is important for the data streams of the sensor subsystems to be temporally aligned (synchronized)*. When data transmission latency is known and jitter null, such alignment is straightforward. However, a smooth data flow is not guaranteed when data are transmitted over a wireless link such as Bluetooth. This is mostly related to buffering in the radio chips and wireless transmission errors. The Bluetooth L2CAP layer ensures that data are retransmitted in case of errors. This leads to a variable latency (jitter) between the moment when the data are acquired by the sensor node and the moment when the data are received by the wearable computer.

We modified the data acquisition logic of the CRN Toolbox to ensure synchronized acquisition of data from multiple sensor subsystems. We use a rate control mechanism based on a FIFO buffer similar to that used by Westeyn et al. (2009). The rate control algorithm aims at keeping the "fill level" of the FIFO buffer at a nominal level that corresponds to the same duration across all sensors domains (typically 0.5 s, i.e., 64 samples at 128 Hz sample rate, or 32 samples at 64 Hz). Data are placed in the buffer immediately upon reception. Data are read out at the guaranteed regular read-out rate, *R out* (i.e., typically higher than or equal to that of the highest-sampling sensor node). The average rate of incoming data, *R in,* is estimated from the incoming data by low-pass filtering the time difference between samples. The data samples in the buffer are read-out R out/R in times. Thus, samples coming from sensors with an incoming rate lower than the read-out rate are upsampled. Data samples are discarded if the read-out rate is lower than that of the connected sensors. A weak integral controller cancels out estimation errors in the incoming sample rate or read-out rate and ensures that the fill level of the buffers remains in average at the nominal level.

### Platform Adaptation for an FOG Assistant

We customized the prototyping platform for our FOG assistant by selecting the appropriate sensors, context-recognition algorithm and user feedback capability. Figure 12.2 shows the complete system worn by a patient. The computing system is attached to the belt placed on the trunk. Raw sensor data and context-recognition logs are stored for offline analysis on a 2 GB MMC card plugged into the wearable computer.

* E.g., when computing the angle between the upper and lower arm by processing the data from two sensor nodes measuring inclination placed on the upper and lower arm.

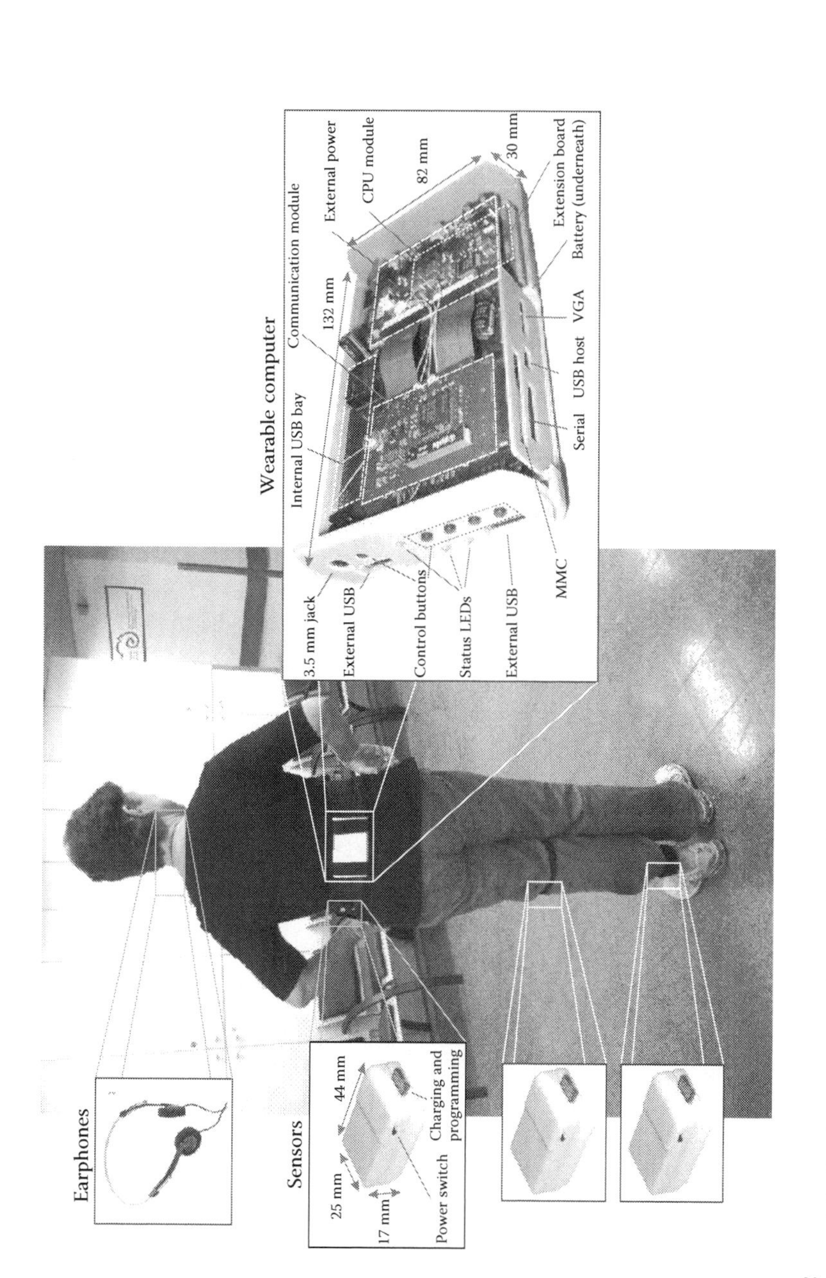


**FIGURE 12.2**
FOG-detection and feedback system worn by a patient. The system consists of a wearable computer, a set of acceleration sensors and headphones. Acceleration sensors are attached to the shank and the thigh using an elasticized strap and Velcro. A third sensor and the wearable computer are attached to the belt. The sensors are wirelessly connected to the wearable computer; the headphones are connected by a cable to the computer.

### *Sensor Selection, Configuration and Placement*

The sensor placement affects context-recognition accuracy and wearability. For an objective and subjective comparative analysis, we investigated multiple on-body locations during the trials in the clinical set-up. Two accelerometer sensors are attached to the shank (just above the ankle) and the thigh (just above the knee) using an elasticized strap and Velcro. A third sensor is attached to the belt of the patient. The acceleration signal is sampled at 64 Hz.

### *Online Context Recognition of Freeze*

We developed an online FOG-detection algorithm based on the principle described by Moore with emphasis on low latency. The flow chart of the algorithm, including the algorithm's parameters, is given in Figure 12.3. Only the shank sensor data have been used for online FOG detection. Sensor data coming from the acceleration sensor are windowed by a 256-sample window (4 s). The windowing is done in steps of 32 samples (0.5 s). On each window block, the first 32 frequency components (0–8 Hz) of the power spectrum of a 256-point Fast Fourier Transformation (FFT) are calculated. The FFT calculation is the main computational part in the whole algorithm.

The energy in the low-frequency part between 0.5 and 3 Hz is summed up, as well as the energy in the higher-frequency part between 3 and 8 Hz. Furthermore, the complete energy between 0.5 and 8 Hz is calculated by the addition of both parts. If the power content of the signal between 0.5 and 8 Hz is above the power threshold (PowerTH), the freezing index (FI) is calculated by dividing the energy in the freeze band by the energy in the locomotor band. For signal parts with a power content below PowerTH (standing parts), FI is set to zero. Finally, FOG is detected whenever the FI exceeds the freeze threshold (FreezeTH). A signal extract, together with the online calculated FI and the FOG-detection signal, is depicted in Figure 12.4.

### *User Feedback*

For the auditory feedback, we integrated a feedback module into the wearable computer plastic case and connected it to the frontal jack. Earphones are used to provide RAS at a rate of 1 Hz to the user when freeze is detected. The RAS is provided to the patient until the patient resumes normal walking.

## Controlled Clinical Proof-of-Concept Study

We conducted a clinical trial to evaluate our wearable assistant with real patients. This clinical trial was performed in the laboratory to test the system's

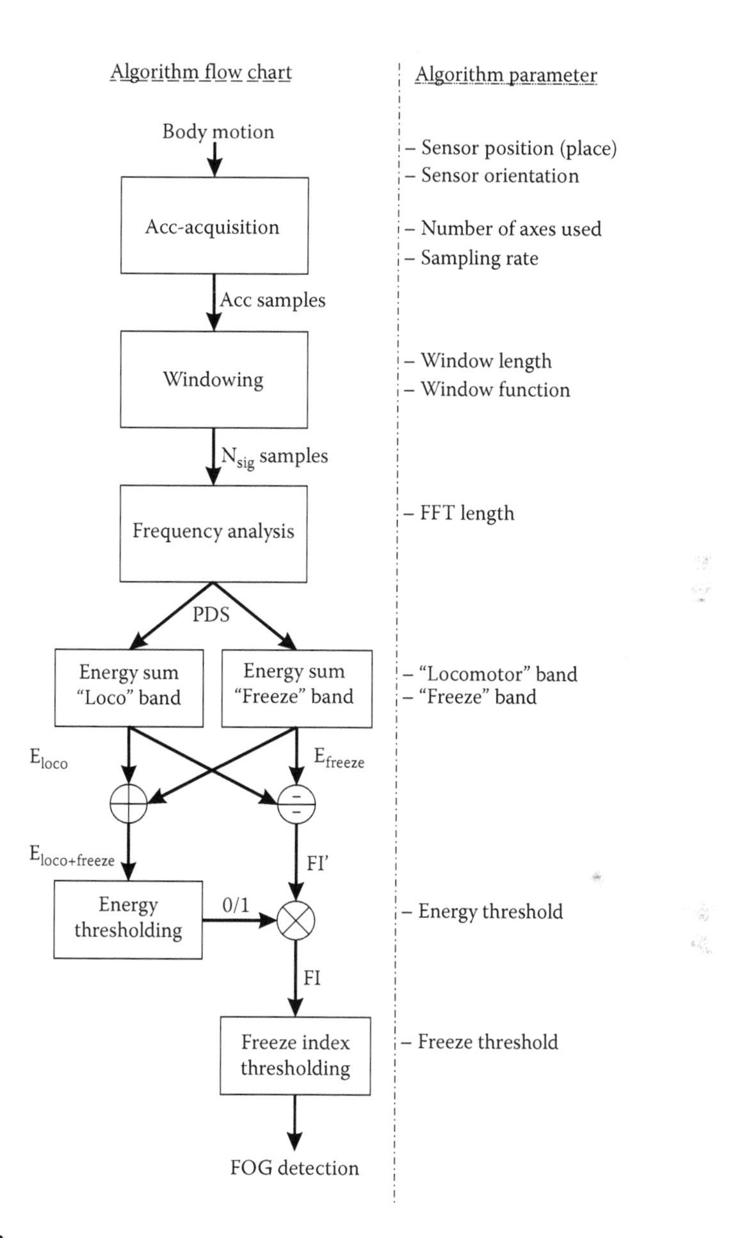


**FIGURE 12.3**
Flow chart describing the FOG-detection algorithm including all parameters.

behaviour in a standardized trial across all subjects who participated. Full annotations were made in real time, and full video and audio documentation was recorded in order to be able to analyse the trial in detail afterwards.

The objectives of this clinical evaluation stage were to understand the technical effectiveness of the system and collect users' subjective feedback about the system. These data were needed for the next design step of the

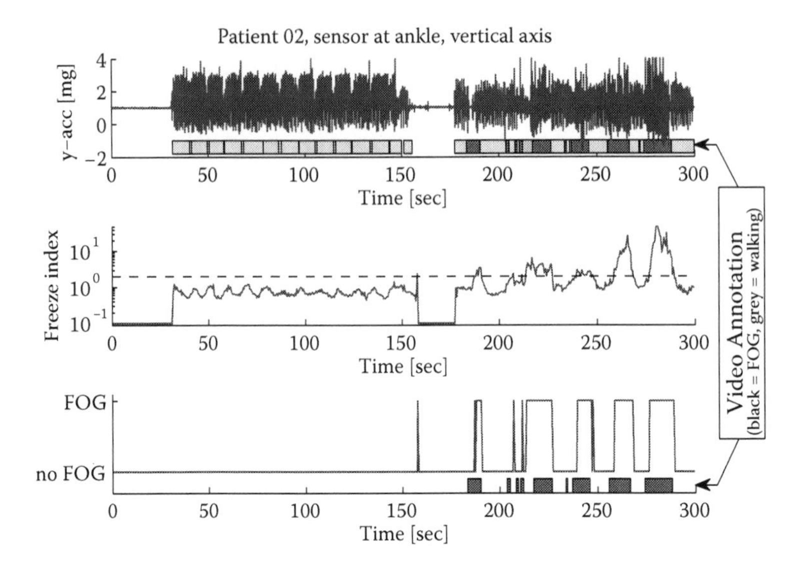


**FIGURE 12.4**
A 5 min signal extract from a PD patient (patient 02) demonstrates the FOG-detection algorithm. *Top,* the raw vertical acceleration of the shank during the study, with the FOG events annotated below it. *Middle,* the freeze index calculated from the power in the freeze band divided by the power in the locomotor band. *Bottom,* automatically detected FOG together with annotation as in the top graph. In this example, most FOG events were identified with little delay. Insertions (time 150 s), fragmentations (time 200 s) and deletions (time 240 s) are also visible.

system towards the free-living set-up, or it could lead to a reconsideration of the sensing and feedback modalities, form factor or concept. If the conclusions were unsatisfactory and modifications were required, the clinical trial would need to be repeated.

## Organizational Steps

The organization of a clinical trial comprises several steps such as the ethical approval and the recruitment process. This can easily take several months and is therefore one big task within the whole development process. In order to give the reader an impression of this task, we explain in this subsection our experience with organizing our clinical proof-of-concept trial.

Three months prior to the planned experiments, we submitted an application to the institutional committee of ethics on experiments with participation of human subjects. This required that all details of the experiment were finalized and agreed upon between both research collaborators, the engineering and medical disciplines, prior to the application. This planning process lasted about 3 months. Drafting the application, including its circulation for internal comments and corrections, took about 2 weeks. One month prior to the planned start date of the experiments, the local Human

Subjects Review Committee approved our study as being in accordance with the ethical standards of the Declaration of Helsinki. The remaining month could now be used for the recruitment of the subjects.

Staff members of the Laboratory for Gait and Neurodynamics recruited 10 patients with PD. Candidates were selected based on existing lists of clinical patients who had expressed their willingness to participate in research efforts in the past. They were first contacted by phone to give them a general description of the experiments and to explain to them the goals of the study. If there was a positive initial reaction, a staff member visited the subject in order to ensure that the subject's clinical condition adhered to the inclusion criteria and that none of the exclusion criteria was relevant to the particular candidate. In addition, detailed explanations were offered about the protocol itself and about potential difficulties that the subject might experience during the experiments. A small, symbolic financial contribution was offered to the subjects as a token of appreciation for their time and efforts, along with full coverage of their travel expenses. Finally, technical details were discussed with the subjects who agreed to participate in the experiments. These included (1) determination of a convenient time; (2) instructions about when they had to take their last anti-parkinsonian medications on the day prior to the experiment, to ensure that all patients would be OFF their medication (at least 12 h since the last uptake); and (3) general instructions regarding comfortable clothing (e.g., comfortable walking shoes for experiments on ambulation).

## Participants

Ten idiopathic PD patients (7 males) with a history of FOG (66.4 ± 4.8 years of age; Hoehn-Yahr score [H&Y] of 2.6 ± 0.65 in ON mode) took part in this study. Patients were excluded in the recruitment process if they had severe vision or hearing loss, dementia or signs of other neurological/orthopaedic diseases. Motor performance among PD patients generally shows large variability. This was also the case among the group of patients who participated in this study. For example, during nonfreezing episodes, some patients maintained regular gait that could hardly be distinguished from that of healthy elderly people, while others had a slow and unstable gait. The patients' characteristics are listed in detail in Table 12.1.

## Validation Protocol

The study was carried out in the Laboratory for Gait and Neurodynamics at the Department of Neurology of the Tel Aviv Sourasky Medical Center (TASMC). Patients were tested in the morning during the OFF stage of the medication cycle (more than 12 h after their last anti-parkinsonian medication intake). The experimental session took about 1 h per patient. Patients arrived at the TASMC at 8:00 or 9:30 a.m. (two recordings per day). First,

**TABLE 12.1**

Patient Characteristics

| Subject ID | Gender | Age (years) | Disease Duration (years) | H&Y Rating in ON | Tested In |
|---|---|---|---|---|---|
| 01 | M | 66 | 16 | 3 | OFF |
| 02 | M | 67 | 7 | 2 | ON |
| 03 | M | 59 | 30 | 2.5 | OFF |
| 04 | M | 62 | 3 | 3 | OFF |
| 05 | M | 75 | 6 | 2 | OFF |
| 06 | F | 63 | 22 | 2 | OFF |
| 07 | M | 66 | 2 | 2.5 | OFF |
| 08 | F | 68 | 18 | 4 | ON |
| 09 | M | 73 | 9 | 2 | OFF |
| 10 | F | 65 | 24 | 3 | OFF |
| Mean | | 66.4 | 13.7 | 2.6 | |
| ± SD | | ± 4.8 | ± 9.67 | ± 0.65 | |

*Note:* H&Y, Hoehn and Yahr scale.

patients were instructed on the experiment, and it was explained to them how they could take advantage of the RAS in case of freezing. The study protocol had two sessions. During the first session the device recorded all the necessary data and performed the online FOG detection; however, the earphones were deactivated. The second session was identical to the first one, but the earphones were activated (feedback system activated). Both sessions consisted of three walking tasks designed to represent normal daily walking. They were (a) straight baseline walking in the laboratory hallway, including several 180° turns; (b) random walking in a reception hall space that included a series of initiated stops and several 360° turns (the examiner spontaneously instructed the subject to turn in different directions, for at least six turns, three in each direction); and (c) walking related to activities of daily living (ADL; the ADL task included entering and exiting rooms, going to the lab kitchen, getting something to drink and going back with the cup of water to the starting room). Each single walking task was performed for about 5–10 min. Participants walked without assistance, but with a therapist close by for safety reasons, at their own pace. At the end of the study patients returned to the examining room, took their medication, had a debriefing with the therapist and filled out a standardized self-report of patient satisfaction and a questionnaire to qualify the systems operation.

Figure 12.5 shows a photograph of the study with two physiotherapists and two assistants who were running and observing the study. All walking trials were recorded on a digital video camera. The leg-movement data were synchronized with the video recordings using three synchronization steps at the beginning of each recording session. One physiotherapist took notes of relevant events during the session. Another assistant annotated the patients'

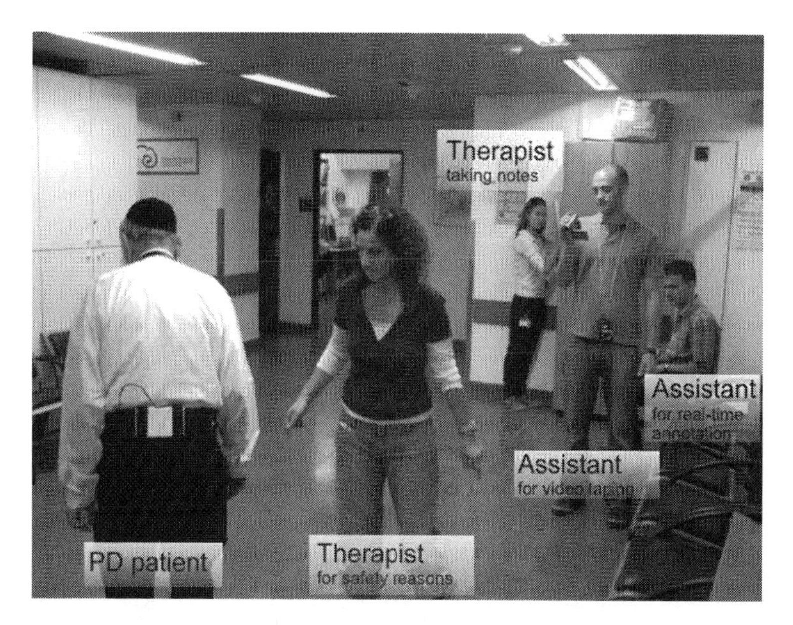


**FIGURE 12.5**
Snapshot of the study, depicting one PD patient, the therapist (near the subject for safety reasons) and the research assistants (further away from the patient) who documented the trials.

current activity (e.g., standing, walking, turning and freezing) in real time by pressing corresponding keys on a laptop computer.

## Technical Validation Results

Our wearable system worked fine throughout all trials, and 8 h 20 min of data were recorded. Eight patients out of the 10 exhibited FOG during our study; two patients did not have any freeze event. In a post hoc analysis, physiotherapists analysed the video recordings to identify FOG events and determine the exact start times, durations and end times. The beginning of an FOG event was detected when the gait pattern (i.e., alternating left-right stepping) was arrested, and the end of FOG was defined as the point in time at which the pattern was resumed. This procedure was similar to an earlier established one (Schaafsma et al. 2003).

The physiotherapists identified 237 FOG events (range 0–66 per subject; mean 23.7 [SD 20.7]) in the video recordings. The length of the FOG events ranged from 0.5 to 40.5 s. (mean 7.3 s [SD 6.7 s]). Half of the FOG episodes lasted less than 5.4 s, and the majority (93.2%) of FOG events were less than 20 s long.

Without user-specific calibration, the frame-based sensitivity and specificity of the online-detected FOG were 73.1% and 81.6%, respectively. The evaluation is based on 0.5 s frames. The reference for all our evaluations is the video annotation by the physiotherapists. A maximum detection delay

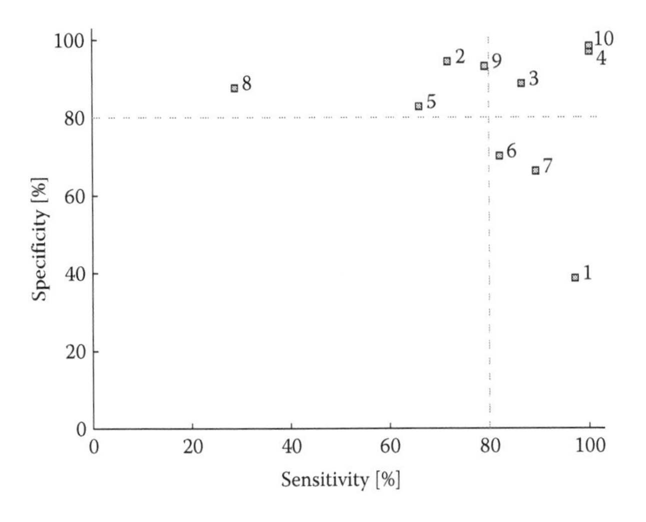


**FIGURE 12.6**
Sensitivity and specificity distribution for the online detection accuracy. Numbers correspond to the patients' identification numbers.

of 2 s is within the detection tolerance and evaluated as correct detection. Figure 12.6 depicts the detection accuracy of the device. For each patient, the sensitivity value (abscissa) and specificity value (ordinate) are plotted. The auditory cueing started properly whenever a FOG episode was detected and stopped again when the patient managed to get out of the freeze and the system detected normal gait.

If such a system becomes a "consumer appliance", provision for an initial calibration would be beneficial. By separating the patients into saccadic and smooth walkers with separate context-recognition parameters, a detection accuracy of 85.9% sensitivity and 90.9% specificity was achieved.

Table 12.2 lists the results for 12 combinations of three sensor positions (ankle, knee and hip) and four combinations of axes. The values are the average performance results of all 10 patients using the algorithm with global parameters (leave-one-out cross-validation). The comparative analysis of on-body sensor locations showed little difference in context-recognition accuracy. The best result was achieved when using the vertical axis of the sensor at the knee. A detailed performance characterization based on the parameter settings is presented and discussed by Bächlin, Roggen, Plotnik, Hausdorff et al. (2009).

## Subjective Validation Results

For a subjective evaluation of the system, we asked the participants to fill out a standardized self-report of patient satisfaction and a questionnaire to qualify

**TABLE 12.2**

Sensor Position Evaluation

| | x | y | z | n |
|---|---|---|---|---|
| *Sensor at ankle* | | | | |
| Sensitivity | 87% ± 16% | 81% ± 14% | 80% ± 13% | 79% ± 15% |
| Specificity | 87% ± 14% | 87% ± 11% | 81% ± 19% | 86% ± 9% |
| *Sensor at knee* | | | | |
| Sensitivity | 76% ± 20% | 85% ± 13% | 82% ± 18% | 82% ± 15% |
| Specificity | 85% ± 16% | 88% ± 13% | 84% ± 20% | 83% ± 13% |
| *Sensor at hip* | | | | |
| Sensitivity | 81% ± 19% | 71% ± 25% | 78% ± 32% | 78% ± 19% |
| Specificity | 84% ± 28% | 79% ± 20% | 79% ± 22% | 80% ± 24% |

*Note:* x = horizontal forward, y = vertical, z = horizontal lateral and n = $\sqrt{x^2 + y^2 + z^2}$ = magnitude of all three axes.

the system's operation. The Visual Analogue Scale (VAS) and the Clinical Global Impression Change (CGIC) scale were used. Furthermore, there was always the possibility to make further comments for each question.

The VAS is a visual sliding scale with two anchor points, one at each extreme. One anchor point is at 0 (i.e., "worst") and the other at 10 (i.e., "best"). Respondents specify their level of agreement with a statement by indicating a position on the VAS between the two end points (Grant et al. 1999). Using the VAS, patients had to grade their walking performance before and after the study, the comfort of the system components and the usefulness of the system for their everyday life.

The CGIC is a 7-point scale that assesses how much the patient's performance or illness improved or worsened relative to a baseline state at the beginning of the intervention (Shelton et al. 2001). For example, in our questionnaire we used the scale to report the change in FOG duration due to using our system. The scale had the following seven anchor points: +3 = much longer; +2, longer; +1, minimally longer; 0, no change; –1, minimally shorter; –2, shorter; or –3, much shorter. Furthermore, the CGIC was used to rate the number of FOG events and whether patients preferred to hear the RAS more or less frequently. To evaluate the experiment from another perspective, the physiotherapists answered a complementary questionnaire at the end of the completed study. The physiotherapists were asked how they would rate the usefulness of the system, the influence on the patients' gait and the suitability for use in everyday life. Furthermore, the physiotherapists were asked whether they saw that patients benefited from the context-aware cueing and used it or whether it was disturbing (Bächlin, Plotnik et al. 2009).

All patients reported that the system was unobtrusive and did not interfere with locomotion. In addition, the physiotherapists did not see any indication that the patients' normal gait was disturbed by the physical size and weight of the sensors and the wearable computer. However, two physiotherapists pointed out that the size of the computing system and the attachment method to the belt should be improved for use in everyday-life studies. Table 12.3 shows the detailed survey questions and their results. Regarding benefits of the device, five out of eight patients who experienced FOG during the study said that they had fewer freezing events with the device. The three other patients could not see any change. Five patients had the impression that their freezing episodes were shorter with the device. Only one thought his episodes were longer than usual, and two could not determine any change. Half of the patients who experienced FOG during the study observed fewer and shorter FOG events. The physiotherapists rated the influence of the automatic identification of FOG events and RAS feedback as beneficial, especially for patients with severe FOG. With respect to the occurrence of the feedback, two patients expressed a preference to hear the RAS less often. In their case, the system was too sensitive, resulting in too many RAS occurrences. Their reaction tends to support the observation that continuous cueing is not appreciated by patients and that RAS should be context-aware. Participants for whom the detection sensitivity was low demanded to have RAS assistance more often. Low sensitivity resulted in missed FOG events in these patients, and therefore they did not always get the auditory assistance when experiencing FOG events. This tends to support the previous observation that patients felt a benefit from RAS. Three participants reported that the feedback occurrence was just right. One participant suggested introducing variations in the audio tone and rhythm to avoid patients becoming used to the system since he believed this could make the RAS feedback even more effective. Similarly, one physiotherapist suggested adjusting the tempo of the RAS according to the walking speed of the patient.

Six participants were optimistic that such a personal assistant could be helpful in their everyday lives. The other four participants said the trial was too short and they could not really judge the usefulness. The physiotherapists also saw potential in the system to support PD patients in their everyday lives. They thought the context-aware automatic RAS would be especially helpful for the PD patients experiencing long FG events. These patients tend to be more able to exploit the RAS and adapt to the rythm. Overall, the self-assessment indicates that some of the patients benefit from the assistive device.

## Lessons Learned and Future Steps

Testing context-aware technology with patients is challenging and costly in many respects: ethical issues must be cleared, patients must be recruited, and a large number of personnel are required during the study. In our case,

**TABLE 12.3**

Detailed Survey Questions and Results

| Questions | Patient ID | | | | | | | | | |
|---|---|---|---|---|---|---|---|---|---|---|
| | 01 | 02 | 03 | 04 | 05 | 06 | 07 | 08 | 09 | 10 |
| How do you grade your walking now? [1 = worst walking / 10 = best walking] | 5 | 8 | 3 | 7 | 5 | 8 | 5 | 5 | 3.5 | 7.5 |
| How do you grade your walking during the experiment? [1 = worst walking / 10 = best walking] | 6 | 6 | 5.5 | 6 | 5.5 | 9 | 5 | 5 | 6.5 | 7.5 |
| Did the belt with the computing device disturb you while walking? [1 = not at all / 10 = very much] | 1 | 1 | 1 | 1 | 1 | 1 | 1 | 1 | 1 | 1 |
| Did the sensor at the leg disturb you while walking? [1 = not at all / 10 = very much] | 2 | 1 | 1 | 1 | 1 | 1 | 1 | 1 | 1 | 1 |
| Did you have fewer/more freezing events with the device? [−3 = much less / 0 = no change / +3 = much more] | 0 | −1 | −1 | _ | 0 | −2 | 0 | −1 | −2 | _ |
| Did you have longer/shorter freezing episodes with the device? [−3 = much shorter / 0 = no change / +3 = much longer] | 0 | −1 | −1 | _ | 0 | −2 | −2 | +1 | −1 | _ |
| Would you prefer to have the assisting RAS more or less often? [−3 = much less often / 0 = no change / +3 = much more often] | −2 | +2 | +2 | _ | +1 | 0 | 0 | −2 | 0 | 0 |
| Would you be able to trigger the assistant yourself by pressing a button on your wristwatch? [No / Sometimes / Yes] | N | Y | Y | Y | Y | Y | S | N | Y | N |
| Do you think such a system could increase your qualtity of life? [1 = not at all / 10 = very much] | 7 | 5 | 8 | _ | 9 | 5 | _ | _ | 8 | 5 |
| Would you accept wearing an acceleration sensor during your everyday life? [Yes / No] | Y | Y | Y | Y | Y | Y | Y | Y | Y | Y |
| Would you accept to ware EMG or ECG sensors glued to your skin during your every day life? [Yes / No ] | Y | Y | Y | Y | Y | N | N | N | Y | Y |

we needed one medical doctor, two physiotherapists and one experimenter. In order to maximize the technical success and ensure that meaningful data are obtained from trials, we present here the lessons learned and best practices from our work developing the FOG assistant.

*Testing procedure:* We extensively tested the whole system and individual components for hardware, software and firmware robustness. Our test procedure included

i. Testing of individual sensor nodes for hardware and firmware reliability by continuous week-long data streaming in static use in the charging station and on the body (recharging at night, on-body during the day). On-body testing allowed us to assess mechanical stability.
ii. Testing of robustness against simulated sensor failure, by purposely switching sensors off and on, resetting them, occluding the antenna and moving sensors far away from the receiver to cause disconnections. This showed that the software framework could continue to record data from the remaining sensors, while attempting to reconnect to the missing sensor without side effects.
iii. Burn-in testing of the complete data acquisition system during a 2-week-long mains-powered recording session with three streaming sensors.
iv. Laboratory testing on actors simulating FOG.
v. Final testing on the location of the trial with patients to ensure no radio interference. This may occur in hospitals where medical devices such MRI scanners are used.

Generally, testing wireless sensors in the target location is recommended whenever suspicions of interference exist, when comparatively high bandwidth is required or when a high density of wireless devices in the same frequency band is present. Furthermore, we favoured reusing sensor nodes that had already proved robust in previous experiments to minimize new risks.

*Sensor nodes:* Wireless sensors were essential for attaching the sensors rapidly to the patients in the clinical set-up and exploring different on-body locations prior to that. We found that our sensors based on the Bluetooth RFCOMM layer allowed a smooth transitions between the desktop computer, laptop and wearable computer, and also across operating systems (MacOS, Linux, Windows), since the sensors appear as regular serial devices. Bluetooth also takes care of error checking and retransmission, which removes many of the difficulties of devising a robust communication protocol. This allowed us to focus on the application.

Since we used the same frame-based communication protocol on top of RFCOMM for all custom sensors, we could develop a single generic data acquisition driver. This allowed us to reuse the same online data visualization software and data acquisition blocks regardless of the sensors.

The sensors were packaged in thermo-retractable plastic during the experiment. This form factor was accepted by the subjects, although they commented on their desire for a professional packaging for later stages of development. We addressed this by developing a rapid prototyping plastic casing afterwards.

*Offline to online algorithm transfer:* Our work is based on initial offline insights from Moore et al. (2008) showing an FI for post hoc analysis. Turning this algorithm into an online version required optimization (decreasing) of the latency of freeze detection to provide feedback quickly after the onset of a freeze. At the same time, the system must be robust against false positives. The initial algorithm had to be improved to avoid the incorrect detection of wilful standing (standing without freeze) as a gait freeze. The modifications required for robust online operation are usually difficult to predict from the literature. Thus, it is important to have computational headroom in the wearable system during the early development phase to improve context-recognition algorithms if needed and make comparative evaluations of different context-recognition principles.

*Software:* We used a C++ code on top of POSIX that allowed easy porting between a desktop PC, a laptop and our wearable computer. C and C++ are portable to many mobile platforms (e.g., the CRN toolbox was demonstrated on the iPhone platform; Bannach and Kunze 2009) and allow for many optimizations when running on resource-constrained devices without resorting to assembly language (Roggen et al. 2006). We achieved rapid prototyping despite the static nature of C++ by using the CRN Toolbox, a data flow–oriented framework based on easy-to-modify scripts representing data processing block interconnections.

*System training and calibration:* Context-recognition algorithms need to be trained and tested from data recorded on users that are in the contexts of interest (here, freezing). Mobilizing patients for this may be difficult due to costs, required manpower and ethical approval requirements. Thus, if available, data recorded in past work from sensors with identical modalities should be used. In our case, we used video material documenting FOG events in PD patients. Actors saw the video material and simulated in the laboratory the walking/freezing of the patients. By recording data from the actors, we could do an initial calibration of the system. During the clinical trial, we had the provision for a first rapid calibration session for the individual patients, based on a preliminary identification of the key algorithm parameters. In practice, we did not make use of this, as the parameters derived from the actors were adequate to recognize FOG in patients.

*Integration and flexibility:* We found that the level of miniaturization and integration was not critical in the clinical set-up, as long as the system could be attached quickly on the body and remained lightweight. In our study, we could attach the system in less than 10 min. The heaviest part, placed on the lower back, where load is easily tolerated, weighed less than 240 g, and sensors placed on limbs, where mobility is important, were less than 30 g.

No stringent requirements regarding the battery life were found, as most users remove wearable assistants at least during the night, when they can be recharged. Our design target was 8 h operation time for all parts, which corresponds to a day's use. By focusing on flexibility and robustness rather than high miniaturization or very low-power operation, we could gain insights into users' and clinicians' perspectives on the technology and its effectiveness, as well as gain valuable experience early in the design phase. This allows making informed decisions in the later technological development of an application-specific miniaturized platform.

*Extensive data recording:* We found it helpful to record data from more sensors than needed for context awareness during the project. This allowed us to identify the best on-body sensor placement locations, and it allowed us to ask patients which placement they prefer. As failures typically increase with the number of system components, it is important to have an architecture that allows for an individual component to fail without affecting the entire system.

The numbers and modalities of sensors should be selected as a trade-off between the gain of information, the risks of failure, the time required to set up the sensor on patients, the ethical approval aspects, user acceptance and system wearability. In our case, we decided not to record heart rate, galvanic skin response or the user's distance to surrounding objects even though this information is relevant to infer gait freeze. This ensured the system remained wearable and rapidly deployable on the patients. Future work may include a single acceleration sensor only and in exchange add another sensor modality.

*Local debriefing:* We purposely reserved time for debriefing every day after the two daily patient trials, as well as three extra days at the end of the complete experiment. By reserving time after the conclusion of the experiment, we could do a detailed debriefing with the physiotherapists, as well as jointly inspect the acquired data to develop a common understanding of the freeze events from the physiotherapist and from our signal-oriented perspective. We found out that such a debriefing is valuable to document the experiment and plan the remainder of the work. Debriefing is best carried out face to face when the experiment is freshly concluded.

*Collecting users' and clinicians' experience:* We found it valuable to perform a debriefing with patients and physiotherapists. There are a number of reoccurring concerns when dealing with wearable technology, such as safety concerns, privacy and ethical issues, wearability and comfort and suitability for daily use. We asked patients specific questions on these points. We also asked open questions to collect patients' comments and ideas on how to improve the technology. We interviewed physiotherapists following a similar protocol. Their complementary perspective was helpful to confirm patients' self-assessment reports.

### Future Steps

As we explained in the section "Identification of Potential Sensor Modalities", there are multiple ways in which to achieve context awareness and provide assistance for PD patients with the FOG symptom. In the preceding sections, we have explained one way. Because the clinical trial showed promising results, there are now two possible directions for the future. First, one might go directly towards the next development steps. Second, one might first want to evaluate further sensor modalities and algorithms in order to determine the best way to recognize the patient's context and react to it. In this case, the next steps would be to perform further clinical trials to evaluate the other possible ways.

If we choose the first direction, towards the next development step, trials in a less controlled environment, such as the user's home, are needed. This semi-controlled study under free-living conditions is necessary because the behaviour of patients can be significantly different in their daily life compared to the clinical trial due to the presence of medical personnel and the constant attention they receive. Trials in less controlled environments bring one closer to free-living conditions. Further, they aim to obtain a more realistic picture of the system's behaviour and the user's perception of the assistant. The main goals of the semi-controlled study are to

a. Assess the system's reliability in more realistic conditions
b. Obtain user feedback on aspects related to acceptance: comfort, wearability, system behaviour, design, etc.
c. Identify issues that may arise during long-term trials in natural conditions

After a successful semi-controlled study, long-term trials would aim at preparing for clinical validation. The goal is to minimize technological failures. Therefore, the robustness and reliability of the assistant need to be assessed during long periods (week- to month-long).

The main goals of the long-term trials are to

a. Obtain realistic user feedback on comfort and wearability
b. Identify any robustness and reliability issues
c. Develop a deployment protocol for clinical trials for technical assistants
d. Optimize costs in view of clinical validation

As the last step before commercialization of the system, the clinical validation must prove the clinical efficacy of the system. Clinical efficacy can be evaluated by deploying the system on a large cohort and assessing patients' status in comparison to a control group. Only this stage will give a clear indication of the efficacy of the system. Assuming positive validation, the main goal is to identify any remaining design issues that may affect production and commercialization.

Finally, after the efficacy of the system is demonstrated and market demand established, the product enters the commercialization phase. Medical certification is sought. Design changes relate to cost optimization and design for production.

In these later stages of the project (from a semi-controlled free-living study to clinical validation), the need for flexibility is less, as the kinds of sensors and algorithms are usually defined. Aspects of comfort, wearability and robustness usually become predominant. Ease of use by patients and technical personnel is also required. Technical personnel rather than developers may take the responsibility for deploying the system. Ideally, the patients themselves should be able to wear the system or recharge it. Technologically, this tends to lead to a miniaturized and optimized platform at hardware and software levels: the number of sensors is reduced to the minimum, algorithms are optimized, and the computational resources and algorithm complexity are matched to minimize power consumption. Some form of partial logging is retained to identify possible problems. Finally, in the commercialization phase, aspects of costs, design for production and device appearance are predominant.

## Conclusion

We described our design flow and our experiences in developing a context-aware wearable health assistant for PD patients with the FOG symptom up to the stage of clinical trials. Because our methodology and the account of lessons learned is in many ways applicable to other kinds of context-aware wearable assistants, this work is also valuable for developers active with other health-related assisstive technologies. The key challenge we faced was satisfying the technological feasibility, user acceptance and clinical efficacy factors at the same time. This multidisciplinary field required collaborative work between technical and medical partners.

From the technological side we showed that developing a wearable context-aware health assistant is not a matter of engineering a single platform from prototype to product. Rather, complementary platforms are required at the different stages of the project, and the platforms' computing resources, sensing resources, wireless communication links, form factors and packaging need to be adjusted accordingly.

From the clinical side we explained the long organization process comprising the ethical approval and the recruitment process as well as the clinical trial itself, including the participants and the detailed study protocol.

Based on the study results we showed the effectiveness of our system, which, to our knowledge, is the first time that FOG has been automatically detected online by a wearable device. Without any user-adaptive training, our system had a sensitivity of 73.1% with a specificity of 81.6%. Because we recorded all data during the study, we could afterwards show that separating the patients into two categories with specific sensitivity already increased the detection performance to a sensitivity of 85.9% and specificity of 90.9%.

Even with a system not fully optimized for wearability, we saw in our study that patients found the system beneficial and were willing to wear it. We did not instruct patients on one specific strategy to take advantage of the assistive technology. Rather, we gave them the freedom to find a suitable personalized approach. For at least two patients we could clearly see visually how they used the metronome sound by balancing their body weight accordingly, to get back from FOG into a normal walking rhythm. Although we could not identify such clear signs for all patients, overall six out of the eight patients that experienced FOG during the study claimed that they benefited from the system.

We explained the next areas of work, which will be investigating additional sensor modalities to recognize further contextual aspects related to FOG. We outlined a few that may help to increase the system's accuracy, decrease its latency and eventually go from freeze detection to freeze preemption. The study of gait characteristics shortly prior to the onset of freeze and changes in physiological parameters such as heart rate or blood pressure are especially promising to reduce recognition latency and improve robustness. Our prototyping platform allows for multimodal sensing. Together with context awareness of complex daily life patterns, this may also lead to new medical insight into factors causal to the occurrence of freezing episodes.

Finally, we discussed the next steps in the development of our context-aware wearable health assistant, including progressively longer trials in free-living situations to study the applicability of such an assistive system for everyday usage and preparation for clinical validation. To this end, a complexity analysis shows that, technically, the system can be miniaturized to the size of a button with the FOG recognition included in the sensor node itself. Complex calculations such as FFTs used in the algorithm can be processed with low power consumption on a device the size of a button (Roggen et al. 2006). Such a system could be entirely integrated into or attached to the normal shoes of the patient, and only the trigger for the external cueing signal would be transmitted to the feedback device. The external cueing signal can be given by a hearing-aid-like device or can even be included in future hearing aids. The feedback from patients in the clinical trial was promising for future work, as many expressed interest in testing our device during their natural daily activities.

## References

Amft, O., Lauffer, M., Ossevoort, S., Macaluso, F., Lukowicz, P., and Tröster, G. (2004). Design of the QBIC wearable computing platform. In *Proceedings of the 15th IEEE International Conference on Application-Specific Systems, Architectures and Processors,* (pp. 398–410). Los Alamitos, CA: IEEE Computer Society.

Bächlin, M., Plotnik, M., Roggen, D., Inbar, N., Giladi, N., Hausdorff, J. M., et al. (2009). Parkinson's disease patients' perspective on context aware wearable technology for auditive assistance. In *Proceedings of the 3rd International Conference on Pervasive Computing Technologies for Healthcare,* IEEE Press.

Bächlin, M., Roggen, D., Plotnik, M., Hausdorff, J. M., Giladi, N., and Tröster, G. (2009). Online detection of freezing of gait in Parkinson's disease patients: A performance characterization. In *Proceedings of the 4th International Conference on Body Area Networks,* Brussels, Belgium: ICST.

Bächlin, M., Roggen, D., Plotnik, M., Inbar, N., Meidan, I., Herman, T., et al. (2009). Potentials of enhanced context awareness in wearable assistants for Parkinson's disease patients with freezing of gait syndrome. In *Proceedings of the 13th International Symposium on Wearable Computers (ISWC),* (pp. 123–130). Los Alamitos, CA: IEEE Computer Society.

Bächlin, M., Roggen, D., and Tröster, G. (2007). Context-aware platform for long-term life style management and medical signal analysis. In *Proceedings of the 2nd Sensation International Conference, Chania, Greece. Citeseer.*

Bamberg, S. J. M., Benbasat, A. Y., Scarborough, D. M., Krebs, D. E., and Paradiso, J. A. (2008). Gait analysis using a shoe-integrated wireless sensor system. *IEEE Trans Inf Technol Biomed,* 12 (4), 413–423.

Bannach, D., Amft, O., and Lukowicz, P. (2008). Rapid prototyping of activity recognition applications. *IEEE Pervasive Computing,* 7 (2), 22–31.

Bannach, D., and Kunze, K. (2009). Building activity recognition applications for the iPhone platform. In Tutorial at Int. Symposium on Wearable Computers.

Bannach, D., Kunze, K., Lukowicz, P., and Amft, O. (2006). Distributed modular toolbox for multi-modal context recognition. In Grass, W., Sick, B., and Waldschmidt, K. (eds). *Proceedings of Architecture of Computing Systems (ARCS)* (pp. 99–113). Heidelberg: Springer.

Bartsch, R., Plotnik, M., Kantelhardt, J. W., Havlin, S., Giladi, N., and Hausdorff, J. M. (2007). Fluctuation and synchronization of gait intervals and gait force profiles distinguish stages of Parkinson's disease. *Physica A,* 383 (2), 455–465.

Benoit, A., Bonnaud, L., Caplier, A., Ngo, P., Lawson, L., Trevisan, D. G., et al. (2009). Multimodal focus attention and stress detection and feedback in an augmented driver simulator. *Personal and Ubiquitous Computing,* 13 (1), 33–41.

Bloem, B. R., Hausdorff, J. M., Visser, J. E., and Giladi, N. (2004). Falls and freezing of gait in Parkinson's disease: A review of two interconnected, episodic phenomena. *Movement Disorders,* 19 (8), 871–884.

Boer, A. de, Wijker, W., Speelman, J. D., and Haes, M. D. (1996). Quality of life in patients with Parkinsons disease: Development of a questionnaire. *Journal of Neurology, Neurosurgery and Psychiatry,* 61 (1), 70–74.

Bonato, P., Sherrill, D. M., Standaert, D. G., Salles, S. S., and Akay, M. (2004). Data mining techniques to detect motor fluctuations in Parkinson's disease. In *Proceedings of the 26th Annual International Conference of the IEEE Engineering in Medicine and Biology Society* (pp. 4766–4769).

Braak, H., Ghebremedhin, E., Rüb, U., Bratzke, H., and Tredici, K. D. (2004). Stages in the development of Parkinson's disease-related pathology. *Cell and Tissue Research,* 318 (1), 121–134.

Bulling, A., Roggen, D., and Tröster, G. (2009). Wearable EOG goggles: Seamless sensing and context-awareness in everyday environments. *Journal of Ambient Intelligence and Smart Environments (JAISE),* 1 (2), 157–171.

Coughlin, L., and Templeton, J. (1980). Hip fractures in patients with Parkinson's disease. *Clinical Orthopaedics and Related Research,* (148), 192–195.

Cubo, E., Leurgans, S., and Goetz, C. G. (2004). Short-term and practice effects of metronome pacing in Parkinson's disease patients with gait freezing while in the "on" state: Randomized single blind evaluation. *Parkinsonism and Related Disorders,* 10 (8), 507–510.

Factor, S. A., Jennings, D. L., Molho, E. S., and Marek, K. L. (2002). The natural history of the syndrome of primary progressive freezing gait. *Archives of Neurology,* 59 (11), 1778–1783.

Fahn, S. (1995). The freezing phenomenon in parkinsonism. *Advances in Neurology,* 67, 53–63.

Giladi, N. (2001). Freezing of gait: Clinical overview. *Advances in Neurology,* 87, 191–197.

Giladi, N., and Hausdorff, J. M. (2006). The role of mental function in the pathogenesis of freezing of gait in Parkinson's disease. *Journal of the Neurological Sciences,* 248 (1–2), 173–176.

Giladi, N., Kao, R., and Fahn, S. (1997). Freezing phenomenon in patients with parkinsonian syndromes. *Movement Disorders,* 12 (3), 302–305.

Giladi, N., McDermott, M. P., Fahn, S., Przedborski, S., Jankovic, J., Stern, M., et al. (2001). Freezing of gait in PD: Prospective assessment in the DATATOP cohort. *Neurology,* 56 (12), 1712–1721.

Giladi, N., Shabtai, H., Simon, E. S., Biran, S., Tal, J., and Korczyn, A. D. (2000). Construction of freezing of gait questionnaire for patients with parkinsonism. *Parkinsonism and Related Disorders,* 6, 165–170.

Grant, S., Aitchison, T., Henderson, E., Christie, J., Zare, S., McMurray, J., et al. (1999). A comparison of the reproducibility and the sensitivity to change of visual analogue scales, Borg scales, and Likert scales in normal subjects during submaximal exercise. *Chest,* 116 (5), 1208–1217.

Hashimoto, T. (2006). Speculation on the responsible sites and pathophysiology of freezing of gait. *Parkinsonism and Related Disorders, 12* (Suppl. 2), S55–S62.

Hausdorff, J. M., Balash, Y., and Giladi, N. (2003). Time series analysis of leg movements during freezing of gait in Parkinson's disease: Akinesia, rhyme or reason? *Physica A: Statistical Mechanics and Its Applications,* 321 (3–4), 565–570.

Hausdorff, J. M., Lowenthal, J., Herman, T., Gruendlinger, L., Peretz, C., and Giladi, N. (2007). Rhythmic auditory stimulation modulates gait variability in Parkinson's disease. *European Journal of Neuroscience,* 26 (8), 2369–2375.

Hoehn, M. M., and Yahr, M. D. (1967). Parkinsonism: Onset, progression and mortality. *Neurology,* 17 (5), 427–442.

Jankovic, J. (2008). Parkinson's disease: Clinical features and diagnosis. *Journal of Neurology, Neurosurgery and Psychiatry,* 79 (4), 368–376.

Johnell, O., Melton, L. J., Atkinson, E. J., O'Fallon, W. M., and Kurland, L. T. (1992). Fracture risk in patients with parkinsonism: A population-based study in Olmsted County, Minnesota. *Age and Ageing,* 21 (1), 32–38.

La Marca, A., Chawathe, Y., Consolvo, S., Hightower, J., Smith, I., Scott, J., et al. (2005). Place lab: Device positioning using radio beacons in the wild. In *Proc. of Pervasive Computing,* 116–133.

Lamberti, P., Armenise, S., Castaldo, V., Mari, M. de, Iliceto, G., Tronci, P., et al. (1997). Freezing gait in Parkinson's disease. *European Neurology,* 38 (4), 297–301.

Lee, S.-W., and Mase, K. (2002). Activity and location recognition using wearable sensors. *IEEE Pervasive Computing,* 1 (3), 24–32.

Lim, I., Wegen, E. van, Goede, C. de, Deutekom, M., Nieuwboer, A., Willems, A.-M., et al. (2005). Effects of external rhythmical cueing on gait in patients with Parkinson's disease: A systematic review. *Clinical Rehabilitation,* 19 (7), 695–713.

Lorincz, K., Malan, D., Fulford-Jones, T., Nawoj, A., Clavel, A., Shnayder, V., et al. (2004). Sensor networks for emergency response: Challenges and opportunities. *IEEE Pervasive Computing,* 3 (4), 16–23.

Macht, M., Krüger, H.-P., Kaussner, Y., Müller, J. C., Stiasny-Kolster, K., Eggert, K. M., et al. (2007). Predictors of freezing in Parkinson's disease: A survey of 6,620 patients. *Movement Disorders,* 22 (7), 953– 956.

Martinez-Martin, P. (1998). An introduction to the concept of "quality of life in Parkinson's disease". *Journal of Neurology,* 245 (S1), S2–S6.

Meidan, I., Plotnik, M., Mirelman, A., Giladi, N., Hausdorff, J. M., and Weise, A. (2009). Autonomic nervous system in Parkinson's disease: New evidence on the relationship to freezing of gait. In *Proceedings of the International Society for Posture and Gait Research (ISPGR).* Bologna, Italy.

Moore, S. T., MacDougall, H. G., and Ondo, W. G. (2008). Ambulatory monitoring of freezing of gait in Parkinson's disease. *Journal of Neuroscience Methods,* 167 (2), 340–348.

Nieuwboer, A., Baker, K., Willems, A.-M., Jones, D., Spildooren, J., Lim, I., et al. (2009). The short-term effects of different cueing modalities on turn speed in people with Parkinson's disease. *Neurorehabilitation and Neural Repair,* 23 (8), 831–836.

Nieuwboer, A., Dom, R., De Weerdt, W., Desloovere, K., Fieuws, S., and Broens-Kaucsik, E. (2001). Abnormalities of the spatiotemporal characteristics of gait at the onset of freezing in Parkinson's disease. *Movement Disorders,* 16 (6), 1066–1075.

Nieuwboer, A., Dom, R., Weerdt, W. de, and Lesaffre, E. (1998). A frequency and correlation analysis of motor deficits in Parkinson patients. *Disability and Rehabilitation,* 20 (4), 142–150.

Okuma, Y. (2006). Freezing of gait in Parkinson's disease. *Journal of Neurology,* 253 Suppl 7: VII27–32.

Otsason, V., Varshavsky, A., La Marca, A., and de Lara, E. (2005). Accurate GSM indoor localization, In Beigl, M., Intille, S., Rekimoto, J., and Tokuda, H. (eds). *Proc. of the 7th Int. Conf. on Ubiquitous Computing* (pp. 141–158). Heidelberg: Springer-Verlag.

Pappas, I. P. I., Keller, T., and Mangold, S. (2004). A reliable gyroscope-based gait-phase detection sensor embedded in a shoe insole. *IEEE Sensors,* 4 (2), 1085–1088.

Pentland, A. (2004). Healthwear: Medical technology becomes wearable. *Computer, 37* (5), 42–49.
Randell, C., Djiallis, C., and Muller, H. (2003). Personal position measurement using dead reckoning. In *Proc. of the 7th IEEE Int. Symposium on Wearable Computers (ISWC'03)* (pp. 166–173). IEEE Computer Society.
Roggen, D., Bharatula, N. B., Tröster, G., Stäger, M., and Lukowicz, P. (2006). From sensors to miniature networked sensorbuttons. In *Proc. of the 3rd Int. Conf. on Networked Sensing Systems (INSS06),* (pp. 119–122). San Diego, CA: Transducer Research Foundation.
Roggen, D., Hamette, P. de la, Tröster, G., and Jenny, R. (2007). Mapping by seeing: Wearable vision-based dead-reckoning, and closing the loop. In Kortuem, G., Finney, J., Lea R., and Sundramoorthy, V. (eds). *Proc. 2nd European Conf. on Smart Sensing and Context (EuroSSC)* (pp. 29–45). Berlin, Heidelberg: Springer.
Roggen, D., Bächlin, M., Schumm, J., Holleczek, T., Lombriser, C., Tröster, G., Widmer, L., Majoe, D., and Gutknecht, J. (2010). An educational and research kit for activity and context recognition from on-body sensors. In *Proc. IEEE Int. Conf. on Body Sensor Networks (BSN)* (pp. 277–282). Los Alamitos, CA: Conference Publishing Services.
Schaafsma, J. D., Gurevich, T., Balash, Y., Bartels, A. L., Hausdorff, J. M., and Giladi, N. (2003). Characterization of freezing of gait subtypes and the response of each to levodopa in Parkinson's disease. *European Journal of Neurology, 10* (4), 391–398.
Schindler, G., Metzger, C., and Starner, T. (2006). A wearable interface for topological mapping and localization in indoor environments. In *Proc. of the 2nd Int. Workshop on Location- and Context-Awareness,* 64–73.
Schumm, J., Bächlin, M., Setz, C., Arnrich, B., Roggen, D., and Tröster, G. (2008). Effect of movements on the electrodermal response after a startle event. *Methods of Information in Medicine, 47* (3), 186–191.
Shelton, R. C., Keller, M. B., Gelenberg, A., Dunner, D. L., Hirschfeld, R., Thase, M. E., et al. (2001). Effectiveness of St John's wort in major depression: A randomized controlled trial. *Journal of the American Medical Association, 285* (15), 1978–1986.
Sixsmith, A., Gibson, G., Orpwood, R., and Torrington, J. (2007). Developing a technology "wish-list" to enhance the quality of life of people with dementia. *Gerontechnology, 6* (1), 2–19.
Stuyven, E., Goten, K. der, Vandierendonck, A., Claeys, K., and Crevits, L. (2000). The effect of cognitive load on saccadic eye movements. *Acta Psychologica, 104* (1), 69–85.
Tröster, G. (2005). The agenda of wearable healthcare. In Haux, R., and Kulikowski, C. (eds). *IMIA Yearbook of Medical Informatics 2005: Ubiquitous Health Care Systems* (pp. 125–138). Stuttgart: Schattauer.
Veltink, P. H., Bussmann, H., Vries, W. de, Martens, W. L. J., and Van Lummel, R. C. (1996). Detection of static and dynamic activities using uniaxial accelerometers. *IEEE Transactions on Rehabilitation Engineering, 4* (4), 375–385.
Westeyn, T., Presti, P., Johnson, J., and Starner, T. (2009). A naive technique for correcting time-series data for recognition applications. In *Proc. 13th Int. Symposium on Wearable Computers,* (pp. 159–160). Los Alamitos, CA: IEEE Computer Society.

# 13

# *Designing a Low-Cost ECG Sensor and Monitor: Practical Considerations and Measures*

**Ahsan H. Khandoker and Brian A. Walker**

**CONTENTS**

Introduction....339
ECG Signals and Diagnosis....341
Design and Construction of ECG Electrodes....344
Electrode Placement....345
The ECG Amplifier: How It Works....347
The Need for an Instrumentation Amplifier....349
The Front End....351
Filtering....353
Automatic Gain Control....355
USB Interfacing....356
USB Configuration and Communication....357
USB Interface Hardware and Device Detection....359
Firmware Operation....361
Displaying the ECG Signals on an PDA or Mobile Phone....362
Discussion and Conclusion....367
References....369

## Introduction

Mobile phones are bridging the digital divide and transforming many economic, social and medical realities, particularly in developing countries. With the penetration of low-cost handsets and the omnipresence of mobile phone networks, tens of millions of people who had never had a computer now use mobile devices. On the other hand, trained health workers and diagnostic testing facilities are a scarce resource in poor countries' rural areas, especially in Africa (World Health Organisation [WHO], 2006). Despite the billions of dollars invested in health in Africa, the shortage of appropriate health workers, particularly in rural areas, in many countries is a major barrier to the coverage of health services for the poor (World Bank 2008). The growing ubiquity of mobile services allows the creation of a new generation of electronic health systems based on mobile computing. *Mobile health (mHealth)* is emerging as an important segment of the field of *electronic health (eHealth)* (Istepanian 2006) that advocates the utilization of mobile technology which supports the next generation of health systems. We suggest that even the simplest solutions would provide a major contribution to health development in these communities. It is possible to create a range of mobile phone applications and low-cost diagnostic devices which will run on health workers' own mobile phones, making them useful for daily activities. The environment imposes severe restrictions, however. First, resource constraints mean that we must avoid introducing new costs, whether capital or recurrent—developing countries' health services have tiny per-capita annual budgets! Also, limitations in the availability of a specialized workforce mean that we cannot rely on distant experts to interpret data or provide the diagnosis. We expect to operate in areas with poor radio coverage, and the applications will be executed on low-end mobile devices. Finally, due to logistic limitations, we must avoid the need for training. Therefore, we argue that a model based on data transmission for distant analysis is unlikely to work.

At present, cardiac disease is a leading cause of death in the developing world as well as the leading cause of death in the developed world. The electrocardiogram (ECG) is the most common method for diagnosing Cardiovascular System Disease (CSD). Prior heart attacks and common abnormal heart rhythms such as in atrial fibrillation can easily be diagnosed using an ECG. The process usually involves attaching a number of electrodes (quite often between 10 and 12 in total) to the arms, legs and chest of the patient to detect the electrical impulses from the heart. The ECG machine displays an ECG trace from the impulses, from which an accurate diagnosis can be made. Finally, the electrodes are removed. The whole process usually takes less than 5 min (Fogoros 2008). It has been stated that 3-electrode ECGs do not provide adequate information for diagnosing diseases compared to 12-electrode ECGs (Mackinnon 2003); however, they do have sufficient sensitivity and specificity (≈80%) when it comes to determining whether an ECG trace is normal one or pathological (Chaikowski et al. 2008). It has even been stated that when used for computer applications, the information content of the 3-electrode ECG could exceed that of the conventional 12-lead one and that recognition rates for diagnosis can be significantly improved (Pipberger 1965).

One of the main concerns in many developing countries is the unreliable electricity supply. In remote communities, there is often no electricity at all. An attempt has already been made to adapt an ECG system for developing countries by powering it off solar energy. However, this product—known as Kadiri—still relies on a lead acid battery for backup, and the costs involved restrict its marketing to large companies (Wegmann 2008). The ECG amplifier developed by our team (Walker et al. 2009) can be powered off a 9 V transistor radio battery, which has its advantages in terms of noise. In fact, one important advantage of the ECG system's use in remote communities where there is no electricity at all is that there is an absence of 50 Hz noise.

In February 2009, Terry Kramer, strategy director at Vodafone, said that there are 2.2 billion mobile phones in the developing world, out of 4 billion in the entire world (Perez 2009). In February 2007, the BBC reported in its news article "Mobile phone lifeline for world's poor" (Anderson 2007) that 85% of new mobile phone subscribers live in emerging markets. Mobile phones have become popular in developing countries as no phone line installation is required (Committee on Disaster Research in the Social Sciences: Future Challenges and Opportunities, (U.S.) National Research Council [CDRSS], 2006). However, these statistics do not apply to smart phones or personal digital assistants (PDAs). Robust cellular phone networks in developing nations mean that phones can be used to communicate directly with doctors and are used in other ways not seen in the developed world (Sprey 2009).

This chapter aims to present step-by-step guidelines and practical considerations in designing, constructing and testing a complete, simple, user-friendly, energy-efficient ECG data logging system that is low cost, aimed at hospitals in developing countries. The ECG system comprises four main components as shown in Figure 13.1: the three electrodes, the ECG amplifier, the USB interface and the software on the phone to display the ECG signals. The ECG system's cost has been minimized and is something which many people in developing countries can afford. This was achieved by using widely available "off-the-shelf" components as much as possible and maintaining the component count to an absolute minimum. Scrap material was used wherever possible. Careful design, construction and testing increased the ECG system's reliability and quality to closely match that of a commercial-grade ECG system. Good-quality ECG signals were observed and have all the distinctive characteristics, which are easily identified even though no conductive gel was applied between the surfaces of the electrodes and the skin, and only unshielded, lengthy interconnecting cables were employed which are highly prone to external interference.

## ECG Signals and Diagnosis

"An electrocardiogram (ECG) is a graphic tracing of the electric current generated by the heart muscle during a heartbeat" (Low et al. 2006, p. 66). This current generates electric potentials on the surface of the skin. Figure 13.2 shows the deflection components making up the ECG signal (Low 2006). The QRS complex is sometimes subdivided into Q, R and S waves and is called a complex because of the usual presence of all three of these waves. The first or second positive waves above the baseline are known as the *R waves*. A negative wave following an R wave is known as an *S wave*, and a negative

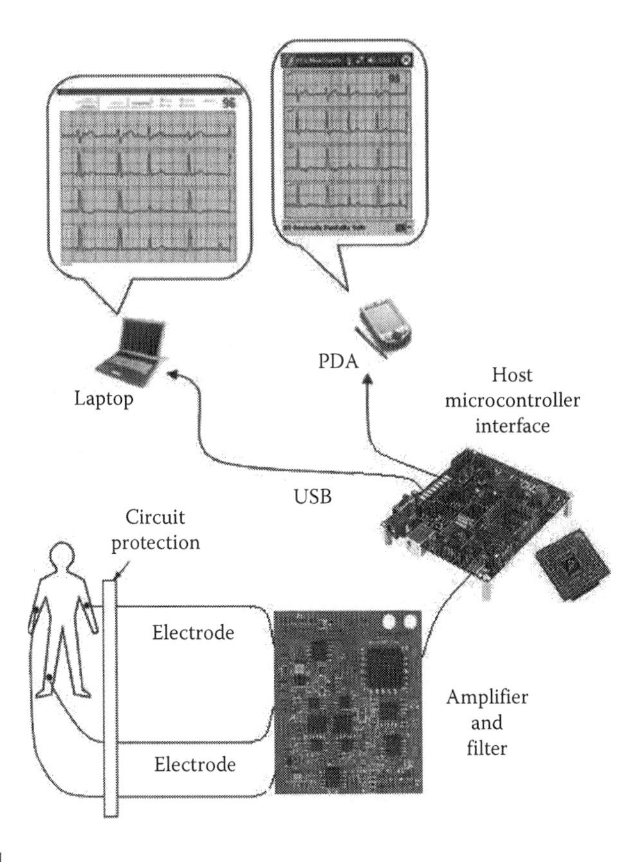


**FIGURE 13.1**
The complete electrocardiogram (ECG) data logging system.

wave which precedes an R wave is known as a *Q wave* (Das 2006). The U wave is sometimes present in the ECG signal and is usually less than one-third of the T wave amplitude. The U wave represents the last phase of ventricular repolarization and is best seen when the heart rate is slow (Huff 2005).

The P wave is caused by atrial depolarization (contraction). The QRS complex represents ventricular depolarization, and the T wave ventricular repolarization (Gardiner 1981). Heart disease can be diagnosed on close inspection of these deflection components. For example, close RR intervals would indicate a high risk of ventricular fibrillation. Stray P waves which do not appear to be part of the cardiac cycle indicate complete heart block, which causes fainting and palpitations. Missing P waves are a strong indication of atrial fibrillation, which increases the risk of stroke. A patient who has atrial fibrillation accompanied by complete heart block would probably require a pacemaker (http://www.ecglibrary.com/). Flat or depressed ST segments could indicate coronary artery disease, which is the leading cause of death worldwide (Carson-DeWitt 2007). Sleep apnoea can be detected using the ECG (Khandoker et al. 2009) by analysing the variation in RR intervals—in

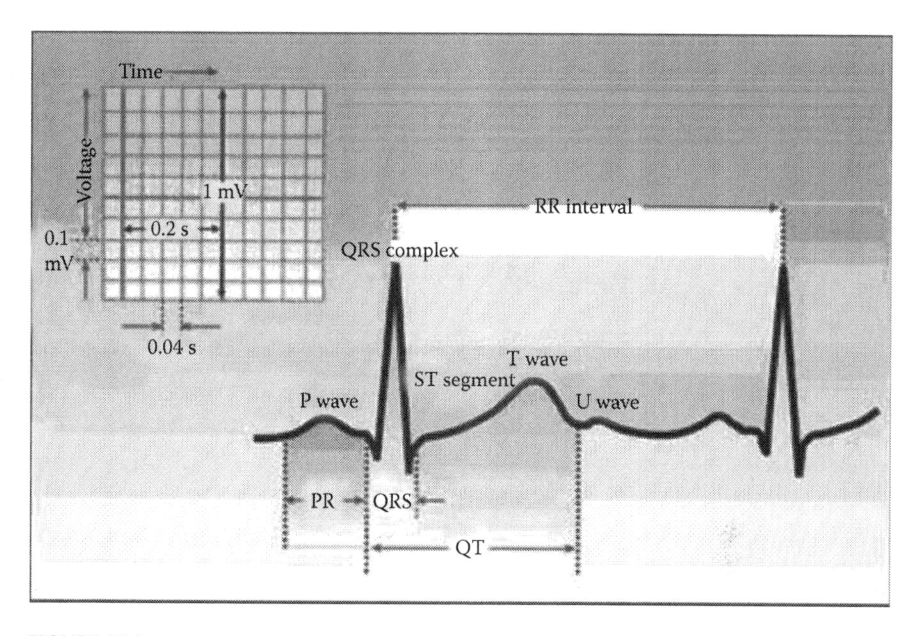


**FIGURE 13.2**
Detailed depiction of the deflection components making up an electrocardiogram (ECG) signal from a healthy patient. The voltage scale is indicated. (Adapted from Yanowitz, F. G., Characteristics of the normal ECG, 2006, University of Utah Spencer S. Eccles Health Sciences Library, retrieved 2010 from http://library.med.utah.edu/kw/ecg/ecg_outline/Lesson3/index.html.)

particular, the RR interval variations can be analysed in the frequency domain using the Hilbert transform (Mietus et al. 2000). Interpreting abnormalities in an ECG signal is not an easy task, as it is very difficult to determine how long the abnormalities have been present. Some people are even born with such abnormalities (Cohen 2005).

Normal heart rates are somewhere between 60 and 90 beats per minute. The PR interval is usually 180–200 ms for an adult and 150–180 ms for a child. The QRS interval is between 70 and 100 ms (Huff 2005). The QT dispersion (QTd) for purposes of diagnosis depends greatly on the heart rate and is given approximately by Bazett's formula (Hegarty 2007). However, Bazett's formula has proven to be inaccurate at high and low heart rates, and other formulas are often used instead. In a healthy patient, QTd is less than or equal to 400 ms (Yanowitz 2006), which sets an upper limit on the QT interval depending on the heart rate. Abnormally long QT intervals result in long QT syndrome, which could lead to ventricular fibrillation and sudden death (Mayo Foundation for Medical Education and Research 2008). One important parameter based on the QT interval is the QT interval dispersion, which is the statistical standard deviation of the QT interval. A study in Venezuela (Fuenmayor et al. 1998) has revealed that ECG signals from malnourished children had higher QT interval dispersions and also exhibited flattened or inverted T waves. Increased QT interval dispersions are also an indication of type 1 diabetes. It is clear that the QT interval is an important parameter used in diagnosis.

## Design and Construction of ECG Electrodes

Designing the electrodes is no trivial matter, and the materials used often distinguish commercial types from homemade ones. In order for an electrode to detect the currents through the body, a current must flow across the interface between the skin and the surface contact of the electrode. For this to happen, electrons must flow. What actually happens is that the metal of the electrode gets oxidized to form positive metal ions (cations), which get discharged into the surface of the skin (the electrolyte). This oxidation process also produces electrons which flow within the metal. Anions from the skin combine with these free electrons in the metal to form a neutral metal atom (reduction). These two chemical reactions together produce a current through the electrode. The resulting half cell potential according to the Nernst equation, due to the uneven distribution of cations and anions, appears on the ECG signal as a DC offset (Lee and Kruse 2008). This DC offset has played an important part in the design of the ECG amplifier. What is more problematic is that even the slightest shifts in the positions of the electrodes with respect to the skin will cause DC baseline drift of the ECG signal due to the varying half cell potentials. This source of noise in the ECG signal is known as the *motion artefact* (Amer 2008).

The best materials to use for electrodes minimize the half cell potential (which causes DC baseline offsets and drifts), are easy to manufacture and are not highly reactive. Materials which produce low half cell potentials increase electrode stability. For example, lithium would not be used for the electrodes as its half cell potential is –3 V, and its reactivity is enough to burn the skin. Another problem to consider is that the half cell potential could vary depending on the amount of current flowing through the electrode. This is known as an overpotential (Hinz 2002). Electrodes which do not have overpotentials are known as nonpolarizable electrodes. Examples of nonpolarizable electrodes include those made from copper and silver chloride (Allaby and Allaby 1999). Normally, commercial electrodes are made from silver with a coating of silver chloride (Amer 2008). The half cell potential of silver chloride is very low, 223 mV, and is ideal for ECG measurement. Copper can also give satisfactory results; it has a half cell potential of 342 mV (Hinz 2002).

Reusable electrodes designed specifically for the limbs are in very common use today. Limb electrodes are used for both historical and practical reasons. Augustus D. Waller used large-area bucket electrodes filled with brine, into which the hands and feet were placed, in his demonstrations

(Morin 2009). Such large-size, large-area electrodes were essential at the time because of the low input impedance of Einthoven's galvanometer, used as the amplifier to receive the ECG signals. As a result, Willem Einthoven used limb electrodes in future experiments. As the input impedance of ECG amplifiers increased in the 1920s due to the advent of vacuum tubes (which, like field-effect transistors (FETs), have high input impedances), the size of the electrodes decreased, and large metal plates were used instead (McAdams 2006). However, though larger electrodes decrease their impedances by having a larger contact area with the skin, they are more prone to 50 Hz noise pickup from the mains.

Metal plates or strips maximize the contact area and minimize electrode impedance. Strips of copper, for example, can encircle the limbs, and clamps available from local hardware stores can be used to fasten the strips to the skin. Homemade electrodes built using strips of metallic material are shown in Figure 13.3.

If metal plates are used, fastening them adequately to the skin is a more involved process, but it can be done using straps of rubber or Velcro which encircle the limbs and attach to the plates (Gomella and Haist 2002). Figure 13.4 shows an example of a limb electrode made from a plate of galvanized iron and a rubber strap attached to the plate. There is a minimum area that the electrode should have in contact with the skin, which is about the size of an Australian 10-cent coin—about the same size as that for a commercial disposable electrode.

## Electrode Placement

Einthoven used what are known as standard bipolar limb leads I, II and III as shown in Figure 13.5 (J. Lee 2000), with the electrodes placed at the extremities of the human body. A *lead* is defined as a stretch between two limb electrodes. The positioning of the electrodes formed what is known as Einthoven's triangle (Pipberger 1965). Each of the leads records the electrical heart activity along a particular axis on the frontal plane (Klabunde 2005). One-lead ECGs cannot perform much in the way of diagnosis as the information obtained will be along only one axis of the frontal plane. However, because lead II lies close to the cardiac axis, it is the most useful lead for the detection of cardiac arrhythmias and provides for the best analysis of P and R waves (J. Lee 2000).

Electrical currents from the heart which flow towards an electrode will record a positive deflection, while currents flowing away will record a negative deflection in the resulting signal (Khan and Cannon 2008). Whether the electrodes are attached at the wrists and ankles or at the shoulders and upper thighs makes no difference to the ECG signals (Klabunde 2005). However,

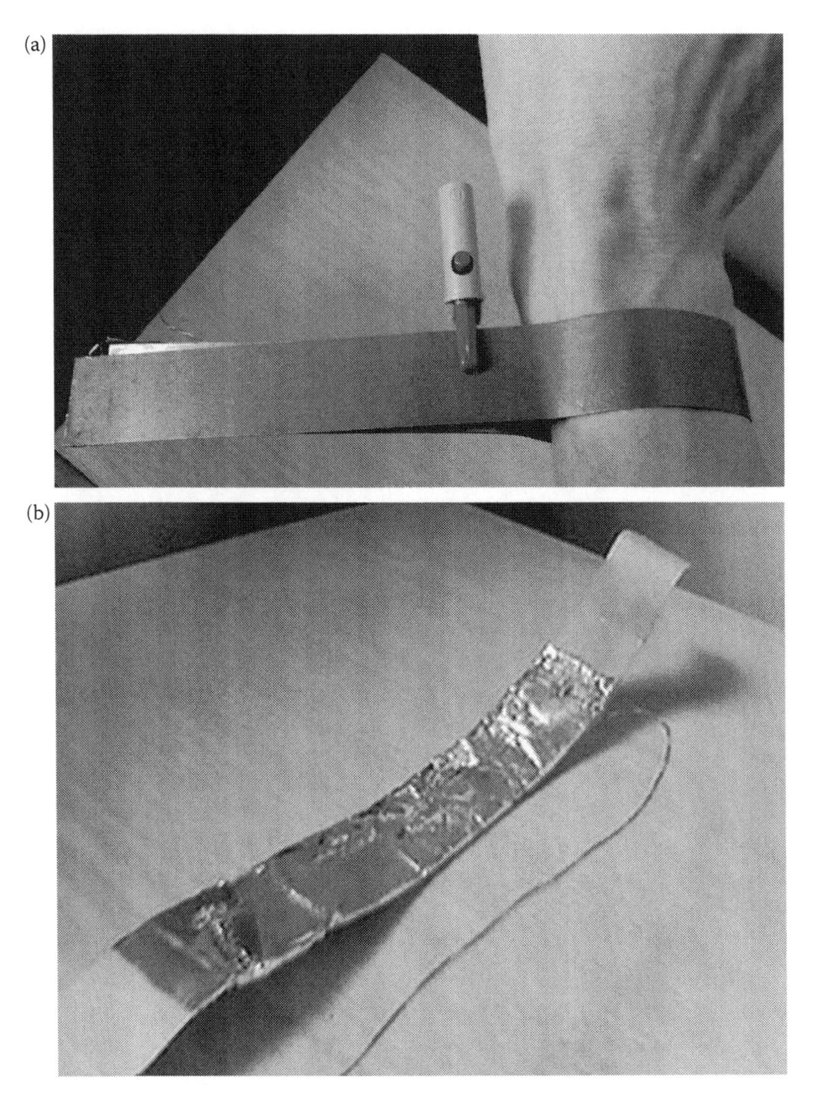

**FIGURE 13.3**
Cheap homemade limb electrodes. Sandpaper was used to ensure good contact, and alligator clips can be used to fasten the electrodes to the limbs. Wires are soldered directly onto the electrode. (a) A copper strip encircling the limb; (b) Aluminum foil from a chocolate bar wrapped around a strip of cardboard.

for best practice, the electrodes should be positioned at the wrists and ankles and at least 15 cm away from the heart to avoid affecting the cardiac axis. The ground reference electrode can be placed anywhere on the body (Bowbrick and Borg 2006), though it is usually placed on the right leg. Electrodes should be placed over bony prominences to minimize electrode movement during respiration and thus baseline wander of the ECG signal (Catalano 2002).

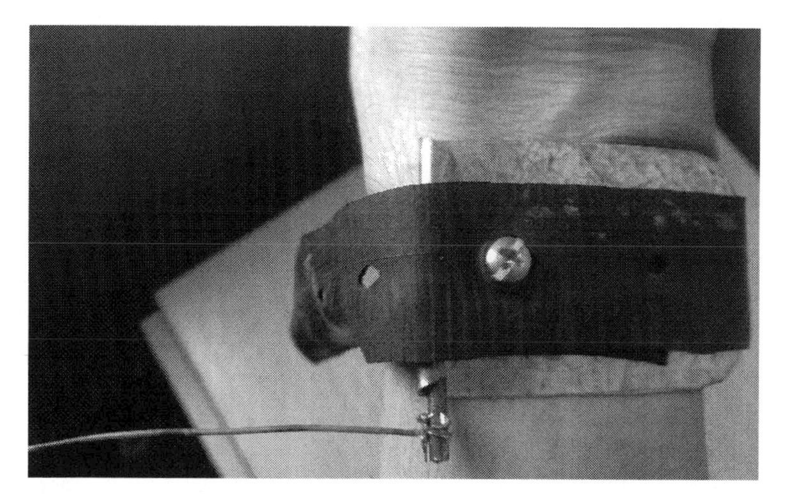

**FIGURE 13.4**
Homemade limb electrode made from a piece of scrap galvanized iron, a strap of rubber from a worn bicycle tyre, a nail which connects the electrode wire to the electrode itself and a self-tapping screw which attaches the rubber to the plate.

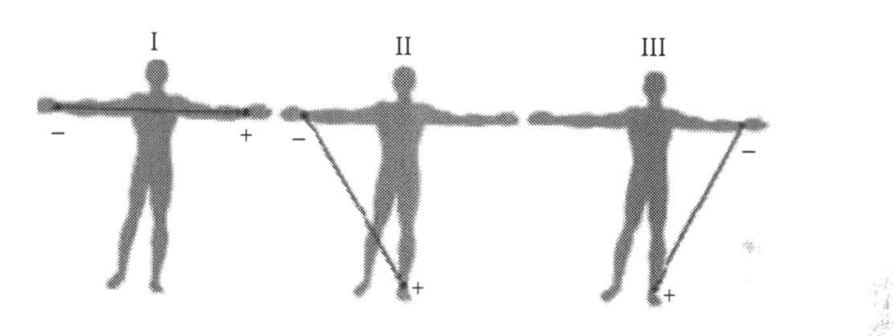


**FIGURE 13.5**
Standard bipolar limb leads originally used by Einthoven. Note that the right leg is the ground reference point to minimize interference. So there are in fact three electrodes.

## The ECG Amplifier: How It Works

Upon inspection of Figure 13.5, we can easily see that a differential amplifier would be most suitable in amplifying the signals and would amplify the voltage difference from two electrodes connected at each input of the amplifier. As stated earlier, amplifying the difference between ECG potentials at two different parts of the body would, in theory, reject electromagnetic interference (EMI). Differential amplifiers are well known for their ability to reject such common mode signals. However, in practice, the parameter which describes this ability to reject such signals, known as the *common mode rejection ratio* (CMRR), is specified on amplifier device data sheets from the manufacturer and is never infinite. Thus, there will always be some residual EMI (mainly

50 Hz noise from the mains) at the output of the amplifier which would need to be filtered. The ground reference electrode is essential because the human body is not at absolute ground potential, and a floating ground would be highly susceptible to noise which would appear between the ground of the amplifier and that of the human body. The electric potential of the human body is higher than at ground, and the amplifier needs a high CMRR so as to reject this potential and the sources of EMI radiated through the loop formed by the human body, the electrode leads and the amplifier itself. By twisting the electrode leads together (minimizing the distance between cables), the EMI in each electrode lead would follow almost the same loop and ideally be cancelled out at the inputs of the amplifier due to its infinite CMRR. No amplifier is ideal, but the CMRR should be at least 100,000 (100 dB). Ideally, it should be at least a million for ECG monitoring (Hemmings and Hopkins 2005).

In addition to the ECG amplifier's CMRR, another parameter to consider is the amplifier's frequency response. The American Heart Association (AHA) Committee on Electrocardiography Standardisation recommends an amplifier bandwidth from 0.05 to 100 Hz, with the response at 0.05 Hz reduced by no more than 30% from the response at 0.14 Hz (Kligfield et al. 2007; Hemmings and Hopkins 2005). The phase response is not stated; however, it is usually adequate if the amplitude response is satisfied. Muscle movement dominates at lower frequencies and could cause significant baseline wander of the ECG signal. Thus, for monitoring purposes only, a cut-off frequency of 0.5 Hz can be used instead of 0.05 Hz for stability. The amplifier should have a flat amplitude response as well as a linear phase response in its passband to prevent distortion. The QRS complex is a feature of high slopes and is thus representative of high-frequency components, whereas the P and T waves are low-frequency components. The ST segment is the lowest-frequency component. Thus, at low frequencies, a nonlinear frequency response could result in ST segment depression, while a nonlinear phase response could result in ST segment elevation. At high frequencies, a cut-off frequency as low as 40 Hz would significantly reduce the amplitude of the R and S waves, and even the whole QRS complex. A third parameter, which appears to be the most important of all, is the amplifier's input impedance. The input impedance of the amplifier determines the size of the electrodes and, as a result, their value and application for diagnosis. One can model an electrode attached to the skin as a voltage source with internal skin–electrode and electrode lead resistance (Thevenin equivalent circuit). The load would then be modelled as the input impedance of the amplifier. Obviously, increasing the amplifier's input impedance or decreasing the skin–electrode resistance will increase the amplitude of the ECG signals at the input of the amplifier. Gel, adhesives and the mechanical construction of the electrodes ensure that the electrodes make very firm contact with the skin in order to minimize skin–electrode impedance. A loose contact will not only decrease the ECG signal level but also increase the coupling of EMI via an increase in the skin–electrode impedance.

## The Need for an Instrumentation Amplifier

An operational amplifier (op-amp) is a DC-coupled high-gain electronic voltage amplifier with differential input and a single ended ouput. In order for the ECG amplifier to have a high input impedance and a high CMRR, it is not sufficient to use a normal one-stage differential operational amplifier (op amp). This is because even the slightest changes in the electrode impedances will affect the differential amplifier's operation and degrade its CMRR (Horowitz and Hill 1989). Noninverting op amps (or unity gain buffers) are known to have extremely high input impedances and serve in many applications to isolate certain sections of circuitry so that the impedance of other parts of the circuitry does not affect the intended operation of these isolated sections. An ECG amplifier consisting of only a standard noninverting op amp has been successfully built and tested by staff at the University of Vienna in Austria (Kirtley 2002). Only two electrodes were used, one of which was the reference ground electrode, which was simply connected to the ground of the amplifier. Commercial disposable electrodes were used, and even a single-pole RC low-pass filter was used at the amplifier's noninverting input in order, as the designer says, "to remove 60 Hz electrical noise picked up from flourescent lights, computers and AC power lines" (Kirtley 2002). This could suggest that perhaps a differential amplifier is redundant, and only two electrodes are necessary. However, the results from this experiment were very poor, due to the ECG signals being "buried" in the huge amount of noise. In order for the ECG amplifier to have any chance to compete with commercial ECGs, the quality of the ECG signals obtained must well exceed many attempts like this conducted by both hobbyists and professionals alike. One of the most important features which determine the quality of ECG signals is the level of noise present. A differential amplifier is absolutely essential to reduce the noise to an acceptable level.

It is quite clear now that what is really needed is a differential amplifier isolated so that changes in electrode impedances do not affect its operation. Adding at least one noninverting op amp amplifier to isolate the standard differential amplifier from the electrodes results in what is known as an instrumentation amplifier. Only the inverting input of the standard differential amplifier needs to be isolated, as its other, noninverting input already offers a high impedance. The inverting input is at a much lower impedance due to negative feedback. Thus, the minimum number of op amps an instrumentation amplifier can have is obviously two. An instrumentation amplifier which uses two op amps, known as the *two op amp instrumentation amplifier*, is shown in Figure 13.6. It can effectively be considered as two noninverting op amp amplifiers in cascade.

The most common instrumentation amplifier in use today uses three op amps—with a noninverting op amp amplifier connected to each of the two inputs of the standard differential amplifier. This clearly would result in

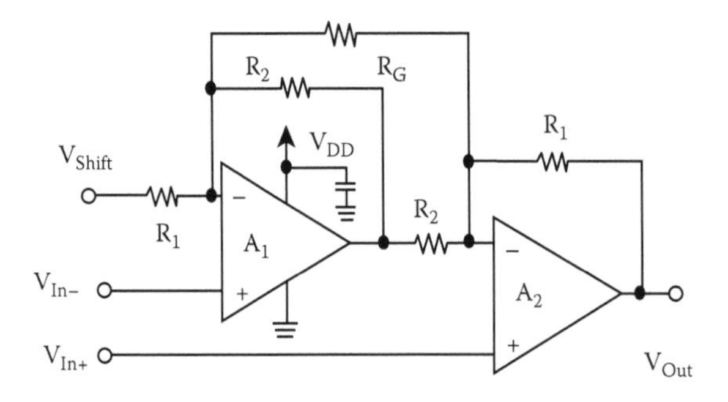


**FIGURE 13.6**
Two op amp instrumentation amplifier. (From Baker, B. C., Operational amplifiers part IV of VI: Working your amplifier inside the single-supply voltage "box", AnalogZONE, 2004, retrieved October 2010 from http://www.analogzone.com/acqt0809.pdf. With permission.)

complete isolation of the differential amplifier from the electrodes, as well as offering complete flexibility in the range of values of the gain which can be achieved, with gains down to unity easily implemented. The symmetric configuration of the three op amp instrumentation amplifier results in a flatter frequency response, particularly at higher frequencies.

However, one of the most important parameters relevant to an ECG amplifier is the amplifier's CMRR, which would determine the amount of common mode noise present at the output of the amplifier. It can be shown using rigorous mathematical analysis (Peyton and Walsh 1993) that at low frequencies, the simple two op amp instrumentation amplifier has increased CMRR compared to that from an amplifier using the three op amp configuration. The two op amp instrumentation amplifier does have two main limitations – one being the minimum amount of gain which can be achieved, which is usually restricted to between 2 and 4 depending on how the resistors are connected (Northrop 2004). Also, at high frequencies, the two op amp instrumentation amplifier has a nonlinear frequency response which results in the bandwidth being one half of that of the three op amp configuration. This is due to signals at one of the amplifier's inputs passing through one more op amp stage than the signals at the other input. However, because an ECG system operates at low frequencies and requires high gain, these limitations are not really of any importance.

The reason for the increased CMRR exhibited in two op amp instrumentation amplifiers has been demonstrated mathematically (Santos 2006), and the results indicate that the CMRR of the standard one op amp differential amplifier used as part of the three op amp instrumentation amplifier degrades the overall CMRR of the instrumentation amplifier. In the two op amp configuration, the overall CMRR is determined only by component mismatch. One other advantage of the two op amp configuration which is often overlooked is the possibility of achieving higher gains. Gains of up to 10,000 can be readily achieved, as the noninverting op amp amplifier stage at the input of the instrumentation amplifier is less prone to saturation (Santos 2006). Thus, with a slight sacrifice on input isolation, the two op amp instrumentation amplifier can be chosen, as it obviously uses fewer components, and it was expected to provide the key in obtaining ECG signals with superior SNR (the signal to noise ratio of the R wave), corrupted with a very low amount of noise. The CMRR is defined as the ratio of the differential gain to the common mode gain. The differential gain simply applies to the difference in the voltage at its two inputs. One of the main advantages of building an instrumentation amplifier from discrete components such as op amps is the ability to tweak component values to obtain better matching, and also to employ ultra-low-noise op amps which could result in lower output noise levels.

## The Front End

Figure 13.7 shows the front end circuitry, connected between the electrodes and inputs to the ECG amplifier, which serves two purposes. First, it serves as protection for the patient as well as for the ECG amplifier itself, and, second, it introduces AC coupling to the inputs of the ECG amplifier.

The ground reference electrode on the patient would ideally be connected to the signal ground potential (about halfway between the two power supply rails) through the output of a noninverting op amp buffer stage to isolate the effects of the human body on the circuit. However, this probably would not work very well in practice without AC coupling, as any difference between the three electrodes would cause a DC offset, as described earlier,

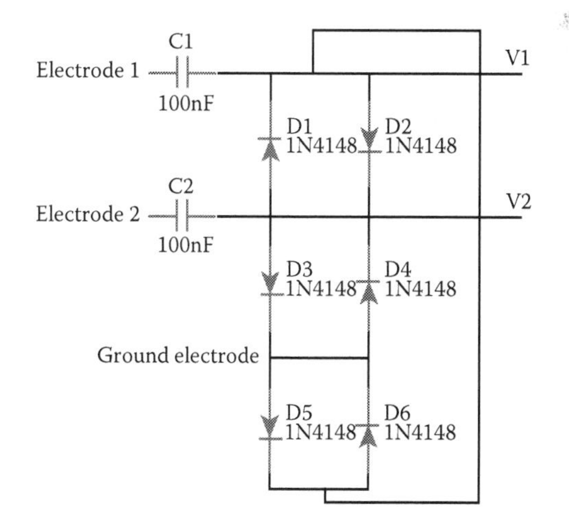


**FIGURE 13.7**
Front end AC coupling network and protection circuitry. V1 connects to input $V_{IN-}$ and V2 to $V_{IN+}$ as indicated in Figure 13.6.

which would saturate amplifiers $A_1$ and $A_2$ in Figure 13.6. This issue has been studied extensively, and the solution is to AC couple the inputs using series capacitors (Pallas-Areny et al. 1989). When this is done, an electrode with a skin–electrode impedance greater than about 86.5 kΩ may affect the frequency response of amplifiers $A_1$ and $A_2$ to the extent of it not meeting the AHA requirements (Valverde et al. 2004). A further problem arises when AC coupling the inputs. The noninverting inputs of amplifiers $A_1$ and $A_2$ must be biased so that there is a bias path for input currents. Floating input potentials will cause amplifiers $A_1$ and $A_2$ to saturate (INA122 data sheet 1997). Introducing series capacitors removes the DC bias path. Thus, a resistive network must be added at the inputs to provide a DC bias path to ground. This would degrade the CMRR (Spinelli et al. 2003). However, there have been proposed improvements to the input end AC coupling network which result in no degradation of the CMRR (Spinelli et al. 2003). As it turns out, the front end network shown in Figure 13.7 using the diodes to provide the DC bias path is one way of introducing effective AC coupling.

The diodes in Figure 13.7 also provide for input protection by clamping the signal to $\approx$ 0.7 V at high signal levels. The diodes act as very high resistances at low signal levels. The series capacitances and the diodes together also act as high-pass filters and remove very low-frequency components in the signals. This front end circuitry is equivalent to using standard balanced AC coupling (Spinelli et al. 2003). The use of AC coupling means that the ground electrode potential is not critical and can simply be the op amp ground reference potential or even the ground of the circuit itself. It is important not to have the bias currents too high; otherwise, saturation of amplifiers $A_1$ and $A_2$ will occur. Bias currents are normally on the order of nanoamperes.

If the ECG amplifier runs off a single supply rail, the ground reference potential (which would ideally have zero impedance so that changes in load currents keeps the potential steady) can be made using a resistive divider connected to the input of a noninverting op amp buffer. The resistor values in the divider depend on the maximum allowable power consumption and noise. Usually, 100 kΩ would be sufficient (Carter 2000). If a change is made to the voltage at the output of the buffer due to load fluctuations, negative feedback will ensure that the difference between the voltages at each of the inputs of the buffer will be zero, and the output voltage will be kept at the same potential as at the noninverting input of the buffer, which would normally be set to half the supply voltage using the voltage divider. However, to minimize the cost, it has been decided to provide for this ground potential at a couple of points throughout the circuit using separate resistor dividers without the use of noninverting op amp buffers, assuming of course that the current to be supplied at the output of each resistor divider is very small in comparison with the total current drawn by the divider—which would be the case if connected to points in the circuit of high impedance. The labelled node in Figure 13.7 indicating where the ground electrode connects to is DC biased using a resistor divider so that the two inputs of the ECG amplifier are DC biased at about half the supply voltage through the use of diodes. The same resistor divider cannot be used to bias other points in the circuit, as load current fluctuations from other parts of the circuit would be superimposed on the bias voltage, and this could lead to saturation of the ECG amplifier due to its extremely high gain; the result could even be an unwanted oscillation due to the amplifier repeatedly coming in and out of saturation.

Some ingenious methods have been used to minimize what is known as *common mode* interference, which is the interference on the reference ground potential common to both inputs of the amplifier, without AC coupling the inputs of the amplifier. One is to use two op amps—one as a buffer and the other as an inverting amplifier—to drive the ground electrode so that a slight positive change in the common mode voltage—usually derived as the sum of the two voltages at both inputs of the differential amplifier—within the circuit of the ECG amplifier would produce a corresponding negative change at the ground electrode (and vice versa). This would then produce a corresponding negative change present at both inputs to the ECG amplifier to compensate for the effect ("Information for Medical Applications", 2004). This is sometimes known as "right leg drive", as the ground electrode is normally attached to the right leg. Another similar technique uses a noninverting op amp configuration which integrates (or low pass filters) the common mode voltage. The output of the noninverting amplifier connects to the input of an inverting amplifier. It is clear that, at very low frequencies, a positive change in the common mode voltage produces a negative change at the output of the inverting amplifier. Thus the output of the inverting amplifier can be used to cancel the effects of DC baseline drift (Sieed et al. 2007). These methods reduce low-frequency common mode voltage variations (such as those caused by muscle movements) but without doing anything about the average DC offset itself. Decoupling capacitors used at the inputs would further minimize common mode interference. However, the DC offset arising from the DC potentials from the electrode–skin interfaces must be blocked before amplification; otherwise, saturation of amplifiers $A_1$ and $A_2$ in Figure 13.6 will occur (Webster 1999). AC coupling is used to block DC offsets and apply high-pass filtering as a low-cost alternative in removing very low-frequency common mode interference.

## Filtering

A passive twin T notch filter can be used to filter the 50 Hz noise. The main advantage of the twin T is the depth of its notch and its simplicity when compared to other types such as the Wien bridge. However, it is often not as easy to tune, due to the need for careful matching of the components (Mims and Helms 2000).

Unfortunately, the passive twin T notch filter would be affected by the load impedance, and thus an op amp buffer is normally used at the output of the filter to isolate the filter from the following output stages in the circuit. Though, in theory, an infinite notch is produced, its low Q means that the filter is likely to cause severe distortion in the filtered signals due to the filter attenuating spectral components of the signals over a very wide range of frequencies. To overcome this problem, an active realization of the filter would have to be designed which enables one to adjust the value for the Q. Figure 13.8 shows the circuit for an active twin T notch filter used in the final design for the low-cost ECG device ("High Q Notch Filter" 1969). The two op amp buffers do not actually play any role in the overall filter response. Increased Q values are obtained using the resistor divider, which provides positive feedback from the output of the filter to the passive notch filter itself. The op amp buffers serve merely to prevent the resistor divider from affecting the overall filter response in any other way than controlling its Q by isolating the resistor divider from the passive notch filter. That way, the resistor divider would not affect the frequency or depth of the notch. Q values of up to around 50 can usually be obtained in this circuit. As positive feedback is used and the buffers would remove losses in the passive components, it is likely that configuring the circuit for higher Q values by use of greater positive feedback would cause the circuit to amplify or even oscillate. Thus, there is an upper limit to how much positive feedback can actually be applied, and simply connecting the output directly back to the passive filter section via the buffer will, in theory, produce an ideal notch with infinite Q. In practice, however, it will cause the filter to oscillate. Mismatches in component values cause the circuit to be increasingly prone to oscillation as well as degrading the depth of the notch.

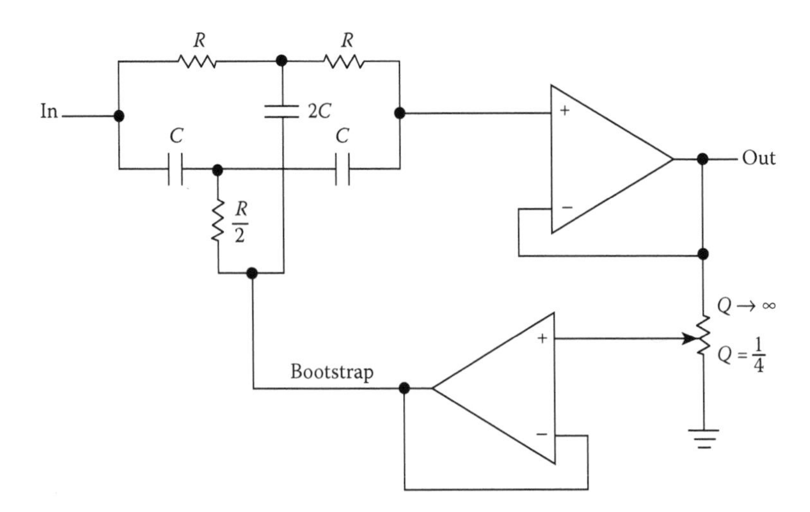


**FIGURE 13.8**
Active twin T notch filter with variable Q. (Horowitz, P. and Hill, W., *The Art of Electronics*, 2nd ed., Cambridge University Press, Cambridge, 1989.)

## Automatic Gain Control

An automatic gain control (AGC) is a control system which uses negative feedback to stabilize the gain of the ECG amplifier in much the same way a phase-locked loop (PLL) stabilizes and locks signal frequencies. An AGC system, like the PLL, is highly nonlinear and complex to analyse. Again like the PLL, the AGC system can be approximated to the first order with a considerable amount of inaccuracy. An AGC system is not the same as a limiter, which does not usually use any form of feedback and merely crops the output waveform so that the required output level is maintained. An AGC system has to maintain the required output level without any form of distortion. An AGC is highly desirable in an ECG system, as smaller ECG signals can be automatically increased and larger ones decreased to the same output level without any adjustment by the user being necessary. This is particularly important when the ECG system is to be used on many patients in quick succession.

Figure 13.9 shows the AGC circuit (George 1998) which became part of the final design of the ECG amplifier after a few changes in the component values. The resistance between the drain and source terminals of the FET is made

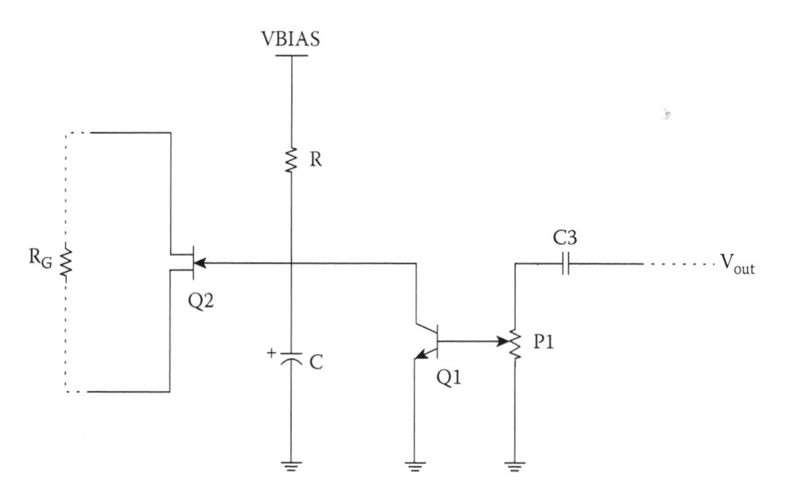


**FIGURE 13.9**
Part of the schematic of an inexpensive AGC amplifier used in the final design of the ECG amplifier. Q2 is part of a variable gain amplifier stage. The output of the amplifier is fed back through C3. It was decided to connect the wiper of P1 to the base of Q1 so as to obtain overall output levels greater than 1.2 V peak to peak. VBIAS = 0.5*VCC. VCC is the supply (or battery) voltage. In other words, the collector of Q1 (and the gate of Q2) is DC biased at half the supply voltage. (Modified from George, J. P., Effective AGC amplifier can be built at a nominal cost, 1998. *Electronic Design*, 46, no. 18, 90–92. Retrieved October 2010 from http://electronicdesign.com/Articles/Index.cfm?AD=1&ArticleID=6272.)

variable by changing the voltage at its gate terminal, provided that the voltage across the FET's drain and source terminals is very small. This poses a challenge with respect to where to place the FET in the ECG amplifier so that a change in the resistance between the FET's drain and source terminals would change the gain of the amplifier. One possible place is where $R_G$ is located in Figure 13.6. As it turns out, the current through $R_G$ is practically zero, and thus the voltage across it is very small, if not zero. Thus, the drain and source terminals of the FET can be connected in place of $R_G$, and a variable gain amplifier results. When the voltage at the base of the bipolar junction transistor (BJT) in Figure 13.9 exceeds a certain threshold, the voltage at the gate of the FET suddenly decreases, and the gain of the amplifier would decrease accordingly. If the voltage at the base of the BJT decreases, the BJT switches off, and the voltage at the gate of the FET slowly increases according to the AGC "response time" set by R8 and C4 in Figure 13.9. The gain of the amplifier would then increase slowly.

The AGC system has to be carefully designed so that the overall phase and gain conditions around its control loop are such that the loop will be asymptotically stable and oscillation will not occur. The AGC system cannot respond immediately in real time to any change in the output level, as severe distortion and even unwanted oscillation will occur. The AGC system's response time is thus limited to a value lower than the lowest useful frequency component of the signal (Martinez 2001). Thus, to control the output R wave levels of the ECG, the AGC response time must be at least about 1 s. However, because the ECG system has to deliver accurate and reliable respiratory rate information which has spectral components centred at frequencies which are about 1/10th of those of the average RR interval (or heart rate), the AGC response time was chosen to be about 10 s.

## USB Interfacing

The decision to use a USB interface came about due to its widespread and almost universal popularity and usage in mobile phones and PDAs. Other ways of getting the ECG signals onto such handheld devices include use of the microphone audio input jack and also Bluetooth, a technology which has increased in popularity in recent times. Infrared communication was popular a few years ago before the advent of Bluetooth. The main problem with infrared communication is its extremely limited area of coverage. Usually, the infrared transmitting device would have to be positioned only a few centimetres away and in direct view of the receiving device, making it impractical for most applications. Wi-Fi is not really an option; the external circuitry required is rather complex, as the interface would need to set up a wireless access point (WAP).

It has been said that the use of the audio input jack would be superior to the use of USB as it is far cheaper to implement, and no extra circuitry is required. However, it appears that many phone manufacturers configure their audio codec chips—dedicated microcontrollers with inbuilt analog-to-digital convertors (ADCs) specifically designed for portable audio that interface directly to the device's main processor—so that high-pass filtering of the input audio signal is switched on by default, and special drivers would need to be installed on these devices to switch this filtering off. For example, the Socket Mobile SoMo 650-M PDA used for testing has a WM 9713 audio codec chip from Wolfson Microelectronics. The driver—a piece of firmware code which runs in the audio chip and which has been loaded onto the chip by the manufacturer—configures the use of the audio chip's hardware functions, one of which is the high-pass filtering function. As it is set on by default, spectral components of the audio signal with frequencies less than about 300 Hz are attenuated by an amount roughly equivalent to a high-pass first-order (20 dB per decade) spectral response with a corner frequency of around 300 Hz. Obviously, given that the spectral components of ECG signals lie between 0.05 and 150 Hz, it is clear that the high-pass filtering would render this option as unsuitable, particularly as the high-pass filtering would have to be switched off in mobile phones on a large scale, which is something that phone manufacturers are highly unlikely to agree to. There are not many actual audio input jacks as such on mobile phones, as the microphone connections are normally inbuilt as connections to the phone's USB port.

Bluetooth is certainly a good option to consider, and given its wireless capabilities and ease of connectability, would have definitely been the one to opt for, provided that the technology could be found in nearly all phones in use in the developing world (which, as yet, is not the case). Another problem is that widely available cheap microcontrollers do not have inbuilt hardware to handle Bluetooth communication, and thus more expensive dedicated chips would need to be used, increasing the cost of the ECG system.

## USB Configuration and Communication

The USB, or universal serial bus, is definitely not new technology, as it was originally developed in 1995 by Intel and Microsoft to meet the need for an inexpensive and widespread solution that would allow connection of a wide range of peripheral devices to their respective PCs (Jungo 2010). One advantage of USB is its plug-and-play feature, which enables the computer to automatically detect the peripheral device upon connection and install any necessary software needed to communicate with the device. USB currently operates at a number of speeds, from low speed (1.5 Mb/s) to high speeds of 480 Mb/s and above. Many cheap, widely available microcontrollers handle USB at what is known as full speed, which corresponds to a rate of 12 Mb/s. Both USB v1.1 and v2.0 can operate at full speed. USB v2.0 is backward-compatible and able to operate at even the lowest speeds. It is rare to find a microcontroller able to handle the highest speeds, for obvious reasons. Microcontrollers with inbuilt dedicated USB hardware can usually handle full-speed USB communication.

It is possible to implement USB communication in ordinary microcontrollers with no dedicated hardware for USB communication by simply implementing it in the firmware. The highest rate in this case is often determined by the maximum number of low-level instruction mnemonics (in assembly language) the microcontroller can execute per second, as well as the number of clock cycles needed per instruction. To handle even the most basic of USB communication requires the use of at least several of these instructions per data bit processed (Starkjohann 2008). Then, it is a case of simple mathematical calculation to work out the minimum clock frequency needed for the processor. The processor's clock frequency usually has an upper limit in the range of tens of megahertz, which would rule out any possibility of high-speed USB communication and would even often rule out the use of full-speed USB communication.

To understand how to configure USB devices, the notions of a USB host and USB client are often introduced. A USB host is often simply a PC; however, some microcontrollers can serve as USB hosts, ones which are classified as on-the-go devices, which means that they are self-contained (i.e., ready to go) and do not need to depend on other devices for their operation. Nearly everything else which connects via a USB port is referred to as a USB client. This includes mice, keyboards, printers and nearly all mobile phones and PDAs. The only exception is a USB hub, which is neither a host nor a client. Unlike in ordinary RS232 serial communication, a USB client cannot work without a USB host. The USB host usually supplies 5 V of power to a USB client and controls many parts of its communication with the client. A system with USB clients connected together (for example, two PDAs connected together) cannot work without interaction with a USB host, and what is probably more surprising is that a system also cannot work which comprises only USB hosts connected together. For example, two PCs connected together cannot communicate over USB! What is needed is a host and a client.

When a USB client connects to a USB host, the configuration process is started, and the host computer acquires this information from the client device. The descriptor is what the USB client device sends to the host so that the host computer knows what kind of device it is. For example, one class of descriptor describes whether the device connected is a human interface device (HID), which means keyboards, mice and so on, or whether it is a mass storage device, which means USB memory sticks. These descriptors are at the device level and also include the product and vendor ID of the device. Note that every USB device in the world has a unique product and vendor ID, which must be obtained from www.usb.org, particularly if the product is to be used commercially. Descriptors at the configuration level obviously describe the configuration of each device, mainly its power consumption. Within each configuration, there is a set of interfaces which in turn have a set of end points. End-point data are at the lowest level, describing things such as the data packet size and transfer type.

End points are used only for USB clients and are simply ports on the device to receive and send data which flow back and forth from host to device. The host has what is known as memory buffers to describe its ports instead of end points. The channels of data sent and received from memory buffer to end point are known as pipes. Data can be transmitted in only one direction at a time. There are four types of transfer of data. Control transfers are usually for configuration and status information. Bulk transfer is for transferring large amounts of data and is the main method of transfer for the ECG signals. Interrupt transfers send and receive small amounts of data infrequently where a maximum service period is guaranteed. Isochronous transfers are for high-speed communication and should not be considered.

For a USB host to connect to a USB client, the USB host must have some knowledge of the various USB descriptors of the device; otherwise, it would not know which device it needs to connect to. If bulk transfers are used to transfer the ECG data, then an appropriate interface descriptor needs to be found which supports this method of transfer. Otherwise, the firmware will not run properly and will send an error message if it cannot locate the correct type of end point. Once the correct end points are found, and the device descriptors match what the host is looking for, then, and only then, will the firmware run and send the ECG data to the device through its USB port.

## USB Interface Hardware and Device Detection

Of course, not just any circuit can connect to the USB port. The very first thing which happens when a USB device is connected to a USB host is the detection of the actual connection. Figure 13.10 shows how the USB data lines should be biased so that the computer is forced to recognize that the circuit connected is a USB device (Hyde 1999). There are only two actual USB data lines, which use non-return to zero (NRZ) differential signalling and are thus labelled D+ and D–. Differential signalling has advantages in terms of crosstalk and coupling of EMI. Normally, no more than 100 mA of current can be supplied to the device; otherwise, an error occurs. If more than 100 mA is required, the device should be powered separately. However, if absolutely necessary, up to 500 mA can be drawn by specifying the appropriate configuration descriptor.

The detection of a rising edge on either D+ or D– indicates that a USB device is attached. The resistors shown bias the lines so that this transition

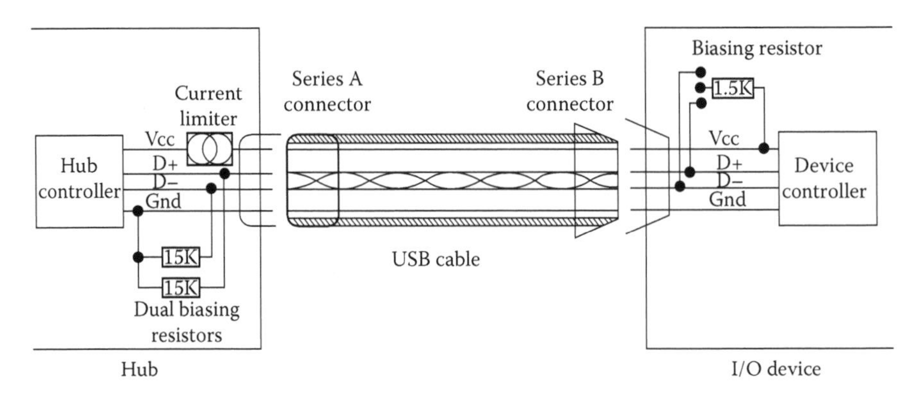


**FIGURE 13.10**
USB cable connection details. (From Hyde, J., *USB Design by Example: A Practical Guide to Building I/O Devices*, John Wiley & Sons, New York, 1999. © 1999 John Hyde. With permission.)

occurs when the device is plugged in. A rising edge on the D– line indicates that the device is a low-speed USB client (1.5 Mb/s), whereas a rising edge on the D+ line indicates a full-speed device (12 Mb/s). The resistor which ties the appropriate line HIGH is switchable to either of the D+ or D– lines. To comply strictly with the USB specification and guarantee that the device will be detected properly when connected to the host, the D+ and D– lines should be regulated to 3.6 V, perhaps by using zener diodes. Alternatively, a 3.6 V regulator can power the entire device. However, the use of zener diodes would be more cost-effective. Usually, series resistors are added along the USB data lines in the device and capacitors which connect each line to ground so as to apply low-pass filtering to the USB signals and remove noise. Inrush currents and flyback voltages mean that decoupling capacitors should be used in the USB device, with values between 1 and 10 μF ("USB in a Nutshell" 2010). The characteristic impedance of the USB bus is about 90 ohms, give or take 15%. The series resistors could be used to match the line's impedance; however, the characteristic impedance should be taken into account when designing USB devices on a printed circuit board (PCB) (Intel Corporation 2001). Once hardware detection has been achieved, the configuration process is started, and data are sent from the host to the device.

Before the USB interface is designed around a specific microcontroller, it is necessary first to see if the microcontroller can actually digitize the ECG signals to an acceptable level; otherwise, another type of microcontroller has to be used. It can be shown (Chan et al. 2003), by some detailed mathematical analysis, that by assuming that the QRS complex is a triangular spike, the SNR required for the error in the heart rate to be within ±1 beat per minute is 36 dB. If one assumes that adequate hardware filtering has been used in the ECG amplifier, then the noise at the output of the USB microcontroller interface is dominated by what is known as *quantisation noise*, which is the noise due to a finite resolution of the ADC and is caused by rounding errors between the actual input signal and the sampled digital one.

If we represent the ECG signal as a square wave having a duty cycle equal to the width of the QRS complex divided by the RR interval, then we arrive at the signal's root mean square (RMS) value of about 0.3 times the square of the R wave amplitude. Assuming that the typical amplitude of the R wave is about 1 V coming out of the ECG amplifier, the ratio of the input signal's RMS value to its peak value would be about –10 dB. This, of course, would be an overestimate, as the QRS complex is not square but triangular. Assuming a triangular shape would probably add another –10 dB or so to the ratio. To satisfy a SNR of 36 dB in the digitized stream, use of a simple mathematical relation between the signal's SNR and the ratio of the signal's RMS to peak value (Lyons and Yates 2005) yields the result that the minimum number of bits for the ADC is likely equal to 9, due to the reasonably large negative decibel value for the ratio and also due to the fact that an ideal ADC has been assumed. The worst-case SNR at the output of the ADC would apply for the minimum sampling rate of about 100 Hz, just satisfying the Nyquist criterion. The SNR can be improved slightly by a technique known as *oversampling* which uses an increased sampling rate, at the expense, however, of using increased processor memory resources. To be on the safe side, and taking into account lower R wave amplitudes, a 10-bit ADC should satisfy the SNR requirement of 36 dB. Fortunately, nearly all commonly available microcontrollers have 10- or 12-bit ADCs.

An example of a microcontroller which could be used for the USB client interface is the very popular and cheap ATTINY45 AVR processor with 10-bit ADCs. In this case, USB communication would be implemented through the firmware as no dedicated USB hardware controllers are inside the microcontroller. Though the USB client interface is clearly quite simple, compact and low cost, the interface cannot work with most PDAs unless the PDA has an inbuilt USB host controller, which not many of them have. However, the client interface should work fine when connected to a PC.

## Firmware Operation

Firmware for the USB interface is normally created using an integrated development environment (IDE) running on a computer, such as the one we used—the Code Warrior IDE from Freescale Semiconductor. Usually, firmware is written either in the C++ language or in assembly language, depending on how critical the timing needs to be for a certain task. Assembly language runs faster as the instructions do not need to pass through a compiler. After the firmware is written, it is programmed, or flashed, into the microcontroller's memory. To make things easier, this is done via a USB port on the computer connected directly to a special-purpose "demonstration kit" or "development board". For the USB interface, we used the Flexis JM Demo development board with the MCF51JM128 microcontroller available from Freescale Semiconductor configured as a USB host.

The actual core operation of the firmware is quite straightforward and involves a never-ending loop, which is run after a procedure of initialization which involves configuring the registers in the microcontroller. Within the loop, the ADC's registers in the microcontroller are accessed after waiting for a time equal to the sampling rate (10 ms, which is equivalent to sampling at 100 Hz), then processing is done to compute from earlier ECG signal data values (normally stored in a variable array) any extra information necessary to be displayed, such as the heart rate, QT interval and PR intervals. Finally, the data are sent through the USB host controller in the processor and out to the USB port on the mobile device, set up as a COM port. The loop repeats at an interval equal to the sampling rate. The USB interface descriptors (in particular, the class, subclass and protocol end-point values) are configured for the intended mobile device during the development stage and not actually changed while the code is run. Figure 13.11 summarizes this in a flow chart. StickOS (Testardi 2009), open-source firmware running in the MCF51JM128 microcontroller configured as a USB client, has been used to display the ECG signal on a PC. A simple program written in BASIC is all that is required to use StickOS.

## Displaying the ECG Signals on an PDA or Mobile Phone

To capture and display the digitized ECG signals sent out from the USB microcontroller interface on a PDA or phone, it is first necessary to determine how a software application can capture such signals. The firmware in the USB interface sends out each sample of the signal 100 times per second via USB and, if necessary, would send out the extra information every 10 s along with the data from the main signal. In the simplest case, a USB client is used and configured as an HID device and connected to a computer running as the USB host. No device drivers on the host computer are necessary for HID devices, and all information from the data retrieved through the USB port can be obtained via what is known as *HID descriptor reports*. Ready-made code from a third party is available ("A USB HID Component for C#" 2007), which makes things even easier by performing all the queries to the data in the reports so that the only thing the software developer needs to do is to handle events such as *usb_OnDataReceived(array memory_buffer),* which is obviously triggered every time data are received through the USB port. The signal data are analysed and displayed by the code inside this event procedure using the array which holds the contents of the host computer's memory buffer at the time the data were received. For example, one statement in the

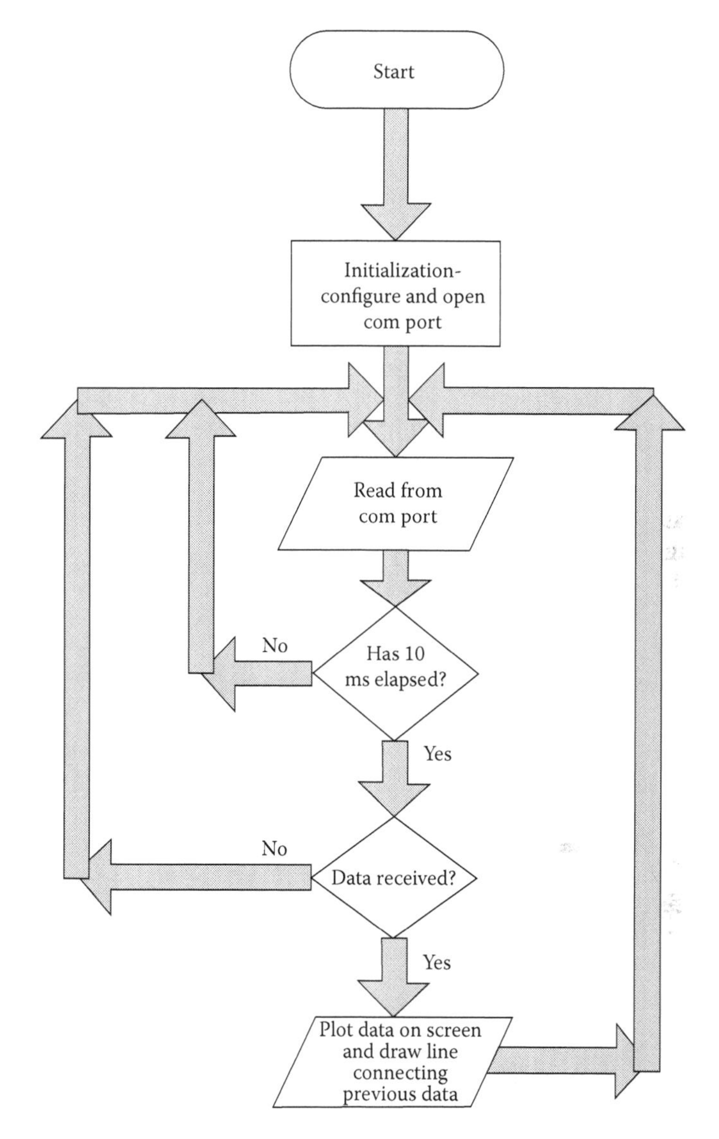


**FIGURE 13.11**
General flow chart for the software to display the ECG signals.

code might be *capture_value = memory_buffer.data[1]*, which would capture the first value of the ECG data. It is quite common to write the code in the C# language using Microsoft Visual Studio, as C# is powerful but simple to use when compared to many other powerful languages such as C++. C# is also relatively easy to learn.

Another way is to associate one of the COM ports in the computer or mobile device to the USB port. Then, the signal data would be converted into serial format, and a software application can be written so as to retrieve the data indirectly through the COM port. When the USB interface is configured as a USB client and connected to the host computer, depending on the class of device the USB interface is configured for, it could tell the computer to install a driver which creates a new COM port associated with the USB port, as is the case when StickOS is used. The computer usually does this by popping up a window to prompt the user to install the driver. This driver is known as a serial over USB virtual COM port driver. Developers achieve this by creating an *inf* file, which can be done easily using a third-party application (Quiroz 2008). The *inf* file contains an entry which is the name of the device connected. The name of the device is in one of the device's USB descriptors, and is part of the firmware in the microcontroller's memory. The entry has to match the one in the *inf* file. When the USB interface is configured as a USB host and connected to a mobile device, provided the appropriate USB interface descriptors have been set up correctly, quite often the device is ready to use as the driver is already installed. If the USB port is being used by another program (for example, some advanced features of ActiveSync continuously monitor the USB port), a COM port can be made available for use by disabling the appropriate functionalities in the settings for the mobile device.

Retrieving the ECG data through the serial COM port is not trivial, as received errors can occur occasionally due to buffer overruns or timing errors. If more than one stream of data is to be received or sent at one time, then a technique known as *multithreading* needs to be implemented to ensure that sent data and multiple received streams of data do not cause data corruption (de Klein 2009). However, the advantage of using COM ports is that connectivity to the device is easily implemented in code and is independent of the class of device.

Once a new COM port is installed and ready to use, Microsoft Visual C++ can then be used to write a program which opens the newly created COM port, configures it and then sends and receives data from it accordingly. The program used to display the ECG signals as an oscilloscope-like trace involves reading data from the COM port at every given short interval, then drawing out the trace on the screen over time using the data retrieved. It is recommended to use Windows 32 API commands in a Visual C++ environment using Microsoft Visual Studio, as these commands would allow the developer complete control and flexibility in the design of the software, and the Windows environment in general is the product of programming in a C language environment.

Using the final design of the ECG amplifier and the USB interface configured as a USB host, the authors' lead II ECG signals were displayed on a Socket Mobile SoMo650-M PDA running the Windows Mobile 6.0 Classic Edition platform. A program was written by the authors using Windows 32 commands in a Visual C++ environment using the Windows Mobile SDK. The ECG signals on the phone are shown in Figure 13.12. The timebase shown is only approximately 100 ms/division. The Visual C++ software reads and displays the ECG data through the USB port on the phone. ECG

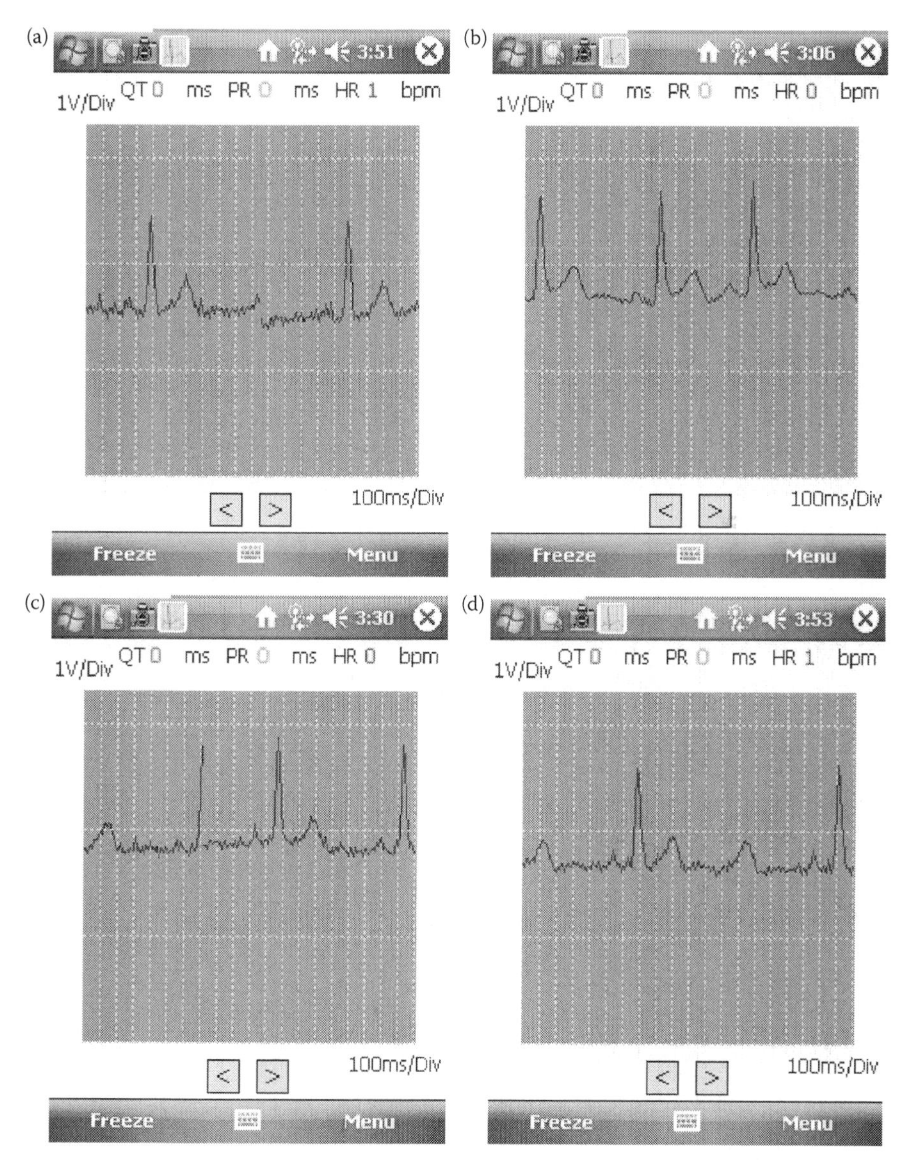


**FIGURE 13.12**
ECG signals obtained on a Socket Mobile SoMo650-M phone using aluminium foil electrodes (a), copper strip electrodes (b), galvanized iron electrodes (c) and commercial electrodes (d).

parameters were also computed and detected, as shown in Figure 13.13, using third-party software (Petrov 2004) modified by the authors for display on a mobile device.

For comparison purposes, lead II ECG signals were also obtained using a CT200C commercial electrocardiograph from Macquarie Medical Systems; these are shown in Figure 13.14.

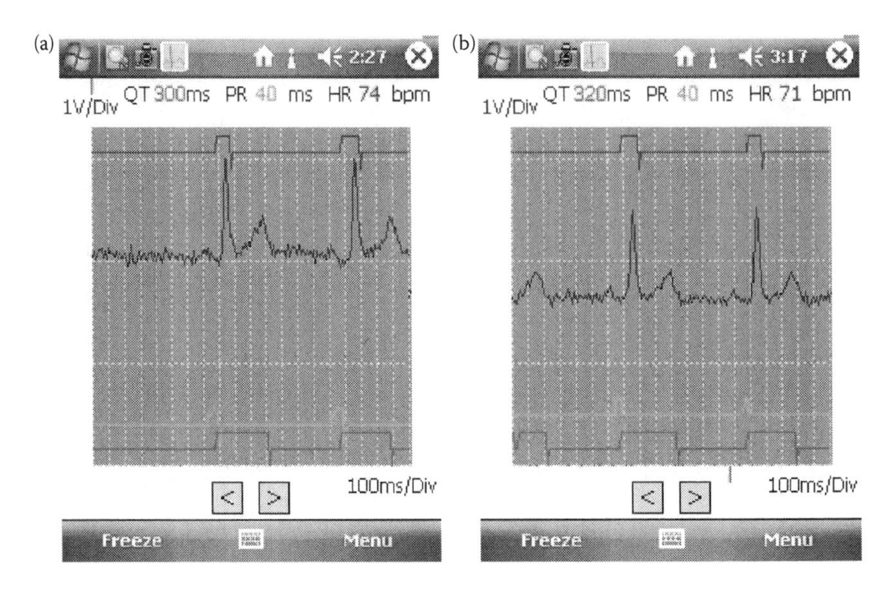


**FIGURE 13.13**
ECG signals on the SoMo650-M phone using copper strip electrodes, complete with automatic computations of the QT, PR and RR (heart rate) intervals. The pulse-like trace above the ECG signals marks out the QRS complexes. The light trace immediately below the ECG signals marks out the PR intervals, and the darker trace at the bottom marks out the QT intervals. (a) Authors' ECG signals and ECG parameters for an average heart rate (HR) of 74 beats per minute (bpm); (b) Authors' ECG signals and ECG parameters for an average HR of 71 bpm.

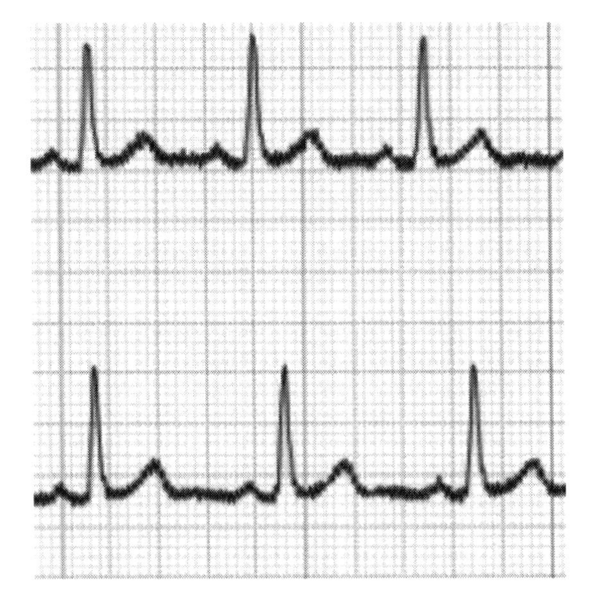

**FIGURE 13.14**
Lead II ECG signals using a commercial electrocardiograph. The timebase is 5 large square divisions per second. Amplitudes are on a scale of 2 large square divisions per millivolt.

## Discussion and Conclusion

Mobile phone technology can provide the basis for a new generation of affordable, easy-to-distribute electronic health solutions for resource-poor communities. We argued in favour of "local applications" as we do not believe that a model based on remote analysis would work in this context. In this chapter, we have provided a way to extend mobile phones' processing and interface capabilities with external sensors to create a low-cost ECG device.

One challenge in providing better healthcare services is how to increase the participation of front-line health workers using limited financial and human resources to deliver assistance to an increasing number of people (Varshney 2007). Mobile and wireless technologies can be effectively utilized by matching infrastructure capabilities to healthcare needs. Research in mHealth scrutinizes ways to take advantage of the wide availability of mobile devices—mobile phones, PDAs, mobile computers—to deliver mobile health solutions for front-line workers. Unlike other mobile ECG devices, our solution provides a near-commercial-quality, cost-effective, energy-efficient way for doctors in developing countries to use their existing mobile phones for cardiac diagnosis, as well as for ECG monitoring of their patients wherever they may be.

Work done by the UN Foundation (2008) shows that governments, companies and nonprofit groups are already developing mHealth applications to improve healthcare. This report presents 51 programs, either currently operating or slated for implementation in the near future, that are taking place in 26 different developing countries. These applications are creating new pathways for sharing health-related information, even in the most remote and resource-poor environments. These projects provide key solutions in (i) education and awareness; (ii) remote data collection; (iii) remote monitoring; (iv) communication and training for healthcare workers; (v) disease and epidemic outbreak tracking; and (vi) diagnostic and treatment support.

In our experience even people on very low incomes in developing countries are acquiring and using mobile phones. They are used for many purposes, including checking market prices and keeping in touch with relatives who migrate to urban areas, rarely for healthcare purposes. Nonetheless, all but the simplest mobile phones now have operating systems, and some have very sophisticated and powerful processors. It seems that implementation of low-cost, high-penetration "analysis, diagnosis and consultation" solutions that explore mobile phones' processing and interfacing capabilities is achievable. Therefore, it is possible to explore this capability and write simple applications that run on front-line health workers' own mobile phones, providing simple "tools" to aid in their daily activities.

Electrodes were constructed, an amplifier was designed and tested, firmware was programmed into a USB microcontroller interface which

**TABLE 13.1**

Bill of Materials: ECG Amplifier Component List

| Quantity | Description | Unit Price (A$) | Total Cost (A$) |
|---|---|---|---|
| 1 | TL074CD SOIC Quad Op-Amp IC | 0.40 | 0.40 |
| 1 | BC 847 SOT23 BJT Transistor | 0.03 | 0.03 |
| 1 | PMBF4416 SOT23 FET Transistor | 0.30 | 0.30 |
| 11 | SMD Resistors | 0.05 | 0.55 |
| 7 | IN4148 DO-35 Small Signal Diodes | 0.02 | 0.14 |
| 7 | SMD Ceramic Capacitors | 0.02 | 0.14 |
| 4 | 0.5 W Trimmers | 0.50 | 2.00 |
| 1 | Zinc Chloride PP3 9 Volt Battery | 0.70 | 0.70 |
| 1 | 9 V PP3 Battery Clip | 0.50 | 0.50 |
| Total project cost (A$) | | | 4.76 |

**FIGURE 13.15**
External view of a prototype of the ECG amplifier built by the authors.

successfully digitized the incoming ECG signals, and, finally, a program in Visual C++ was written to display the ECG signals on a mobile device. Comparing the results obtained using our final design of the ECG amplifier shown in Figures 13.12 and 13.13 with the results from a commercial ECG system shown in Figure 13.14, it is clear that the ECG amplifier produces ECG signals of a quality very close to that obtained from a commercial ECG system. The SNR of the ECG signals obtained from the ECG amplifier appears to have exceeded the theoretical requirement of 36 dB, for a maximum error in the heart rate of 1 bpm under certain conditions, depending on the amount of noise present. It also appears that the use of copper electrodes gives better-quality ECG signals due to the lower resistance between the skin and the surface of the electrodes, as copper is a better conductor of electricity. It was also found that DC baseline drift was more of a problem when using electrodes with contact surfaces of smaller area. The use of aluminium foil electrodes gave by far the worst results.

Upon inspection of the ECG parameters shown in Figure 13.13, it appears that the QT and RR intervals are within theoretical limits, but the PR interval shows some inaccuracy due to its sensitivity to the amount of noise present. The PR interval is underestimated. However, the pulse-like marker waveforms clearly indicate that the appropriate parts of the ECG signal are correctly identified.

Typical commercial ECG machines today cost in excess of $3000, and given that the average income in Mozambique (where the ECG monitor in its final form is to be sold) is A$854 per person per year (Education Services Australia 2009), this is an extortionate price to pay, as it is in many other developing countries. The poor maintenance of existing commercial equipment in developing countries has resulted in the development of many low-cost ECG testers, one of which was built using A$28 worth of components and sells for A$65 (Hill 2006). The target selling price of the ECG monitor for this project has never been precisely stated. Engineering World Health normally sets good targets, and they have set a target for such low-cost ECG devices for developing countries as being limited to A$50 for a prototype and A$4 each when sold in batches of 500 or more (Bermani et al. 2004). Table 13.1 provides the bill of materials for a prototype of the ECG amplifier as shown in Figure 13.15. Prices apply for quantities purchased in bulk (500 plus) from element 14. Note that the bill of materials does not include board fabrication and costs for assembly. It also does not include the cost for the USB interface. However, in the final product some of the components listed in the table will be replaced with cheaper alternatives—in particular, adjustable components will be replaced with fixed ones. The electrodes would presumably come from scrap materials. From the total cost stated in the table, it is possible that even the entire ECG system—including the USB interface, which would contain a microcontroller chip costing only a couple of dollars when bought in bulk—will cost only around A$4.

## References

Allaby, A., & Allaby, M. 1999. *A Dictionary of Earth Sciences*, 2nd ed. Oxford: Oxford University Press.

Amer, M. B. 2008. Lecture 7 Biopotential Electrodes (Ch. 5). King Saud University, Riyadh, Saudi Arabia, Introduction to Biomedical Equipment BME 201 Lecture Notes, 2008 (unpublished).

Anderson, T. 2007. Mobile phone lifeline for world's poor. *BBC News*. Retrieved 2009 from: http://news.bbc.co.uk/2/hi/business/6339671.stm

Avans Hogeschool. 2007. A USB HID component for C#. *The Code Project*. Retrieved October 2010 from http://www.codeproject.com/KB/cs/USB_HID.aspx?msg=2283285.

Baker, B. C. 2004. Operational amplifiers part IV of VI: Working your amplifier inside the single-supply voltage "box". AnalogZONE. Retrieved October 2010 from http://www.analogzone.com/acqt0809.pdf.

Bermani, B., Dandekar, B., & Ingemi, P. 2004. *EvalECG: Low Cost ECG Tester.* ECE 2799 Technical Report. Worcester MA: Worcester Polytechnic Institute.

Bowbrick, S., & Borg, A. N. 2006. *ECG Complete*. Oxford: Churchill Livingstone.

Carson-DeWitt, R. 2007. Coronary artery disease (Coronary heart disease). *EBSCO Publishing*, Ipswich, MA, USA. Retrieved October 2010 from *The Mount Sinai Medical Center*, New York, USA. http://www.mountsinai.org/Other/Diseases/Coronary%20artery%20disease

Carter, B. 2000, November. *A Single-Supply Op-Amp Circuit Collection*. SLOA058 Application Report. Texas, TX: Texas Instruments.

Catalano, J. T. 2002. *Guide to ECG Analysis*, 2nd ed. Philadelphia, PA: Lippincott Williams & Wilkins.

Chaikowski, I., Stromeyer, H., Bukhman, V. & Kovalenko, O. 2008. Value of the single-lead ECG in comparison with 12-lead ECG. *Ukrainian Journal of Telemedicine and Medical Telematics*, 6, no. 1, pp. 25–27.

Chan, P., Cunjalo, M., McKenzie, B., Monsurate, R., & Nelson, S. 2003. *Design Specification for a Personalised Medical Emergency and Distress System*. School of Engineering Science, Simon Fraser University, Burnaby, B. C., Canada.

Cohen, J. S. 2005. Electrocardiogram. *eMedicine Health*. Retrieved October 2010 from http://www.emedicinehealth.com/electrocardiogram_ecg/page7_em.htm

Committee on Disaster Research in the Social Sciences: *Future Challenges and Opportunities. 2006. Facing Hazards and Disasters: Understanding Human Dimensions. (U.S.) National Research Council*, Washington, DC: National Academies Press.

Das, S. 2006. The normal ECG and the 12-lead system. *AnaesthesiaUK*. Retrieved October 2010 from http://www.frca.co.uk/article.aspx?articleid=100681

De Kleink, R. 2009. Serial library for C++. Retrieved October 2010 from http://www.codeproject.com/KB/system/serial.aspx.

The ECG Library website. (2011) Retrieved October 2010 from http://www.ecglibrary.com/ecghome.html

Education Services Australia. 2009. The Global Education website. *Mozambique.* Retrieved October 2010 from http://www.globaleducation.edna.edu.au/globaled/page643.html

Fogoros, R. N. 2008. The electrocardiogram (ECG). *About.com Heart Health Center.* Retrieved October 2010 from http://heartdisease.about.com/cs/ekgecg/a/ECG.htm

Fuenmayor, A. J., Fuenmayor, A. M., Mora, R. E., & Fuenmayor, A. C. 1998. QT-interval dispersion in malnourished children. *Clinical Cardiology* 21, no. 3, 201–205.

Gardiner, J. 1981. *The ECG—What Does It Tell*? Cheltenham, Gloucestershire, UK: Nelson Thornes.

George, J. P. 1998. Effective AGC amplifier can be built at a nominal cost. *Electronic Design*. Retrieved October 2010 from http://electronicdesign.com/Articles/Index.cfm?AD=1&ArticleID= 6272.

Gomella, L. G., & Haist, S. A. 2002. *Clinician's Pocket Reference*, 9th ed. New York: McGraw-Hill.

Hegarty, D. 2007. Why Bazett's formula? *Anesthesia & Analgesia*, 105, 535-a.
Hemmings, H. C., & Hopkins, P. M. 2005. *Foundations of Anesthesia: Basic Sciences for Clinical Practice*, 2nd ed. Philadelphia, PA: Mosby.
High Q notch filter. National Semiconductor Application Note, March 1969.
Hill, D. 2006. Graduate Zach Jones dedicated to medical devices for developing nations. *Pratt School of Engineering at Duke University*. Durham, NC, USA. Retrieved October 2010 from http://www.pratt.duke.edu/news/?id=436
Hinz, R. 2002. Biopotential electrodes. Lecture notes, Medical Engineering Summer Course 2002, Polytechnic University of Timisoara, Timisoara, Romania.
Horowitz, P., & Hill, W. 1989. *The Art of Electronics*, 2nd ed. Cambridge: Cambridge University Press, Cambridge.
Huff, J. 2005. *ECG Workout: Exercises in Arrythmia Interpretation*, 5th ed. Philadelphia, PA: Lippincott Williams & Wilkins.
Hyde, J. 1999. *USB Design by Example: A Practical Guide to Building I/O Devices*. New York: John Wiley & Sons, New York.
INA122 Single Supply, MicroPower Instrumentation Amplifier. 1997. Datasheet No. PDS-1388B. *Burr-Brown Corporation* USA.
Information for medical applications. 2004. Texas Instruments Applications Guide. *Texas Instruments Incorporated*, Fort Worth, TX, USA.
Intel Corporation. 2001. *High Speed USB Platform Design Guidelines*. Santa Clara, CA, USA.
Istepanian, R., Laxminarayan, S. & Pattichis, C., ed. 2006. M-Health: Emerging Mobile Health Systems. *Kluwer Academic / Plenum*, New York, USA: Springer. p. 623.
Jungo Ltd. 2009. WinDriver USB User's Manual v10.10. Retrieved October 2010 from http://www.jungo.com
Khan, M. G., & Cannon, C. P. 2008. *Rapid ECG Interpretation*, 3rd ed. Totowa, NJ: Humana Press.
Khandoker, A. H., Karmakar, C. K., & Palaniswami, M. 2009. Automated recognition of patients with obstructive sleep apnoea using wavelet-based features of electrocardiogram recordings. *Computers in Biology and Medicine*, 39, no. 1, 88–96.
Kirtley, C. 2002. Operational amplifier electrocardiogram. Biomedical Engineering BE 513 Course Notes (unpublished), University of Vienna, Vienna, Austria.
Klabunde, R. E. 2005. *Cardiovascular Physiology Concepts*. Philadelphia, PA: Lippincott Williams & Wilkins.
Kligfield, P. et al. 2007. Recommendation for the standardization and Interpretation of Electrocardiogram. *Circulation*, 115, no. 10, 1306–1324.
Lee, J. 2000. ECG monitoring in theatre. *World Federation of Societies of Anaesthesiologists*, no. 11, article 5.
Lee, S., & Kruse, J. 2008. Biopotential electrode sensors in ECG/EEG/EMG systems. *Analog Devices*, Norwood, MA, USA. Retrieved October 2010 from http://www.analog.com/static/imported-files/tech_docs/ECG-EEG-EMG_FINAL.pdf
Low, Y. F., Mustaffa, I. B., Saad, N. B. M., & Bin Hamidon, A. H. 2006, June. Development of PC-based ECG monitoring system. *4th Student Conference on Research and Development*, SCOReD 2006. pp. 66–69.
Lyons, R., & Yates, R. 2005. Reducing ADC quantization noise. *Microwaves & RF*. Retrieved October 2010, from http://www.mwrf.com/Articles/ArticleID/10586/10586.html

Mackinnon, L. T., Ritchie, C. B., Hooper, S. L. & Abernethy, P. J. 2003. *Exercise Management: Concepts and Professional Practice*. Champaign, IL: Human Kinetics.

Martinez, G. I. 2001. Automatic gain control (AGC) circuits: Theory and design. ECE1352 Analog Integrated Circuits I Course Notes, University of Toronto, Canada.

Mayo Foundation for Medical Education and Research. 2008. Long QT syndrome. Retrieved 2010 from http://www.mayoclinic.com/health/long-qt-syndrome/DS00434/DSECTION=complications

McAdams, E. 2006. Ch. Bioelectrodes. In J. G. Webster, Ed., *Encyclopedia of Medical Devices and Instrumentation*, 2nd ed. New York: John Wiley & Sons. pp. 120–166.

Mietus, J. E., Peng, C. K., Ivanov, P. C. H., & Goldberger, A. L. 2000. Hilbert transform based sleep apnea detection using a single lead electrocardiogram. *Computers in Cardiology*, 27, 753–756.

Mims, F. M., & Helms, H. 2000. *Mims Circuit Scrapbook*, Vol. II Ed. Eagle Rock, VA: Newnes.

Morin, E. 2009. A short history of bioelectricity and the ECG. ELEC408 Lecture Notes, Queen's University, Kingston, Ontario, Canada.

National Semiconductor. 1969. High Q notch filter. National Semiconductor Linear Brief 5 (LB-5). Arlington, TX, USA.

Northrop, R. B. 2004. *Analysis and Application of Analog Electronic Circuits to Biomedical Instrumentation*. CRC Press, Boca Raton, FL.

Pallas-Areny, R., Colominas, J., & Rosell, J. 1989. An improved buffer for bioelectric signals. *IEEE Transactions on Biomedical Engineering*, 36, no. 4, 490–493.

Peacock, C. 2010. USB in a NutShell. *Beyond Logic*. Retrieved October 2010 from http://www.beyondlogic.org/usbnutshell/.

Perez, S. 2009. Mobile phones to Serve as Doctors in Developing Countries. *ReadWriteWeb*, Retrieved 2009 from: http://www.readwriteweb.com/archives/mobile_phones_to_serve_as_doctors_in_developing_countries.php

Petrov, G. 2004. ECG Recording, Storing, Filtering, and Recognition. *The Code Project*. Retrieved October 2010, from http://www.codeproject.com/KB/cpp/ecg_dsp.aspx

Peyton, A. J., & Walsh, V. 1993. *Analog Electronics with Op Amps: A Source Book of Practical Circuits*. Cambridge University Press, Cambridge.

Pipberger, H. V. 1965. Advantages of three lead cardiographic recordings. *Annals of the New York Academy of Sciences*, 126, no. 2, 873–881.

Quiroz, S. 2008. *How to Make a Graphical User Interface for Your USB Application*. Freescale Semiconductor, Jalisco, Mexico.

Santos, B. 2006. *Precision Instrumentation Amplifiers in Single Supply Systems*. Intersil Corporation, Milpitas, CA.

Sieed J., Bulbul, A. N. M. A., Ali, S. N., Rashid, S. M. S., & Roy, A. 2007. *PC Based ECG System*. EEE 426 Technical Report. Dhaka, Bangladesh: Bangladesh University of Engineering and Technology.

Spinelli, E. M., Pallas-Areny, R., & Mayosky, M. A. 2003. AC-coupled front-end for biopotential measurements." *IEEE Transactions on Biomedical Engineering*, 50, no. 3, 391–395.

Sprey, K. 2009. The CellScope: Transforming the cell phone into a mobile microscope. *Gizmag*. Retrieved October 2010 from http://www.gizmag.com/cellscope-mobile-microscopes/ 11463/

Starkjohann, C. 2008. V-USB. *Objective Development Software*. Retrieved October 2010 from http://www.obdev.at/avrusb/

StickOS [Firmware]. Retrieved 2010 from http://www.cpustick.com.

Testardi, R. 2009. StickOS for Freescale MCFJM128 v1.82 [Firmware]. CPUStick.com. Boulder, CO, USA. Retrieved October 2010 from http://www.cpustick.com

USB in a nutshell. n.d. Retrieved 2010 from http://www.beyondlogic.org/usbnutshell/

Valverde, E. R., Arini, P. D., Bertran, G. C., Biagetti, M. O., & Quinteiro, R. A. 2004. Effect of the electrode impedance in improved buffer amplifier for bioelectric recordings. *Journal of Medical Engineering & Technology*, 28, no. 5, 217–222.

Varshney, U. 2007. *Pervasive Healthcare and Wireless Health Monitoring*. Vol. 12, pp. 113–127.

V-USB. Retrieved 2010 from http://www.obdev.at/avrusb/

UN Foundation. 2008. *mHealth for Development*. s.l. : Vodafone Foundation, 2008.

Walker, B., Khandoker A. H., & Black, J. 2010. Low cost ECG monitor for developing countries. In Palaniswami, M., Marusic, S., & Law, Y. W. (Ed.) *Proceedings of Fifth International Conference on Intelligent Sensors, Sensor Networks and Information Processing December 7-10 2009, IEEE Press, NY, USA*. pp. 195–199. Melbourne, Australia.

Webster J. G., ed. 1999. *The Measurement, Instrumentation and Sensors Handbook*. Boca Raton, FL: CRC Press, Springer-Verlag and IEEE Press.

Wegmann, S. 2008. ETH Zurich student devises solar ECG. Retrieved October 2010 from http://www.ethlife.ethz.ch/archive_articles/080710_solar_E KG/index_EN.

World Bank. 2008. *Health Workers Needed: Poor Left without Care in Africa's Rural Areas*. Washington, DC.

World Health Organisation. 2006. *The Global Shortage of Health Workers and Its Impact*. Fact sheet 302. Geneva, Switzerland: World Health Organisation.

Yanowitz, F. G. 2006. Characteristics of the normal ECG. University of Utah Spencer S. Eccles Health Sciences Library. Retrieved October 2010 from http://library.med.utah.edu/kw/ecg/ecg_outline/Lesson3/index.html

# 14

# *Sensors, Monitoring and Model-Based Data Analysis in Sports, Exercise and Rehabilitation*

**Jurgen Perl, Daniel Memmert, Arnold Baca, Stefan Endler, Andreas Grunz, Mirjam Rebel, and Andrea Schmidt**

**CONTENTS**

Introduction ........ 375
Data Monitoring ........ 376
  Position-Detection Sensors and Devices ........ 376
  Motion-Detection Sensors and Devices ........ 377
  Force Sensors ........ 378
  Physiological Sensors and Devices ........ 379
  Sensor Networks for Monitoring Physical Activity ........ 379
Model-Based Simulation ........ 381
Neural Network–Based Process Analysis ........ 384
  Neural Network–Based Motor Analysis ........ 385
  Neural Network–Based Analysis of Game Tactics ........ 391
Net-Based Analysis of Rehabilitation Processes ........ 394
  Case Study ........ 396
Conclusion and Outlook ........ 401
References ........ 402

## Introduction

During the last 20 years, computer-based data analysis has become a central task in sports, exercise and rehabilitation in order to optimize performance, prevent disorder and support rehabilitation. Consequently, data monitoring has become a central need, and a lot of high-tech methods and devices have been developed which support automatic data monitoring even in complex processes like games. In the first section of this chapter, one of the authors (Baca) gives an overview of current sensor monitoring techniques and applications.

However, very often the problem arises that the useful information is not directly available from the monitored data. Instead, single pieces or complex patterns of information have to be derived from the data by means of analysis tools. If the relevant information is known at least in principle, it can be detected by means of data mining procedures. If the information is unknown, modelling methods such as antagonistic metamodels or artificial neural networks can be helpful for adaptation analysis or pattern recognition. The second section introduces the basic ideas of adaptation analysis based on dynamic models. The third section deals with pattern recognition in motor and game processes. Preventing overload and recognizing dangerous patterns in monitored data are of particular importance in the case of rehabilitation processes. Therefore, the last section develops ideas for the use of the presented methods in the area of healthcare.

## Data Monitoring

Technological solutions enabling the monitoring of human motion and physiological signals during sports and exercise are gaining increased attention as tools for preventing overload and for supporting rehabilitation in movement activities. Wireless technologies employing small sensors are particularly beneficial since they allow monitoring of kinematic, kinetic and physiological data without affecting individuals in executing their motions (Armstrong 2007; Baca 2008; Baca et al. 2009; Chi 2008).

In a typical set-up, sensors are carried by the human, mounted onto the sports equipment or deployed in the environment. They acquire the relevant motion and performance data like heart rate, velocity or reaction forces. The digitized signals are then transmitted to a mobile client (e.g., a mobile phone) via short-range protocols (e.g., Bluetooth, ZigBee, ANT). The data are either analysed directly at the client or sent to an Internet server (e.g., via EDGE, UMTS, or HSDPA/HSUPA), where they are further processed. Based on the results of the analysis, feedback can be given, or specific measures can be set (e.g., in the case of fall-detection devices). Parameters characterizing performance and sports activity can thus be supervised continuously.

Variables such as package dimensions, mass, costs and, in particular, accuracy are key factors for the selection of sensors and have to be considered for any measurement set-up. The reduction in the sizes of both sensors and data loggers allows vast amounts of field data to be collected during almost any physical or sporting activity.

### Position-Detection Sensors and Devices

In order to roughly estimate position (and speed) in outdoor sports activities over long distances (e.g., running or mountain biking), global positioning system (GPS)–based systems are well suited (e.g., Eskofier et al. 2008).

Differential GPS (dGPS) systems, which use a network of ground-based reference stations with known fixed positions for obtaining correction data, provide a (more expensive) alternative to achieve better accuracy (e.g., Larsson and Henriksson-Larsén 2005).

Position-detection sensors, which may also be used indoors (or outdoors, even under cloudy weather conditions), are based on radio or microwave technology. Objects to be tracked are equipped with one or several lightweight tags transmitting electromagnetic waves. Several receiver stations in the surrounding environment under investigation determine the time of arrival (TOA) and (optionally) the angle of arrival (AOA) of the emitted signals. One master station calculates the position of the tag(s) from the different values. Systems of that kind (see Stelzer, Pourvoyeur, and Fischer 2004 for an example) allow the monitoring of not only the positions of individuals over time but, moreover, the positions of groups of individuals, such as the members of a team.

RFID (radio frequency identification) chips transmit a unique signal, which is identifiable by short-distance readers. This technology is used, for example, in marathon races (cf. Malkinson 2009). Whenever a runner wearing such a chip crosses fixed mats throughout the race, the position is identified, and the time is recorded.

Video-based motion analysis systems are based on optical sensors (cameras) deployed in the environment. They are used to identify the position of the human body and/or how its individual limbs are configured during a motion. Using markers, the pose is defined by the three-dimensional (3D) position of marker points attached to the human body. Optoelectronic real-time motion capture systems identify the image coordinates of the marker points by processors that are part of the cameras. In many situations of everyday life, however, there are neither markers attached to the human body, nor is there intention or justification to do so. Image sequences acquired simultaneously from multiple views may form the basis for reconstructing 3D joint data at each instant without the use of markers. Applying optimization algorithms, the kinematic pose parameters of a human body model are calculated such that the synthesized shapes obtained are most similar in appearance to the actual shapes (edges, silhouettes, contours, etc.) of the real subject in the multiview camera images. Poppe (2007) gives a good survey of the current state of the art. Complex environments, noise, occlusions and shading complicate tracking of human segments and negatively affect accuracy. However, some promising results have also been reported when tracking individuals in challenging indoor and outdoor environments using off-the-shelf handheld video cameras (Hasler, Rosenhahn et al. 2009; Krosshaug and Bahr 2005) and when tracking subjects wearing loose or heavy clothing (Hasler, Stoll et al. 2009).

## Motion-Detection Sensors and Devices

Accelerometers are widely used to measure changes in position or to detect activities (e.g., Che-Chang and Yeh-Liang 2010; Ermes et al. 2008).

Theoretically, an acceleration signal that is mathematically integrated once gives the velocity, and one that is integrated twice gives the change in position. In practice, very accurate signals requiring a very exact compensation for the acceleration due to gravity and for the drift of the respective signal are needed to do this. A 3D tracking device usable for practical purposes, often referred to as an *inertial measurement unit* (IMU), therefore comprises rotation sensors (gyroscopes) in addition to three accelerometers (see Mayagoitia, Nene, and Veltink 2002 for a system applicable for long-term recordings in clinical, sport and ergonomics settings and Ghasemzadeh et al. 2009 for an application in the sport of golf). Sophisticated algorithms rely on all signals acquired to determine changes in position. There are also solutions where IMUs and (three-axial) magnetometers are integrated in order to obtain even more accurate results. The IMUs may be small in weight and size; the measured data may be transmitted using wireless technologies. Popular applications lie in the area of virtual reality in general and virtual training in particular. Computer–human interaction systems have been designed for different sports. The motions of users are captured and used as input for controlling the virtual run in order to improve the realism and the enjoyment during the game experience. Head-, hand- and body-tracking devices are utilized.

IMUs and inertial sensors (accelerometers and gyroscopes) based on wireless technologies are well suited to monitor human body segment orientations during activities in natural environments in real time. Ghasemzadeh et al. (2009), for example, describe an application for providing feedback for improving the golf swing. A body sensor network (BSN) consisting of several sensor nodes equipped with inertial sensors (a triaxial accelerometer and a biaxial gyroscope) placed on the body and the golf club is used to capture the movements of the golf swing.

## Force Sensors

Several kinds of force-measuring devices are used when monitoring human motion. The two most important devices are probably the force platform (or force plate) and pressure sensor arrays. A force plate is a rigid platform that provides the forces and torques applied to the surface of the platform in all three dimensions as well as the point of application of the resulting force vector. Force plates are typically mounted in the floor or ground. Major manufacturers—among others—are Kistler (Winterthur, Switzerland; www.kistler.com) and AMTI (Advanced Mechanical Technolgy, Inc., Watertown, Massachusetts, USA; www.amti.biz). Due to miniaturization of sensors, arrays of small force sensors have been developed that allow measurement of the pressure distribution on a body segment (e.g., the foot). By utilizing shoe insoles with these sensors, the pressure distribution under a person's feet can be determined continuously. There are commercial systems available (e.g., Pedar from Novel, Munich, Germany; www.novel.de) which allow connecting those insoles to a small portable data acquisition device that is able to communicate wirelessly with compatible personal digital assistants (PDAs), notebooks and standard PCs. This makes them applicable in monitoring force data in nonlaboratory conditions as well. Pressure distribution under the feet is measured in a variety of sports activities (e.g., in skiing).

If force sensors are integrated into sports equipment or fitness devices, additional benefit for the users can be achieved (cf. Stevens et al. 2006). Strain gauges can be applied to selected positions of the equipment or device to measure the mechanical stress, which corresponds well to applied forces or torques. Intelligent fitness devices, for example, can automatically adapt to the users' needs, thereby protecting them from conducting exercises in a way that would injure their bodies.

## Physiological Sensors and Devices

In addition to positions, velocities, accelerations and forces, human activities are also characterized by physiological signals, which may be acquired and transmitted using physiological sensors with miniaturized electronics to condition, digitize, process and transmit them wirelessly. A great variety of sensors have been developed for this purpose. Among these, in particular heart rate monitors have gained wide popularity. However, the spectrum of signals to be monitored is much larger. There are devices for mobile measurement of electrocardiographic (ECG), electroencephalographic (EEG) and electromyographic (EMG) signals, as well as of blood pressure, brain and body temperature, respiratory function and skin conductance, just to mention some important examples.

Wearable physiological monitoring systems appear to be particularly promising for collecting many of these signals in various applications. They consist of an array of sensors embedded into the clothing of the wearer to continuously monitor selected physiological parameters and transmit the acquired data wirelessly to a remote monitoring station. In addition, such wearable computing devices may also be used to identify motions. Coyle et al. (2010) present a number of case studies in the fields of healthcare, rehabilitation and sports performance. Knitted stretch sensors which are part of textiles are able to change their electrical resistance when stretched. Thereby, motions or respiration activity can be recognized (e.g., Paradiso, Loriga, and Taccini 2005). In the case of arm and upper-body motions, for example, certain gestures can be identified in this way.

## Sensor Networks for Monitoring Physical Activity

In order to assist or advise individuals during their physical activity, systems have been designed and developed which are able to monitor and transmit a combination of position, motion, force and physiological signals. As an example, one such approach (Baca and Kornfeind 2007; Preuschl et al.

2010) is briefly presented. The system (a prototype has already been developed) provides interactive communication technology to athletes/exercising students and coaches, physical education teachers or experts and helps to evaluate and improve certain performance parameters with respect to the individual's performance level. Characteristic parameters of the performance of individual persons or of a group of athletes/students can be supervised continuously.

In this way coaches/teachers/experts are able to advise athletes/students individually. The individual athletes/students get feedback on the quality of their motion. Persons may thus be assisted in interpreting their bodies' reactions to physical strain. Their performances are recorded, and the bodily changes during a certain time may be documented. Positive effects of physical exercise are highlighted. This might help to promote health-conscious behaviour and to enhance the willingness to engage in physical exercise. In particular, a sophisticated regulation of physiological stress can prevent persons from demotivating experiences due to fatigue.

An overview of the system is depicted in Figure 14.1. Sensors, either carried by the person or mounted onto the sports equipment, are used to measure different parameters like the heart rate, velocity or reactive forces of the exercising person. Wireless sensors that use the ANT+ connectivity solution (extension of the ANT protocol; Dynastream Innovations, Cochrane Alberta, Canada) for communication may be integrated. Moreover, by using a NEON (Spantec, Linz, Austria), a sensor platform with ANT+ interoperability (DAQ device in Figure 14.1), for the acquisition of analogue or digital sensor signals (e.g., from accelerometers or strain gauges), the range of supported sensor types can be broadened. By utilizing this wireless body area network (WBAN) based on the ANT protocol (cf. Kusserow, Amft, and Tröster 2009, who use similar components) the measured sensor data are transmitted to a mobile device (e.g., a smart phone). The required ANT interface to the mobile

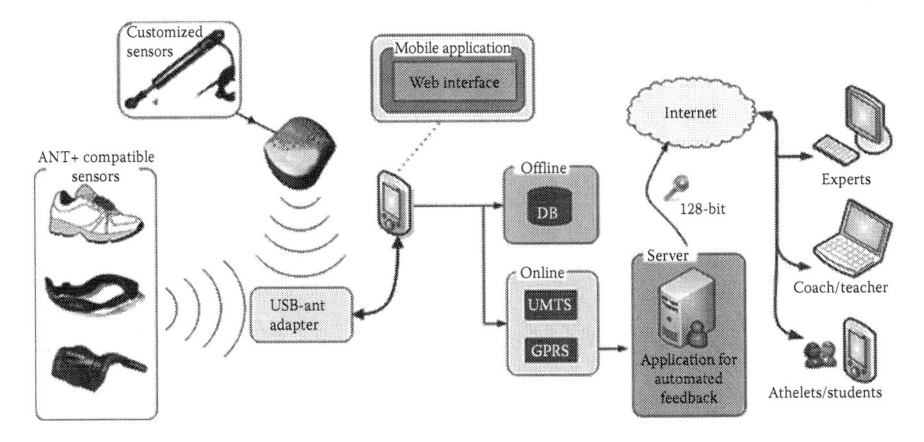


**FIGURE 14.1**
Overview of a mobile coaching system.

device may be provided by a USB-to-ANT adapter. The (preprocessed) data are then transmitted to an application server using wireless communication technologies (UMTS, HSUPA). From these data, feedback and/or exercising instructions are generated and sent back to the exercising person. The feedback may thus be based on a variety of parameter values characterizing the motion technique and the individual performance. It is either automatically generated by a server application or individually provided by an expert, a coach or a teacher. Coaches viewing this information on a laptop can give appropriate instructions; experts (e.g., doctors) make critical decisions on the person's safety. Feedback may be given not only at the training site but also from remote locations.

For the technical implementation of the proposed system, different sensor and wireless technologies have to be adapted to the sports' movements. From performance tests with the prototype implementation, it can be concluded that the system (with up to four sensors) is applicable in sports like running, cycling and mountain biking. However, there are still some limitations in data transmission which need to be addressed in further developments.

The realization of the system enables the evaluation of the application scenarios for ubiquitous technologies in sport and the acceptance of such applications not only in elite sport but, moreover, also in fitness and exercise programmes. Furthermore, the right timing and method of giving feedback in terms of performance enhancement may be investigated.

One particularly challenging task in implementing the system for individual sports is the generation of automatic feedback. Extraction of the relevant information from all the data streams acquired is a prerequisite for success. To achieve this aim, it may be necessary to identify patterns in the multidimensional time functions of the parameters acquired. Recommendations can then be based on the pattern(s) found. Another approach would be to find models which adequately describe dependencies between the parameters (e.g., load–performance interaction) and to give recommendations for the time course of the load.

## Model-Based Simulation

One relatively simple example of translating data to information is that of running speed and heart rate (or, more generally, load input and performance output): common devices like heart rate and speed monitors, GPS watches and bike computers can be used for measuring speed and heart rate data. Also, heart rate variability (HRV) can give important data on the status of an athlete. Information taken from those data could be take the form "Heart rate is too high, so reduce speed" or "Heart rate variability analysis indicates the need for recovery." Companies like Polar, Firstbeat Technologies and Black Tusk meanwhile provide a variety of devices including wireless data transfer to cloud computers for online data analysis. However, as is well known, things are more complicated due to physiological adaptation dynamics. Therefore, for instance, capacities, flow rates and delays have to be taken into consideration in order to answer questions like "*When* must the speed be reduced—and *how much*?"

Antagonistic models can help to simulate the physiological system dynamics and to optimize performance and prevent overload—provided that, of course, sensors are available for monitoring load and performance data like weight, speed, positions and heart rate.

One successful approach is given by the Performance Potential Meta-Model PerPot, which is based on the ideas of System Dynamics, simulating the time-dependent load–performance interaction iteratively by means of potentials, rates and delays: the basic idea of PerPot is that the time series of load input (e.g., measured in kilometres per hour or watts) does not influence performance directly but effects changes in internal buffers—which can be interpreted as fatigue or recovery potentials—from which the performance potential is changed by a delayed negative rate (fatigue potential – decreasing performance) and a delayed positive rate (recovery potential – increasing performance). This combination of negative and positive effects from just one source is what is called *antagonistic* in modelling. Antagonism is a rather simple and natural way of process control, e.g., in the case of focusing motions. However, it is sometimes not quite easy to describe the process dynamics by means of static functions, as is done in the fitness fatigue approach (Banister et al. 1975). The reason is that antagonistic processes normally do not stay in just one phase but dynamically change their states—which can be better represented by iteratively working level-rate models like those developed in the field of System Dynamics in order to simulate the behaviour of dynamic systems. (For more information see Perl 2002a, Perl and Endler 2006, and Perl 2008b; for a comparison of the fitness fatigue approach and the PerPot approach, see Pfeiffer 2009).

In its application, PerPot has to be adjusted to the athlete, which can be easily done by a 20 min load test, e.g., running at 5 different speed levels for 4 min each. From this calibration the PerPot software gets the central parameters like start values and delays and is then able to simulate the time-dependent performance response (like heart rate) corresponding to load input (like speed). Doing it this way, one can recognize, just by simulating, whether the speed profile of a marathon run will be successful or not (Perl and Endler 2006). The results are quite fascinating as the following marathon example may illustrate. One of the authors (Endler) used the PerPot software for training optimization as well as for speed-profile scheduling. The simulation calculated an optimal schedule and promised a result of 2:57:00. He followed the schedule recommendation and finished in 2:56:54.

While in those first steps the results were more or less prototypical, there are unpublished studies from 2010 evaluating PerPot on a broader base of data, particularly demonstrating that even a simulative calculation of the anaerobic threshold is possible with the help of PerPot. Obviously, the presented approach is helpful not only for marathons but also for all kinds of sports where load input data correspond to performance output data. Moreover, it can be used in the area of healthcare, for example, in order to simulate specific load effects of illness and therapy or to prevent overload. Briefly summarized, the described application can be characterized by the following process:

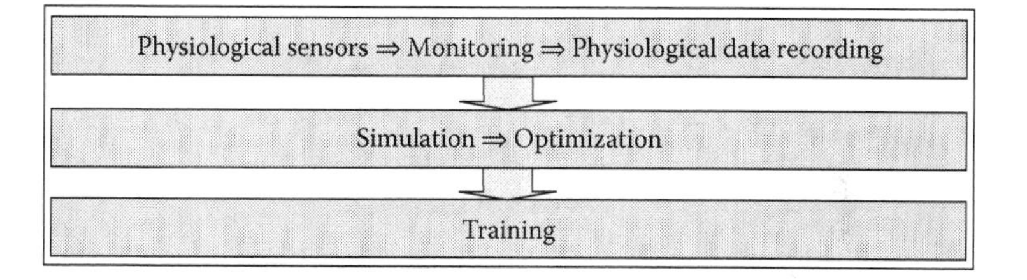


A typical case study then has the design sketched in Table 14.1.

The situation is obviously more complicated if the data are not simple-related quantities like speed and performance but build complex patterns as in motions, games or rehabilitation processes. Here, the problem arises that monitoring provides just raw material where the needed information is often hidden in the huge amount of data.

**TABLE 14.1**

Design of a Typical Case Study on Load–Performance Interaction

| | |
|---|---|
| Aim of the study | Analysis of load–performance adaptation processes and calculation of characteristic parameters. |
| Data | Load input and performance output data from training processes and/or competitions, in particular, load data (km/h, watts) and performance data (heart rate) from running and biking. |
| Processing | 1. Time series of load and performance data are used for model calibration and parameter calculation.<br>2. Load data are used for simulation of performance data by means of calibrated parameters. Simulated performance data are<br>– compared to original ones for evaluation<br>– used to predict future performance data |
| Results | Parameters like delay values or anaerobe thresholds as well as future performance profiles can be calculated or predicted with high precision.<br>Examples:<br>The influence of special events like training camps can be predicted quantitatively with high precision.<br>Results in marathons can be predicted with a deviation of less than 1 min. |

Another limitation of the current PerPot approach is caused by the one-dimensional stream of input data, which makes it difficult to simulate the effects of concurrent loads as in triathlons or decathlons or treatment optimization in medical applications. Currently, ideas are being developed to combine PerPot and artificial neural networks in order to recognize and analyse patterns of concurrent processes. There are first positive results from such an approach of combining PerPot and neural networks in order to optimize clinical treatment processes, where the patterns of simulated processes were analysed by means of networks to find the best-fitting treatment with the lowest stress for the patient, including early recognition of threatening critical situations (Perl 2007).

## Neural Network–Based Process Analysis

A main aspect of process analysis in sport is the analysis of time series. In the case of games, those time series normally contain the positions of the players and the ball. In the case of motions, the time series usually contain biometric data. Characteristic of such analyses are the complexity and the dynamics of the data, which make it at least difficult to use conventional methods like mathematical-statistical approaches for the respective pattern analyses. In turn, experiences with self-organizing maps or Kohonen Feature Maps (SOM, KFM: see Kohonen 1982, 1995; Dynamically Controlled Networks [DyCoN]: see Perl 2001a, 2002a, 2008a) prove that they can reduce complexity without reducing contained information and, in the case of process patterns, dynamics as well. In particular, those artificial neural networks can help to determine the type of a given process. Furthermore, if semantic qualities of processes are asked for, the answer normally is encoded in the complex data patterns and difficult to analyse by means of conventional approaches. In contrast, networks like SOM offer a means of calibration: the neurons of the trained network are checked and evaluated by experts once. A data set recognized by such a calibrated network can thus be classified regarding its type as well as its semantic meaning.

The idea of how such networks work can be briefly described as follows: suppose the data sets to be classified are from a motion; they may, for instance, contain speed values of limbs as well as coordinates and angles of articulations. During the training phase, the network is offered those data sets. Step by step it decides which neuron corresponds to which input, adapting the neurons to the input and so eventually representing the space of multi-dimensional inputs by a two-dimensional matrix of correspondingly prepared neurons. That is, in the end each neuron contains a generalized data set representing a collection of data sets which are sufficiently similar to each other. Moreover, adjacent neurons can have similar entries and so form clusters which represent types of input data sets. In a basketball free throw, for example, the first data sets may be of the type "concentration and preparation", the following ones of the type "reach back", and so on.

## Neural Network–Based Motor Analysis

In the area of motor analysis, the data from motion processes like steps, hits or throws can be analysed under the aspects of stability, and diverging patterns can be detected and analysed regarding their counterproductive effects. The corresponding process in the case of net-based analysis is given by

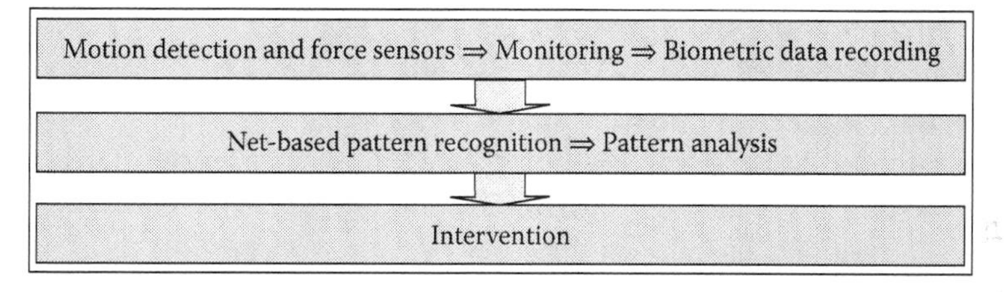


The example in Figure 14.2 briefly demonstrates how the net-based process analysis works. Assume an 11-step time series of biometric data. For each time step the resulting data set is mapped to the corresponding neurons of the net, and if the neurons are connected by edges in the order of the data sets, then the result is a two-dimensional simplified mapping of the process.

Such trajectories are simple two-dimensional transformations which contain most of the relevant information from the original motions and therefore make comparisons between them much easier. However, even those rather

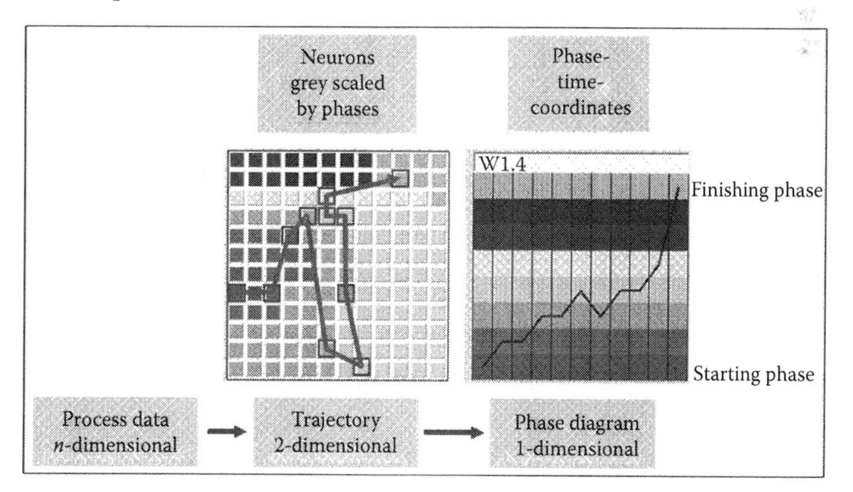


**FIGURE 14.2**
Grey scale-encoded semantic clusters with process trajectory (left) and corresponding phase diagram (right) of a motion. (From Grunz, A., Memmert, D., and Perl, J., *International Journal of Computer Science in Sports*, 8(1): 22–37, 2009.)

simple patterns can cause problems—if, e.g., a trajectory stays on the same neuron for more than one step or if synchronous steps of two trajectories are mapped to quite different neurons of the cluster. In both cases (and similar ones), it is difficult to see whether the trajectories are similar or not.

This problem can be solved by transforming the two-dimensional trajectories into one-dimensional phase diagrams, where *phase* here is meant to be a piece of the motion with specific semantics (like "preparation" or "reach back"). This is done by matching the phase-representing clusters with a semantic scale, which allows the neurons to be replaced by the respective corresponding semantic phase. In Figure 14.2 shades of grey are used to encode the phases and simplify the representation. The following example of basketball free throws, which is taken from the doctoral project of one of the authors (Schmidt), demonstrates the advantages of the phase diagram approach with regard to process analysis (see Table 14.2).

As has already been explained, net-based analysis of motor patterns is a method providing researchers with information inherent in large data sets

**TABLE 14.2**

Overview of the Free-Throw Study

| | |
|---|---|
| Research questions | The aim of the study was to explore the information that can be drawn from the movement pattern, with particular focus on individual differences and the athletes' skill level. Compensatory variability as a criterion of expertise is assumed. These specific research problems can be solved by measuring the *stability* of the body movement. These measurements not only encapsulate isolated features (from isolated parts of the body), but the kinematic chain of the whole body movement must also be analysed to investigate these questions. |
| Data | Angular displacements and angular velocities of the joints |
| Analysis | Kinematic analysis<br>Pattern recognition with DyCoN:<br>1. Training of the neural network of the first layer (shooting data network)<br>2. Trajectory analysis<br>3. Identification of the movement phases<br>4. Transformation of the shooting trajectories on the phase diagram<br>5. Training of the neural network of the second layer (trajectories network)<br>6. Semantic clustering of the trajectories network<br>7. Typification of the free-throw shooting patterns<br>8. Test of intra-individual stability<br>Mixed-methods design with conventional qualitative and quantitative analyses in the form of case studies |
| Results | Clearly defined individual characteristics found in the movement pattern data |

*Source:* From Schmidt, A., *Forum Sportwissenschaft*, 19. Hamburg, Czwalina, 2010; Schmidt, A., *Movement Pattern Recognition in Basketball Free-Throw Shooting*, 2011, submitted.

which cannot be extracted by means of conventional methods like statistics. With DyCoN (a special type of KFM; see Perl 2004a, 2004b) it is possible to reduce the complexity of such data sets by calculating a complex feature which consists of several data sets and their relationship. For example, in a recent research study on the basketball free throw, the net was trained with data captured using motion analysis techniques. These data consisted of the angular displacements and velocities from the articulations of the whole kinematic chain (wrist, elbow, shoulder, hip, knee and ankle joints). A total of 21 shooters were tested, each of whom executed five shots, resulting in a total of 105 data sets. (Unlike conventional KFMs, a DyCoN can successfully be trained even with such a low number of training data; also see Rebel's explanations in the section "Net-Based Analysis of Rehabilitation Processes".)

When the net is trained, tests could be accomplished with the data sets from the individual shots. The results of these individual tests are trajectories passing specific neurons of the network. Each trajectory represents the kinematic chronology of the corresponding shot. It is the complex feature combining the items (angle measures) which are in interaction with each other (see Figure 14.3).

With a semantic identification of the network regions, the movement phases have been deliberately detected, as shown in Figure 14.4. Thus, it is possible to detect the series of movement phases. Figure 14.5 shows this succession for the shots from participant 09 (T41–T45, advanced shooter) and participant 17 (T81–T85, expert shooter) in phase diagrams. Individual characteristics like the skill level of movement patterns have clearly been detected. The phase diagrams reveal specific movement patterns corresponding to specific structures of movement organization according to particular shooters. A common functional pattern is the launching action, each with one measurement point in the last three movement phases. Novice and advanced shooters typically did not hold this pattern consistently for five shots. Nevertheless, some shooters, like, e.g., participant 19, showed a different launching pattern, scoring very successfully. It is concluded that their movement (pattern) is marked by an individual functionality. The shooting action of the most successful participant (he scored on all five shots, four times without touching the rim) is characterized by a minor variability in the preparatory movement phases and a high stability in the launching phases (see Figure 14.5).

The variability in the preparatory phases is supposed to be a kind of functional variability. Beginning with a consistent starting position, his throwing movements are built up continuously over all the phases. He omits only the fifth phase, which has been identified as dysfunctional.

Indeed, several different modes of movement organization have been revealed, suggesting functional individual differences. In two separate stages, the stability of the respective motor patterns is tested, and the throwing patterns are classified. These stages are realized by training a second

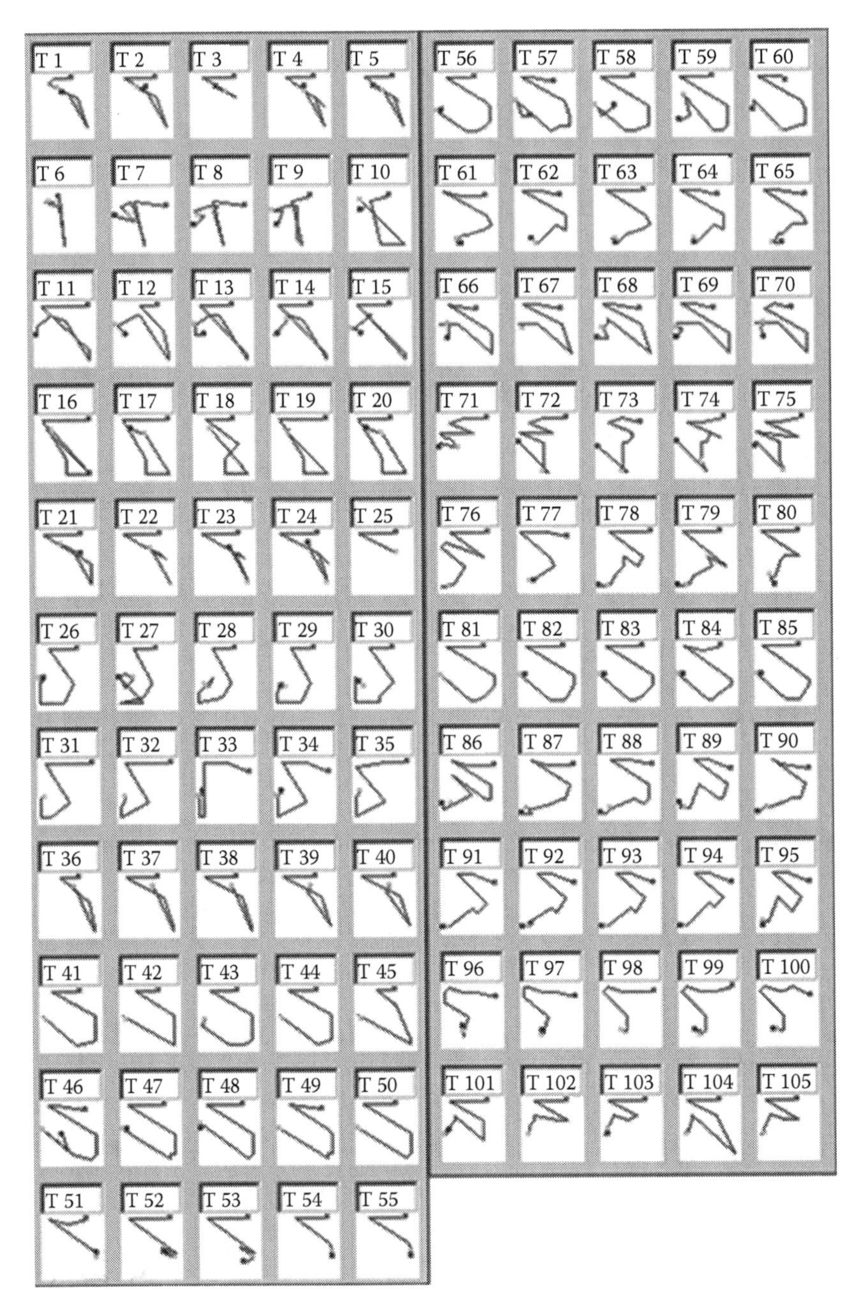


**FIGURE 14.3**
The trajectories of the 105 shots analysed. All five trajectories in each half row represent the free-throw movements of one participant. The trajectories are arranged in order of skill level, from those of participant 01 (T1–T5), a novice, to participant 21 (T101–T105), an expert shooter.

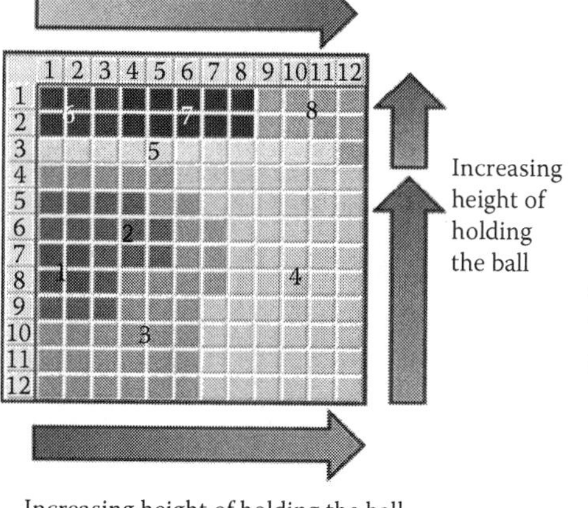


**FIGURE 14.4**
Representation of the shooting data net, showing the regions where the specific movement phases are localized.

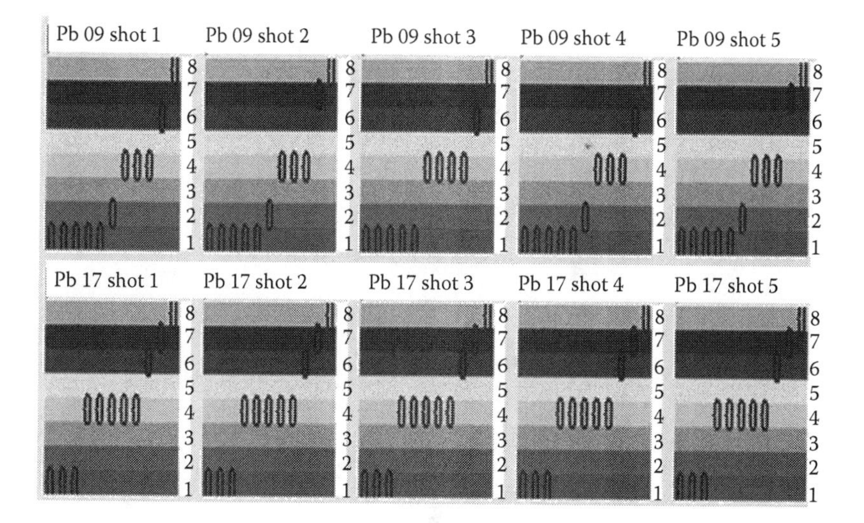


**FIGURE 14.5**
Phase diagrams of the five shots from participant 09 (T41–T45) and participant 17 (T81–T85).

DyCoN with the data of the phase series (representing the movement patterns of the throwing action).

The procedures described in the preceding have been verified using several methods of analysis (e.g., qualitative) on the same free-throw patterns in order to ensure the validity of the DyCoN method (see Schmidt 2010).

For example, a fruitful avenue for future research may entail analyses of the basketball free throw with a focus on the most successful shooters. Such an approach may provide the opportunity to precisely differentiate highly functional movement patterns between skilled players. Moreover, a long-term objective of this programme of research is to develop an automatic analysis system for free-throw performance. Using DyCoN, further research is currently being conducted on the analysis of the golf drive. This example offers the development of modelling procedures for highly complex movement tasks.

Other applications of net-based motor analysis include the modelling of complex interactions (Schöllhorn and Perl 2002d), individual differentiation of dynamic foot pressure patterns (Schöllhorn et al. 2002), tactical analysis in squash (Perl 2001b, 2002a, 2002b), Perl 2004a (load–performance interactions), rehabilitation processes (Rebel 2004b) and motor learning (Raab, Perl, and Zechnall, 2003). Lamb, Bartlett and Robins (2010) used a similar approach (also based on KFM) assessing the coordination stability of the golf chip shot.

In the presented example—and even more in the case of soccer in the following section—it is not only the recognition of the phases of a process that is of importance. It is also of central interest to recognize similarities and dissimilarities between processes as well as striking patterns which are different from expectations. Although Figure 14.3 gives an idea how stable the shots of a player are, it is difficult to say whether the shots of two of the players are similar. The phase diagrams in Figure 14.5 make those decisions much easier, as will be discussed based on Figure 14.6 (see Grunz, Memmert, and Perl 2009):

Figure 14.6 shows the phase diagrams of the five free shots of four selected players. (Note that the lines or tracks in Figure 14.6 have the same meaning

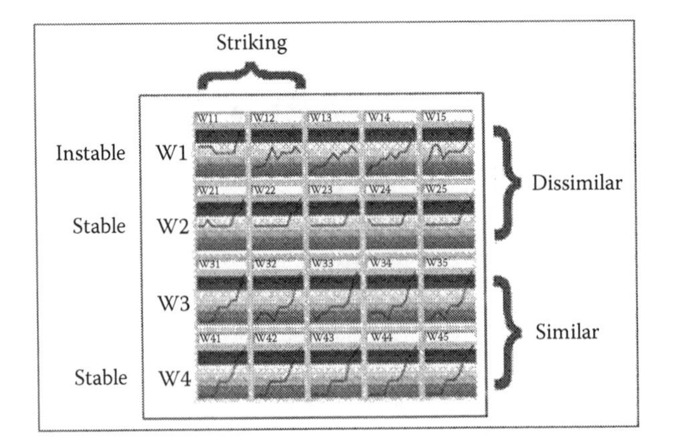


**FIGURE 14.6**
Phase diagrams of five free shots of four players W1, W2, W3 and W4. (From Grunz, A., Memmert, D., and Perl, J., *International Journal of Computer Science in Sport*, 8(1): 22–37, 2009.)

as the sequences of 0s in Figure 14.5.) Regarding Figure 14.4, the shades of grey represents the standard shot process from the start phase 1 (bottom) to the final phase 8 (top). Therefore, tracks from bottom left to top right are to be expected, as are shown by W3 and W4, while the tracks of W2 are unexpected and striking. Moreover, there are differences regarding inter-individual similarity and intra-individual stability: the shot types of W3 and W4 are quite similar to each other but different to those of W1 and W2, which are different to each other, too. Also, the shots of W1 are rather unstable compared to those of the other players.

On the basis of phase diagrams, such striking feature analyses also can be done and have been done automatically by means of neural networks on a second level. To this aim, the sequences of phase shades, which encode the shots, are replaced by sequences of phase numbers, which then can be taken for training a shot net. Each neuron of such a shot net then corresponds to a type of shot.

The situation in the case of games is much more complicated because of tactical aspects as well as group activities and interactions with the opponent team. Some results from a project currently run by four of the authors (Endler, Grunz, Memmert and Perl) are briefly sketched in the following section. Some earlier approaches can be found, for instance, in the works of Lames and Perl (1999), Perl (2002b) and Jäger, Perl and Schöllhorn (2007).

## Neural Network–Based Analysis of Game Tactics

The basic idea of network-based game analysis is that groups of players form patterns of positions, which can be analysed by means of networks in the same way as biometric data of a motion. This means that, first of all, the position data have to be recorded from the game. This is, of course, a high demand on the regarding monitoring techniques which range from sensor recording to computer-based video analysis. The corresponding process in the case of net-based analysis is therefore given by

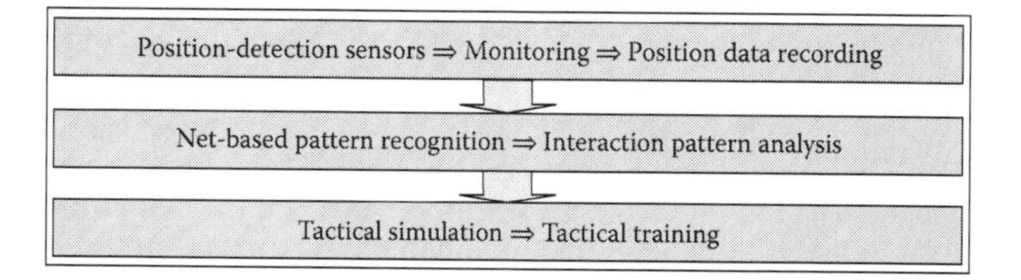


A typical example of a case study is presented in Table 14.3.

In detail, the process steps can be described as follows. The usual way of getting information about activities and interaction in a game is that of monitoring the position data of the players and the ball. In most team games, however, it makes no sense to follow the movements of single players if

**TABLE 14.3**

Design of a Case Study Dealing with Net-Based Game Analysis

| | |
|---|---|
| Aim of the study | Recognition of tactical concepts by means of network-based pattern analysis of group constellations |
| Data | Position data for the players and the ball |
| Method and steps of analysis | Position data for the players and the ball are recorded from the game.<br>Position data of groups of players are combined to form constellations.<br>Constellation data are analysed by means of neural networks, resulting in constellation types.<br>Constellation types and ball data are combined to represent interaction.<br>Interactions are analysed by means of neural networks resulting in interaction types. |
| Results | Time series of the types of<br>– behaviour of tactical groups<br>– interaction between opponent groups<br>Evaluation of the success of types of tactical concepts and interaction |

the strategic plans of a team are of interest. Instead, it makes more sense to observe the interactive behaviour of groups like "offence" and "defence" to understand the strategic plans behind the actions (Grunz, Memmert, and Perl 2009; Memmert, and Perl 2009a, 2009b).

Hence, the most interesting question from the point of view of a coach is: What can be learned about the interaction of the opponent teams? A first step in those analyses is to compare the tactical patterns of opponent groups like the offence group of team A and the defence group of team B. To this aim, the specific actions of both teams have to be trained to specific individual nets, where the actions form trajectories of group constellations. These trajectories can be transformed into phase diagrams like those from basketball free throws in Figure 14.6. Finally, experts can give the different phases (i.e., the clusters of the network) semantic meanings, as has been shown for basketball in Figure 14.4. The resulting phase diagram then also provides a stepwise description of the game activities, i.e., a game protocol.

Figure 14.7 shows an example of the correspondence of French offence activities and Italian defence activities in the World Cup final in 2006. In order to make the interactive "rhythms" more obvious, the diagrams are reduced to the parts which the processes run through.

The respective top lines show the time intervals, and the text boxes in the middle show the semantic interpretations that can be taken from the automatically generated protocol.

Taking the corresponding phase diagrams from both teams together with the ball positions enables the training of a further net with that interaction information, as shown in Figure 14.8. Such an interaction network is able to detect and specify types of interaction. With such a net, interactions can be classified—i.e., under the aspects of the advantage or success of one team

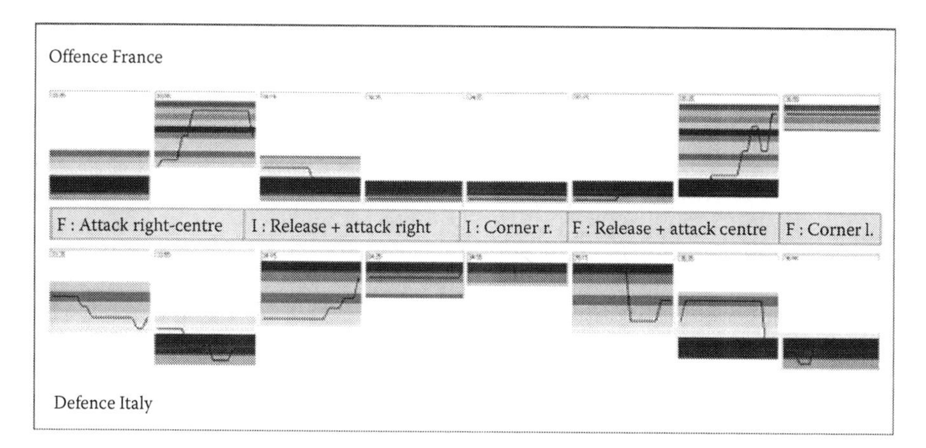


**FIGURE 14.7**
Synchronous phase diagrams of French offence (*top*) and Italian defence (*bottom*) (Grunz, A., Memmert, D., and Perl, J., *International Journal of Computer Science in Sports*, 8(1): 22–37, 2009). The process-representing lines in the diagrams are the results of a two-step network analysis. In the first step the types of the respective group constellations are recognized and transferred to the stripe grey shades of the diagrams. In the second step the time series of constellation types (i.e., the process trajectories) are mapped to the diagrams, forming a one-dimensional mapping of the tactical behaviour. (The data were offered by a project partner without information about the sensors or recording techniques used.) (From Grunz, A., Memmert, D., and Perl, J., *International Journal of Computer Science in Sport*, 8(1): 22–37, 2009.)

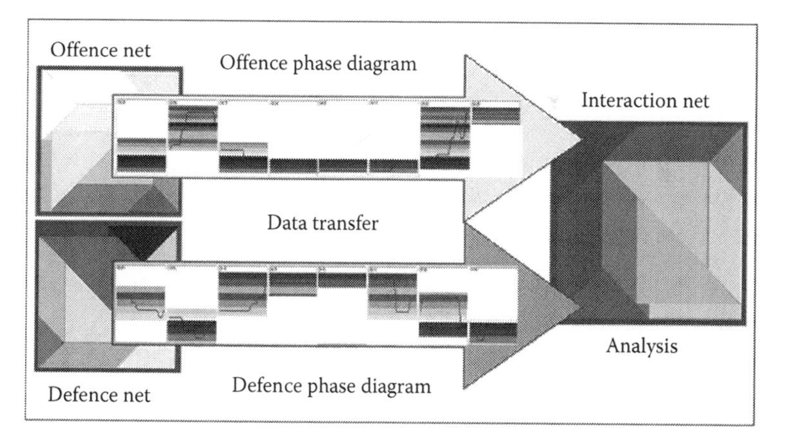


**FIGURE 14.8**
Interaction net trained with offence and defence information of the opponent teams. (From Grunz, A., Memmert, D., and Perl, J., *International Journal of Computer Science in Sport*, 8(1): 22–37, 2009.)

or the other. This way, it seems to be possible not only to recognize specific tactical behaviour but also to support recognition and evaluation of weak and strong actions. Together with statistical information about the frequencies of activities, this allows for simulation of possible tactical processes and their success.

The idea which Memmert and Perl (2009a, 2009b) have mentioned is to support tactical training by recognizing typical or particular creative behaviour as well as by simulating and evaluating tactical variants. Obviously, net-based prognosis of the development and success of processes can be helpful for rehabilitation as well, as is demonstrated in the following section.

## Net-Based Analysis of Rehabilitation Processes

In the area of rehabilitation, after a clinical intervention the patient is monitored periodically in order to prevent critical situations. The recorded data, however, often are difficult to interpret because they can be redundant as well as inconsistent and even contradictory. Net-based pattern recognition can help to detect striking patterns as well as predict and prevent critical developments.

The first example is taken from Perl and Dauscher (2006; also compare Perl 2007, p. 311). They give some results regarding the postoperative rehabilitation of risk patients: "after the operation, psychological items were recorded from the patients for 7 days (3 times a day 10 values). Among others, these items regarded their pain and their anxiety, in order to analyse and improve the rehabilitation process. The problem was that the answers were quite subjective, the values depending on the complex situation and feeling. Therefore, the combination of a rather high number of items and the small number of repetitions seemed to prevent the detection of characteristic patterns."

The first step of the net-based approach, therefore, was a net training with Monte Carlo–generated data, which compensated for the low amount of original data (compare Schmidt's case study in the section "Neural Network–Based Motor Analysis"). The data representation was then cut into segments of 1 day in length, where the three neurons that represented the respective three daily data sets were connected by a "daily trajectory" (see Figure 14.9).

As in the examples of basketball and soccer, clusters that are the same shade of grey contain neurons of equal or similar meaning regarding the possible patient states. Figure 14.10 shows one typical result: during the third, fourth and fifth days, the processes are quite similar, running "from right to left" (positive during the days and reset to a negative state during the respective nights). During the sixth and the seventh day, the processes run inversely, "from left to right", indicating a dramatic change to a negative state.

In comparison to Figure 14.10, Figure 14.11 shows the example of another patient with an obviously quite different rehabilitation process. The day-night rhythm starts to change already after the fourth day. But in contrast to Figure 14.10, this change indicates the start of stabilization, which is finished after the seventh day in a positive state.

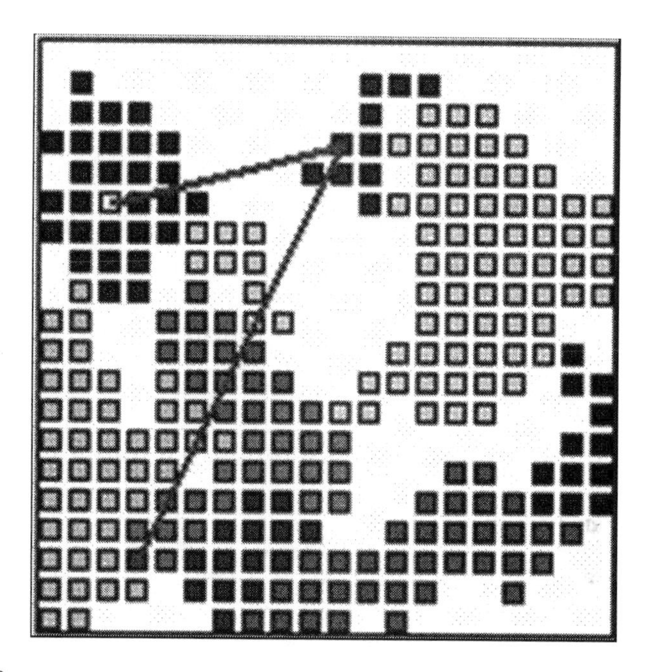

**FIGURE 14.9**
Example of a daily trajectory, starting in the neuron at bottom left and ending in the neuron at top left. (From Perl, J. and Dauscher, P., *Computational Intelligence for Movement Science*, 299–318, Idea Group, Hershey, PA, 2006. With permission.)

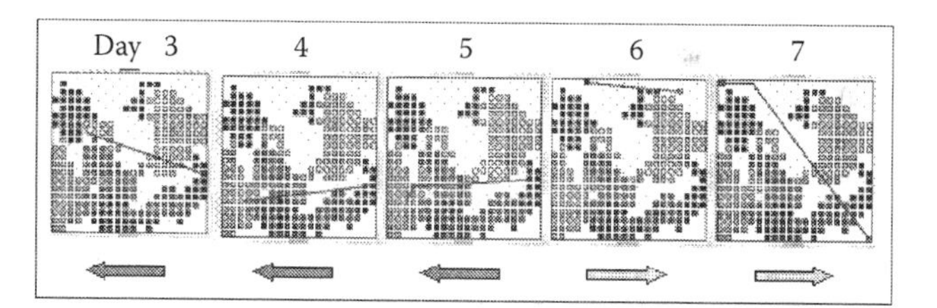


**FIGURE 14.10**
Daily trajectories of the patient's state. The change of trajectory directions between the fifth and the sixth days indicates a change to a negative state (Also see Perl and Dauscher 2006): while during days 3, 4 and 5 the trajectories run from right to left—i.e., from worse to better—(and back in the night), the situation changes dramatically in the night of the 6th day, when the daily trajectory has moved to a negative area of the network and then starts to move from worse to worst. (From Perl, J. and Dauscher, P., *Computational Intelligence for Movement Science*, 299–318, Idea Group, Hershey, PA, 2006. With permission.)

As Perl and Dauscher (2006, p. 312) have pointed out,

> the problem of analyzing human behavioral processes is that they form time-dependent structures of situations, which differ inter-individually and intra-individually, depending on types and contexts of activities. Processes can be described as time-dependent trajectories of single patterns of situations, where the trajectories are patterns on a higher level

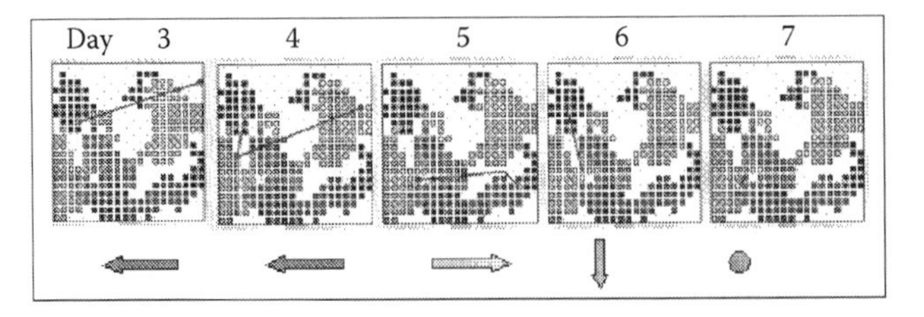


**FIGURE 14.11**
Trajectories of a process that is starting to stabilize after the fourth day. Days 3 and 4 reproduce the behaviour shown Figure 14.10. Day 5 shows a complete turn in the direction of the trajectory, which, moreover, has moved to a positive area of the network and there becomes stable during days 6 and 7. (From Perl, J. and Dauscher, P., *Computational Intelligence for Movement Science*, 299–318, Idea Group, Hershey, PA, 2006. With permission.)

(with the situation patterns as components) and so can be analyzed using the methods mentioned above. This approach is particularly useful if the complex situation patterns can be replaced by their much simpler type specifications. This way, the high-dimensional original process can be replaced by a low-dimensional representing trajectory, which is much easier to handle and to analyze.

Although a net-based pre-processing of patterns improves the trajectory handling the problem remains if and how trajectories can be compared for intra- and inter-individual analysis. It turns out that due to technical reasons similarity of movement-trajectories is difficult to evaluate, as will be discussed in the following section. Therefore, the idea of taking such trajectories as meta-patterns and again using a net for clustering can work but needs some preparation by means of conventional data processing methods, which are known, for instance, from voice recognition.

Meanwhile, the approach of using phase diagrams has been developed successfully also in the field of processes in sport, as has been explained in the preceding section.

## Case Study

The closing example was run in 2004 by one of the authors (Rebel). It shows in a complete case study how rehabilitation processes can be supported by means of the neural networks approach:

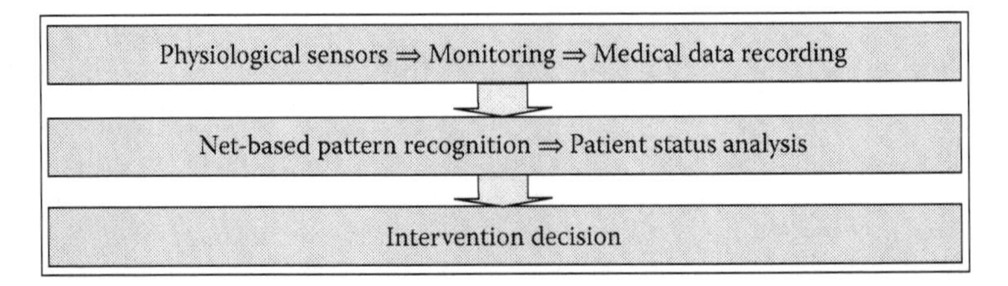

Mirjam Rebel's doctoral thesis, "Analyse von Rehabilitationsverläufen nach Kreuzbandrekonstruktionen" (An analysis of rehabilitation processes after reconstructive surgery of the cruciate ligaments; Rebel 2004a), examined the impacts of emotional, cognitive and behavioural parameters on the recovery process after reconstruction of the anterior cruciate ligaments (ACL). Therapists often report on typical rehabilitation processes in the ACL reconstruction after treatment. Therefore, Rebel's research strategy followed two research objectives: On the one hand, the identification together with a detailed description and typification of the rehabilitation processes was examined. On the other hand, an interpretative explanation of the preceding processes was provided.

The research design of the work, an overview of which is given in Table 14.4, was based on Petermann's (1992) concept of the "controlled practice (Kontrollierten Praxis)". That is, routine cases from the therapeutic practice as well as their critical incorporation according to appropriate rules were documented. The focus was placed on a comprehensive diagnostic evaluation, follow-ups and a systematic comparison of individual cases which allowed an arrangement in groups. For development of the research question, an adequate spot check rather than a mainly representative sample was chosen. The number of patients examined was limited to 30 at the beginning of the study to enable a wide variety of possible rehabilitation processes but also due to financial constraints. Thus, 30 ACL outpatients of the sports medicine training centre of the ATOS Clinic in Heidelberg were part of the study.

A mainly qualitatively oriented longitudinal analysis was chosen. For the illustration and modelling of individual and typical rehabilitation processes, criterion variables were defined which describe the rehabilitation processes after an ACL operation in detail. The particular measurement points for the collection of the criterion variables were scheduled in a 3-week period. This period was chosen in order to precisely depict the rehabilitation process with

**TABLE 14.4**

Research Design of the Work

| | |
|---|---|
| Aim of the study | The explorative study focused on the detailed description and characterization of rehabilitation processes of patients after a surgical reconstruction of the anterior cruciate ligaments.<br>No explicit research hypotheses could be formulated at the beginning of the study. |
| Data | Daily physiological data of 30 ambulant patients |
| Analyses Methods | 1. Modelling of rehabilitation processes by means of neural networks (of type DyCoN), based on criterion variables<br>2. Characterization of grouped rehabilitation processes by means of qualitative data analysis software, based on explanatory variables |
| Results | Nonoptimal behaviour with regard to pain, regeneration and motion in daily life and at work proved to be a major risk factor after surgery of the anterior cruciate ligaments. Those particular types of behaviours reflect the cognitive processing of and emotional reaction to the handicaps caused by the operation. |

as much longitudinal data as possible but without stressing the patients with additional expenditure of time. In order to trace the rehabilitation processes as realistically as possible, no predecision on the number of measurement points was made. The definition of the criterion variables was a result of a subjective knee score (Höher et al. 1995) and the subjective information from the interview data. The criterion variables can be related to the following five categories: (i) functionality, (ii) performance ability, (iii) pain, (iv) training therapy standard and (v) satisfaction.

The difficulty in the identification of typical rehabilitation processes was apparent in the data material. To display the rehabilitation processes on the basis of the quantitatively collected criterion variables, the neural network modelling method was chosen. The main advantage of using the neural network modelling method for this study is the data reduction. Conventional methods do not have the capacity to display the high amount of data with 12 criterion variables per measurement point, whereas network modelling enables an illustration without a considerable loss of information. The specific goal was to transfer the structure and the phase development of the 12 criterion variables which were collected during longitudinal research onto a net. The sequence of the identified neurons of the rehabilitation process in a trajectory showed a specific, individual pattern.

In this study, the DyCoN (see Perl 2001b) was used with a resolution of 20 × 20 neurons. As opposed to other nets, each neuron of such a net can learn continuously under individual adaptation control, which was taken from the PerPot approach (see the section "Model-Based Simulation"). This kind of control enables a stepwise training, for instance, starting with stochastic data and continuing with original data. Thus, a specific learning pattern can be impressed on a net (Perl 2001b) with few training steps (<100). In a first step, following special data processing, a net was stochastically pretrained which contained all the basic information (dimensionality and change-over possibilities) from the defined criterion variables. As a result of the training, the net developed neuron clusters which were consistent with similar attribute structures. Specific data sets were assigned to adjacent neurons of the net in a certain way so that areas of adjacent neuron clusters represented similar line patterns.

After the network training, each neuron contained a specific data structure of the 12 criterion variables, although the neurons were displayed as two-dimensional position coordinates in the illustrations. When activated, a respective neuron of the net displays the attribute combination for which it is responsible. The network, which is trained with specific information, forms the basis for a subsequent illustration of the rehabilitation processes. For this purpose the attribute combinations were inputted as learning stimuli, and the so-called winner neuron was identified as the one most similar to the stimulus and thus indicating its responsibility. In order to identify the responsible neuron, the pretrained net required between 30 and 100 learning steps. When the data structure showed a high similarity for several measurement points, the same neuron could be activated.

The different measurement points are connected to a trajectory in which an O signals the beginning and an X signals the end of the rehabilitation process. In this way, the 30 individual rehabilitation processes were displayed on their very own net and saved for further data processing. For neural networks there is no possibility to calculate significances or probabilities of error, which are common in quantitative statistics. The comparison of the rehabilitation processes, which were illustrated as trajectories, is based on qualitative criteria. The most important qualitative criterion was the trajectory process in relation to the reference point. The points (O = rehabilitation start; X = rehabilitation end) are chronologically connected in the trajectory. The reference point, represented by a neuron containing the highest values of all criterion variables, signals the 100% rehabilitation success. A spatial approximation towards the reference point implies an improvement in the rehabilitation result, whereas increasing distance from the reference point implies deterioration. Particular attention was given to the variation of progressive and recessive processes. Table 14.5 and Figure 14.12 exemplify a trajectory and the corresponding percentage data.

Due to the characteristic process of the trajectories, we arranged the rehabilitation processes into groups for analysis. The analysis resulted in several process types. Process type I (for example, see Figure 14.13) was characterized by a constant progressive development. In the beginning, the rehabilitation process trajectory approached the optimum rehabilitation result. The review of the underlying data showed that all identified responsible neurons built upon each other in terms of a positive development. The rehabilitation processes of the process type III (see Figure 14.14) were characterized by a positive, progressive development followed by a decline. In the beginning, the trajectory of the rehabilitation process approached the reference point before dropping away and concluding.

Because this was a case study, no evaluation or comparable study was carried out. The method of artificial neural networks was new in the field of clinical processes and somewhat sophisticated for the nonspecialist. Therefore—although it was introduced and briefly described—the method was not discussed in depth with patients or medical staff.

The results, however, were discussed with patients, medical doctors and physiotherapists. It turned out that subjective estimations, clinical test data

**TABLE 14.5**

Percentage Data Corresponding to the Trajectory in Figure 14.12

| Rehabilitation start 50% | | | | |
|---|---|---|---|---|
| Functionality | Pain | Performance ability | Satisfaction | Training therapy standard |
| 57% | 40% | 70% | 50% | 17% |
| Rehabilitation end 61% | | | | |
| Operability | Pain | Performance ability | Satisfaction | Training therapy standard |
| 67% | 50% | 80% | 30% | 50% |

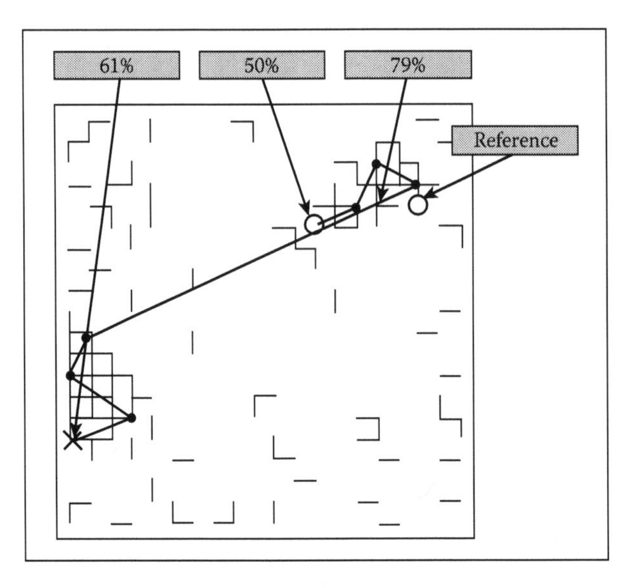


**FIGURE 14.12**
Neural network with a trajectory process and reference point (O). Phases of process development: increase – development towards the reference point – decline – stabilizing and rehabilitation result. Interchange from progressive to recessive development takes place between the fourth and the fives measurement point.

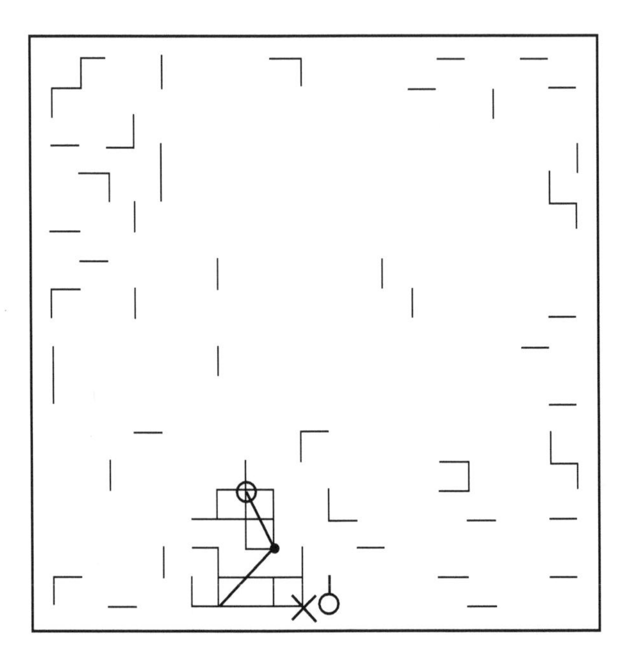

**FIGURE 14.13**
Example of a progressive trajectory development. See also the explanations in the text.

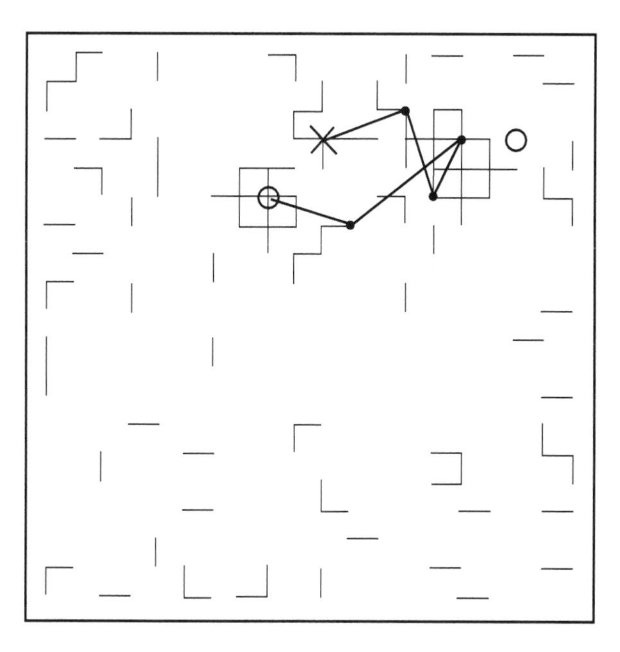

**FIGURE 14.14**
Example of a progressive development followed by a decline. See also the explanations in the text.

and net-based trajectory analyses strongly converged and corresponded with each other. In sum, the neural network modelling method turned out to be an adequate medium to visualize individual rehabilitation processes and proved to be an adequate basis for a qualitative analysis.

## Conclusion and Outlook

The idea of this chapter was to demonstrate how data sensors, recording and monitoring can fruitfully work together with net-based pattern analysis in order to recognize specific situations in time and prevent problems. This obviously is of particular importance in the area of healthcare where bad pattern recognition can prevent success even in the case of excellent monitoring. In turn, even the best pattern analysis does not help if sensors, monitoring or recording supply insufficient data.

Numerous experiments and projects in sports improved the experience in the field of net-based recognition of patterns in automatically recorded data. The case studies from the last section demonstrate how the results can be transferred to the area of healthcare. Future analyses will show that it is not only individual processes that can be handled in the described way. Phenomena like epidemics or pandemics can also be the subject of monitoring and net-based pattern analysis in order to enable early recognition of the first indicators (Perl 2007).

## References


Armstrong, S. 2007. Wireless connectivity for health and sports monitoring: A review. *British Journal of Sports Medicine* 41 (5): 285–289.

Baca, A. 2008. Feedback systems. In *Computers in Sport*, ed. P. Dabnichki and A. Baca, 43–67. Southampton: WIT Press.

Baca, A., Dabnichki, P., Heller, M., and Kornfeind, P. 2009. Ubiquitous computing in sports: A review and analysis. *Journal of Sports Sciences* 27 (12): 1135–1346.

Baca, A., and Kornfeind, P. 2007. Mobile coaching in sports. In *Adjunct Proceedings of Ubi-Comp 2007*, ed. J. E. Bardram et al., 172–179. Innsbruck, Austria.

Banister, E., Calvert, I., Savage, M., and Bach, I. 1975. A system model of training for athletic performance. *Australian Journal of Sports Medicine* 7: 57–61.

Che-Chang, Y., and Yeh-Liang, H. 2010. A review of accelerometry-based wearable motion detectors for physical activity montoring. *Sensors* 10: 7772–7788.

Chi, E. H. 2008. Sensors and ubiquitous computing technologies in sports. In *Computers in Sport*, ed. P. Dabnichki and A. Baca, 249–268. Southampton: WIT Press.

Coyle, S., Mitchell, E., Ward, T. E., May, G., O'Connor, N. E., and Diamond, D. 2010. Textile sensors for personalized feedback. In *Proceedings of the Information Access for Personal Media Archives Workshop (IAPMA2010), Milton Keynes, U.K,* online doras.dcu.ie/15373/

Ermes, M., Parkka, J., Mantyjarvi, J., and Korhonen, I. 2008. Detection of daily activities and sports with wearable sensors in controlled and uncontrolled conditions. *IEEE Transactions on Information Technology in Biomedicine* 12 (1): 20–26.

Eskofier, B., Hartmann, E., Kühner, P., Griffin, J., Schlarb, H., Schmitt, M., et al. 2008. Real time surveying and monitoring of athletes using mobile phones and GPS. *International Journal of Computer Science in Sport* 7 (1): 18–27.

Ghasemzadeh, H., Loseu, V., Guenterberg, E., and Jafari, R. 2009. Sport training using body sensor networks: A statistical approach to measure wrist rotation for golf swing. In *Proceedings of the 4th International Conference on Body Area Networks* (Los Angeles, California, April 1–3, 2009). Brussels, Belgium: ICST.

Grunz, A., Memmert, D., and Perl, J. 2009. Analysis and simulation of actions in games by means of special self-organizing maps. *International Journal of Computer Science in Sport* 8 (1): 22–37.

Hasler, N., Rosenhahn, B., Thormählen, T., Wand, M., Gall, J., and Seidel, H.-P. 2009. Markerless motion capture with unsynchronized moving cameras. In *Proceedings of the IEEE Computer Society Conference on Computer Vision and Pattern Recognition 2009*, 224–231. Piscataway, NJ: IEEE Service Center.

Hasler, N., Stoll, C., Rosenhahn, B., Thormahlen, T., and Seidel, H.-P. 2009. Estimating body shape of dressed humans. *Computers and Graphics* 33 (3): 211–216.

Höher, J., Münster, A., Klein, J., Eypasch, E., and Tiling, T. 1995. Validation and application of a subjective knee questionnaire. *Knee Surgery, Sports Traumatology, Arthroscopy* 3: 26–33.

Jäger, J., Perl, J., and Schöllhorn, W. 2007. Analysis of players' configurations by means of artificial neural networks. *International Journal of Performance Analysis of Sport* 3 (7): 90–103.

Kohonen, T. 1982. Self-organized formation of topologically correct feature maps. *Biological Cybernetics* 43: 59–69.

Kohonen, T. 1995. *Self-Organizing Maps*. Berlin, Heidelberg, and New York: Springer.

Krosshaug, T., and Bahr, R. 2005. A model-based image-matching technique for three-dimensional reconstruction of human motion from uncalibrated video sequences. *Journal of Biomechanics* 38 (4): 919–929.

Kusserow, M., Amft, O., and Tröster, G. 2009. BodyANT: Miniature wireless sensors for naturalistic monitoring of daily activity. In *Proceedings of the 4th International Conference on Body Area Networks* (Los Angeles, California, April 1–3, 2009). Brussels, Belgium: ICST.

Lamb, P., Bartlett, R., and Robins, A. 2010. Assessing coordination stability by means of a second SOM. Lecture at the XI. Workshop on Computer Science in Sport "Sportinformatik trifft Sporttechnologie", September 15–17, Darmstadt, Germany.

Lames, M., and Perl, J. 1999. Identifikation von Ballwechseltypen mit Neuronalen Netzen [Identification of rally types by means of neural networks]. In *Dimensionen und Visionen des Sports*, ed. K. Roth, T. Pauer, and K. Reichle, 103. Hamburg, Germany: Szwalina.

Larsson, P., and Henriksson-Larsén, K. 2005. Combined metabolic gas analyser and dGPS analysis of performance in cross-country skiing. *Journal of Sports Sciences* 23 (8): 861–870.

Malkinson, T. 2009. Current and emerging technologies in endurance athletic training and race monitoring. In *Proceedings of the IEEE Toronto International Conference—Science and Technology for Humanity (TIC-STH)*, 581–586. Toronto, Canada: Institute of Electrical and Electronic Engineers (IEEE).

Mayagoitia, R., Nene, A. V., and Veltink, P. H. 2002. Accelerometer and rate gyroscope measurement of kinematics: An inexpensive alternative to optical motion analysis systems. *Journal of Biomechanics* 35 (4): 537–542.

Memmert, D., and Perl, J. 2009a. Analysis and simulation of creativity learning by means of artificial neural networks. *Human Movement Science* 28: 263–282.

Memmert, D., and Perl, J. 2009b. Game creativity analysis by means of neural networks. *Journal of Sport Science* 27: 139–149.

Paradiso, R., Loriga, G., and Taccini, N. 2005. A wearable health care system based on knitted integrated sensors. *IEEE Transactions on Information Technology in Biomedicine* 9 (3): 337–344.

Perl, J. 2001a. Artificial neural networks in sport: Concepts and approaches. *International Journal of Performance Analysis of Sport* 1: 10–15.

Perl, J. 2001b. DyCoN: Ein dynamisch gesteuertes neuronales Netz zur Modellierung und Analyse von Prozessen im Sport [On a dynamically controlled neural network for modelling and analysis of processes in sports]. In *Sport und Informatik 5*, ed. J. Perl, 85–98. Cologne: Sport und Buch Strauß.

Perl, J. 2002a. Adaptation, antagonism, and system dynamics. In *Perspectives—The Multidisciplinary Series of Physical Education and Sport Science*, ed. G. Ghent, D. Kluka, and D. Jones, 4, 105–125. Oxford: Meyer, and Meyer Sport.

Perl, J. 2002b. Game analysis and control by means of continuously learning networks. *International Journal of Performance Analysis of Sport* 2: 21–35.

Perl, J. 2004a. Artificial neural networks in motor control research. *Clinical Biomechanics* 19 (9): 873–875.

Perl, J. 2004b. A neural network approach to movement pattern analysis. *Human Movement Science* 23: 605–620.

Perl, J. 2007. Neuronales Netz im medizinischen Bereich [Neural networks in the field of medical therapy]. *Klinik* 6 (7): 10–13.

Perl, J. 2008a. Physiologic adaptation by means of antagonistic dynamics. In *Encyclopedia of Information Science and Technology* (2nd ed.), ed. M. Khosrow-Pour, VI, 3086–3092. Hershey, New York: Information Science Reference.

Perl, J. 2008b. Modelling. In *Computers in Sport*, ed. P. Dabnichki, and A. Baca, 121–160. Southampton: WIT Press.

Perl, J., and Dauscher, P. 2006. Dynamic pattern recognition in sport by means of artificial neural networks. In *Computational Intelligence for Movement Science*, ed. R. Begg and M. Palaniswami, 299–318. Hershey, London, Melbourne, and Singapore: Idea Group.

Perl, J., and Endler, S. 2006. Training- and contest-scheduling in endurance sports by means of course profiles and PerPot-based analysis. *International Journal of Computer Science in Sport* 5 (2): 42–46.

Petermann, J. 1992. Controlled practice (Kontrollierte Praxis). In *Psychologische Diagnostik*, ed. R. S. Jäger and F. Petermann, 147–154. Weinheim, Germany: Psychologie-Verlags-Union.

Pfeiffer, M. 2008. Modeling the relationship between training and performance—a comparison of two antagonistic concepts. *International Journal of Computer Science in Sport* 7: 13–32.

Poppe, R. 2007. Vision-based human motion analysis: An overview. *Computer Vision and Image Understanding* 108 (1–2): 4–18.

Preuschl, E., Baca, A., Novatchkov, H., Kornfeind, P., Bichler, S., and Boeckskoer, M. 2010. Mobile motion advisor—A feedback system for physical exercise in schools. *Procedia Engineering* 2 (2): 2741–2747.

Raab, M., Perl, J., and Zechnall, D. 2003. The mapping of intrinsic and extrinsic information in continuous visuo-motor control. *E-Journal Bewegung und Training*. http://www.sportwissenschaft.de/index.php?id=291

Rebel, M. 2004a. Analyse von Rehabilitationsverläufen nach Kreuzbandrekonstruktionen [An analysis of rehabilitation processes after reconstructive surgery of the cruciate ligaments]. PhD diss., University of Heidelberg, Germany.

Rebel, M. 2004b. *Wenn der Kopf in die Knie geht—Analyse von Rehabilitationsverläufen nach Kreuzbandrekonstruktionen* [When the head moves into the knees—analysis of rehabilitation processes after reconstructions of the crucial ligament.] Bonn, Germany: Kovac.

Schmidt, A. 2010. Bewegungsmustererkennung anhand des Basketball Freiwurfes [Movement pattern recognition by the means of the basketball free throw]. *Forum Sportwissenschaft* 19. Hamburg, Germany: Czwalina.

Schmidt, A. 2011, submitted. Movement pattern recognition in basketball free-throw shooting. *Human Movement Science*.

Schöllhorn, W., and Perl, J. 2002. Prozessanalysen in der Bewegungs- und Sportspielforschung [Process analysis in motion and sports game research]. *Spectrum der Sportwissenschaften* 14 (1): 30–52.

Schöllhorn, W. I., Schaper, H., Kimmeskamp, S., and Milani, T. L. 2002. Inter- and intra-individual differentiation of dynamic foot pressure patterns by means of artificial neural nets. *Gait and Posture* 16: 159.
Stelzer, A., Pourvoyeur, K., and Fischer, A. 2004. Concept and application of LPM: A novel 3-D local position measurement system. *IEEE Transactions on Microwave Theory and Techniques* 52 (12): 2664–2669.
Stevens, G., Wulf, V., Rohde, M., and Zimmermann, A. 2006. Ubiquitous fitness support starts in everyday's context. In *The Engineering of Sport 6*, vol. 3, ed. E. F. Moritz and S. Haake, 191–196. New York: Springer.

# 15

# *Robust Monitoring of Sport and Exercise*

**Andrew J. Wixted**

**CONTENTS**

Introduction ........ 408
Current Sports Monitoring Systems: The Systems, Their Users, Their Outputs and Issues ........ 409
Current Monitoring Systems ........ 409
The Data Consumer ........ 409
The Core Monitor Outputs ........ 410
Issues for Monitoring Systems ........ 411
Sports Monitoring Research: The Purpose and Example Projects ........ 412
Purpose of Research Monitoring ........ 412
Examples of Research Monitoring Projects ........ 413
Sensors for Sports Monitoring: The Sensors, Their Outputs, Their Limitations and Their Uses ........ 416
Sensor Devices ........ 416
Using MEMS Sensor Outputs for Monitoring Sport and Exercise ........ 417
Useful Outputs from Accelerometers ........ 417
Useful Outputs from Gyroscopes ........ 418
Limitations of MEMS Accelerometers and Rate Gyroscopes ........ 418
Sensor Limits ........ 419
Utilizing Accelerometer Outputs in Sport Monitoring ........ 420
Orientation ........ 420
Frequency ........ 421
Consistent Repetitive Action ........ 421
Other Useful Outputs of MEMS Inertial Sensors ........ 422
Signal Processing for Sports Monitoring: Tools and Techniques for Extracting Information from Sensor Outputs ........ 422
Analysis Tools ........ 422
Kalman Filters and Neural Networks ........ 426
Sensor Hardware, Synchronization, Networking and Mounting: Putting It All Together ........ 427
Sensor Hardware ........ 427
Sensor Synchronization ........ 428
Sensor Networking ........ 429
Sensors on Athletes ........ 430

Current and Future Research ........ 431
Summary ........ 432
References ........ 433

## Introduction

The perception of the term *monitoring of sport and exercise* depends on the viewpoint of the individual, and this viewpoint is shaped by his or her activities and interests. Physiologists will have a different viewpoint of monitoring to that of a biomechanist or a strength and conditioning coach. Athletes will have a multilayered viewpoint that encompasses a broad spectrum of monitoring activities. An iconic view of sports monitoring is the coach with a stopwatch timing an athlete in trials before the big meet. This is still an important aspect of monitoring, but monitoring has moved far beyond this particular paradigm.

Modern monitoring of sport and exercise is being driven by developments in technology, particularly electronics and microelectromechanical systems. An athlete's diary, in which training sessions are recorded, has migrated from an exercise book through the personal computer to now reside in a memory chip in a watch or mobile phone. Electronic monitoring pervades many aspects and levels of sport and exercise. When exercising for fitness on the local sidewalks or on the running machine at the gym, the individual can choose to monitor heart and respiration rate, speed, temperature and distance covered. All of this information can be automatically and seamlessly logged and reviewed at a later time.

In a sports science laboratory the athlete can be monitored for almost any conceivable activity. The force plate resolves applied forces in three dimensions, and motion capture systems can be used to resolve complex dynamic systems into sets of numbers that define the optimal operation of the biomechanical system. Lactate levels, blood oxygen levels and energy expenditure can all be measured in the sports science laboratory.

The advances in miniaturized technology are pushing towards moving many laboratory-based activities out into the field. This assists in shifting towards a more ecologically valid assessment of sporting activity. Some laboratory-based assessment tools are ingrained in the athlete development process and form part of a standardized testing and recording system. This can lead to the development of lightweight, wearable electronic monitoring systems that mimic the laboratory instrument; however, in some cases the wearable system itself could produce a far richer analysis. Conversely, attempting to replicate some laboratory measures may be out of the reach of current technology, and this can lead to the rejection of a whole gamut of technological options based on the failure to achieve one particular aim. To optimize the monitoring options it is necessary to identify the strengths and weaknesses of different systems in particular situations. In sport there is such a wide range of activity that could be monitored that something that will appear obvious in 10 years has not yet been considered. At the time of writing, the "hawk-eye" system used to enhance cricket coverage and assist in tennis umpiring has been in use for only nine years, but many of the next generation of athletes will not remember a time without it.

## Current Sports Monitoring Systems: The Systems, Their Users, Their Outputs and Issues

### Current Monitoring Systems

The "standard" of sport and exercise monitoring is the heart rate monitor. This crosses all sport boundaries and includes built-in monitors in modern exercise equipment, sports watches, cycle computers and other technologies. The mainstay of scientific monitoring of population exercise levels is the accelerometer, used to generate estimates of energy expenditure in daily activity. A more recent arrival is the sports global positioning system (GPS) device. The sports GPS is used in many sports including running, cycling, sailing and various football codes. With the continued advances in semiconductor technology, these monitoring systems are merging with wearable combined GPS and heart rate monitors that can wirelessly connect to external sensors and transmit to a base unit.

Once you move past the generic sensors, the next level of commercial sensors is the sport-specific technologies. Cycling uses sophisticated technologies, monitoring not just speed and distance but also power through the crank, chain or wheel, depending on the brand, along with altitude and incline monitoring. Parachuting utilizes sensors for safety in the form of automatic activation devices monitoring the altitude and rate of descent. There are also sophisticated skydiving computers and data loggers that record each jump.

The inputs and outputs of the sensor systems are either simple, like pedometers with a reset button and a display of step counts, or complex, where the onboard computer generates a continuum of statistics covering physiological load, strength, distances, timing and any statistic that may be considered remotely useful.

### The Data Consumer

The data consumer of the sports monitoring device, usually the athlete and the support staff, determines the sophistication, the ease of use and the system outputs. The output of a device is either real time, providing instantaneous feedback, or some form of logged output for later retrieval and recording in a database. For some sports, athlete monitoring is performed for the media. The advent of wireless GPS devices allows live statistics for team players. It also assists coaches in identifying, in real time, which players need to be brought in so a fresher player can take the field. Sports administrators also use the output of some systems as a way of managing events or gathering statistics for planning for future events. For all these groups, the devices are commercially available, ruggedized devices with well-documented user interfaces for ease of use.

Researchers are data consumers of a different type. Researchers perform complex monitoring of athletes for a myriad of physiological and biomechanical evaluations. The devices used by researchers may be commercial devices, but at the leading edge of sports research the devices are often ad hoc monitoring systems built in-house or developed in conjunction with another research group. The output of these devices is the raw sensor data, which researchers then analyse, looking for useful activity signatures. The sensors that monitor the athlete may need to be accurately attached, the data collected synchronously from multiple systems and the outputs processed over a long period of time.

## The Core Monitor Outputs

The core monitors are the sports monitoring devices that are in widespread use, either across a broad spectrum of sports or in particular sports. Heart rate monitors and pedometers are popular for many reasons, but the primary reasons are that they are simple to operate and simple to understand, and they provide an immediately useful output. The heart rate monitor might be a complex heart rate monitor, fitness logger and calorie counter, but in the end it puts out a simple number. If the heart rate monitor is used as a real-time monitor, then the output figure is used to moderate the effort. This could be on a treadmill or other exercise machine where the person is trying to keep his or her heart rate in the appropriate zone—for burning fat or increasing fitness—depending on his or her personal goals.

Pedometers are similarly easy to use, in that the pedometer produces a simple number that can be recorded at the end of a session or the end of a day. The user can record this number in a diary or more commonly via a website. The accuracy of the pedometer is often not strictly relevant; the number is just a number that an individual can use as a comparison for past and future activity. An upmarket pedometer might record steps hour by hour and keep a log of each day, but in the end it produces a simple, easy-to-understand number.

Cycling computers have an increased level of technology, but this is linked to the sport, where, in endurance racing, the cyclist travels up and down hills of various grades, and the altitude can change by thousands of metres over the duration of a day. The combination of the grade and altitude can determine the power output the cyclist can safely sustain. While it is necessary for the computer to use sensors to measure the power delivered to the wheel and to capture speed, altitude and heart rate, the result is a real-time output that guides the effort of the cyclist in a particular situation.

Even though some sports have monitoring technology, other sports where the technology requirements are similar do not yet have access to the technology. For swimming, a prime candidate for automated monitoring to track the long training sessions, no commercial device currently exists. Accelerometer-based devices can track the stroke type, the lap times and the stroke rate (Davey et al. 2005, 2008), but this research has not made the jump to a commercial device.

This is not meant to be an exhaustive review, as there are many niche applications for technology. The very lightweight, unpowered radio frequency identification (RFID) tags are a common low-cost technique used to track a large field of athletes in long-distance events. Orienteering uses electronically equipped control points, and the athletes carry a specialized memory stick in lieu of the original control point record card. Other sports have their sport-specific systems. In some of these cases, the systems are akin to the pool or track timing system.

### Issues for Monitoring Systems

Electronic system monitoring is often assumed, by the uninitiated, to be 100% correct. For some systems the accuracy of the system is far greater than the approximations that have previously been considered acceptable. In the case of heart rate, an error of ±1 beat per minute for an electronic monitor compares favourably to an experienced individual monitoring his or her heart rate for 15 s with an estimate within ±4 beats per minute of the correct value.

Where an absolute measure is necessary, the error in some electronic systems would be considered unacceptable. Using an athlete-mounted GPS device to measure the time for an athlete to cover a course would be acceptable, but using that time to select a race winner would be unacceptable. While the timing accuracy of the GPS is extreme, since it is derived from multiple atomic clocks carried in the GPS satellites, the identification of position is sufficiently inaccurate that it could not be used reliably to discriminate between close times. Many sensor-based systems suffer from the same problem of having no absolute measure in some domain. A swimming monitor can detect a dive and reliably detect the turn at the end of each lap, but accurately determining when the swimmer touches the wall is difficult. So a swimming monitor could give very useful information on lap times and stroke rates but be totally inadequate as a race timing device. Despite these issues, if the monitoring technology is appropriately chosen and matches the requirements, the sensor systems supply valuable information to their users.

## Sports Monitoring Research: The Purpose and Example Projects

### Purpose of Research Monitoring

A plethora of monitoring devices have been developed, and a range of research activities have been performed in attempts to develop appropriate monitoring for a wide variety of athlete- or sport-related measures. Research focuses on a specific research question, usually associated with physiological, biomechanical or performance assessment (Figure 15.1). Ultimately, sports research should feed research outcomes back into the sporting community.

In biomechanics, there are many variables of interest to researchers. There are also different research goals; some research is performed to improve improvements in technique and performance, while other research is aimed at identifying injury predictors. In running, sports biomechanists are interested in everything from the performance contribution of the little toe through to the contribution of the head. In any sport, every link in the system is studied and at various depths of understanding. The baseball pitcher is studied at a basic level that measures ball speed and body rotation and at the complex in-depth level of joint-by-joint operation of muscles and ligaments. The sport of rowing has numerous opportunities to implement monitoring technologies on everything, including the foot stretchers, oarlocks, oars, seats and the athletes themselves. Virtually anything that can be monitored is useful in understanding the existing complex actions and interactions and in determining where those actions can be improved, either to enhance performance or to reduce injury risk. The types of sensors vary and include different types of strain gauges, accelerometers, gyroscopes, angular encoders and any other pieces of hardware that the researcher finds useful.

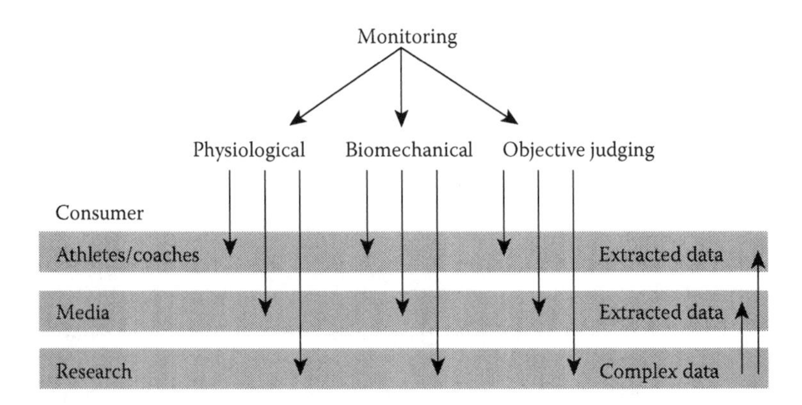

**FIGURE 15.1**
Monitoring occurs for different disciplines and purposes and for different user groups.

For untethered, noninvasive physiological monitoring, the number of monitored parameters is small. Heart rate, blood oxygen saturation, skin temperature and respiration are the basic monitored values. Core body temperature can be monitored using an ingested capsule. Lactate and glucose monitoring has been investigated over time, and research by the American military has looked at implantable monitors for lactate monitoring. Finally, some research is aimed at developing objective measurement systems for sports that rely on subjective measures of competition scoring.

## Examples of Research Monitoring Projects

There are, and have been, numerous research projects operating over a wide spread of technologies in the area of sports and exercise monitoring. The following projects include various physiological, biomechanical and objective assessment research projects using sensor technology in some form. They are also interesting as they cover a diverse set of sports, but the research outcomes are applicable across different disciplines. This research includes systems such as the boxing suit (Partridge et al. 2005), in-sole pressure sensors (Billing et al. 2004, 2006), instrumented snowboarders (Harding et al. 2005, 2007, 2008), instrumented swords (James 2008; James et al. 2005), instrumented rowing sculls and physiological monitoring (Wixted et al. 2007).

Boxing has been and, along with many other combative sports, continues to be a popular sport. As in other combative sports, athletes engaged in boxing can sustain serious injuries, and this can be exacerbated by the relatively subjective scoring system. Due to the subjective scoring system, some punches, although not technically higher scoring, tend to be given a greater weighting because of their visibility. In particular, a head shot can be observed and scored more easily than a flurry of body shots. The research of Partridge et al. (2005) investigated the development of an objective boxing scoring system that could identify and score all legal and illegal shots. By recording legal body and head shots equally, the incentive to target the head is reduced, potentially reducing the injury rate. The researchers developed a boxing suit with piezoelectric polymer sensors incorporated into the scoring and nonscoring areas of the suit and gloves. These sensors were networked via local electronics and Bluetooth wireless units to a scoring system that could capture and synchronize the data from both boxers within milliseconds. This research has culminated in the boxing suit being commercially available (PWP Designs) as the *Automated Boxing Scoring System*. Research into the boxing suit has continued, with collaboration between the Australian Institute of Sport (AIS) and Australia's Commonwealth Scientific and Industrial Research Organisation (CSIRO).

The work of Billing et al. (2004, 2006) is an example of the development of wearable "laboratory instrumentation", in this case the force plate. The force plate is a large, in-floor device that collects ground contact data from subjects as they walk or run over the plate. This may require the athlete to make repeated passes over the force plate to get a "natural" contact with the plate. The force plate output is the time-varying vertical and horizontal forces during the contact. This can be recorded graphically as the three-dimensional forces or used dynamically to visualize the changing force vector during contact. The force plate is often used in conjunction with motion capture systems in the study of athlete biomechanics. Billing and associates developed in-sole pressure sensors that collect contact pressure from four key points on the foot. Used alone, these sensors cannot resolve the three-dimensional contact forces. By using a trained artificial neural network to combine the output from these sensors with three-dimensional accelerometry, Billing and associates were able to reproduce the three-dimensional contact forces of the force plate. This type of wearable instrumentation allowed the collection of laboratory-type data with the environmental validity of an athlete on the track. It is also an example of wearable instrumentation that provided new opportunities for the in situ study of the biomechanics of running.

The instrumented snowboarding research of Harding et al. (2005, 2007, 2008) is another example of research that investigates a sport where there is a level of subjective skill assessment. Harding and associates, in these and other papers, investigated not only the development of sensor technology for the snowboarding athlete but also the basis for the scoring used by judges in competition. This is not a sport like diving or gymnastics where there are clearly identified set pieces and transitions with specific levels of difficulty but a sport where judging is linked to a culture that values style and innovation. From their research, Harding and associates identified some key performance parameters, such as total air time and average degree of rotation, which in combination had good correlation to scores. By combining various signal-processing techniques and using these to analyse the output of body-mounted three-dimensional (3D) accelerometers and gyroscopes, Harding and associates were able to reliably extract the key performance parameters identified previously. In outlining further research Harding identified some possible areas of investigation such as capturing parameters relating to single pieces of the total performance. This technology would then be applicable to some of the existing sports that include complex aerial and rotational activities.

A number of sports require the use of an implement such as a hockey stick, golf club, cricket bat or, in the case of James et al. (2005), a training sword, or *bokken*. Disciples of a particular sport develop their skill through a range of training activities including solitary practice and skill acquisition sessions with an experienced practitioner. Solitary practice is valuable but relies on the developing athlete correctly recalling and practicing the appropriate sequence of body placements and moves. Experienced athletes or coaches are not always available to correct a developmental athlete's training. In this and other research (James 2008), James and associates used three-dimensional accelerometers to record the movements of a bokken wielded by an experienced practitioner. The resultant data were wirelessly transmitted in near real time to a computer where they were processed and displayed visually in a format that was interpretable by the athletes. The data from the experienced practitioner were used as a template, and the developmental athletes were able to compare and correct their movements in a continuous feedback cycle during their training. This allowed the developmental athletes to more quickly acquire particular technical skills. This same sensor-based biofeedback technique is being used in a variety of sports.

Rowing is a high-tech sport with considerable research into all aspects of the equipment including the shape of the shell. Modern rowing shells are often carbon fibre composites carefully designed to minimize drag. For many years the most common piece of electronic technology used with rowing boats was the impeller. This was a small propeller attached under the boat at a carefully chosen position where it would not increase the drag of the boat. The rotation rate of the impeller, and hence the speed of the boat, was detected by a magnetic sensor fitted inside the shell immediately above the impeller. Although the impeller itself had a negligible contribution to drag, usually less than 0.1% of total drag, the impellers catch water-borne material, sometimes increasing the drag to a detectable level. The impeller can also get clogged and not rotate freely or even come off and be lost. Research into the use of alternate sensors initially considered accelerometers as a replacement for the impeller. As discussed later in this chapter, this was not a particularly appropriate use of the technology, as extracting velocity was and is problematic. In researching this problem the acceleration signal was recognized as a valuable resource. When the accelerometer output was first synchronized with video, coaches immediately recognized the value of this signal as it allowed them to see the power pulses and deceleration of the boat. With the use of the video, the causes of unexpected decelerations were identified, and changes to the rowing style were implemented. Ultimately, accelerometers were combined with other technology, in particular the GPS, to create a rowing unit. This technology has been used by elite Australian rowing teams since prior to the 2004 Athens Olympics. (Land Victoria 2005; Prime Minister's Science, Engineering and Innovation Council 2004).

Physiological monitoring is considered by many people to be the forte of the heart rate monitor, but in many large studies the most common parameter of interest is the estimated energy expenditure of the subject. The tool selected for this task is the accelerometer. In this context, *accelerometer* refers to a purchased item that outputs a number, or "count", that is directly related to the estimate of energy expenditure. These devices can log the data and output the number over specific time frames, e.g., counts per minute or per 5 minutes or per hour. These devices are reasonably accurate for low-intensity activity but will often have zero or even negative correlation for high-intensity activity. Because sport and exercise are usually related to more intense activity, the traditional form of accelerometer "counts" as an energy estimator was not useful. The research of Wixted et al. (2007) investigated the mechanism of the effect of running on the "count" style of the estimator. As an alternative, it was found that, for submaximal running, an individual's step rate correlated highly with running speed. For a group, the step rate modified by leg length also correlated highly with running speed. Existing research had already shown that running speed modified by body mass correlated highly with energy expenditure. By combining step rate with leg length and mass, the energy expenditure of running athletes could be estimated ($R^2$ of 0.81 for mixed-gender recreational athletes). Step rate can be robustly and reliably estimated from the raw accelerometer signal and is impervious to calibration and overload errors.

## Sensors for Sports Monitoring: The Sensors, Their Outputs, Their Limitations and Their Uses

### Sensor Devices

Currently, the bulk of monitoring technology projects appear to be in the area of kinematic monitoring using microelectromechanical systems (MEMS) inertial sensors. These are very small devices that include a mechanical component and support electronics on a tiny silicon chip (Figure 15.2). The common MEMS inertial sensors are the accelerometer and the rate gyroscope. Today, a MEMS inertial sensor operating on three orthogonal axes (3D) comes in a package of 4 × 4 × 0.9 mm. The small size of these devices and their low cost have led to a proliferation in their use in sports monitoring.

In 1991 Analog Devices developed the MEMS accelerometer for use in vehicle airbag deployment systems. By 1999, thanks to the commercial pressures of the automotive industry, it was easy to obtain ±2 g dual-axis MEMS accelerometers at a reasonable price and in a small 5 × 5 × 3 mm package. Getting triple-axis accelerometers required a second accelerometer device carefully aligned and mounted on edge. For sports monitoring the 2 g accelerometers were quickly adopted. Within a few years integrated triple-axis (3D) accelerometers (Figure 15.2) became available in a single package with similar dimensions to the dual-axis device. The 3D accelerometers could also be obtained with switchable outputs of up to 18 g. In 2010, dual-axis MEMS accelerometers of 70 g were available. 3D accelerometers are widely used in game consoles and are incorporated in numerous mobile telephones.

MEMS gyroscopes became commercially available in the mid 1990s, and by the mid 2000s the use of 300°/s gyroscopes in sports monitoring was not uncommon. By 2009, 3D gyroscopes were being included in a wide range of mobile phones and other mobile and game platforms. 3D gyroscopes with rates of 2000°/s were commercially available from InvenSense in a 4 × 4 × 0.9 mm device.

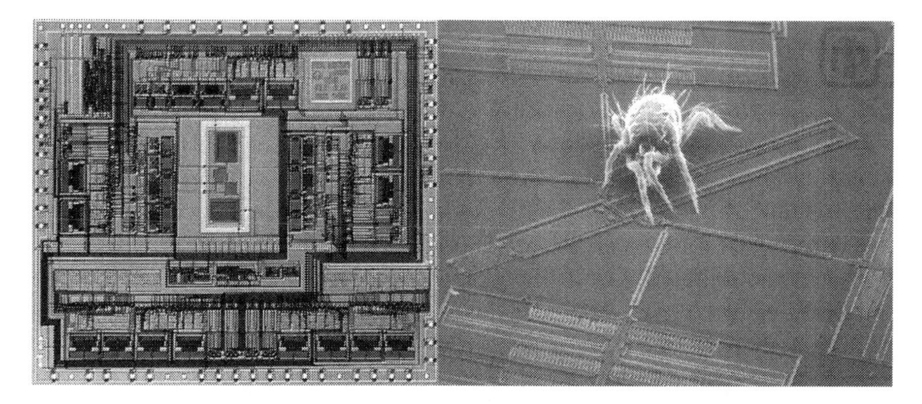

**FIGURE 15.2**
*Left*, a three-axis Sandia MEMS accelerometer (fits in a 5 × 5 mm footprint). *Right*, for a sense of scale, a spider mite on the MEMS mirror assembly. (Courtesy of Sandia National Laboratories, SUMMiT(TM) Technologies, www.mems.sandia.gov.)

## Using MEMS Sensor Outputs for Monitoring Sport and Exercise

The use of MEMS sensors depends on the scenario under investigation. Accelerometers measure acceleration, and gyroscopes measure angular velocity. If the physics of the activity and the locations of the monitoring points mean that either of these devices, or both together, can be used to extract useful information, then an attempt will be made to gather that information. Some of the typical useful outputs of accelerometers and gyroscopes are identified here.

### *Useful Outputs from Accelerometers*

- Stationary 3D MEMS accelerometers can detect their own orientation relative to gravity.
- Accelerometers on moving objects can detect the movement of the object. If the object is moving with a repetitive action, then the frequency of that action can be extracted.
- If the repetitive action is consistent, such as the trunk movement or shoe movement of a running athlete, then a pattern of the movement can be extracted and the particular movement modelled. This is often done in conjunction with motion capture data.
- If the pattern is consistent, then a change of pattern can be detected.
- If an impact occurs, then the accelerometer can detect the intensity and direction of the impact.
- As for the impact, the magnitude and direction of an applied acceleration can be detected.

### *Useful Outputs from Gyroscopes*

- Gyroscopes output the rate of rotation (angular velocity). To obtain an angle of rotation requires the mathematical integration of the angular velocity.
- Outputs from linked gyroscopes can be compared and used to identify joint movement.

## Limitations of MEMS Accelerometers and Rate Gyroscopes

Anyone with a high school understanding of physics will immediately recognize that if we have a stationary start position or a known starting velocity or acceleration, then we should be able to integrate the acceleration to get velocity and integrate the velocity to get position. In that case, we can build a 3D representation of life and track all our movements and wanderings using 3D accelerometers.

Unfortunately, life is not this simple for a number of reasons. The predominant reason is that MEMS accelerometers detect gravity, and the constantly changing orientation of the sensor mixes gravity with a dynamic movement-related acceleration signal and creates a chaotic signal. Most of the other reasons relate to the nature of semiconductor materials, the manufacturing process and digital representations of the analogue world (Figure 15.3a, b and c).

Other problems arise if the accelerometer is rotating about an axis. If the axis of rotation is through the device, the rotation will not be detected; if the axis is offset, then the acceleration will be a combination of centrifugal and tangential acceleration, which can be indiscernible from an applied linear acceleration.

Accelerometers need to be calibrated. A simple calibration technique is to align each axis with ± gravity, and then a scaling factor and offset value are calculated for each axis. This method suffers from inaccuracies due to fractional misalignments with gravity. A better method is the six-point method of Lai et al. (2004), where any misalignment is catered for in the calculation. This method, although an improvement, cannot deal with any nonlinearity of the sensor (Figure 15.3d). Gyroscopes, while not affected by gravity in the same way as accelerometers, also suffer from errors due to the technology, the manufacturing process and the inability to digitally represent an analogue world.

The final issue for accelerometers and gyroscopes is that they very often have no absolute reference. In timing a race there is an absolute start and end position that exist in a particular frame of reference. An inertial sensor mounted on an athlete or on a piece of portable sporting equipment knows almost nothing about the external world. A stationary 3D accelerometer knows its orientation relative to gravity but knows nothing else about

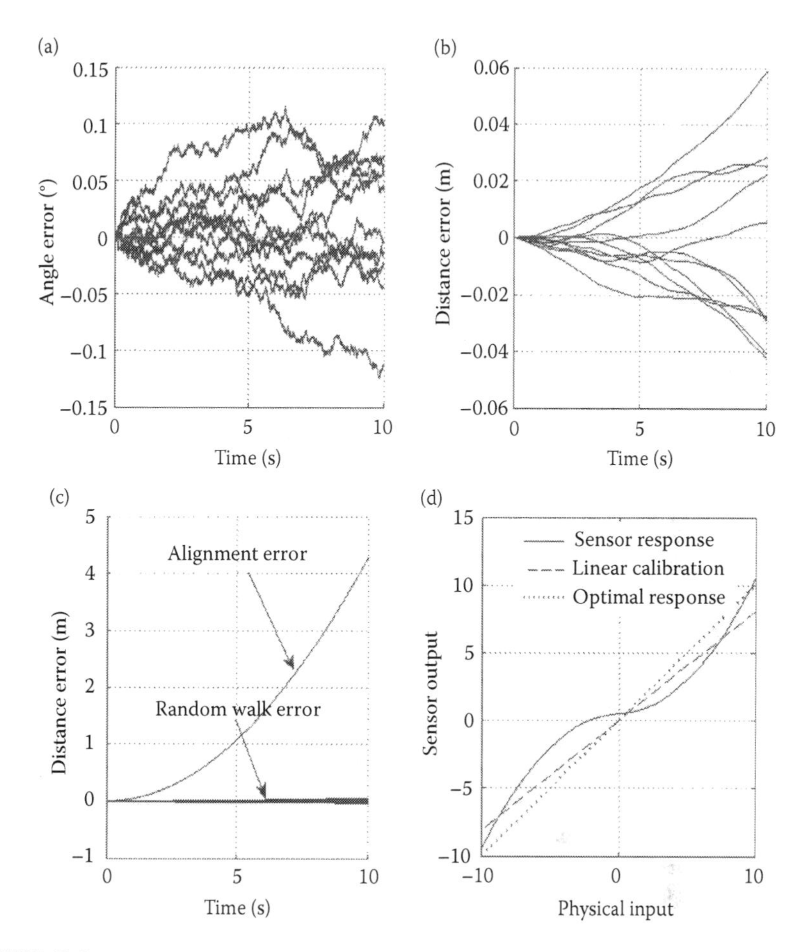


**FIGURE 15.3**
(a) Random walk errors for gyroscope output after integration to angle. (b) Random walk errors for accelerometer output after double integration to distance. (c) Comparison of random walk errors from (b) to error due to 0.5° misalignment of horizontal accelerometer. (d) Sensor bias, non-linearity and calibration errors.

the world around it. All information derived from inertial sensors must be obtained by extracting some relevant activity signature. There is ongoing research into the use of accelerometers in combination with other devices such as rate gyroscopes and MEMS magnetometers for inertial navigation systems (INS).

### *Sensor Limits*

MEMS accelerometers and gyroscopes have provided many research opportunities, and as newer devices with larger ranges become available, the opportunities increase. The range and frequency of human movement may

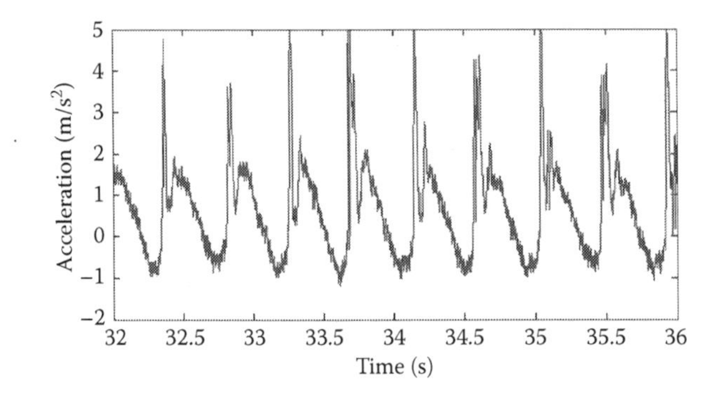


**FIGURE 15.4**
Overloading of a ±2 g vertical accelerometer during walking. The overload occurs when the foot strike is transmitted up to the trunk-mounted accelerometer. Sample rate 1000 Hz, filter frequency 227 Hz.

appear small when compared to the mechanized world, but when sport is involved, the intensity of the activity can exceed the range of many devices. Accelerometers rated at ±2 g often have a useful range exceeding 2 g, but impacts, such at the foot contact impact transmitted up to the trunk of a runner, can easily overload the accelerometer (Figure 15.4).

The 300°/s gyroscopes were useful for some sports, such as monitoring body roll in swimming, but even full-body rotation could not be captured in sports such as parachuting *style and accuracy*, where an elite athlete can start and stop a full 360° rotation in less than 1 s.

Analysis of motion capture data from different sports had identified limits that the latest versions of accelerometers and gyroscopes may just be capable of measuring. For cricket fast bowlers, the centrifugal acceleration on the forearm near the wrist exceeded 70 g, and the wrist rotation rates during bowling exceeded 3500°/s (Wixted, Spratford et al. 2010). Actual measurements using 100 g accelerometers on the wrist have returned values as high as 85 g for subelite fast bowlers. Analysis of baseball pitchers indicates rotation rates of approximately 4500°/s.

## Utilizing Accelerometer Outputs in Sport Monitoring

### *Orientation*

Orientation is a key output for some activities. If a researcher desired to detect the orientation of a baseball bat or cricket bat during a particular movement or in response to a particular pitcher or bowler (or type of pitch or delivery), then a 3D accelerometer would be ideal. If a researcher wanted to estimate the movement of a bat through a response, then the momentary stationary orientations before and after the stroke could be used in conjunction with gyroscopes and appropriate software in the modelling of this response. Orientation is useful for other activities, and for some activities the athlete does not need to be stationary. The difference between swimming backstroke and using any other stroke is clearly discerned in the level of the accelerometer output. For a football player, the difference between vertical and horizontal can be easily identified. For a continuous repetitive movement such as running or swimming, the acceleration due to the repetitive action can be filtered off, leaving just the orientation signal. For a more chaotic movement, such as a rugby maul, very short moving average filters can be used to detect when the player is momentarily stationary, at which point the orientation can be extracted.

### *Frequency*

Frequency of activity is a directly usable output for swimmers, rowers, runners and any other athletes who perform a repetitive activity. For running, the step rate or stride rate can be used for extracting the estimated energy expenditure (Wixted et al. 2007) and for estimating the runner's speed. For swimmers and rowers, the stroke rate is an important parameter for estimating both physiological load and speed.

### *Consistent Repetitive Action*

With any repetitive action such as trunk movement while running, accelerometers generate a signal that contains a repetitive component. This component contains a mix of information that may be due to changes in orientation introducing a gravity signal into the horizontal accelerometers plus a signal due to movement along the various axes. In the running action there is a period of free fall where total acceleration should approach zero, but it may not, since the sensor is monitoring only a body component, not the whole body. The result is a pattern that needs interpretation. Sometimes this pattern is augmented with motion capture data or data from other sensors that help in interpreting the signal and how the acceleration signal relates to the athlete's biomechanics. Regardless of the cause of the signal, either orientation or motion changes, the resultant signal is able to be framed and interpreted to identify biomechanical factors of interest.

Even though a repetitive action might appear consistent, the accelerometer's data from each swimming stroke or running stride are different in some way to those from the preceding or subsequent one. It may not be possible to directly compare individual running strides from one point in time with those from another point in time. In this case it is useful to generate a representative picture of the stride. This can be done by taking a number of successive strides and overlaying them to generate a pattern representing an average stride at that time and place. This pattern removes transient effects or interstride variability and leaves the fundamental acceleration pattern and hence allows interpretation of the underlying biomechanical activity.

The signal will often have a left-right asymmetry indicating the particular idiosyncrasies of the athlete. In the case of swimming a freestyle stroke, the athlete rotates along his or her longitudinal axis, and, for some swimmers, a hip-mounted medio-lateral–oriented accelerometer can oscillate almost between ±1 g due to the hip roll. This signal can give a good indication of how far the hips roll in each direction, and any asymmetry of the motion is quickly identified. For some runners, the representative stride acceleration pattern can indicate the contact and flight phases and direction of applied forces (Wixted, Billing, et al. 2010). For athletes who strike their heels hard at initial contact, a trunk-mounted accelerometer can show a clear contact signal, while for an athlete landing with a flatter foot, the contact signal is muted. The harder heel strike is the most likely explanation of the large spikes in Figure 15.4, as not all subjects caused this overload while walking.

### *Other Useful Outputs of MEMS Inertial Sensors*

Multiple MEMS sensors can be combined in a system to extract interesting and useful information. Some sports activities, such as baseball pitching, tennis serving or cricket bowling, involve rotation of the spine. Excessive rotation of the spine has been identified as an indicator of future injury (Portus et al. 2004). Normally, this type of measure would be performed in a laboratory setting using a full motion capture system, with the athlete undergoing numerous trials. With multiple synchronized gyroscopes mounted along the spine, using the spine as the axis of rotation, the timing and magnitude of the rotation can be measured in the training and competition environments. The same technique can be used to measure the relative rotation of the leg or arm segments as in the tennis serve analysis of Ahmadi et al. (2010). MEMS gyroscopes are also often used to detect the orientation changes that occur to accelerometers, and then the output of the accelerometer can be mathematically de-rotated to generate a cleaner output. As noted earlier, there are many sports and many things that can be monitored in sport. The preceding examples are a sampling of monitoring research. Table 15.1 compares some of the sensor outputs with some of the parameters of sport where the sensor outputs have application.

## Signal Processing for Sports Monitoring: Tools and Techniques for Extracting Information from Sensor Outputs

### Analysis Tools

The main tools used for processing sensor data include many trigonometric functions, high- and low-pass filters of various lengths and various forms of

**TABLE 15.1**

Some Combinations of Sensor Outputs and Their Uses

| Sensor System Outputs | Examples of Uses |
|---|---|
| Repetitive action patterns | Biomechanical assessment of technique<br>Skill acquisition for developing athletes |
| Timing between identifiable events | Speed, transmission delays in biomechanical linkages |
| Frequency (rate) of repetitive activity | Physiological load, energy expenditure, running speed, boat speed, etc. |
| Stationary orientation | Body position, e.g., prone, upright<br>Implement position (bat, club, hockey stick) |
| Dynamic orientation | Swimming stroke identification<br>Biomechanical assessments<br>Signal processing for feature extraction |
| Intensity of acceleration | Impacts (between players)<br>Physiological load (non-intense activities)<br>Measuring intensity of forces acting on the spine, the head, etc. in different sports |
| Intensity of angular velocity | Skill acquisition (tennis, cricket batting, cricket bowling, snowboarding, etc.) |
| Relative angle of rotation | Injury prevention (identifying twisting of body segments like the spine) |
| Joint angle | Skill acquisition and assessment, biomechanical assessment of running, jumping, cricket bowling, rowing, etc. |

transformation tools. There are also self-correcting processing technologies that are appropriate for working with patterns and are often used in conjunction with inertial sensors.

Digital filters with long durations can be used to extract orientation from data from athletes engaged in a repetitious activity such as running or swimming. The duration of the filter is proportional to the period of repetition. For instance, a runner with a step frequency of 3.2 Hz (3.2 steps per second) has a stride frequency of 1.6 Hz. The output of a low-pass filter set at 1.4 Hz will have virtually no power in the signal, indicating that little or no activity is occurring below the stride frequency. The filter duration would typically be chosen to cover a number of strides; in this case a filter length of around 4 s would cover 6 complete strides or 12 steps. The actual filter lengths are a trade-off between what is technically desirable and what produces a useful outcome. In swimming the stroke rate and the number of strokes that occur in one lap of the pool may preclude using longer filters.

The extracted orientation information can be further processed using various trigonometric relationships to extract the angle of the sensor relative to gravity, and this angle can be used in a Euler rotation matrix to remove any long-term orientation shift. In the example in Figure 15.5, an athlete has a

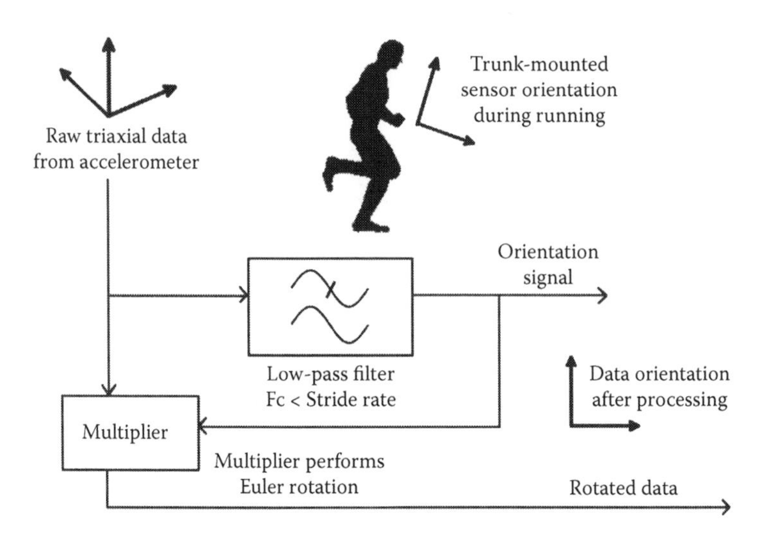


**FIGURE 15.5**
Signal processing to remove quasi-static orientation of an athlete's trunk during running and to generate an output signal for a vertically aligned sensor.

sensor mounted on the lower back. If the athlete leans forward as he or she runs, the 3D sensor is tilted forward, and the vertical- and anterior-posterior–oriented axes both receive a mix of vertical and forward-back acceleration. By mathematically rotating the sensor data so that the vertical accelerometer aligns with gravity, a cleaner, more useful output is obtained.

Relatively short filters are useful for identifying short-duration low- and high-intensity activity. In the case of a game of rugby, a short rolling mean, rectangular filter can be used to identify the short-duration high-impact activity that occurs in a hit up between two players or during a tackle. At the other end of the scale, a player involved in a tackle may be momentarily stationary; therefore, a very short filter can also be used to detect this event.

The fast Fourier transform, or FFT, is a tool often used in signal processing to extract the frequency components present in a signal. The output of an FFT changes in discrete intervals (Figure 15.6), which is not usually a problem for high-frequency systems such as radio, but at the frequencies of interest for athlete monitoring, these discrete intervals are sometimes too far apart to be useful. In this situation the frequency information, such as step rate or stroke rate, is extracted in the time domain by zero crossing detection. For a runner, the accelerometer's vertical axis has a very clear oscillatory signal, and this is the easiest point to extract the running step rate. The zero crossing detector is combined with a gravity-removing high-pass filter (Figure 15.7), and the rate is extracted. This method achieves a step rate output with an accuracy that is more than sufficient for monitoring purposes. In Figure 15.6, the output of a zero crossing detector operating on 15 Hz data is compared to the FFT used with 150 Hz data. The jitter in the zero crossing detector can

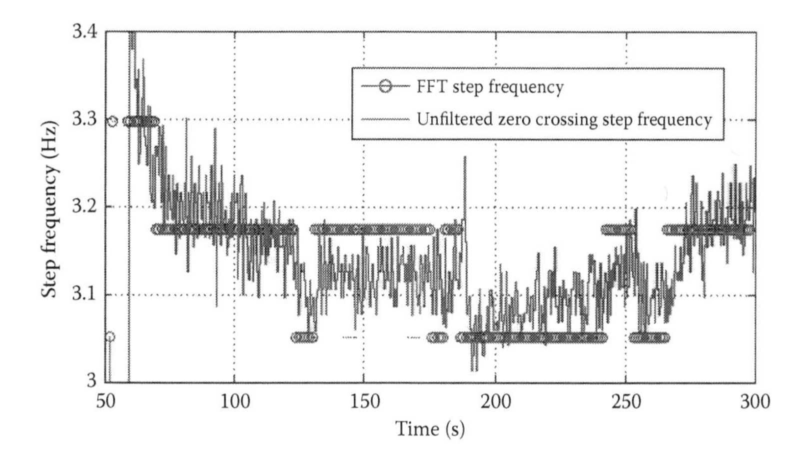


**FIGURE 15.6**
Comparison of fast Fourier transform (FFT) output as an estimator of step frequency with 150 Hz sampled data as the input and a linear interpolated zero crossing detector operating on 15 Hz sampled data. The FFT outputs only at discrete intervals.

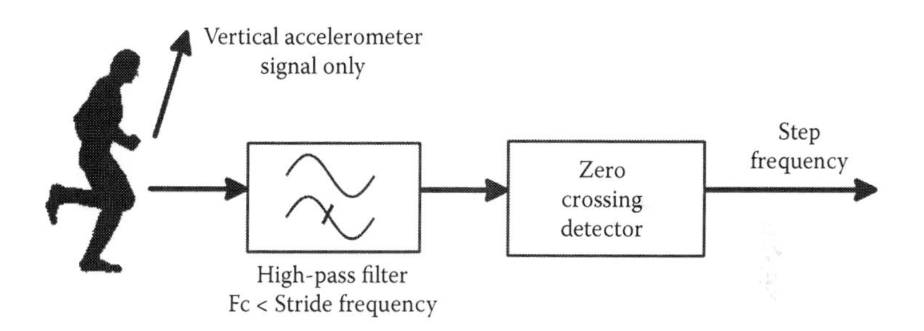


**FIGURE 15.7**
Extracting step frequency from running using a single accelerometer. The high-pass filter removes the gravity offset from the accelerometer signal.

be removed by using a higher sample rate or by averaging two or three successive steps. The FFT is a processor-intensive calculation and not generally suitable for athlete-mounted systems. The zero crossing detector can operate very efficiently using very few processor cycles and is therefore ideal for lightweight, low-power, athlete-mounted monitors.

Multiple devices are often linked to measure a complex system. Using multiple gyroscopes on a limb, joint rotation can be detected. In Figure 15.8, three gyroscopes were mounted on an arm, one on either side of the axis of the elbow joint and one at the wrist. If the arm was moving while the elbow was straight, the gyroscopes tracked together, but once the elbow began to flex, the gyroscopes mounted on the forearm tracked away from the upper arm–mounted gyroscope. At a rudimentary level, if the gyroscopes are

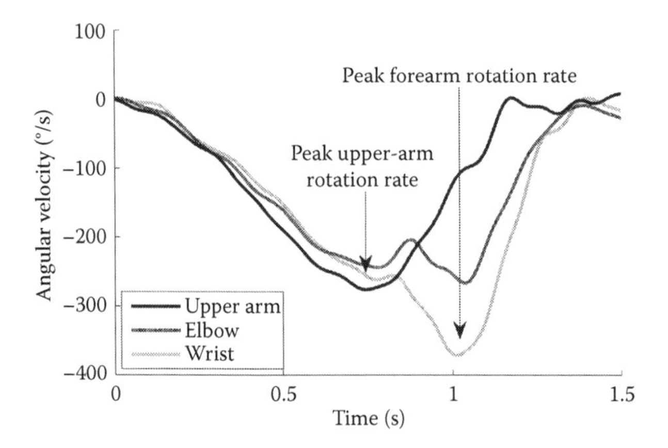


**FIGURE 15.8**
Using multiple gyroscopes to monitor joint movement.

correctly calibrated, then the angle between them can be extracted. At a more sophisticated level, more sensors can be combined with the original sensors to create a system that detects segment orientation and that can automatically self-calibrate.

## Kalman Filters and Neural Networks

Both these techniques can be used to take multiple sources of data from multiple sensors and combine them to create an output that is useful in research. In the case of artificial neural networks, the sensor system, which can consist of many sensors, forms the input to the system. The system output may be required to mimic a reference system, for instance, the output of a laboratory-based system. The artificial neural network, which contains a number of hidden processing layers between the inputs and outputs, "learns" how to combine the input signals to produce the same output as the reference system. The system can be trained by concurrently collecting data from the sensor systems and from the laboratory equipment and exposing the system to the required output. The effectiveness of the system is tested by collecting data from both systems and then comparing the outputs of the two systems. The work of Billing et al. (2006) is an example of a wearable sensor system using an artificial neural network to mimic the output of a standard piece of biomechanics laboratory equipment, the force plate.

The Kalman filter's strength is the ability to take noisy data and create an output that appears less noisy. The Kalman filter predicts future values of the system state and uses the difference between the predicted value and the measured value to improve the predictions. This necessitates an external input to the system that can be used to correct the predicted values. As noted earlier, the inertial sensor systems have almost no knowledge of the outside world, but by looking at specific circumstances, some references can be identified. The obvious reference is the stationary orientation, but others exist. For instance, a foot-mounted sensor system on a runner can use a predictor of foot height above the running surface. While the foot is flat on the surface, the height must be zero. Therefore, if the system can cleanly detect the contact phase, then the Kalman filter can operate. If the height estimate is working efficiently, then, provided the model is appropriate, the other outputs of the system will also be operating effectively.

## Sensor Hardware, Synchronization, Networking and Mounting: Putting It All Together

### Sensor Hardware

All monitoring projects require electronic hardware to power the MEMS sensor devices and sample the signal produced by the sensors. This hardware is basic microprocessor technology with consideration of the requirements of athlete monitoring. This means that sensors potentially need to operate over long periods and be of very light weight and small size. Over recent years, the advances in technology have made these objectives more easily achievable. The monitoring systems used for some of the projects detailed earlier, such as the early rowing monitoring hardware, the wearable force plate and several other projects, were based around the real-time operating system of Wixted et al. (2002) and various versions of hardware systems including those of James, Davey & Gourdeas 2004 and James, Davey & Rice 2004 and later commercialized systems.

The basic monitoring system comprised the power supply, the microprocessor, the data logging memory and the sensors. For systems with no user controls and just basic logging of sensor data, this hardware was fitted completely on a circuit board that was 2 × 1.5 cm with the components on one side and the micro-SD flash card for data logging mounted beneath. The newer lithium polymer batteries also assisted in reducing the size of the sensor hardware. The first iteration of hardware used 32 MB SD card flash memory. Even this amount of memory, modest by today's standard, could save 24 h of continuous three-channel, 10 bit-per-sample, 100 Hz sampled data. If the data were compressed, this could rise to more than 80 h of data. For a system with an inbuilt wireless and a user interface with a multi-line display, the size increased to about 4 × 2.5 × 0.7 cm. A number of commercial sports monitoring products now exist that combine multiple sensors including triaxial accelerometers, gyroscopes and magnetometers. These can also include GPS.

Battery life depends on the battery size, the system sample rate (the number of times per second the sensors are read), the number of channels of data and the sensors themselves. The higher the sample rate, the shorter the battery life; the more channels of data, the shorter the battery life; and the more sensor devices switched on, the shorter the battery life. For some athlete-monitoring applications, the hardware must be very small and light but capture a large quantity of data, so the battery life could be as short as 30 min. In other circumstances, the same size battery could last several hours.

The useful life cycle of basic sensor systems is reasonably long, with many research uses for older sensor systems. As research progresses, more applications for sensors are identified. As the sensor technology improves, more monitoring options become available, and therefore even more research options become available. The result of this cycle is that older sensor systems continue in use, while newer research devices are continuously being built. To minimize the overheads in continuously designing new hardware, it is possible to reuse designed components of the system. For basic logging-based systems (as opposed to wireless), the core component is the logger, a combination of the microprocessor, battery, charging circuit and a flash memory card. Communications take place via a USB interface, which can also double as the battery-charging circuit. As newer sensor devices become available, these can be mounted on a daughterboard and connected to the logger via a small connector. This arrangement results in hardware that is not optimal for size but saves a considerable amount of time and expense in designing new sensor platforms. Today, because of the interest in MEMS sensors, these devices are available in specialist electronic shops. If size is not a major issue, a hobbyist can develop a system to the same specifications as these research devices. Miniature data loggers and miniature sensor boards can be purchased over the Internet for as little as A$100, and many hobbyists are developing software to use the sensors to control radio-controlled planes and cars and other models. Some suppliers make the hardware design and controlling software available under open-source style licenses, allowing anyone to base new hardware on the same design and to copy the operating system software.

## Sensor Synchronization

When multiple sensors are monitoring multiple limb segments or multiple separate locations as happens in a rowing boat, these sensors need to be synchronized so that events happening to one sensor can be directly related to events happening at another sensor. Sensors can be synchronized by connecting them together with wiring, or they can be synchronized wirelessly. Post hoc synchronization is a commonly used process where logged data from multiple sensor systems are aligned (or synchronized) by matching external event signatures across all the sensors. For example, an athlete fitted with multiple sensors might be asked to do a "jump" before and after a trial. This jump is identified in the data from each sensor, and the data are aligned.

Post hoc synchronization can work very well for many applications, but when the timing of events needs to be accurate to within a few samples, the synchronizing event signature must be very specific. For these applications the synchronizing event can be generated electronically. A sharp magnetic pulse can be used to insert a signature on magnetometer channels, or a strong infrared pulse can be detected on an infrared receiver. Another common method of generating the synchronizing pulse is the use of a wireless receiver and synchronizing pulse generator. The beauty of post hoc synchronization is that unsophisticated logging-only sensor systems can be used individually and independently of each other or together in a complex monitoring environment. As long as the data collection protocols include the generation of the synchronizing event, the data can be synchronized when they are processed.

Wired or cabled networks are intrinsically synchronized where some master device provides a synchronizing clock signal that controls all other systems in the network and where all the sampled data are fed by cable directly to the master unit for logging. Wireless networks can have a range of synchronizing techniques depending on the network topology and network technology. Some *wireless* networks are intrinsically synchronized, while others require additional *sensor* network protocols to manage the sensor synchronization.

## Sensor Networking

Sensors are usually networked where data from multiple sensors need to be logged in a synchronized format. Networked sensors can be connected via cables to a base unit or have their data transmitted wirelessly. Wireless networks appear to be able to handle vast amounts of data, but the organization of the network and the number of sensors in the network can limit the network throughput. If sensors are small with limited battery power and using a Wi-Fi wireless network, they need to maximize their data throughput to minimize overall power consumption. This is because there is a certain amount of preamble and wireless synchronizing that occurs on every transmission. To maximize throughput, data should be transmitted in larger blocks. This amortizes the cost of setting up the connection over more data bytes, reducing the overall power cost per byte. In this situation data packets from different sensors need timing information transmitted in the packets so the data in the packets can be correctly reassembled in relation to the packets from other sensors. Wi-Fi–based networks have other limitations. As Wi-Fi uses a carrier sense multiple access, collision avoidance (CSMA-CA) mechanism, you cannot guarantee when a particular sensor will get an opportunity to transmit. If there are too many sensors in the network, the network can become saturated because the sensors are continuously trying to transmit but getting blocked by other transmitters. If the sensors are being used in a public venue, they will also be competing for airtime with the numerous wireless devices in the venue. In a Bluetooth network, the transmission time slots are guaranteed, but the packet sizes are much smaller, and the overall data rate is also much lower compared to the potential data rates of a Wi-Fi network.

Modern wireless technology is highly complex and, in the case of Bluetooth, incorporates very tight synchronization. Using a prebuilt wireless unit simplifies many networking issues but can bring in different issues due to the existing fixed network topologies. For instance, using a technology like Bluetooth makes some sensor networks into simple plug-together technology but makes other network topologies unworkable. If it is necessary to monitor two or three sensors in close proximity, then a Bluetooth solution is ideal. If it is necessary to monitor 40 football players including on-field and bench players, the problem is more complex (see Fraser et al. 2006). Many sensor networks use their own wireless network protocols in order to get more flexibility in the networking than some of the existing technologies provide. Using a home-grown solution can also bring a price advantage because adding a wireless chip to a sensor board during design is low cost. Many wireless chip manufacturers also provide the computer code necessary to program and use their chip, and this simplifies the creation of the programming necessary to run the sensor platform.

## Sensors on Athletes

Athletes, particularly elite athletes, can be sensitive to the presence of sensors on their bodies or on their equipment. If an athlete wearing a sensor comes in fractionally behind an athlete not wearing a sensor, then the question in the athlete's mind will be *"Did that sensor slow me down?"*

This is a legitimate question. Some early attempts at mounting sensors did cause problems. Initial attempts at attaching sensors on swimmers with straps had some unexpected effects with sensors getting caught in the water flow and lifting away from the athlete's body like a wing. Other sensors flipped away from the body and acted like a brake. The solution for swimmers was to have the swimsuits modified to include a small pouch so that the sensors were held firmly in place inside the swimsuit. A similar technique was used with sports shorts with a small pouch inside the waistband.

Today, many athletes wear sensors as a matter of routine. Some professional sporting organizations require their athletes to wear sensors during all their team training sessions. These can include heart rate monitors and commercial GPS devices. These sensors can be quite large compared to research sensors. By making the wearing of sensors routine, athletes become desensitized to their presence and are more accepting of wearing them. When all athletes on a team wear sensors, no particular athlete feels singled out, and this also assists in desensitizing the athletes to the existence of the sensors.

## Current and Future Research

At the Centre for Wireless Monitoring and Applications (CWMA; Griffith University, Australia), there are a number of ongoing sports monitoring projects covering sports including tennis, cricket, hockey and swimming. There is also research into the use of mobile phones as the sensor platform of the future.

The tennis research has been looking at the tennis first serve (see Ahmadi et al. 2006, 2010). Initially, this research used sensors to monitor the rotation rates of specific body and arm segments during the tennis serve. It quickly became apparent that the available sensors could not capture the high rotation rates that were occurring. Ahmadi and associates then developed their data using virtual sensors, where motion capture data were reprocessed to give the acceleration and rotational velocity just *as if* a sensor was collecting the data. These rotation rate data were effective in identifying key performance variables related to the tennis serve. More recently, newer rate gyroscope MEMS devices have become available with ranges up to ±6000°/s. These new devices have allowed the collection of rotation rate information directly from the subject and confirmed the data from the virtual sensors. This topic achieved significant international attention with a story in *New Scientist* early in 2010 ("Motion Sensors Could Help").

The sport of cricket, currently one of the top money-earning sports and one of the most popular sports in the world, is involved in a considerable amount of sports science research. In the area of biomechanics, this research predominantly looks at two aspects: performance and injury prevention. The research at the CWMA looks at a third, and seemingly inconsequential, aspect related to the laws of cricket, e.g., the straightening of the arm during a bowling delivery (also known as *chucking*). This research of Wixted, Spratford et al. (2010) first investigated the bowling arm action using a library of existing motion capture data. Similar to the tennis research, this project used virtual sensors generated from the existing motion capture data. The output from this process became the input to the design of prototype sensors that could capture the very high accelerations and rotation rates experienced by the bowling arm during a bowling delivery. In a nutshell, and ignoring numerous complications, the basic premise of this research is that two joined limb segments moving relative to each other will experience different rotation velocities (Figure 15.8). For example, an arm that is flexed at the start of the bowling action and straightens during the bowling action will result in diverging rotational velocities on a particular axis. Two appropriately located linked gyroscopes can detect this divergence. Some of the numerous complications include knowing that the arm is in fact bowling, or knowing when the arm action starts and when the ball is released. Other complications occur if the bowler has an unusual arm, for instance, an arm that hyperextends (bends backwards at the elbow) or an arm that can abduct or adduct (bend sideways at the elbow). Each of these aspects is currently being researched.

Stamm et al. (2009) has been building on the work of Davey et al. (2005, 2008) and other researchers in expanding knowledge in the area of swimming. Davey and associates used accelerometers to create a system that could recognize swimming strokes and turns and accurately capture stroke rates and lap timing. Stamm's work investigates swimming with a finer granularity, taking specific actions, such as the push-off, and quantifying them. Present research also includes quantifying the forces and pressures that occur within each swimming stroke. To assist in the swimming research, other researchers at the CWMA have been developing dual-medium antennas capable of operating in the water, in the air and at the air–water boundary. The dual-medium antennas enable the operation of wireless sensor networks on swimmers in the pool.

Over the years, researchers have custom-made sensor devices or purchased expensive commercial sensors. Because of the ubiquitous nature of mobile phones, the increasing list of features of phones and their low cost, mobile phones are becoming a useful research tool. The current iPhone/iPod Touch comes with all the necessary sensors and networking hardware including accelerometers, gyroscopes, light sensors, GPS, Wi-Fi and Bluetooth. The iPhone and iPod also come with sufficient processing power, memory and graphical display that all necessary data processing can occur on-board with feature-rich results displayed directly on the phone. Rowlands and James (2010) characterized the wireless performance of athlete-mounted iPods, finding that they could maintain full Wi-Fi network connectivity at 50 m from the access point regardless of orientation or occlusion by the athlete's body. They also assessed the viability of the iPod Touch as both an individual sensor and a networked sensor, with researchers able to clearly identify athlete activity in a number of different tests. The mobile phone looks as if it could be the future for a range of sports monitoring scenarios, from single-sensor individual monitoring with on-board processing and logging, full team monitoring with a client/server networked architecture and even remote monitoring via the cellular network for endurance events. The price of an iPod Touch compares very favourably with those of many commercial sport monitors with similar sensor technology.

## Summary

There are many research projects into sports and exercise monitoring, and these cover many technologies, many sports and many aspects within a sport. Generally, this research falls into three categories: biomechanical, physiological and skill assessment. The monitoring technologies use whatever transducers are available that suit the specific application. Some technologies are widely used and accepted and have a long history of use. Generally, the outputs are simple real-time numbers that can be used for feedback or for logging.

There is considerable research interest in using MEMS technologies in sports monitoring, and the important aspect of implementing these technologies is to match the available useful outputs to the problem. There are some useful outputs that can be robustly extracted. Orientations can be extracted from stationary, momentarily stationary or repetitively moving sensors. Activity intensities over various intervals and activity rates are extractable. Multisensor systems can use differences in the signal levels or differences in signal timing to estimate various parameters. The patterns of signals extracted from MEMS devices can be usefully interpreted in terms of the sport being monitored. Systems that have some sort of reference or reset point can use this information to improve estimates over time.

Current research is expanding into many more fields, and as sensor technology and electronics technology advance, many more possibilities exist for all forms of sports monitoring.

## References

Ahmadi A, Rowlands DD, James DA, 2006, Investigating the translational and rotational motion of the swing using accelerometers for athlete skill assessment. In *Proc IEEE Sensors Conference*, Daegu, South Korea, 22–25 Oct 2006, pp. 980–983.

Ahmadi A, Rowlands DD, James DA, 2010, Development of inertial and novel marker-based techniques and analysis for upper arm rotational velocity measurements in tennis, *Sports Engineering*, 12 (4), 179–188.

Billing DC, Hayes JP, Harvey EC, Baker J, 2004, Measurement of ground reaction forces using unobtrusive, on-athlete instrumentation. In *Proc IEEE International Conference on Intelligent Sensing and Information Processing*, Chennai, India, Jan 4–7, pp. 218–221.

Billing DC, Nagarajah CR, Hayes JP, Baker J, 2006, Predicting ground reaction forces in running using micro-sensors and neural networks, *Sports Engineering*, 9 (1), 15–27.

Davey NP, Anderson ME, James DA, 2005, An accelerometer-based system for elite athlete swimming performance analysis, *Proc. SPIE*, 5649, pp. 409–415.

Davey NP, Anderson ME, James DA, 2008, Validation trial of an accelerometer-based sensor platform for swimming, *Sports Technology*, 1 (4), pp. 202–207.

Fraser MJ, James DA, Thiel DV, 2006, Innovative techniques for extending the range and node limits in Bluetooth-based wireless sensor networks, *Proc. SPIE* 6035, pp. 60350E: 1–12.

Harding JW, Mackintosh CG, Martin DT, Hahn AG, James DA, 2008, Automated scoring for elite half-pipe snowboard competition: important sporting development or techno distraction? *Sports Technology*, 1 (6), 277–290.

Harding JW, Mackintosh CG, Martin DT, Rosemond D, James D, Dowlan S, 2005, Detection of key performance variables during half-pipe snowboard competitions and the validation of an accelerometer-based (AIS ROVERTS) algorithm to calculate air-time during elite halfpipe snowboarding, Paper presented at the National Elite Sports Council, Athlete Services Forum 2005, Canberra, Australia.

Harding JW, Small JW, James DA, 2007, Feature extraction of performance variables in elite half-pipe snowboarding using body mounted inertial sensors. In *Proc. SPIE* 6799, pp. 679917: 1–12.

Land Victoria, 2005, It's a home-grown athletic support, *Landmark,* no. 21, March, pp. 1–3, Land Victoria Division of the Department of Sustainability & Environment, Melbourne Aust,

James DA, 2008, A biofeedback system for swing skill acquisition in implement sports, Paper presented at the Japan Society for Mechanical Engineering, Joint Symposium on Sports Engineering and Human Dynamics, Akita, September 9–11.

James DA, Davey N, Gourdeas L, 2003, An integrated platform for micro sensor data acquisition, storage and analysis, *Proceedings of SPIE,* 5274, pp. 371–378.

James DA, Davey N, Gourdeas L, 2004, A modular integrated platform for microsensor applications, Proc. SPIE, 5274, pp. 371–378.

James DA, Davey N, Rice T, 2004, An accelerometer based sensor platform for insitu elite athlete performance analysis. In *Proc IEEE Sensors Conference*, Vienna, Austria, 24–27 Oct. 2004, pp. 1373–1376.

James DA, Uroda W, Gibson T, 2005, Dynamics of swing: A study of classical Japanese swordsmanship using accelerometers, in *The Impact of Technology on Sport,* Subic A, Ujihashi S: Eds. pp. 355–360, ASTA, Japan.

Lai A, James DA, Hayes JP, Harvey EC, 2004, Semi-automatic calibration technique using six inertial frames of reference, *Proc. SPIE*, 5274, pp. 531–542.

Motion sensors could help tennis players serve an ace, 2010, *New Scientist,* 206 (2754) p. 15.

Partridge K, Hayes JP, James DA, Hahn A, 2005, A wireless-sensor scoring and training system for combative sports, *Proc. SPIE,* 5649, pp. 402–408.

Portus MR, Mason BR, Elliott BC, Pfitzner MC, Done RP, 2004, Technique factors related to ball release speed and trunk injuries in high performance cricket fast bowlers, *Sports Biomechanics,* 3(2), 263–284.

Prime Minister's Science, Engineering and Innovation Council, 2004, Routine monitoring of rowing performance—The 'Rover', Report on Science and Technology in Sport, June 1, p. 13, Canberra, Australia: Prime Minister's Science, Engineering and Innovation Council, available online at http://www.innovation.gov.au/Science/PMSEIC/PMSEICMeetings/Documents/STinSportReport.pdf Accessed 25 May 2011.

PWP Design. Automated Boxing Scoring System, available online at http://www.pwpdesigns.com.au/index.php?page = sports-technology. n.d.

Rowlands DD, James DA, 2010, An architecture for iPod/iPhone applications in field Sports. *Procedia Engineering* 2 (2), p. 3493.

Stamm A, Thiel DV, Burkett B, James DA, 2009, Roadmapping performance enhancement measures and technology in swimming, in *The Impact of Technology on Sport II*, ed. Fuss A, Subic A, Ujihashi S, pp. 713–715, Taylor & Francis, London.

Wixted AJ, Billing D, James DA, 2010, Validation of trunk mounted inertial sensors for analysing running biomechanics under field conditions, using synchronously collected foot contact data, *Sports Engineering,* 12, 207–212, available online at http://dx.doi.org/10.1007/s12283-010-0043-2.

Wixted AJ, James DA, Thiel DV, 2002, Low power operating system and wireless networking for a real time sensor network, in *Proceedings of IEEE International Conference on IT and Applications,* Bathurst Australia, 25–29 Nov. paper 112.

Wixted AJ, Spratford W, Davis M, Portus M, James DA, 2010, Wearable sensors for on-field near real-time detection of illegal bowling actions, in *Proc. Biannual Conference of Science, Medicine & Coaching in Cricket*, 1–3 June 2010, pp. 165–168, Cricket Australia, Melbourne, Australia.

Wixted AJ, Thiel DV, Hahn A, Gore C, Pyne D, James DA, 2007, Measurement of energy expenditure in elite athletes using MEMS based inertial sensors, *IEEE Sensors Journal,* 7 (4), pp. 481–488.

# Index

A

Acceleration, components of, 23–24
Accelerometer board embedded in heel of boot, 234
advantages, 234
Accelerometers (acceleration sensors), 41, 174–175; *see also* Inertial sensors; Microelectromechanical systems (MEMS) accelerometers
advantages and disadvantages, 28, 260
clinical applications, 37–39, 287
osteoarthritis, 193–195
definition and nature of, 287, 415
gyroscopes and, *see under* Gyroscopes
research on, 22
schematic representation of biaxial accelerometer based on mass-springer-damper model, 23
triple-axis (3D), 416, 418
Accelerometry
estimation of body inclination, balance control, and postural transitions, 25–28
future trends, 40–41
AC component of acceleration, 24
Accountability (countermeasure in home healthcare system), 124–126
Activities, level of functioning and, 166
Activities of daily living (ADLs), 322
Activity counts, 30, 32
Activity monitoring, 206–208
Activity monitors (AMs), 30
Actuators, 12–13
Adduction, 187
Administration (countermeasure in home healthcare system), 124–126
ADUC7022, 261
ADXL325, 260
Akinesia, 272
Ambient assistive living (AAL) technologies, 283
Ambient intelligent (AmI) systems, 284
AmbiMax, 88
Amplifiers, *see* Electrocardiogram (ECG) amplifier
Analog Devices ADUC7022, 261
Analog Devices ADXL325, 260
Analog-to-digital convertors (ADCs), 357, 360–361
Angular velocity, *see* Gyroscopes
ANT (network), 380–381
Antagonism (modelling), 382
Anterior cruciate ligament (ACL), 397
Arthritis, 163; *see also* Osteoarthritis
Arthroplasty assessment, 39
Artificial neural networks (ANNs), 29
Assurance (countermeasure in home healthcare system), 124–126
Atmel AT32UC3B0256 microcontroller, 262
Atmel AVR32, 262
Authorization (countermeasure in home healthcare system), 124–126
Automatic gain control (AGC), 355
Automatic gain control (AGC) amplifier, 355, 356
schematic of a, 355–356
Availability (countermeasure in home healthcare system), 124–126

B

Balance assessment, 39
Battery life, 427–428
Bazett's formula, 343
Biaxial acceleration sensor, 23
Bioacoustic sensors, 7
Biochemical sensors, 7
Bioelectric sensors, 7

Biomagnetic sensors, 7; *see also* Electromagnetic field sensors; Magnetic sensors
Biomechanical sensors, 7
Biomedical engineer, 253
Biomedical sensors
  classification, 7
  requirements of, 8
Biooptical sensors, 7
Bipolar junction transistor (BJT), 356
Bluetooth, 9, 10, 117, 263, 328
  RFCOMM layer, 313, 315, 328
Body area networks (BANs), 2–4, 9, 11, 51; *see also* Wireless body area networks
Body-centric networks, *see* Body area networks
Body-coupled communication (BCC), *see* Intrabody communication
Body functions, 166
Body sensor networks (BSNs), 129, 130
  *vs.* home sensor networks (HSNs), 117–118
Body structures, 166
Boots, accelerometer board embedded in heel of, 234
Bradykinesia, 255; *see also* Parkinson's disease
  defined, 249
  in Parkinson's disease, 38, *see also* Kinesia system: system validation to clinical standards
  tremor and, 254–255

## C

Capacity and performance (human activity), 166
Cardiovascular system disease (CSD), 340; *see also* Electrocardiogram
Carnot's rule, 90
Carrier sense multiple access/collision avoidance (CSMA/CA), 97
Certified Information Systems Security Professional (CISSP), 123
"Christmas tree effect," 32
Chucking, 431
Clinical Global Impression Change (CGIC), 325
Coaching system, mobile, 380
CodeBlue, 306
Coefficient of variation (CV), 273
  of activity, 32–33
COGKNOW, 288
Cognitive radios, 18
Co-learning approaches, 37
Common mode interference, 353
Common mode rejection ratio (CMRR), 347–352
Complementary filtering approach, 28
COM (hardware interface) ports, 362–364
Conditional random fields (CRF), 37
Conjunctiva, transcutaneous oxygen monitoring at, 145–147
Context-aware wearable assistant, *see under* Parkinson's disease (PD) patients
Contextual factors (health conditions), 166
Controlled practice, 397
CRN Toolbox, 315–316, 329
Cross P-fold validation, 36
Curse of dimensionality, 34
Cyclic Redundancy Check (CRC), 119
Cycling computers, 410–411
Cylindrical muscle model, 55–57

## D

DAPHNet project, 313
DC baseline offset and drifts, 344
DC bias, *see* DC baseline offset and drifts
DC component of acceleration, 24
Decision support system (DSS), 290, 291, 294
Degrees of freedom (DOF), 260
Dementia, *see* NOCTURNAL assistance technology; Nocturnal sensing and interventions
Diabetes mellitus type 1, 148, 164
Diabetes mellitus type 2, 148, 164
Differential global positioning system (dGPS), 377
Diffie–Hellman-based EKE protocol, 131
Dipole antenna, 63, 65
Dissolved oxygen (DO), 143–144

Dynamically Controlled Networks (DyCoN), 386, 387, 389–390, 398
Dyskinesias, 249, 254
assessment and monitoring, 38

## E

Electric field distributions, 57–60, 62, 64
Electrocardiogram (ECG), 340
Electrocardiogram (ECG) amplifier, 340, 367–369
automatic gain control (AGC) amplifier, 355, 356
schematic of a, 355–356
component list, 347
external view of prototype of, 368
front end circuitry and, 351–353
history, 345
how it works, 347–348
instrumentation amplifier
need for an, 349–351
two op, 349–351
Electrocardiogram (ECG) electrodes
design and construction, 344–347
homemade limb, 345–347
placement, 345–346
Electrocardiogram (ECG) sensor systems, 53, 76, 340, 367–369
data logging system, 341, 342
electricity supply in developing countries and, 340–341
filtering, 353–354
firmware operation, 361–363
front end circuitry, 351–353
AC coupling network and protection circuitry, 351
signals and diagnosis, 341–344
USB cable connection details, 359, 360
USB configuration and communication, 357–359
USB interface hardware and device protection, 359–361
USB interfacing, 356–357, 362–364
Electrocardiogram (ECG) signals
displaying them on a PDA or mobile phone, 362–366
flow chart for software to display, 362, 363
from healthy patient, deflection components making up, 341, 343
lead II ECG signals using a commercial electrocardiograph, 365, 366
signal-to-noise ratio (SNR), 351, 360, 361, 368
Electrogoniometers, 192
Electromagnetic field sensors, AC and DC, 192–193
Electromagnetic interference (EMI), 347–348
Electromagnetic vibration-based microgenerator devices, 92, 93
Electronic health (eHealth), 340
Electrooculography (EOG), 311
Elliptic Curve Cryptography (ECC), 131–134
ENABLE (Enabling Technologies for People with Dementia), 288
Encrypted Key Exchange (EKE) protocol, 131
Energy expenditure (EE) estimation, 30
approaches to improving the accuracy of, 30–33
Energy harvesting (EH)
EH opportunities and demonstrated capabilities, 86–87
EH sources and their energy harvesters, 85
EH technologies for healthcare WSNs, prospect of, 96–97
overview, 85–87
Energy harvesting (EH) systems, research on, 87–95
Environmental factors, 166–167
Ethylene-propylene, fluorinated, 143
Everlast, 88
Extraction algorithms, 34

## F

Fall detection, 39–40
pre-impact, 40
systems of, 40
Fall-detection sensor, 103–104
Fast Fourier transform (FFT), 318, 424–425

Feature selection algorithms, 34
Field-effect transistors (FET), 345, 355–356
Field metabolic rate, 226
Finger-worn motion sensor, 259–261
Flexible biosensors, 148–154; *see also* Flexible sensors
Flexible devices (for healthcare networks)
 for biomonitoring, 140–142
  for information systems, 140
Flexible oxygen sensors for transcutaneous oxygen monitoring, 142–147
 structure, 143
Flexible sensors; *see also* Glucose sensors
 in body area network, 141
Fluorinated ethylene-propylene (FEP), 143
Force plate, 413–414
Free-throw study, 386
Freeze index (FI), 309
Freezing of gait (FOG), 304–305, 312
Freezing of gait questionnaire (FOG-Q), 304
Freezing of gait (FOG) treatment
 limitations of, 305
 state of the art in, 305–306
Frequency-hopping spread spectrum (FHSS), 10
Functional Electrical Stimulation (FES), 13
Functioning and disability, level of, 166

## G

Gait; *see also* Walking
 temporal and spatial parameters of, 29–30
Gait acceleration measurement, 238–239
Gait analysis
 clinical paradigm, 169, 171
 and monitoring walking, 167–172
 technology, 167–168
Gait monitoring, *see* Idiopathic toe walking (ITW) children
Galvanic coupling, 53
Geometrical approach (classification methodology), 37
Global positioning system (GPS), 176, 206, 224, 225, 285, 310
 differential, 377
 sports, 409–411
Glucose dynamics after oral glucose tolerance test, 152–154
Glucose monitoring
 continuous, 148–149
 tear glucose monitoring at the eye, 152–154
Glucose oxidase (GOD), 149–152
Glucose sensors
 flexible, 149–152
  calibration curve, 149, 151
  structure, 149, 150
 transparent, 149–151
GNOME platform, 119
Ground reaction force (GRF), 186–190
Gyroscopes (gyros), 195, 200, 260, 378, 420; *see also* Inertial sensors
 *vs.* accelerometers, 259, 260
 accelerometers combined with, 27, 195–197, 199, 201, 203–205, 223, 248
 advantages and disadvantages, 28
 bradykinesia and, 272, 273
 3D, 416, 418
 and magnetic sensors in wearable sensor systems, 26–28
 magnetometers, integrated sensors, and, 26–27, 175–176
 MEMS, 260–261, 416, 417, 422
 osteoarthritis and, 195–197, 199, 201–205
 Parkinson's disease and, 259, 260, 266, 268–270, 272, 273
 rate, 416, 418
  limitations of, 418–419
 strap-down integration approach and, 27
 triaxial, 27, 38, 40
 triple-axis (3D), 416
 used to monitor joint movement, 425–426
 useful outputs from, 418

## H

Healthcare sensor networks (HSNs), 1–4, 106
applications of, 4
adults, 5
elderly, 5–6
the young, 4–5
commercial challenges and barriers to successful implementation, 15–17
engineering and technical challenges, 6
actuators, 12–13
network architectures and telecommunications, 9–11
power management, 11–12
security, 12
sensor fusion algorithms and models, 9
sensor hardware, 7–8
future work and future directions, 17–19
human-centric challenges, 13–15
schematic of, 3
Healthcare sensor systems, 115–117
compatibility issues between different environments, 117–118
implementations of, 116
limitations with power and security, 118–120
Healthware, 306
Heart rate variability (HRV), 130, 381–382
Heel acceleration signals, graphical representation of, 239, 240
Hidden Markov models (HMMs), 37, 228
HMAC-MD5, 132, 133
Hoehn and Yahr (H&Y) scale, 304, 305n
Home health controller (HHC), 112–113
Home sensor networks (HSNs), 117
*vs.* body sensor networks (BSNs), 117–118
Human activity classification, 33–36
classification methodologies, 36–37
Human interface device (HID), 358
Human interface device (HID) descriptor reports, 362
Hybrid systems, 197–198
Hypokinesia, 272

## I

Idiopathic toe walking (ITW)
causes of, 232
consequences of, 232
gait assessment, 232–233
nature of, 231–232
Idiopathic toe walking (ITW) children, ambulatory gait monitoring in, 230–231
challenges in, 233
behavioural issues, 233
sensor data interpretation, 235–237
sensor selection and placement issues, 233–235
differences in gait features in toe walking and normal stride, 237–238
experiments, algorithm development and statistical analysis, 242–243
acceleration measurement methods, 238–239
algorithm development for identifying strides, 239
analogue output connector, 241
battery charger, 242
external on/off switch, 242
hardware description, 241
serial RS232 interface, 242
system using sensors to monitor and assess gait remotely, 239–240
lack of an objective method for, 233
Incline estimation, 29–30
INDEPENDENT, 306
Independent components analysis (ICA), 36
Individual search (algorithm), 34
Individual validation, 36
Industrial, scientific and medical (ISM) band, 10

Inertial measurement unit (IMU), 27, 260, 378
Inertial navigation system (INS), 419
Inertial sensors, 39, 193, 378; *see also* Accelerometers; Gyroscopes
*inf* file, 364
Information and communication technology (ICT), 288–289
Information assurance, security, and privacy threats, 112–115, 120
  countermeasures to threats, 123–124
    environment information, 129–130
    KDC schemes, 128–129
    key management, 125, 127
    mapping the countermeasures, 124–126
    pairwise key establishment, 127
    privacy, 124
    random key establishment, 127–128
    using physiological data to establish keys, 130–132
  denial of service, 122
  difficulty in using complex technology, 122
  disclosure of sensitive data, 121
  efficiency issues and experimental evaluations, 132–134
  expectation of real-time communication, 123
  expectation of reliability, 123
  future directions, 135
  impersonation of service, 121
  impersonation of user, 120–121
  inability to keep track of changing technology, 122
  lack of trust in the system, 122
  modification of data, 121
  modification of software, 121
  repudiation, 122
Infrared detectors, *see* Passive infrared detectors
Infrared sensors, *see* Passive infrared (PIR) sensors
Infrastructure area networks (IANs), 2, 3, 8
Integrated development environment (IDE), 361
Integrated sensors, 175–176
Intelligent Device for Energy Expenditure and Activity (IDEEA), 222–223
*International Classification Functioning, Disability, and Health (ICF),* 165–168
Interpulse interval (IPI), 130
Intrabody communication, 51–52, 72
  configuration of electrodes, 60–65
    arrangement of input electrodes of receivers, 60, 63
    electric field distributions, 60, 62
    signal voltages received by receiver on wrist, 60–61, 64
  distance between human body and circuit board, 67
  impedance matching of electrodes, 68, 70–72
  spacing between contact electrodes, 66–67
  transmission characteristics of on-body devices, 67–69
  transmission model, 52–59
    carrier frequency and, 59–60
    electric field distribution including an off-body device, 57–59
    numerical simulation, 55–57
  transmitter model, 57
  using capacitive coupling, 59
  using contact electrodes, 53–55
iPod Touch, 432

## J

Joint movement, using multiple gyroscopes to monitor, 425–426

## K

Kalman filtering, 27, 176
Kalman filters and neural networks, 426–427
Key distribution centre (KDC), 125
Key distribution centre (KDC) schemes, 128–129
Key establishment protocols, 125, 127
Key management, 125, 127

Kinesia system
  benefits, 256, 278–279
  challenges to widespread clinical use, 276–279
  components, 258
  development, 248, 254, 256–258, 267
    test engineering, 265–266
  FDA clearance, 277
  nature of, 248
  Parkinson's disease and, 248, 253, 254, 256–258
  sensor unit specifications, 259
  software interface, 264
  specifications, 259
  system validation to clinical standards, 266–267
    automated bradykinesia assessment compared to clinical standard, 267–274
    automated tremor assessment compared to clinical standard, 267–272
    patient acceptance, 275
    quantitative and independent bradykinesia feature extraction, 274–275
  UPDRS and, 267–274, 278–279
  use at home, 278–279
Knee adduction moment (KAM), 186, 187, 189–191, 209
  absolute, 190
  net external, 185–187
  peak, 187
Knee arthritis, *see* Osteoarthritis: prospective motion sensor technologies for knee OA
Knee arthroplasty assessment, 39

**L**

Lead (stretch between limb electrodes), 345
Leave-1-out (validation approach), 36
Levodopa (LD), 305
Light-dependent resistor (LDR), 288
Light-emitting diode (LED), 288
Load–performance interaction, 383
Longitudinal arrangement (electrode arrangement), 53

**M**

Machine learning approach, 41
Magnetic sensors, 7, 26–28; *see also* Electromagnetic field sensors
  advantages and disadvantages, 28
Magnetometers, 175–176
Mass-spring-damper model, 23
Maximum entropy Markov models (MEMM), 37
Medical telemetry service, wireless, 10
Medi-Link dialer, 16
Medium access control (MAC) protocol, 97
MedRadio band, 10
Mercury tilt switch, 287
Message Authentication Code (MAC) value, 119–120
Metabolic equivalent of task (MET), 223, 226–228
Microelectromechanical systems (MEMS), 8, 416
  MEMS sensor outputs for monitoring sport and exercise, 417
    useful outputs from accelerometers, 417
    useful outputs from gyroscopes, 418
Microelectromechanical systems (MEMS) accelerometers, 23–25
  limitations, 418–419
    sensor limits, 419–420
  measure specific force projection along sensitive axis, 23–24
  three-axis Sandia, 416
Microelectromechanical systems (MEMS) gyroscopes, 260–261, 416, 417, 422
Microelectromechanical systems (MEMS) inertial sensors, 22
  useful outputs of, 422
    combinations of sensor outputs and their uses, 422, 423
Mobile coaching system, 380
Mobile health (mHealth), 340
Mobile phones, 339–341, 357, 367
  displaying ECG signals on, 362–366

Mobility problems; *see also* Walking
causes of, 162–163
prevalence of, 162–163
age and, 161–162
Mobi system, 314
Modified Bradykinesia Feature Extraction (MBRS), 274, 275
Modified Bradykinesia Rating Scale (MBRS), 274, 275
Motion artefact, 344
Motion detection, 287
Motion measures not derived from laboratory measures, field-based, 205–208
Motion-sensing/tracking technologies
categories of, 191
characteristics of ideal, 208–209
Movement classification, features for, 34–36
signal features for activity-classification tasks, 34, 35
Multithreading, 364

**N**

Neural network-based analysis of game tactics
design of case study dealing with, 392
Neural network-based analysis of rehabilitation process, case study of, 396–401
research design, 397
Neural network (NN) models, 228
Neural network with trajectory process and reference point, 399, 400
NOCTURNAL (Night Optimized Care Technology for UseRs Needing Assisted Living) assistance technology, 284, 290, 297
addressing user acceptance, 295–297
architecture, 293
sensing/intervention platform, 291–295
Nocturnal sensing and interventions, peculiarities of, 288–291
Non-insulin-dependent diabetes mellitus (NIDDM), 148

**O**

Osteoarthritis (OA), 163, 183–184; *see also* Arthritis
laboratory-based motion measures in
impact forces, 187–188
kinematic differences, 188–189
net external knee adduction moment (KAM), 185–187
motion sensors in, 208–209
characteristics of ideal, 208–209
movement analysis in, 184–189
precedents and prospects for use of motion sensors, 198–205
prospective motion sensor technologies for knee OA, 189–191
motion sensors, 191–198
Oxygen, dissolved, 143–144
Oxygen sensors for transcutaneous oxygen monitoring, flexible, 142–147

**P**

Pairwise schemes, 125
Parkinson's disease (PD), 247–250, 252–253, 304; *see also* Unified Parkinson's Disease Rating Scale
monitoring motor fluctuations in, 38
Parkinson's disease (PD) motor symptoms, ambulatory and remote monitoring of; *see also* Kinesia system
clinically driven design input specifications, 253
clinician characteristics, 256–257
patient characteristics, 253–255
technology development, 258–259, *see also* Kinesia system
finger-worn motion sensor, 259–261
human factors, 265
software development, 264–265
wireless data telemetry, 263
wrist-worn command module, 258, 261–263

Parkinson's disease (PD) patients
walking, 164
wearable context-aware assistant for, 307, 308, 326, 332–334
collecting users' and clinicians' experience, 330–331
development stages of, 307, 308
extensive data recording, 330
future steps, 331–332
integration and flexibility, 329–330
local debriefing, 330
offline-to-online algorithm transfer, 329
sensor nodes, 328–329
software, 329
system training and calibration, 329
testing procedure, 328
Partial pressure of oxygen in arterial blood ($PaO_2$), 142, 146
Participation (involvement in life situations), level of functioning and, 166
Passive infrared detectors (PIDs), 285, 287, 291, 292
Passive infrared (PIR) sensors, 285–286, 294
PDAs, *see* Personal digital assistants
Pedometers, 30, 173–174, 224, 225, 410
osteoarthritis patients and, 205–209
Peer Intermediaries Key Establishment (PIKE) scheme, 129
Performance and capacity (human activity), 166
Performance Potential Meta-Model PerPot, 382–384
Personal digital assistants (PDAs), 115, 116
PDA assessment, 29–33
Personal health assistant (PHA), 306
Personal *vs.* environmental factors (capacity and performance), 166–167
Phase diagrams, one-dimensional, 386
Phase-locked loop (PLL), 355
Physical activity (PA) monitoring, 222–223
challenges in, 224–225
behavioural, 228–230
data interpretation, 226–228
hardware and software, 225–226
in children, reasons for, 224
Piezoelectric generators, 92, 93
Piezoelectric-powered RFID system, 92, 93
Polydimethylsiloxane (PDMS), 150–152
Polydimethylsiloxane (PDMS) membrane, 141–142
Poly(2-MPC–co-DMA) (PMD), 150–152
Position-detection sensors, 377
Position sensors, 176
Principal components analysis (PCA), 34, 36
Probabilistic approach (classification methodology), 36–37
Prosthesis, 39

## Q

QRS complex (electrocardiography), 341–342, 348, 360, 361
QT dispersion (QTd), 343
QT interval (electrocardiography), 343–344
Quality-of-service (QoS) provisions, 78, 79
Quantisation noise, 360–361
Q wave (electrocardiography), 342

## R

Radio-frequency identification (RFID) chips, 377
Radio-frequency identification (RFID) detectors, 285, 287
Radio-frequency identification (RFID) system, piezoelectric-powered, 92, 93
Radio-frequency identification (RFID) tags, 285, 287, 411
Radio frequency (RF) spectrum, 10, 18
Radio frequency (RF) transmission, 9–11
*RANDKEY*, 130
Random key establishment, 127–128

Random key predistribution schemes, 125–128
RC5, 132, 133
Real-time motion measurement, 204–205
Rehabilitation; *see also* Sports, exercise, and rehabilitation
objective skill evaluation for, 38–39
Rehabilitation process, neural network-based analysis of, 394–396
case study, 396–401
research design, 397
Resistor–capacitator (RC) model, 59
Rhythmic auditory stimulation (RAS), 306, 312, 326
"Right leg drive," 353
Rigidity, 249
Rivest, Shamir, and Adleman (RSA) password protocols, 131
Rowing, 415
RR intervals (electrocardiography), 342–343
RSA, 132–134

S

Sandia MEMS accelerometers, three-axis, 416
Secure environment values (SEVs), 129, 131–132
Security Protocols for Sensor Networks (SPINS) protocol, 128–129
Seebeck's theory, 102
Sensor calibration, 238
Sensor-enabled smart home, 112–113
Sensor network, 118, 429
Sensors (in healthcare), 284–288
types and characteristics of, 285, 286
Sensor system outputs, 422, 423
Sequential classifier, 36
Sequential search (algorithm), 34
Sequential supervised learning, 37
Signal processing (for sports monitoring)
tools and techniques for extracting information from sensor outputs
analysis tools, 422–426
Kalman filters and neural networks, 426–427
Signal-to-noise ratio (SNR) of ECG signals, 351, 360, 361, 368
Single-contact-electrode transmitter, 61, 63–65
Single-frame classifier, 36
Sit-to-stand (STS) postural transition, 28
SKIPJACK, 132, 133
Sleep, *see* NOCTURNAL assistance technology; Nocturnal sensing and interventions
Snowboarding, 414
Socket Mobile SoMo650-M phone, ECG signals on, 364–366
Solar energy harvesting (SEH) systems, 88–89
Spatial parameters, 29
Spatio-temporal parameters of gait, 203
Sports, exercise, and rehabilitation, 375–376, 401–402; *see also* Rehabilitation
model-based simulation, 381–384
neural network-based process analysis, 384–394
neural network-based analysis of game tactics, 391–394
neural network-based motor analysis, 385–391
Sports and exercise monitoring, 376, 408–409, 432–433
for different disciplines, purposes, and user groups, 412
force sensors, 378–379
motion-detection sensors and devices, 377–378
physiological sensors and devices, 379
position-detection sensors and devices, 376–377
sensor devices, 416, *see also* Microelectromechanical systems
on athletes, 430
hardware, 427–428
networking, 429–430
synchronization, 428–429

sensor networks for monitoring physical activity, 379–381
utilizing accelerometer outputs in consistent repetitive action, 421–422
frequency, 421
orientation, 420–421
Sports and exercise monitoring research
current and future, 431–432
example projects, 413–416
purpose, 412–413
Sports GPS, 409–411
Sports monitoring systems, 409
core monitor outputs, 410–411
the data consumer, 409–410
issues for, 411
SQRT application, 133
Standard Microsystems LPR5150AL, 260
Step frequency from running, extracting
using single accelerometer, 424, 425
Strap-down integration approach, 27
ST segment (electrocardiography), 348
Supercapacitor (supercap), 83
Swimming, 432
System dynamics, 382
System validation *vs.* verification, 267

## T

Tear glucose level, measurement methods of, 152, 153
Telemetry service, wireless medical, 10
Template-matching algorithm, 37
Temporal parameters, 29
Thermal energy harvesting (TEH)
from human warmth for WBANs in medical healthcare system, 97–105
experimental test results, 105–106
TEH sensor node for fall detection, schematic diagram of, 102–103
TEH structure with thermoelectric generator (TEG), 100–102
TEH systems, 89–92
examples of, 90–92
power management circuit, 102–103
Thermoelectric generator (TEG), 90, 92
prototype of, 102
thermal analysis of, 101
Time-division multiple access (TDMA) technique, 97
TinyOS, 119, 132, 134
Toe walking, *see* Idiopathic toe walking
Tracking systems, 285, 287
Trajectory development, progressive, 399, 400
followed by a decline, 399, 401
Transcutaneous gas at body surfaces, 142–143
Transcutaneous oxygen monitoring
at conjunctiva, 145–147
flexible oxygen sensors for, 142–147
Transcutaneous partial pressure of oxygen ($TcPO_2$), 142, 144–147
Transmission line matrix (TLM) method, 55
Transmitter boxes of different heights, 57–58
electric field distributions for, 57, 58
Transversal arrangement (electrode arrangement), 53
Tremor; *see also* Parkinson's disease
assessment, 267–272
bradykinesia and, 254–255
Trusted third party, 128
limitations with using sensor node as, 129
Two-contact-electrode transmitter, 61, 63–65

## U

Ubiquitous computing, 76
Ultra low-power (ULP) radio, 11–12
Ultrasonic sensors, 192
Ultra wideband medical (UWB) band, 10
Unified Parkinson's Disease Rating Scale (UPDRS), 250–252, 256
kinesia and, 267–274, 278–279

Universal synchronous/asynchronous receiver/transmitters (USART), 262
Uptimers, 173

**V**

Validation process, implementation of, 36
Vibration-based energy harvesting (VEH) research projects, 92–94
Vibration energy harvesting (VEH) systems, 92–94
Vibration-powered wireless sensor node, 92, 93
Visual Analogue Scale (VAS), 325
Visual C++ , 364

**W**

Walking
- with disease, 161–162
- for health, 159–161
- health conditions affecting, 162–164
- monitoring, 176–177
  - how to monitor walking, 173–176
  - reasons for, 159–165
  - what to monitor, 165–172

Walking speed, 29–30
Wearable context-aware assistant, *see under* Parkinson's disease (PD) patients
Wearable devices, 140
Whole-body model, high-resolution, 55, 56
Wi-Fi, 9, 429–430
Wind energy harvesting (WEH) systems, 94–95
- examples of, 94–95

Wind turbine generators (WTGs), 94
Wireless access point (WAP), 356
Wireless body area networks (WBANs), 96–100
Wireless medical telemetry service (WMTS), 10
Wireless networks, 429
Wireless personal area network (WPAN), 9
Wireless sensor networks (WSNs), 76, 117–118, 128
- architecture, 77–78
- motivation for healthcare, 76
- protocol stack, 78–79
- wireless sensor nodes, 79–80

Wireless sensor nodes (WSNs), 76, 79–80
- block diagram of, 79–80
- energy-harvesting solutions for, 84–95
- problems in powering, 80–81
  - high power consumption, 81–82
  - limits on energy sources, 82–84

Wrist-worn command module, 258, 261–263

**X**

Xbow sensor nodes, 81–82
Xor application, 134

**Z**

Zero-frequency (DC) component of acceleration, 24
Zigbee protocol, 10